João Canning-Clode (Ed.)

Biological Invasions in Changing Ecosystems

Vectors, Ecological Impacts, Management and Predictions

This work is dedicated to the memory of my father

João Canning-Clode

João Canning-Clode (Ed.)

Biological Invasions in Changing Ecosystems

Vectors, Ecological Impacts, Management and Predictions

Managing Editor: Katarzyna Michalczyk

Associate Editor: Anssi Vainikka

Language Editor: Blake Turner

DE GRUYTER OPEN

Published by De Gruyter Open Ltd, Warsaw/Berlin

Part of Walter de Gruyter GmbH, Berlin/Munich/Boston

ISBN: 978-3-11-043865-9
e-ISBN: 978-3-11-043866-6

Bibliographic information published by the Deutsche Nationalbibliothek
The Deutsche Nationalbibliothek lists this publication in the Deutsche Nationalbibliografie; detailed bibliographic data are available in the Internet at http://dnb.dnb.de.

Managing Editor: Katarzyna Michalczyk

Associate Editor: Anssi Vainikka

Language Editor: Blake Turner

www.degruyteropen.com

Cover illustration: Commomn lionfish © lilithlita, Cane toad © Capstoc

Contents

Part III. **Management and Control of Biological Invasions**

Part IV. **Predictions and New Tools in Biological Invasions**

Preface

Adaptive radiation, allopatric speciation and local adaptation, alongside sexual selection, are the key evolutionary processes that have generated the current biodiversity. Globally, steadily increasing trade and the movement of humans across continents have broken the original distribution ranges of species by increasing the rates at which biological organisms enter non-native locations. Certain species have also been introduced on purpose with the goal of increasing the variety of fished, hunted or otherwise utilized species. Biological homogenization due to the intentional and unintentional movement of species is practically an irreversible process and one of the greatest threats to global biodiversity.

When leaving the original native environment, species are often released from their natural enemies as described by the enemy release hypothesis. On the other hand, an invasive species typically reaches new areas as only a small subset of individuals, causing a genetic bottleneck. To explain the great success of many invasive species in new environments, we must understand the ecological interactions between native and invasive species, niche structures and environmental characteristics in relation to the requirements of the invading species. We also need to understand if and how the invasive species adapt to their new environment and how the surrounding ecological environment adapts to the presence of the new species. At more subtle levels, we need knowledge on the evolutionary processes that determine if two related species or forms will interbreed and lose their special characteristics due to hybridization. Such fine-scaled processes and their consequences cannot be understood without support from modern genetics.

Sometimes, invasive species have detrimental ecological effects on local ecosystems by out-competing native species or by predating on endogenous species that severely suffer or become extinct as witnessed in Lake Victoria after the introduction of the Nile perch in 1954. Unfortunately, invasive species rarely travel alone, but are often accompanied by a wide variety of parasites and diseases. For example, diseases that arrived in Europe along with imported North American crayfish have driven most of the native European crayfish populations to extinction. Global climate change is altering the distribution of hosts, but also the distribution of their parasites. As some of the invasive hosts act as disease vectors for human diseases, invasions also increase human health concerns. Some invasive species are agricultural pests or weeds, and induce substantial economic losses. In rare cases, invasive species improve the ecosystem functioning by providing a new direct or indirect resource for the existing fauna.

This open-access, printed-on-demand book, edited by Dr. João Canning-Clode and written by the leading authors in their discipline, provides an up-to-date overview of the central themes in invasion biology and ecology. The open access publishing format, still a novel one for scientific books, will make science accessible not only to scientists but also to journalists, decision makers, and the general public. More importantly, the present open access format also guarantees full no-cost accessibility to students, thus significantly increasing the efficiency of education across the globe.

Efficient transfer of knowledge from the scientific community to the rest of society is central to successfully preventing and controlling the harmful impacts of biological invasions. The crucial distinction between native and non-native species by ordinary citizens will often have strong conservational implications, as the problems cannot be solved without recognizing them. For example, well-intended release of pets such as aquarium fish and crayfish to natural waters forms a serious disease risk for native species.

This book starts by describing the vectors and history of biological invasions in different systems and within different taxonomic groups. Understanding the past and current mechanisms of how invasive species have been transferred to new areas is crucial for the planning of any management measures taken to prevent invasions. The second section focuses on the impacts of biological invasions, especially in host-parasite systems. Environmental parasitology is a rapidly developing research field, and of importance for predicting both the outcomes of biological invasions and global climate change. Section III deals with practical examples of how biological invasions can be managed in both terrestrial and aquatic systems. Finally, section IV reviews contemporary modelling and DNA-based methods that can be used to study both the mechanisms of invasions and their predicted future outcomes. This last section also discusses how climate change might interfere with invasions, and how aquatic communities might reach new assemblage structures due to invasions by new species.

As a whole, this book provides illustrative examples of biological invasions, synthesizes the current knowledge by identifying general patterns and factors that impact the resilience of biological systems, and gives insights into practical management problems. As such, I anticipate this work will have valuable use as a reading material for university students and anyone interested in learning more about the ecology and biology of invasions. The book has a strong conservational message: the once-invaded species cannot often be eradicated. Introducing a new species into an ecosystem is a one-way choice that needs to be based on prior, careful, holistic, and precautionary impact assessment. Thus, any unintended species translocation is to be avoided, and developing barriers to stop unintentional species migrations is a timely management challenge.

Anssi Vainikka, in Joensuu, 10.12.2014

List of Contributors

Preface

Anssi Vainikka
University of Eastern Finland
Department of Biology
Finland
anssi.vainikkka@uef.fi

General Introduction

João Canning-Clode[1,2,3]
[1]MARE – Marine and Environmental Sciences Centre, Estação de Biologia Marinha do Funchal, Portugal
[2]Centre of IMAR of the University of the Azores, Department of Oceanography and Fisheries, Horta, Azores, Portugal
[3]Smithsonian Environmental Research Center, Edgewater, USA
canning-clodej@si.edu

Parts I, II, III and IV

João Canning-Clode[1,2,3]
[1]MARE – Marine and Environmental Sciences Centre, Estação de Biologia Marinha do Funchal, Portugal
[2]Centre of IMAR of the University of the Azores, Department of Oceanography and Fisheries, Horta, Azores, Portugal
[3]Smithsonian Environmental Research Center, Edgewater, USA
canning-clodej@si.edu

Filipa Paiva[1,2]
[1]GEOMAR - Helmholtz Centre for Ocean Research Kiel, Department of Experimental Ecology, Germany
[2]MARE – Marine and Environmental Sciences Centre, Estação de Biologia Marinha do Funchal, Portugal
fpaiva@geomar.de

Chapter 1

James T. Carlton
Williams College - Mystic Seaport
Maritime Studies Program
USA
james.t.carlton@williams.edu

Gregory M. Ruiz
Smithsonian Institution
Smithsonian Environmental Research Center
USA
ruizg@si.edu

Chapter 2

Phillip Cassey
University of Adelaide
School of Biological Sciences
Australia
phill.cassey@adelaide.edu.au

Miquel Vall-llosera Camps
University of Adelaide
School of Biological Sciences
Australia
miguel.vall-lloseracamps@adelaide.edu.au

Ellie Dyer
Zoological Society of London
Institute of Zoology
United Kingdom
ellie.dyer@ioz.ac.uk

Timothy Blackburn
University College London
Genetics, Evolution & Environment
United Kingdom
t.blackburn@ucl.ac.uk

Chapter 3

Mark D. Sytsma
Portland State University
Center for Lakes and Reservoirs
USA
mark.sytsma@pdx.edu

Toni Pennington
Tetra Tech, Inc.
Water Division
USA
Toni.Pennington@tetratech.com

Chapter 4
George K. Roderick
University of California, Berkeley
Environmental Sciences, Policy, and Management
USA
roderick@berkeley.edu

Maria Navajas
INRA
CBGP (INRA/IRD/Cirad/Montpellier SupAgro)
France
maria.navajas@inra.fr

Chapter 5
Pam L. Fuller
U.S. Geological Survey
Nonindigenous Aquatic Species Program
USA
pfuller@usgs.gov

Chapter 6
Christina M. Romagosa
University of Florida
Department of Wildlife Ecology and Conservation
USA
cmromagosa@ufl.edu

Chapter 7
April M.H. Blakeslee
East Carolina University
Department of Biology
USA
blakesleeap14@ecu.edu

Chapter 8
Japo Jussila
The University of Eastern Finland
Department of Biology
Finland
japo.jussila@uef.fi

Al Vrezec
National Institute of Biology
Slovenia
al.vrezec@nib.si

Jenny Makkonen
The University of Eastern Finland
Department of Biology
Finland
jenny.makkonen@uef.fi

Raine Kortet
The University of Eastern Finland
Department of Biology
Finland
raine.kortet@uef.fi

Harri Kokko
The University of Eastern Finland
Department of Biology
Finland
harri.kokko@uef.fi

Chapter 9

Name Laura Härkönen
University of Oulu/University of Eastern Finland
Department of Ecology/Department of Biology
Finland
laura.harkonen@oulu.fi

Arja Kaitala
University of Oulu
Department of Ecology
Finland
arja.kaitala@oulu.fi

Chapter 10

Ane Timenes Laugen
Åbo Akademi University and Novia University of Applied Sciences
Aronia Coastal Zone Research Team
Finland
Ane.Laugen@abo.fi

Johan Hollander
Lund University
Department of Biology
Sweden
Johan.Hollander@biol.lu.se

Matthias Obst
University of Gothenburg
Department of Marine Sciences
Sweden
Matthias.Obst@marine.gu.se

Åsa Strand
University of Gothenburg
Department of Marine Sciences
Sweden
Asa.Strand@gu.se

Chapter 11
Antonella Petrocelli
Italian National Research Council (CNR)
Institute for Coastal Marine Environment (IAMC)
Italy
antonella.petrocelli@iamc.cnr.it

Ester Cecere
Italian National Research Council (CNR)
Institute for Coastal Marine Environment (IAMC)
Italy
ester.cecere@iamc.cnr.it

Chapter 12
Frederico Cardigos[1,2]
[1]MARE – Marine and Environmental Sciences Centre, University of the Azores
Department of Oceanography and Fisheries, Portugal
[2]Centre of IMAR of the University of the Azores, Department of Oceanography and Fisheries, Horta, Azores, Portugal
frederico@uac.pt

Jorge Fontes[1,2]
[1]MARE – Marine and Environmental Sciences Centre, University of the Azores

Department of Oceanography and Fisheries, Portugal
[2]Centre of IMAR of the University of the Azores, Department of Oceanography and Fisheries, Horta, Azores, Portugal
fontes@uac.pt

João Monteiro[1,2]
[1]MARE – Marine and Environmental Sciences Centre, University of the Azores Department of Oceanography and Fisheries, Portugal
[2]Centre of IMAR of the University of the Azores, Department of Oceanography and Fisheries, Horta, Azores, Portugal
jmonteiro@uac.pt

Paola Parretti[1,2]
[1]MARE – Marine and Environmental Sciences Centre, University of the Azores Department of Oceanography and Fisheries, Portugal
[2]Centre of IMAR of the University of the Azores, Department of Oceanography and Fisheries, Horta, Azores, Portugal
pparretti@uac.pt

Ricardo Serrão Santos[1,2]
[1]MARE – Marine and Environmental Sciences Centre, University of the Azores Department of Oceanography and Fisheries, Portugal
[2]Centre of IMAR of the University of the Azores, Department of Oceanography and Fisheries, Horta, Azores, Portugal
rssantos4@me.com

Chapter 13

Richard Shine
University of Sydney
Department School of Biological Sciences
Australia
rick.shine@sydney.edu.au

Chapter 14

Aníbal Pauchard
Universidad de Concepción, Institute of Ecology and Biodiversity (IEB)
Facultad de Ciencias Forestales
Chile
pauchard@udec.cl

Rafael García
Universidad de Concepción, Institute of Ecology and Biodiversity (IEB)

Facultad de Ciencias Forestales
Chile
ragarcia@udec.cl

Sergio Zalba
Universidad Nacional del Sur. Bahía Blanca
Departamento de Biología, Bioquímica y Farmacia
Argentina
szalba@criba.edu.ar

Mauro Sarasola
Instituto Nacional de Tecnología Agropecuaria (INTA)
Argentina
msarasola@bariloche.inta.gov.ar

Rafael Zenni
Universidade deBrasília
Brazil
rafaeldz@gmail.com

Silvia Ziller
The Horus Institute for Environmental Conservation and Development
Brazil
sziller@institutohorus.org.br

Martín Nuñez
INIBIOMA, CONICET, Universidad Nacional del Comahue
Laboratorio Ecotono
Argentina
nunezm@gmail.com

Chapter 15

Mark S. Minton
Smithsonian Environmental Research Center
Marine Invasion Research Laboratory
USA
mintonm@si.edu

A. Whitman Miller
Smithsonian Environmental Research Center
Marine Invasion Research Laboratory
USA
millerw@si.edu

Gregory M. Ruiz
Smithsonian Environmental Research Center
Marine Invasion Research Laboratory
USA
ruizg@si.edu

Chapter 16

Michelle Leishman
Macquarie University
Department of Biological Sciences
Australia
michelle.leishman@mq.edu.au

Rachael Gallagher
Macquarie University
Department of Biological Sciences
Australia
rachael.gallagher@mq.edu.au

Chapter 17

Mário Santos
Universidade de Trás-os-Montes e Alto Douro
Laboratory of Applied Ecology, CITAB - Centre for the Research and Technology of Agro-Environment and Biological Sciences
Portugal
mgsantos@utad.pt

Rita Bastos
Universidade de Trás-os-Montes e Alto Douro
Laboratory of Applied Ecology, CITAB - Centre for the Research and Technology of Agro-Environment and Biological Sciences
Portugal
ritabastos@utad.pt

Joana Vicente
Universidade do Porto
Centro de Investigação em Biodiversidade e Recursos Genéticos & Faculdade de Ciências (Departamento de Biologia)
Portugal
jsvicente@fc.up.pt

Uta Berger
Technische Universität Dresden
Institute of Forest Growth and Computer Sciences
Germany
Uta.Berger@tu-dresden.de

Britaldo Silveira Soares Filho
Universidade Federal de Minas Gerais
Centro de Sensoriamento Remoto
Brazil
britaldo@csr.ufmg.br

Hermann Rodrigues
Universidade Federal de Minas Gerais
Centro de Sensoriamento Remoto
Brazil
hermann@csr.ufmg.br

Joaquim Alonso
Instituto Politécnico de Viana do Castelo
Escola Superior Agrária
Portugal
malonso@esa.ipvc.pt

Carlos Guerra
Instituto Politécnico de Viana do Castelo
Escola Superior Agrária
Portugal
carlosguerra@esa.ipvc.pt

João Martins
Universidade do Porto
Centro de Investigação em Biodiversidade e Recursos Genéticos & Faculdade de Ciências (Departamento de Biologia)
Portugal
joaofilipepm@gmail.com

João Honrado
Universidade do Porto
Centro de Investigação em Biodiversidade e Recursos Genéticos & Faculdade de Ciências (Departamento de Biologia)
Portugal
honradoj@gmail.com

João Alexandre Cabral
Universidade de Trás-os-Montes e Alto Douro
Laboratory of Applied Ecology, CITAB - Centre for the Research and Technology of Agro-Environment and Biological Sciences
Portugal
jcabral@utad.pt

Chapter 18

Frédérique Viard[1,2]
[1]Sorbonne Universités, UPMC Univ Paris 06
UMR 7144, Adaptation et Diversité en Milieu Marin, Station Biologique de Roscoff
France
[2]CNRS, UMR 7144, Adaptation et Diversité en Milieu Marin, Station Biologique de Roscoff, France
frederique.viard@sb-roscoff.fr

Thierry Comtet[1,2]
[1]Sorbonne Universités, UPMC Univ Paris 06
UMR 7144, Adaptation et Diversité en Milieu Marin, Station Biologique de Roscoff
France
[2]CNRS, UMR 7144, Adaptation et Diversité en Milieu Marin, Station Biologique de Roscoff, France
thierry.comtet@sb-roscoff.fr

Chapter 19

Theo Light
Shippensburg University
Department of Biology
USA
tsligh@ship.edu

Peter B. Moyle
University of California, Davis
Wildlife, Fish, and Conservation Biology
USA
pbmoyle@ucdavis.edu

João Canning-Clode

General Introduction – Aquatic and Terrestrial Biological Invasions in the 21st Century

Motivation and Book Structure

In the preface of his 2007 book, Wolfgang Nentwig (Nentwig, 2007) starts by posing an obvious question: "*Yet another book on biological invasions?*". In fact, the market already provides several valuable volumes focusing on biological invasions in both aquatic and terrestrial systems (e.g. Mooney *et al.*, 2005; Lockwood *et al.*, 2006; Davis, 2009; Rilov & Crooks 2009a; Richardson, 2011; Simberloff, 2013). However, invasion science is a dynamic discipline and new detection tools, modeling techniques, and eradication and management strategies have been developed and updated in recent years. For example, in the past 20 years, molecular genetic approaches have been increasingly used to investigate the origin and rapid evolution of non-indigenous species (NIS) and other biological invasion processes (Geller *et al.*, 2010; Darling, 2015). These DNA tools have recently been used to detect NIS, in taxonomic identification, to elucidate sources and vectors of introductions, to monitor NIS expansions in invaded regions, and to describe the consequences of introductions for native communities (Razgour *et al.*, 2013; Darling, 2015). Furthermore, modeling techniques have been developed in recent years to forecast spatial and temporal patterns of NIS distributions in future climate change scenarios (see also Chapter 17).

In this context, the present book integrates with the current questions and hypotheses being postulated in invasion science in a timely approach (Sax *et al.*, 2005; Rilov & Crooks 2009a; Galil *et al.*, 2011). It is generally accepted that a better understanding of the processes controlling the success of invasive species and what processes will influence their persistence in space and time will contribute to improved tools for environmental managers seeking to reduce or prevent invasions of new species.

In this book, a collection of efforts from 57 renowned worldwide invasion scientists covers our current knowledge of biological invasions as well as their impacts, patterns and mechanisms in marine, freshwater and terrestrial ecosystems. The book presents a multidisciplinary approach to biological invasions with key study cases from different biogeographic regions structured into four sections. In Part I – 'Biogeography and vectors of biological invasions', six chapters characterize several vectors of introduction of non-indigenous species as well as spatial and temporal scale patterns of invasions across different ecosystems and taxonomic groups.

Part II – 'Biological invasions in aquatic ecosystems and in host parasite systems', comprised of five chapters, is dedicated to invasions in aquatic ecosystems but also to the significance of parasites in the context of biological invasions. Furthermore, the four chapters included in Part III – 'Management and control of biological invasions' have a strong emphasis on different behaviors of well known invasive species and further discuss possible management tools and future advice for preventing and controlling this global environmental threat. Finally, the closing four chapters included in Part IV – 'Predictions and new tools in biological invasions' looks into the future of biological invasions. In this final section, contributions emphasize emerging molecular tools, climate change, and modeling techniques.

Brief Discipline History

Naturalists have likely been observing and thinking about biological invasions in some form or other for centuries. Indeed, Charles Darwin, Charles Lyell and Frank Egler all referred to invasive species in their work (Richardson, 2011). However, it was the pioneering and influential work of British ecologist Charles Elton (Elton, 1958) that first drew attention to the phenomena of biological invasions and, more importantly, their impacts on local communities and ecosystems. Since Elton's milestone work, a growing number of volumes across ecosystems and biogeographic regions have been contributing to strengthen the study of biological invasions (e.g. Mooney *et al.*, 2005; Lockwood *et al.*, 2006; Nentwig, 2007; Davis, 2009; Rilov & Crooks 2009a; Richardson, 2011; Simberloff, 2013). Likewise, a few scientific journals (e.g. *Biological Invasions, Aquatic Invasions, NeoBiota* or *BioInvasions Records*) and conferences (e.g. *International Conference on Marine Bioinvasions* or *NeoBiota - European Conference on Biological Invasions*) are exclusively dedicated to the study of biological invasions.

Moreover, and for example in the marine system, NIS have received great attention from scientists, managers and policy makers. As a result, this global problem has been addressed from different angles in various EU-funded FP6 and FP7 projects (ERNAIS, 2001; ALARM, 2008; DAISIE, 2008; VECTORS, 2011; INVASIVES, 2013). Valuable work has also proceeded on several regional concepts focused on databases and management (NEOBIOTA, 1999; REABIC, 2001; NOBANIS, 2012; AquaNIS, 2013). Recently, the European Marine Strategy Framework Directive (MSFD) has also included NIS among key descriptors required for assessing and setting a qualitative target - Good Environmental Status (GES).

Finally, the interest of the general public in this global problem is increasing and this is reflected in the intensification of media coverage on biological invasions, including television shows, magazines, newspapers and blogs, and many others.

The Invasion Process

When organisms are deliberately or accidentally introduced into a new ecosystem a biological invasion may take place. Whether in land or water, to successfully invade an ecosystem or a community, these so-called 'invasive species' have to pass through a series of stages known as the 'invasion process' (Lockwood *et al.*, 2006). The first stage of this process can be seen as the 'transport phase' where a particular species is transported (intentionally or unintentionally) from their native range to a new area and released into the wild. Second, arriving into a new environment, individuals must establish a viable population ('establishment phase'), or the population becomes extinct. Third, this now-established population of invasive individuals needs to increase in abundance and expand its geographic range ('spread phase'), otherwise it remains in low abundances and with only a small local distribution. Finally, once this invasive population is able to establish and spread its distribution range, the population will then alter the invaded community causing ecological and economic impacts ('impact phase'). The capacity of a given invader to pass each of these stages dictates the success of an invasion.

Nowadays, there are several cases of successful invasive species both in aquatic and terrestrial systems, that due to their impacts (ecological and/or economic) have become famous worldwide. A good example in the marine system is the well-documented invasion of the Indo-Pacific lionfish *Pterois volitans* and *Pterois miles* (featured in this book cover) into the Western Atlantic and Caribbean. The first sighting of this marine invader was reported in Florida in 1985, and since their establishment in the Bahamas in 2004, they have colonized more than 7 million km^2 in the Western Atlantic and Caribbean (Côté *et al.*, 2013). In the terrestrial system, the cane toad *Bufo* [*Rhinella*] *marina* is probably considered the most famous and iconic invader (see Chapter 13). These toads are native to South and Central America and have been moved to several countries to (unsuccessfully) control insect pests. These toads were introduced to Australia in 1935 and have since spread across several regions of the continent with severe ecological and economic impacts (Chapter 13).

Challenges in the 21st Century

Once successful, biological invasions by animals, plants or pathogens are one of the greatest environmental and economic threats and, along with habitat destruction, a leading cause of global biodiversity loss (Mack *et al.*, 2000). In addition, the rate of detected invasions in terrestrial and aquatic ecosystems has increased significantly in the last two decades for several reasons: 1) search effort, which follows the establishment of specialized research groups in different continents and across different disciplines; 2) the development and use of new emerging molecular approaches; and 3) climate change.

The growing interest in invasion science in the 21st century is reflected in a simple and original literature survey I performed for this introductory account. To verify the growing interest in biological invasions in this century I conducted a simple search at "Web of Science" for articles published between 2000 and 2014 that included the keywords 'invasive' OR 'invasion' OR 'non-native' OR 'alien' as the main topic. To avoid any bias with medicine and cancer research due to the use of the terms 'invasive' and 'invasion', I restricted the search to the following research areas: ecology, biology, limnology, fisheries, forestry, plant sciences, marine freshwater biology, biodiversity and conservation, environmental sciences, oceanography, entomology and zoology. The results of this literature survey show that a total of 36344 articles on biological invasions were published from 2000 to 2014 with a growing tendency over the years (Figure G1A). Lockwood *et al.* (2006) performed a simple meta-analysis where they also show an increasing pattern on the number of citations returned from a search on the Science Citation Index between 1975 and 2004 within the field of invasion ecology. The results I show here corroborate the same tendency after 2004 but it seems the number of papers appear to stabilize after 2011 at a rate of a little below 4000 papers per year (Figure G1). Interestingly, the United States of America is the country that contributes by far the most scientific articles on biological invasions, followed by Australia and Canada (Figure G1B). China and European countries such as England, France, Germany, Spain and Italy complete the top 10 territories that contribute the most to the scientific body of knowledge on this phenomenon. Finally, the majority of these contributions are in the field of Environmental Sciences and Ecology with more than 16000 papers in the course of 15 years (Figure G2). Marine and Freshwater Biology, as well as Plant Sciences, are also well represented with approximately 6000 contributing papers between 2000 and 2014.

For the coming years, I expect that the interest in biological invasions will continue, with the production of more scientific papers and an increase in citations. New records and detections of non-indigenous species will probably continue in the next decades in several biogeographic regions, both in aquatic and terrestrial ecosystems. New emerging techniques such as molecular approaches and spatial and temporal modeling will play a key role in producing a significant amount of papers in coming years. However, the slope of increase in publication rate will probably decrease in future years as can already be seen in Figure G1A after 2010.

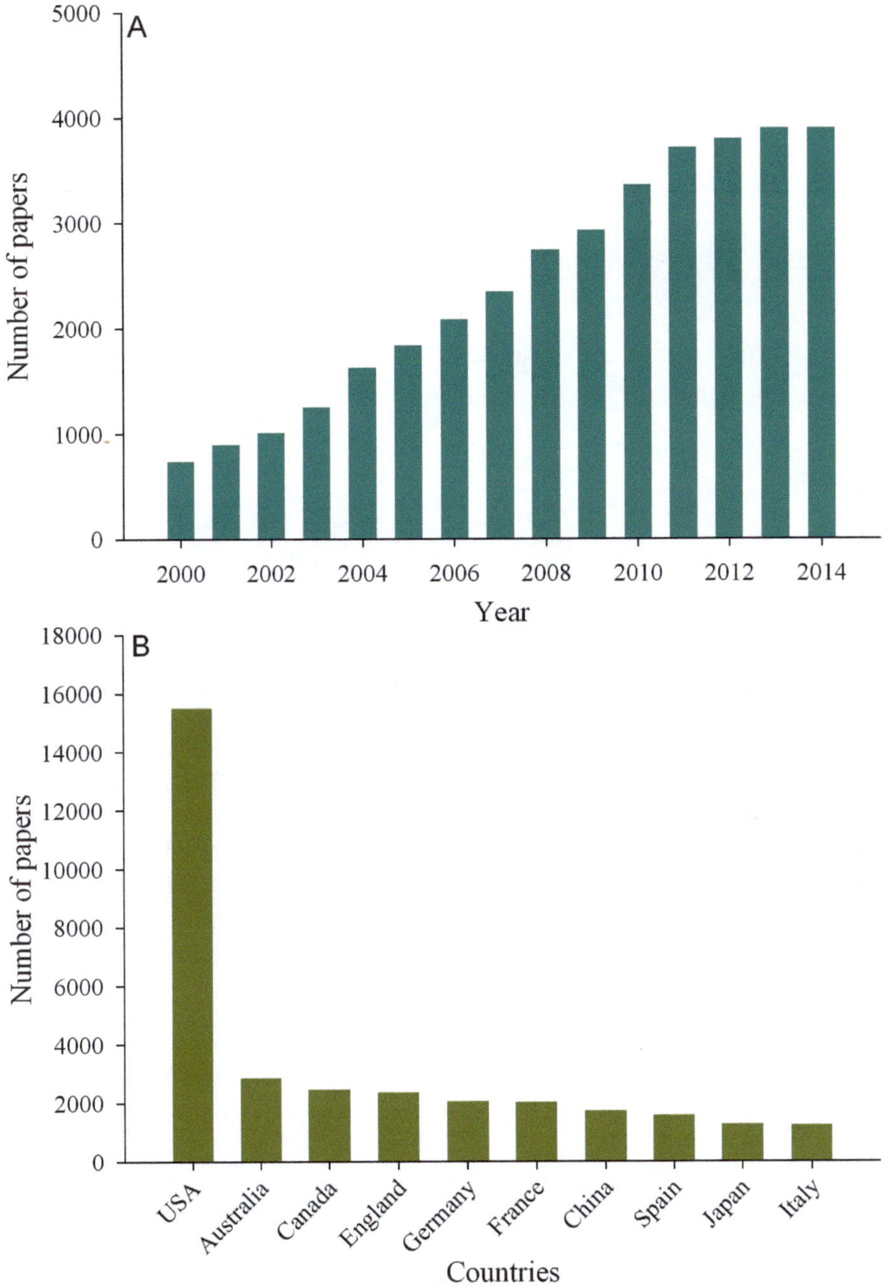

Fig. G1: Number of articles published in invasion science from 2000 to 2014 for this literature search (A) and most represented countries contributing to the field in the 21st century (B).

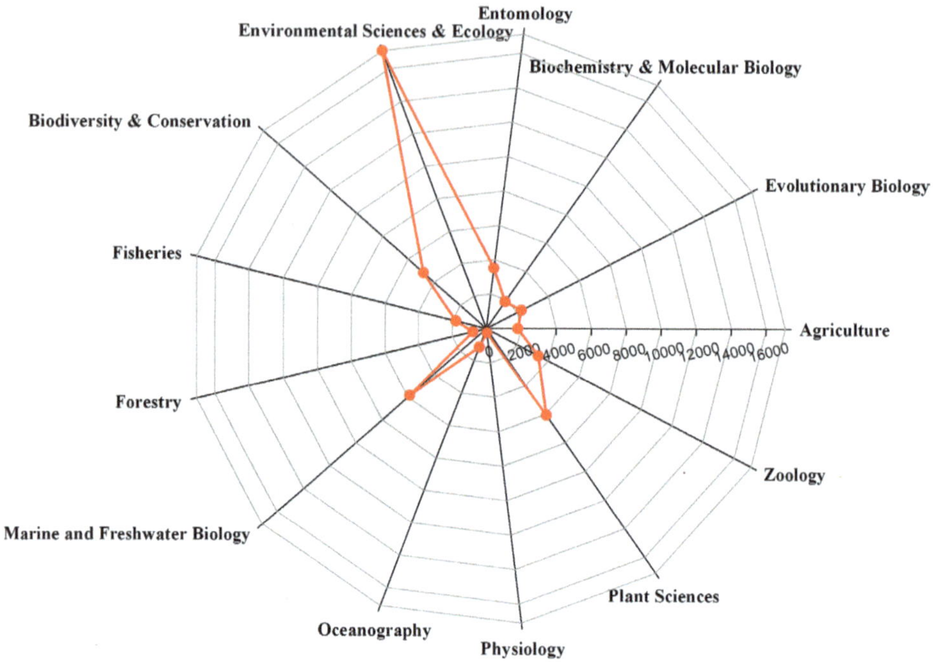

Fig. G2: Most represented research areas in papers published between 2000 and 2014 in biological invasions from a literature survey I performed in "Web of Science".

A Final Note on Definitions and Invasion Terminology

In recent years there have been several efforts trying to achieve harmony and consistent usage in terminology within invasion biology and ecology (e.g. Occhipinti-Ambrogi & Galil, 2004). This much-needed consistency in bioinvasion terms would likely facilitate debate about this issue for the scientific community, but also for policy makers, managers and the general public. Below I outline a few key terms now being used in invasion science:

Non-Indigenous Species (NIS): I personally have preference for this designation as it describes a given species that was moved outside its usual geographical range via anthropogenic actions (this could be intentional or accidental), irrespective of its impact on native species and native ecosystems.
Invasive species: an invasive species by definition must be a NIS, but one that has caused demonstrable impact, both in ecological and economic terms.
Cryptogenic species: this is a species of unknown origin or a species that is neither undoubtedly native nor NIS.

Native species: this term refers to a species that occurs naturally in a given area/ecosystem/region. These species were not introduced via human actions, either intentionally or accidentally.
Biological invasion or bioinvasion: This is a very broad term that refers to the introduction of NIS into new ecosystems/area/regions via human actions but also considers natural range expansions.
Propagule pressure: this term can be seen as the introduction effort, i.e. the pool of individuals introduced in a new ecosystem/area/region and the number of times it is released.

However, concepts currently used in invasion science are highly unlikely to ever reach national or international uniformity because they vary among scientific disciplines, countries, and linguistic borders (Carlton, 2002; Occhipinti-Ambrogi & Galil, 2004; Rilov & Crooks, 2009b). In future years, I expect invasion biologists and scientists will continue to employ i) the same word that probably means distinctive things to different workers; but also ii) different words that mean the same thing. A good example of this is the word '*invasive*', which could imply a species with documented ecological and economical impact in a certain region or simply suggest nothing more than an '*exotic*', '*alien*' or '*non-indigenous species*'. A further example would be the definition of '*propagule pressure*' which, for certain scientists, simply defines the term quantitatively as the number of individuals introduced in a new area and the number of times an invader is released. In contrast, to other authors '*propagule pressure*' could imply a much broader concept, including propagule viability and other parameters such as stress tolerance. In this context, I decided not to attempt any consistency in invasion terms for the current volume and, as such, all authors in this book have used invasion-related terms of their own choosing.

Bibliography

ALARM (2008). Assessing Large Scale Risks for Biodiversity with Tested Methods (ALARM). European Commission Sixth Framework Programme (FP6).
AquaNIS (2013). Editorial Board. Information system on Aquatic Non-Indigenous and Cryptogenic Species. World Wide Web electronic publication. http://www.corpi.ku.lt/databases/aquanis. Version 2.36+. Accessed 2015-01-26.
Carlton, J. (2002). Bioinvasion Ecology: Assessing Invasion Impact and Scale. In E. Leppäkoski, S. Gollasch S. Olenin (Eds.), Invasive Aquatic Species of Europe. Distribution, Impacts and Management (pp. 7-19), Netherlands: Springer.
Côté, I. M., Green, S. J., Hixon, M. A., (2013). Predatory fish invaders: Insights from Indo-Pacific lionfish in the western Atlantic and Caribbean. Biological Conservation, 164, 50-61.
DAISIE (2008). DAISIE European Invasive Alien Species Gateway (http://www.europe-aliens.org).
Darling, J. (2015). Genetic studies of aquatic biological invasions: closing the gap between research and management. Biological Invasions, 17, 951-971.
Davis, M. A. (2009). Invasion Biology. Oxford: Oxford University Press.
Elton, C. (1958). The ecology of invasions by animals and plants. Chicago: University of Chicago Press.

ERNAIS (2001). European Information and Research Network on Aquatic Invasive Species (ERNAIS). http://www.reabic.net/ERNAIS.aspx. Last access November 2015.
Galil, B. S., Clark, P. F., Carlton, J. T. (2011). In the wrong place – alien marine crustaceans: distribution, biology and impacts. Berlin: Springer.
Geller, J. B., Darling, J. A., Carlton, J. T. (2010). Genetic Perspectives on Marine Biological Invasions. Annual Review of Marine Science, 2, 367-393.
INVASIVES (2013). Invasive seaweeds in rising temperatures: impacts and risk assessments. SEAS-ERA. ERA NET. European Commission Seventh Framework Programme (FP7).
Lockwood, J., Hoopes, M., Marchetti, M. (2006). Invasion Ecology. Malden, USA: Blackwell Publishing.
Mack, R. N., Simberloff, D., Lonsdale, W. M., Evans, H., Clout, M., Bazzaz, F. A. (2000). Biotic invasions: Causes, epidemiology, global consequences, and control. Ecological Applications, 10, 689-710.
Mooney, H. A., Mack, R., McNeely, J. A., Neville, L. E., Schei, P. J., Waage, J. K. (2005). Invasive Alien Species: A New Synthesis. Washington, DC: Island Press.
Nentwig, W. (2007). Biological Invasions. Ecological Studies, Vol. 193. Heidelberg: Springer Verlag.
NEOBIOTA (1999). European Group on Biological Invasions - NEOBIOTA. http://www.oekosys.tu-berlin.de/menue/neobiota. Last access November 2015.
NOBANIS (2012). European Network on Invasive Alien Species - NOBANIS. http://www.nobanis.org, Last access November 2015.
Occhipinti-Ambrogi, A., Galil, B. S. (2004). A uniform terminology on bioinvasions: a chimera or an operative tool? Marine Pollution Bulletin, 49, 688-694.
Razgour, O., Juste, J., Ibáñez, C., Kiefer, A., Rebelo, H., Puechmaille, S. J., Arlettaz, R., Burke, T., Dawson, D. A., Beaumont, M., Jones, G. (2013). The shaping of genetic variation in edge-of-range populations under past and future climate change. Ecology Letters, 16, 1258-1266.
REABIC (2001). Regional Euro-Asian Biological Invasions Center (REABIC). http://www.reabic.net/About.aspx. Last access November 2015.
Richardson, D. M. (2011). Fifty Years of Invasion Ecology: The Legacy of Charles Elton. Berkely: Wiley-Blackwell.
Rilov, G., Crooks, J. (2009a). Biological Invasions in Marine Ecosystems: Ecological, Management, and Geographic Perspectives. Ecological Studies Series. Berlin: Springer.
Rilov, G., Crooks, J. (2009b). Marine Bioinvasions: Conservation Hazards and Vehicles for Ecological Understanding. In G. Rilov, J. Crooks (Eds.), Biological Invasions in Marine Ecosystems (p.3-11). Berlin, Heidelberg: Springer.
Sax, D. F., Stachowicz, J. J., Gaines, S. D. (2005). Species invasions: insights into ecology, evolution and biogeography. Sunderland, MA: Sinauer Press.
Simberloff, D. (2013). Invasive Species: What Everyone Needs to Know. Oxford: Oxford University Press.
VECTORS (2011). Vectors of Change in Oceans and Seas Marine Life, Impact on Economic Sectors (VECTORS). European Commission Seventh Framework Programme (FP7).

Part I. **Biogeography and Vectors of Biological Invasions**

João Canning-Clode, Filipa Paiva

Summary of Part I

Biological invasions have become one of the most significant threats to global biodiversity as they have severe consequences for ecosystems. The movement of aquatic and terrestrial non-indigenous species worldwide, either intentional or accidental, has triggered a homogenization of biota by breaking down natural barriers. Numerous vectors responsible for the introduction of these non-indigenous species have been identified and described over the years. Several studies have been conducted exclusively to understand and quantify the role of vectors and the impacts of invasions.

The first section of this book includes six chapters characterizing several vectors of introduction as well as spatial and temporal scale patterns of invasions across different ecosystems and taxonomic groups. The section opens with an account on vectors of marine invasions, where James Carlton and Greg Ruiz (**Chapter 1**) present a general framework for vector science, expanded from a previous 10-year-old model. In the marine system, several studies conducted over the years have contributed to the current understanding of species movements on our planet. However, over time, the causes, routes, corridors, vectors, propagule pressure, and vector strength of these movements have changed, leading to a fair number of unanswered questions. In this opening chapter, Carlton and Ruiz leave the term 'pathway' behind and give life to 'vector science', a framework that could encompass the 4 stages of the standard models of invasion (transport, introduction, establishment and spread). The authors further review the role of several vectors that contribute to the spread of marine invasive species.

Chapter 2 by Phillip Cassey and co-authors reviews the biogeography of bird invasions, with particular emphasis on the history and market trade of birds. Despite their great capacity for flying long distances, bird species have also been influenced by humans in their movement across continents. Interest in birds has been increasing over time as a source of food, hunting sports, and also for pets and ornamentation, contributing to a breaking down of the remaining barriers. This chapter identifies two eras that had a major impact on the spread of non-native birds: i) the era of the Acclimatisation Societies between the eighteenth and twentieth centuries; and (ii) the era of the international trade in wild birds for bird-keeping from the late-twentieth century to the present. The authors refer to several examples from different geographic locations and elaborate on future trends in avian invasions.

Mark Sytsma and Toni Pennington provide an overview of the pathways and vectors involved in the anthropogenic and natural dispersal of freshwater vascular plants in **Chapter 3**. Non-native vascular plants are known to cause ecological damage when transplanted to other ecosystems. In this chapter, the authors categorize vectors as primary (vectors for initial introduction to a new habitat) and secondary (natural processes for spread following establishment) and illustrate these two categories with several examples. In this account, authors attribute key importance to vector management, but suggest that new effective predictive models of non-native species dispersal will require a combination of better knowledge of organism biology, probability of introduction, and site suitability.

Insects and other arthropods, too, are considered relevant invasive species both in magnitude and in their economic impact. **Chapter 4** by George K. Roderick and Maria Navajas outlines the major pathways and vectors for insect invasions and also provides novel insights on new emerging tools, such as molecular population genetics, computational methods, climate modeling, and collection science. The authors further highlight the significance of policy and risk assessment in the management of invasive arthropods.

As a result of human actions, freshwater fishes and invertebrates are capable of traveling between and across ecosystems, contributing to a more homogeneous biodiversity. In freshwater ecosystems, aquaculture and aquarium release appear to be the main sources of introduction, but stocking and ornamental pet industry have high importance. Through time, many factors that influence invasion by invertebrates and fishes have changed due to reasons ranging from improvements of transport to the change of peoples' interest in different animal groups. In **Chapter 5** of this work, Pam Fuller reviews the causes and vectors of invasions in freshwater fishes and invertebrates with several North American examples and also provides basic solutions for these problems.

Finally, in **Chapter 6** Christina Romagosa describes the live animal trade, the vectors through which animals are transported, and how each vector has contributed to invasions on a global scale. The transport of live animals involves the exchange of several millions of individuals worldwide and is a growing issue. The intentions of the transport of live animals are numerous and human-driven, such as zoos, tourist attractions, and religious purposes. In addition, certain species are exploited for other purposes such as food, traditional medicine, or fishing bait.

James T. Carlton, Gregory M. Ruiz

1 Anthropogenic Vectors of Marine and Estuarine Invasions: an Overview Framework

1.1 Introduction

We present here a simple general framework for vector science, modified, expanded and updated from earlier discussions a decade ago in Carlton and Ruiz (2005). Vector science encompasses the phenomena involved in the passive movement of species by human activities and the resulting propagule pressure, as well as the number of invasions related to a given vector over time. Vector science thus captures portions of four of the stages in classic invasion models: *transport*, *introduction*, *establishment* and *spread* (Blackburn *et al.*, 2011; Lockwood *et al.*, 2013).

Numerous papers have summarized and classified the many vectors by which human activity moves marine and estuarine biota (including viruses, bacteria, protists, fungi, animals, and plants) within and between oceans (Carlton & Scanlon, 1985; Carlton *et al.*, 1995; Hayward, 1997; Cohen & Carlton, 1997; Campbell & Hewitt, 1999; Chapman *et al.*, 2003; Carlton & Cohen, 2003; Ribera, 2003; Padilla & Williams 2004; Lewis *et al.*, 2005; Minchin *et al.*, 2005; Minchin *et al.*, 2006; Cook *et al.*, 2008; Davidson *et al.*, 2008; Cohen & Zabin, 2009; Chang *et al.*, 2009; Hewitt *et al.*, 2009; Davidson *et al.*, 2010; Haydar & Wolff, 2011; Ruiz *et al.*, 2011; Frey *et al.*, 2014; Ashton *et al.*, 2014). These vectors include, for example, the movement of species with marine products (such as aquaculture stocks, live seafood, and bait, as well as the epi- and endo-biota associated with these) and with ships, dry-docks, barges, drilling platforms, and other watercraft, and the release of species for purposes such as fisheries enhancement, marsh or seagrass restoration, and as aquarium discards.

Quantification of the scale and tempo of these and many other vectors varies hugely over time and space. While some vectors have been extensively studied, many others remain poorly known, resulting in, at best, a largely sparse and fragmented data picture for most vectors, especially in terms of informing modern-day management strategies (Williams *et al.*, 2013). Adding to this menu are other emerging vectors that also remain poorly assessed, such as the increasing amount of ***marine debris*** (long-lasting floating substrates, such as plastic, styrofoam, and fiberglass, which may significantly alter the natural transoceanic dispersal of many species (Barnes, 2004; Barnes & Milner, 2005; Godwin *et al.*, 2008; Gregory, 2009). A recent example is the debris field generated by the 2011 Tōhoku Earthquake and Tsunami, leading to the dispersal of Japanese species across the North Pacific Ocean to North America and the Hawaiian Islands (Calder *et al.*, 2014). Other vectors arise as unintended consequences of various endeavors (such as species attaching to and being dispersed on tracking bands of migratory birds (Tottrup *et al.*, 2010) and marine mammals (Reisinger *et al.*, 2009), as well as the movement of scientific sampling gear between hydrothermal vents (Voight *et al.*, 2012).

Rather than reviewing these many individual vectors, we take a broad view and ask how the range of phenomena involved in vector science can be rigorously framed and quantified.

1.2 A Framework for Vector Science

Table 1.1 presents a model that focuses on the definition and examples of quantification of six elements of vector science: cause, route, corridor, vector, propagule pressure, and vector strength. The word *pathway* is abandoned (as proposed by Carlton & Ruiz, 2005), as it is used uncritically to refer to any of the first four of these distinct phenomena. While "pathway" is in widespread use in management and regulatory frameworks, it is exactly for this reason that we argue for replacing the word with more fine-grained, distinct concepts, to clarify and focus management goals and regulatory enforcement. As noted, this suggestion is not new, but was proposed 10 years ago. We note that virtually all aspects of the phenomena and vector elements discussed here may be modified by global climate change (Pederson *et al.*, 2011), including warming sea temperatures (which may alter species survival during transit), storm activity, sea level rise (impacting port and harbor conditions), and so forth. These scenarios are beyond the scope of the present chapter, but are a critical foundation for predictive and risk assessment models for vector management.

Tab. 1.1: A framework for vector science and its quantification.

PHENOMENON	EXAMPLES OF QUANTIFICATION [per location (such as a Port or Port System) in a defined time period (daily, annual, decadal, etc.)]
CAUSE Why a species is transported, either **accidental [AC]** (unintentional, inadvertent, escape) or **deliberate [DB]** (intentional, planned, planted)	* # deliberate vectors * # accidental vectors * # of species assignable to AC or DB release or both
ROUTE The geographic path over which a species is transported from the **donor area** (origin; may be defined as Last Port of Call [LPOC] to the **recipient area** (destination or target), which may include one or more corridors	* # and location of donor regions (may be as # of LPOC, or # of general biogeographic regions) * # and location of recipient regions * # and location of geographic routes (= total range of different regimes through which vector may pass during transit, such as a vessel moving solely through temperate waters, tropical waters, or combinations thereof)

continued **Tab. 1.1 :** A framework for vector science and its quantification.

PHENOMENON	EXAMPLES OF QUANTIFICATION [per location (such as a Port or Port System) in a defined time period (daily, annual, decadal, etc.)]
CORRIDOR The physical conduit over or through which the vector moves within a route, such as roads, highways, canals, sea lanes, and railroad beds	* # and location of corridors (for example, the number of sea lanes in use, now or historically; a ***route*** may be a combination of different sea lanes) * total # of non-native introductions associated with a corridor (for example, the # of successful invasions believed to have passed through the Suez Canal into the Mediterranean Sea)
VECTOR How a species is transported, that is, the physical means or agent	* # vectors (such as shipping and its subcomponents) * size and rate: quantity (in appropriate units) / time * duration: length of transit time from donor to recipient area * timing: for example, the season when a vector is active * total # of non-native introductions associated with a vector
PROPAGULE PRESSURE The quantity, quality, and frequency of propagules (such as spores, eggs, larvae, or adults) released in a given location	* diversity: # species/vector/time period * density (D): # individuals/unit area or volume/vector * frequency (rate): D/vector/time * quality: viability and/or reproductive capability (for example, % ovigerous individuals/species)
VECTOR STRENGTH The established invasions in a region related to a given vector	* # established non-native species / vector / unit of time

1.2.1 Causes of Invasion

In the long history of human-assisted invasions, species have been moved both **accidentally** (often termed "hitchhikers") and **deliberately** (often linked to a specific purpose, motive, or rationale). Distinguishing between these two may in many circumstances (for both historical and modern invasions) not be possible, although genetic analyses linked to geographic origin of certain introduced populations may assist in rejecting one or the other (for example, the improbability of known vectors moving a species accidentally from one point to another). Many species have been moved both intentionally and unintentionally, and deliberate releases may be legal or illegal. Quantification of causes thus consists of assigning species to one of three bins: *accidental, deliberate* or *both.*

1.2.2 Routes of Invasions

Invasion routes – the geographic paths – from potential **donor** to potential **recipient** (destination or target) areas have changed vastly over the past several thousand years. Intraoceanic travel by vessels may date back tens of 1000s of years, transoceanic travel by 1000s of years, and interoceanic travel by centuries (Riley *et al.*, 1971; Hurles *et al.*, 2003; Matisso-Smith & Robins, 2004; Balter, 2007; Hattendorf, 2007; Anderson, 2009). Transport of species both in and on ships has proceeded for so long to all shores of the world that the earliest invasions, linked to early voyages of discovery, have not yet been detected. That interoceanic invasions were well underway long ago is manifested by the ship-mediated presence of the Pacific Ocean green alga *Halimeda opuntia* in the Caribbean by the 1600s (Kooistra & Verbruggen, 2005).

It is thus critical to involve maritime historians in any analysis of the invasion history of a particular region. Many shipping routes, including routes of exploration, colonization, and trade, are now extinct. These passages provided key linkages between distant areas of the globe that no longer exist by sea routes today, and without doubt transported a great many species whose modern-day distribution is best explained by now-defunct paths. In turn, modern-day global shipping assessed in terms of network analysis has proven to be a fruitful approach in terms of understanding the modern scale of connectivity (Kaluza *et al.*, 2010; Keller *et al.*, 2010; Seebens *et al.*, 2013).

While most literature has concentrated on the carriage of species from a donor area to the recipient area, surprisingly little work has focused on the accumulation of species *en route* by a vector, with ships providing an obvious experimental platform in this regard. As reviewed by Carlton (2011), Darwin (1854) noted that a vessel travelling from Britain to West Africa acquired the barnacle *Megabalanus tintinnabulum* in the region of Namibia, and then sailed to South America, where *Balanus psittacus* settled on the *Megabalanus*; on the return voyage, additional *M. tintinnabulum* then settled on the *B. psittacus*. Similarly, Pilsbry (1916) described a ship being colonized in the Caribbean by the barnacle *Balanus trigonus*, upon which *M. tintinnabulum* then settled, followed by the barnacle *Newmanella radiata*. On return to New England, young *Amphibalanus eburneus* barnacles were acquired.

Carlton & Hodder (1995), using an experimental approach, describe the accumulation of species on a vessel travelling along the Pacific coast of North America. Results included the long-distance movement of both introduced and native species, the acquisition of benthic species (which boarded the vessel when it settled into harbor mud at one point), and the decapitation of the reproductive polyps of hydroids, leading to the potential broadcast dispersal of the latter as the ship moved down the coast. Chapman *et al.* (2013) describe the accumulation and dispersal of barnacles and oysters on a vessel with a long Atlantic history moving through the Panama Canal to the Pacific coast of North America.

Marine buoys are likely candidates for similar accumulation scenarios, but few such incidents have been described. Abbott (1961) reported finding a stranded buoy on a beach in the Carolinas (United States) which had become fouled by a tropical assemblage of species, including the bivalves *Chama*, *Spondylus*, and *Pinctada*; then, having been carried north, the buoy acquired a "circular band" of small blue mussels (*Mytilus edulis*).

Quantification of routes consists of assessing the number of well-resolved geographic paths to a given region, ideally followed by an assessment of how the source and number of routes have changed over time.

1.2.3 Invasion Corridors

Corridors and routes are often confused in the literature. For example, the Panama Canal may be referred to as a *route* of invasion, whereas it is better framed as the physical conduit through which ships (the *vector*) pass from one port to another (the *route*, as defined above and in Table 1). Certain sea corridors are, fortunately, far easier to assess through historical time than are many sea routes. Thus, the opening of modern-era sea level canals can be dated to the time of the first ship that passed through. A terrestrial analogy is the opening of long-distance railroads, and the first train to pass along the rail bed.

Important to consider is the detailed history of a corridor over time, including how it may have changed environmentally. Species accumulations by vectors in corridors have rarely been studied, although presumably passages are typically not of long-lasting duration. In a remarkable case of long-corridor residency, 14 ocean-going ships became trapped in the Suez Canal for 8 years between 1967 and 1975 during the "Six Day War;" these ships became known as the "Yellow Fleet," due to the yellow desert dust that eventually covered them (Ben-Eliahu & ten Hove, 2011). Lengthy residency time of a vessel in port or at anchor presumably leads to increased potential for already-present fouling species to release propagules or to become detached from the hull, although this does not appear to have been studied with the Yellow Fleet. Instead, however, the Fleet "acquired a massive biofouling aggregation" (which a Norwegian expedition sampled in January 1975), with vessels either then leaving the Canal with these aggregations under their own power (Ben-Eliahu & ten Hove, 2011) or being towed away.

Quantification of corridors consists of measurements of the number of species passing along or through a corridor over a given period of time.

1.2.4 Vectors

Vectors are the conveyance in or on which a species is transported. As noted above, an extensive literature now exists on the plethora of human-mediated activities and agents moving marine and estuarine species around the world in relatively short periods of time. Far less attention has been paid to a detailed exploration of historical vectors, such as solid ("dry") ballast used by ships for 1000s of years (Carlton, 2007). Much fruitful work remains here, especially when linked to historical routes. Such is the diversity of both historic and modern vectors that it is often a challenge to unequivocally assign a particular introduction to a particular vector. As noted by Carlton & Ruiz (2005), such cases are often referred to as being due to an "unknown vector," when polyvectism is actually meant. Species for which there is no known vector to clearly explain their appearance in a new region (i.e. cryptovectic species) are extremely rare.

Of no small importance is to understand the diversity of modern-day vectors moving along routes in an increasingly polyvectic world. Considerable effort is required to assess which vectors are in play in a given region. For a rapid assessment approach, *vector blitzes* have been suggested (Carlton, 2009; Williams *et al.*, 2013). In its simplest manifestation, a vector blitz would consist of deployment of a team of professionals and students to survey—in one urban region in a short period of time—ports, fishing docks, seafood vendors (including grocery stores), bait shops, aquarium stores, colleges, schools, and other venues for the presence of live non-native animals and plants.

Vectors are quantified in a number of ways to assess patterns in time and space: by *vector diversity* (the number of distinct vectors in a given location within a given time period), *vector size* (quantity, such as gallons of ballast water, or ship hull surface area), *vector duration* (the time of the voyage from A to B, relative to increasing (faster voyages) or decreasing (slower voyages) survival of entrained or attached species), and *vector timing* (such as the season when the vector is active).

1.3 Propagule Pressure

Propagule pressure consists of not only the *number* (*density*) of propagules released into a region over a given period of time (the *inoculum*), but the *diversity* of species involved, as well as the *viability and reproductive capacity* of those species upon release (Wonham *et al.*, 2005; Verling *et al.*, 2005; Colautti *et al.*, 2006; Carlton & Ruiz, 2011). Measuring propagule pressure is thus no small challenge. Ruiz *et al.* (2013) noted that vector metrics such as number of arriving vessels or volume of ballast water discharged may be poor proxies for propagule supply (see also Carlton & Ruiz, 2011, for additional discussion).

In terms of propagule diversity, taxonomic expertise is often severely limited, with a consequent focus on larger and better known species, resulting in what may be an extensive under-estimate of inoculum diversity. Often ignored or marginally identified in vector sampling are protists (which may include scores of species), sponges, hydroids, nematodes, flatworms, small polychaetes, copepods, and a large number of other microscopic or difficult-to-identify taxa, with mention of these lacunae being rarely made in vector studies. Many species may also be overlooked because of the limited knowledge of those analyzing samples, with taxa such as folliculinid ciliates or the rhizopod *Gromia*—both of which can be common if not abundant in many samples—simply being passed over, unrecognized as living organisms. While next-generation sequencing of bulk samples may be able to identify the spectrum of species being carried by a given vector at a given time, the result is currently likely to produce an amalgamation of a large number of unidentifiable OTUs (operational taxonomic units), as well as both weak and strong GenBank "hits", thus still leaving the identification of many species uncertain or unknown.

Strictly speaking, propagules are only propagules if they are capable of surviving in the environment after release or discharge. Those individuals that are severely physiologically compromised upon arrival such that they are not capable of reproduction, or species that die immediately upon entering a new environment (due, for example, to a severe temperature or salinity shock) would not be members of the "pressure" component of a propagule equation.

Toward this end, and using ballast water as a model system, a number of investigators have attempted direct and indirect assessments of during-transport or post-transport viability. Examples include Hamer *et al.* (1998) looking at the viability of the megalopa larva of the American crab *Cancer irroratus*; Smith *et al.* (1999), who collected polychaetes, bivalve mollusks, copepods, and mysids from ship ballast water and assessed survivorship against temperature and salinity characteristics of Chesapeake Bay; and Bailey *et al.* (2003) who examined the viability of cladoceran, copepod, and rotifer diapausing eggs in ballast water sediment. Kang *et al.* (2009) examined phytoplankton viability in ballast water in growth media of different temperature regimes, while Steinberg *et al.* (2011) looked at the efficacy of different vital stains to determine protist viability. Villac & Kaczmarska (2011) examined the viability of diatoms in ballast water using the vital stain fluorescein diacetate, protoplasm integrity, and chlorophyll autofluorescence, as well as growing phytoplankton samples in culture media (which further revealed species not otherwise detected). Wang *et al.* (2014) have described a portable rapid-detection system for live microalgae in ballast water.

These and other experimental studies point to a general conclusion that a great many species both survive ballast water transport and are capable of reproduction. Earlier work (Carlton & Geller, 1993), which had the advantage of a marine laboratory a short drive from an active port—thus permitting examination of a large number of live ballast samples and concomitant culture and grow-out (by D. Carlton) of a wide range of taxa—similarly demonstrated on-arrival viability and reproductive capacity of many invertebrate and diatom species.

For ballast water and other vectors, additional evidence of post-transport viability (and thus propagule pressure) is the presence of individuals that are reproductive, either by having mature gonads, by being ovigerous, or by the presence of incubated larvae or juveniles (such as in peracarid crustaceans). Experimental approaches can also provide insight into the potential scope of viability of individual species under mimicked transit conditions (for example, Schaffelke & Deane, 2005; Nyberg & Wallentinus, 2009; Carney *et al.*, 2011).

Quantification of propagule pressure thus involves as thorough an assessment of species diversity as is feasible and practicable, estimates of propagule density and, when possible, estimates of the number of individuals and species that are reproductive upon arrival.

1.3.1 Vector Strength

Vector strength is the number of species believed to have been introduced with a given vector over a given period of time. Vector strength is typically underestimated—sometimes extensively so—by the need to assign many species to a polyvectic status, obscuring detailed patterns of any particular vector over space and time.

Quantification of vector strength consists of calculating the number of species per vector per unit time.

1.4 Epilogue

Causes, routes, corridors, vectors, propagule pressure, and vector strength have varied vastly over centuries along all shores of the world. Despite a vastly burgeoning interest in invasion science in the last 25 years, a surprising number of gaps exist in our knowledge and understanding of how vectors operate, offering fruitful areas for research at every level. For example, we know of no region in the world—no port, harbor, bay, or estuary—where all aspects of vector science as described above have been characterized. Individual components—the causes, routes, corridors, vectors, propagule pressure, and vector strength—have been described and occasionally quantified in various regions, but no one location has been the benefit of a complete historic and modern assessment of vector pressure. This type of lacuna—which should, in theory, be addressable—in our understanding sets forth clear and important goals for marine invasion science.

In a nutshell

- Scores of human-mediated vectors are in play today that are capable of moving marine species within a matter of hours or days to any shore of the world. Vector science attempts to describe the full range of phenomena in these transport events.
- Invasions may be due (*cause*) to accidental or deliberate movements along geographic paths (*routes*) from donor to recipient regions. Between source and target areas, additional species may also be acquired by the vector.
- Vectors and the species they carry may move along well-defined *corridors* – such as sea lanes and through canals – which themselves may influence dispersal history. Many *vectors* (*polyvectism*) may move one species along many routes, and many vectors deliver millions of propagules to a region in a single episode.
- Propagule pressure is thus measured as the diversity of delivered species (although the species-level identify of many taxa may be extraordinarily difficult to determine) and their density, along with an assessment of the reproductive viability of the post-transported individuals.
- *Vector strength* is the number of species introduced with a given vector, but is typically underestimated, at times extensively so, by the need to assign many species to a polyvectic status, obscuring detailed patterns of any particular vector over space and time. No region of the world has been adequately described at the beginning of the 21st century in terms of overall vector pressure.

1.5 Bibliography

Abbott, R. T. (1961). How to know the American marine shells. New York: The New American Library.

Anderson, A. (2009). The rat and the octopus: initial human colonization and the prehistoric introduction of domestic animals to Remote Oceania. Biological Invasions, 11, 1503-1519. Ashton, G., Davidson, I., Ruiz, G. (2014). Transient small boats as a long-distance coastal vector for dispersal of biofouling organisms. Estuaries and Coasts, 37, 1572-1581.

Bailey, S. A., Duggan, I. C., van Overdijk, C. D. A., Jenkins, P. T., MacIsaac, H. J. (2003). Viability of invertebrate diapausing eggs collected from residual ballast sediment. Limnology and Oceanography, 48, 1701-1710.

Balter, M. (2007). In search of the world's most ancient mariners. Science, 318, 388-389.

Barnes, D.K.A. (2004). Natural and plastic flotsam stranding in the Indian Ocean. In J. Davenport, J. L. Davenport (Eds), The Effects of Human Transport on Ecosystems: Cars and Planes, Boats and Trains (pp. 193-205). Dublin: Royal Irish Academy.

Barnes, D.K.A., Milner, P., (2005). Drifting plastic and its consequences for sessile organism dispersal in the Atlantic Ocean. Marine Biology, 146, 815-825.

Ben-Eliahu, M. N., ten Hove, H. A. (2011). Serpulidae (Annelida: Polychaeta) from the Suez Canal – from a Lessepsian migration perspective (a monograph). Zootaxa, 2848, 1-147.

Blackburn, T. M., Pysek, P., Bacher, S., Carlton, J. T., Duncan, R. P., Jarosik, V. *et al.* (2011). A proposed unified framework for biological invasions. Trends in Ecology and Evolution, 26, 333-339.

Calder, D. R., Choong, H. H. C., Carlton, J. T., Chapman, J. W., Miller, J. A., Geller, J. (2014). Hydroids (Cnidaria: Hydrozoa) from Japanese tsunami marine debris washing ashore in the northwestern United States. Aquatic Invasions, 9, 425-440.

Campbell, M. L., Hewitt, C. L. (1999). Marine Biological Invasions of Port Phillip Bay. Vectors, shipping, and trade. In C. L. Lewis *et al.* (Eds), Centre for Research on Introduced Marine Pests, Technical Report 20 (pp. 45-60). Hobart : Centre for Marine Research.

Carlton, J. T. (2007). Ballast. In J. B. Hattendorf (Ed.), The Oxford Encyclopedia of Maritime History (pp. 249-251). Oxford, New York: Oxford University Press.

Carlton, J. T. (2009). Aquatic Invasive Species Mid-Atlantic Workshop: Vector Management: A Prevention Solution. Baltimore, MD, USA: Sea Grant, Maryland.

Carlton, J. T. (2011). The Global Dispersal of Marine and Estuarine Crustaceans. In: B. S. Galil, P. F. Clark, J. T. Carlton (Eds), In the wrong place: alien marine crustaceans – distribution, biology and impacts (pp. 3-23). Dordrecht: Springer.

Carlton, J. T., Cohen, A. N. (2003). Episodic global dispersal in shallow water marine organisms: the case history of the European shore crabs *Carcinus maenas* and *Carcinus aestuarii*. Journal of Biogeography, 30, 1809-1820.

Carlton, J. T., Geller J. B. (1993). Ecological roulette: The global transport of nonindigenous marine organisms. Science, 261, 78-82.

Carlton, J. T., Hodder, J. (1995). Biogeography and dispersal of coastal marine organisms: experimental studies on a replica of a 16th-century sailing vessel. Marine Biology, 121, 721-730.

Carlton, J. T., Reid, D. M., van Leeuwen, H. (1995). Shipping Study. The role of shipping in the introduction of non-indigenous aquatic organisms to the coastal waters of the United States (other than the Great Lakes) and an analysis of control options Report Number CG-D-11-95, The National Sea Grant College Program/Connecticut Sea Grant Project R/ES-6. Washington, D.C. and Groton, Connecticut: Department of Transportation, United States Coast Guard.

Carlton, J. T., Ruiz, G. M. (2005). Vector science and integrated vector management in bioinvasion ecology: conceptual frameworks. In H. A. Mooney, R. N. Mack, J. A. McNeely, L. E. Neville, P. J. Schei, J. K. Waage (Eds), Invasive Alien Species: A New Synthesis (pp. 36-58). Covelo, California: Island Press.

Carlton, J. T., Ruiz, G. M. (2011). Assessing the relationship between propagule pressure and invasion risk in ballast water. Committee on Assessing Numeric Limits for Living Organisms in Ballast Water, National Research Council. Washington, D.C.: The National Academies Press.

Carlton, J. T., Scanlon, J. A. (1985). Progression and dispersal of an introduced alga: *Codium fragile* ssp. *tomentosoides* (Chlorophyta) on the Atlantic coast of North America. Botanica Marina, 28, 155-165.

Carney, K. J., Delaney, J. E., Sawant, S., Mesbahi E. (2011). The effects of prolonged darkness on temperate and tropical marine phytoplankton, and their implications for ballast water risk management. Marine Pollution Bulletin, 62, 1233-1244.

Chang, A. L., Grossman, J. D., Spezio, T. S., Weiskel, H. W., Blum, J. C., Burt, J. W. *et al.* (2009). Tackling aquatic invasions: risk and opportunities for the aquarium fish industry. Biological Invasions, 11, 773-785.

Chapman, J. W., Miller, T. W., Coan, E. V. (2003). Live seafood species as recipes for invasion. Conservation Biology, 17, 1386-1395.

Chapman, J. W., Breitenstein, R. A., Carlton, J. T. (2013). Port-by-port accumulations and dispersal of hull fouling invertebrates between the Mediterranean Sea, the Atlantic Ocean, and the Pacific Ocean. Aquatic Invasions, 8, 249-260.

Cohen, A. N., Carlton, J. T. (1997). Transoceanic transport mechanisms : The introduction of the Chinese mitten crab, *Eriocheir sinensis*, to California. Pacific Science, 51, 1-11.

Cohen, A. N., Zabin, C. J. (2009). Oyster shells as vectors for exotic organisms. Journal of Shellfish Research, 28, 163-167.

Colautti, R.I., Grigorovich, I. A., MacIsaac, H. J. (2006). Propagule pressure: a null model for biological invasions. Biological Invasions, 8, 1023-1037.

Cook, E. J., Ashton, G., Campbell, M., Coutts, A., Gollasch, S., Hewitt, C. *et al.* (2008). Non-native aquaculture species releases: implications for aquatic ecosystems. In M. Holmer, K. Black, C. M. Duarte, N. Marba, I. Karakassis (Eds.) Aquaculture in the Ecosystem (p.155-184). Berlin: Springer.

Darwin, C. (1854). A monograph on the sub-class Cirripedia, The Balanidae, or sessile cirripedes, the Verrucidae. London: The Ray Society.

Davidson, I. C., McCann, L. D., Fofonoff, P. W., Sytsma, M. D., Ruiz, G. M. (2008). The potential for hull-mediated species transfers by obsolete ships on their final voyages. Diversity and Distributions, 14, 518-529.

Davidson, I. C., Zabin, C. J., Chang, A. L., Brown, C. W., Sytsma, M. D., Ruiz, G. M. (2010). Recreational boats as potential vectors of marine organisms at an invasion hotspot. Aquatic Biology, 11, 179-191.

Frey, M.A., Simard, N., Robichaud, D. D., Martin, J. L., Therriault, T. W. (2014). Fouling around: vessel sea-chests as a vector for introduction and spread of aquatic invasive species. Management of Biological Invasions, 5, 21-30.

Godwin, S., Harris, L., Charette, A., Moffitt, R. (2008). The marine invertebrate species associated with the biofouling of derelict fishing gear in the Papahanaumokuakea-Marine National Monument: a focus on marine non-native species transport. Kaneohe, Hawaii: Hawaii Institute for Marine Biology.

Gregory, M. R. 2009. Environmental implications of plastic debris in marine settings -- entanglement, ingestion, smothering, hangers-on, hitch-hiking and alien invasions. Philosophical Transactions of the Royal Society B, 364, 2013-2025.

Hamer, J. P., McCollin, T. A., Lucas, I. A. N. (1998). Viability of decapod larvae in ships' ballast water. Marine Pollution Bulletin, 36, 646-647.

Hattendorf, J. B. (Ed.) (2007). The Oxford Encyclopedia of Maritime History. Oxford: Oxford University Press.

Haydar, D., Wolff, W. J. (2011). Predicting invasion patterns in coastal ecosystems: relationship between vector strength and vector tempo. Marine Ecology Progress Series, 431, 1-10.

Hayward, B. W. (1997). Introduced marine organisms in New Zealand and their impact in the Waitemata Harbour, Auckland. Tane, 36, 197-223.

Hewitt, C. L., Gollasch, S., Minchin, D. (2009). The vessel as a vector – biofouling, ballast water and sediments, In G. Rilov, J. Crooks (Eds.), Biological invasions in marine ecosystems (p.117-131). Heidelberg: Springer-Verlag.

Hurles, M. E., Matisso-Smith, E., Gray, R. D., Penny, D. (2003). Untangling oceanic settlement: the edge of the knowable. Trends in Ecology and Evolution, 18, 531-540.

Kaluza, P., Kolzsch,A., Gastner, M. T., Blasius, B. (2010). The complex network of global cargo ship movements. Journal of the Royal Society Interface, 7, 1093-1103.

Kang, J.-H., Hyun, B.-G., Shin, K. (2009). Phytoplankton viability in ballast water from international commercial ships berthed at ports in Korea. Marine Pollution Bulletin, 60, 230-237.

Keller, R. P., Drake, J. M., Drew, M. B., Lodge, D. M. (2010). Linking environmental conditions and ship movements to estimate invasive species transport across the global shipping network. Diversity and Distributions, 17, 93-102.

Kooistra, W. H. C.F., Verbruggen, H. (2005). Genetic patterns in the calcified tropical seaweeds *Halimeda opuntia, H. distorta, H. hederacea,* and *H. minima* (Bryopsidales, Chlorophyta) provide insights in species boundaries and interoceanic dispersal. Journal of Phycology, 41, 177-187.

Lewis, P. N., Riddle, M. J., Smith, S. D. A. (2005). Assisted passage or passive drift: a comparison of alternative transport mechanisms for non-indigenous coastal species into the Southern Ocean. Antarctic Science, 17, 183-191.

Lockwood, J. L., Hoopes, M. F., Marchetti, M. P. (2007). Invasion ecology. Oxford: Blackwell Publishing

Matisso-Smith, E., Robins, J. H. (2004). Origins and dispersals of Pacific peoples: evidence from mtDNA phylogenies of the Pacific rat. Proceedings of the National Academy of Sciences, 101, 9167-9172.

Minchin, D., Floerl, O., Savini, D., Occhipinti-Ambrogi, A. (2006). Small craft and the spread of exotic species. In J. Davenport, J. L. Davenport (Eds.), The Ecology of Transportation: Managing Mobility for the Environment (p.99-118). Berlin: Springer.

Minchin, D., Gollasch, S., Wallentinus, I., (Eds.) (2005). Vector pathways and the spread of exotic species in the sea. ICES Cooperative Research Report 271. Copenhagen, Denmark: ICES.

Nyberg, C. D., Wallentinus, I. (2009). Long-term survival of an introduced red alga in adverse conditions. Marine Biology Research, 5, 304-308.

Padilla, D. K., Williams, S. L. (2004). Beyond ballast water: aquarium and ornamental trades as sources of invasive species in aquatic ecosystems. Frontiers in Ecology and the Environment, 2, 131-138.

Pederson, J., Mieszkowska, N., Carlton, J. T., Gollasch, S., Jelmert, A., Minchin, D. *et al.* (2011). Climate change and non-native species in the North Atlantic. In P. C. Reid, L. Valdes (Eds.), ICES Status Report on Climate Change in the North Atlantic/ICES Cooperative Research Report, No. 310 (p.174-190). Copenhagen, Denmark: ICES.

Pilsbry, H. A. (1916). The sessile barnacles (Cirripedia) contained in the collections of the U. S. National Museum: including a monograph of the American species. Bulletin of the United States National Museum, 93, 1–366.

Reisinger, R. R., McIntyre, T., Bester, M. N. (2009). Goose barnacles hitchhike on satellite-tracked southern elephant seals. Polar Biology, 33, 561-564.

Ribera Siguan, M. A. (2003). Pathways of biological invasions of marine plants. In G. M. Ruiz, J. T. Carlton (Eds.), Invasive species: vectors and management strategies (p.183-226). Washington, D.C.: Island Press.

Riley, C. L., Kelley, J. C., Pennington, C. W., Rands, R. L. (Eds.) (1971). Man across the sea. Problems of pre-Columbian contacts. Austin: University of Texas Press.

Ruiz, G. M., Fofonoff, P. W., Ashton, G., Minton, M. S., Miller, A. W. (2013). Geographic variation in marine invasions among large estuaries: effects of ship and time. Ecological Applications, 23, 311-320.

Ruiz, G. M., Fofonoff, P. W., Steves, B., Foss, S. F., Shiba, S. N. (2011). Marine invasion history and vector analysis of California: a hotspot for Western North America. Diversity and Distributions, 17, 362-373.

Schaffelke, B., Deane, D. (2005). Desiccation tolerance of the introduced marine green alga *Codium fragile* ssp. *tomentosoides* -- clues for likely transport vectors. Biological Invasions, 7, 557-565.

Seebens, H., Gastner, M. T., Blasius, B. (2013). The risk of marine bioinvasion caused by global shipping. Ecology Letters, 16, 782-790.

Smith, L.D., Wonham, M. J., McCann, L. D., Ruiz, G. M., Hines, A. H., Carlton, J. T. (1999). Invasion pressure to a ballast-flooded estuary and an assessment of inoculant survival. Biological Invasions, 1, 67-87.

Steinberg, M. K., Lemieux, E. J., Drake, L. A. (2011). Determining the viability of marine protists using a combination of vital, fluorescent stains. Marine Biology, 158, 1431-1437.

Tottrup, A. P., Chan, B. K. K., Koskinen, H., Hoeg, J. T. (2010). 'Flying barnacles': implications for the spread of non-indigenous species. Biofouling, 26, 577-582.

Verling, E., Ruiz, G. M., Smith, L. D., Galil, B., Miller, A. W., Murphy, K. R. (2005). Supply-side invasion ecology: characterizing propagule pressure in coastal ecosystems. Proceedings of the Royal Society B, 272, 1249-1257.

Villac, M. C., Kaczmarska, I. (2011). Estimating propagule pressure and viability of diatoms detected in ballast tank sediments of ships arriving at Canadian ports. Marine Ecology Progress Series, 425, 47-61.

Voight, J. R., Lee, R. W., Reft, A. J., Bates, A. E. (2012). Scientific gear as a vector for non-native species at deep-sea hydrothermal vents. Conservation Biology, 26, 938-9421.

Wang, J., Chu, H., Sun, J., Song, Y., Jiang, J., Pan, X., Sun, Y., Li, D. (2014). Chip-based portable device for detection of microalgae viability in ballast water. Applied Mechanics and Materials, 522-524, 3-6.

Williams, S. L., Davidson, I. C., Pasari, J. R., Ashton, G. V., Carlton, J. T., Crafton, R. E. *et al.* (2013). Managing multiple vectors for marine invasions in an increasingly connected world. Bioscience, 63, 952-966.

Wonham, M. J., Lewis, M. A., MacIsaac, H. J. (2005). Minimizing invasion risk by reducing propagule pressure: a model for ballast-water exchange. Frontiers in Ecology and Evolution, 3, 473-478.

Phillip Cassey, Miquel Vall-Llosera, Ellie Dyer, Tim M. Blackburn

2 The Biogeography of Avian Invasions: History, Accident and Market Trade

2.1 Introduction

Birds are the most accomplished dispersers of all the terrestrial vertebrates. Most bird species are volant (capable of flight) and birds have colonised and occupied all major landforms, oceans, and islands of the world. However, the approximately 9,993 extant species are not regularly distributed, in either number or identity (Jetz *et al.*, 2012), and there exist distinct biogeographic patterns that have developed over their evolutionary history in both their diversity (species richness) and endemism (evolutionary uniqueness). Most hotspots of bird species richness are located in the tropics and in mountainous areas of mainland continents, whereas hotspot regions of endemism tend to be on large islands and/or island archipelagos (Orme *et al.*, 2005). These natural distributions of bird species and the patterns of movement that led to them have subsequently been greatly affected by the actions and movement of people.

There is sparse but consistent evidence that people have been moving animals, either purposefully or inadvertently, for thousands of years, and across a wide range of regions and civilisations (Lever, 1979; Yalden, 1999). Ancient civilisations translocated both domesticated and non-domesticated species as sources of food, for ornamentation (of people or landscapes), for game hunting, or as pets (Hughes, 2010; Tella, 2011). One of the earliest domesticated (and liberated) bird species was the domestic fowl, the descendant of the Red Jungle Fowl (*Gallus gallus*). Jungle Fowl are a native of southeast Asia and archaeological and palaeoclimatic data suggest that they were transported to northern China c. 8,000 years ago (West & Zhou, 1989). They were subsequently introduced as poultry to other parts of Asia (China, Indonesia) around 5400 BC (Miao *et al.*, 2013) to Pacific islands by Polynesian colonisers in 1000 BC (Steadman *et al.*, 1990), and possibly to South America before the arrival of Europeans between 1321-1407 AD (Storey *et al.*, 2007). Domestic fowl were also transported across Europe during the Bronze and Iron Ages and were well established by the time of the Roman Empire's fall (Sykes, 2012). The principal motivation for their spread from Asia was to enhance the human diet, but they were also valued for their song, their eggs and feathers, and for the sport of cockfighting. In Western Africa, domestic fowl only became widespread after 1000 AD.

In the ancient Mediterranean region, there was keen interest in exotic species for exhibition in menageries and gardens, and for slaughter in the arenas. Many Greek and Roman temples had sacred groves that provided shelter to birds, fish, reptiles and mammals, among which were exotic species originating from far-off countries. Paradises in the ancient world, particularly areas set aside by affluent individuals, contained a variety of wild and exotic species. Alexander the Great's expedition from Greece to Persia and the north of the Indian subcontinent led to the importation of many species from India and other lands. Around 270 BC, Alexandria had zoological and botanical gardens that included many exotic species (and humans) from India as well as parts of Africa. A trade developed in particular species that could be maintained as pets. Among those kept in private households (or garden collections) were many bird species, including peacocks, pheasants, parrots, cranes, storks, flamingoes, rails, crows, starlings, magpies, thrushes and nightingales, that were housed individually or in aviaries; some were valued for their song, and others because they could be taught to talk (Jennison, 1937). The Common Pheasant (*Phasianus colchicus*), which takes its name from the River Phasis in the Transcaucasus, was first introduced to Europe by ancient Greeks and Romans in c. 1300 BC, and was subsequently introduced on many more occasions during the Middle Ages (Mason, 1984). Large-scale releases of pheasants continue in Europe to this day. The Helmeted Guineafowl (*Numida meleagris*) from Africa is first mentioned in Athens by Sophocles in the 5th century BC, while pheasants are mentioned in Aristophanes' play, 'The Birds,' first performed in 414 BC.

In pre-Columbian America, it has been proposed that there was extensive use of, and traffic in, exotic birds (Haemig, 1978). Before the arrival of Europeans, American cultures are believed to have bred and raised exotic birds in aviaries to supply the needs of the feather industry and pet trade. Long-distance merchants could have transported birds hundreds of kilometres outside their natural ranges, and these birds would have been a potential source for accidental or deliberate introductions. The best documented early example of an introduction in the Americas is that of the Great-tailed Grackle (*Quiscalus mexicanus*) to Mexico (Haemig, 1978; 2012), but there is some evidence that other species were introduced at the same time, including the Tufted Jay (*Cyanocorax dickeyi*) introduced to Mexico (Haemig, 1979; but see Bonaccorso *et al.*, 2010), and the Red-legged Thrush (*Turdus plumbeus*) introduced to the West Indies (Ricklefs & Bermingham, 2008).

Despite the long history of movements of bird species, the real 'golden age' of translocations did not start until the middle of the nineteenth century, at which point there was a step-change in the rate at which bird species introductions occurred worldwide (Blackburn *et al.*, 2009). Since then, we have identified two distinct periods of major activity in the transportation of bird species that overshadow all others in terms of their influence on bird introductions and invasions. We have defined these two periods as: (1) the era of the Acclimatisation Societies during the great European diaspora between the eighteenth and twentieth centuries; and (2) the era of the international trade in wild birds for bird-keeping from the late-twentieth century to the present. In the following sections we describe the activities that define these two eras in detail, and discuss the very different influences that they have had on the biogeography (through establishment and spread, *sensu* Blackburn *et al.*, 2011) of exotic bird species.

2.2 Avian Translocations in the Age of Discovery

Since their dinosaurian origin (Zhou, 2004), bird species have (under their own effort) swum, walked, and flown around the globe. Given that birds are such a highly mobile and naturally widespread taxon, one might ask how much influence the re-distribution of species by humans can really have. The simple answer is: a lot!

Like birds, humans are also great colonisers. The greatest period of colonisation in human history, in terms of numbers of people moving their permanent home from one region to another distant region, was the European diaspora of the 19th and early-20th centuries. These movements of people sparked the transportation of many species, both large and small, in both directions across the oceans (Crosby Jr, 1972). Unfamiliar with their new colonized land, Europeans introduced plants and animals to make the alien environment feel more like home, to beautify their gardens, provide sport for hunters, and 'aggrandise' the colony. Above all, however, they wanted to make the land sustainable and economically productive. These motivations led to the first great period of activity in the deliberate transportation and introduction of birds: the founding of the Acclimatisation Societies (McDowall, 1994). The British were especially prominent in this activity. For example, c. 40% of all known bird introductions occurred as a result of activity relating to the British occupation of just four geopolitical regions: Hawaii, New Zealand, Australia, and the continental USA (Blackburn *et al.*, 2009). After 1863, New Zealand alone was home to more than half of the world's Acclimatisation Societies, with coverage of almost the entire country. The influence of the British (and other Europeans) in New Zealand is illustrated by the origin of birds introduced, with many species (and many successful species) coming from Europe and from New Zealand's closest colonial neighbour, Australia (Figure 2.1).

It is possible to quantitatively compare the rate of natural colonisation of areas by birds with the rate of anthropogenic introductions in the era of the Acclimatisation Societies, particularly for oceanic islands. For example, St Helena is a volcanic island in the South Atlantic Ocean that was first discovered by the Portuguese in 1502. By 1588 St Helena had experienced its first avian introductions (*Phasianus colchicus* and *Alectoris chukar*; Lever, 2005). Since then, at least 35 bird species have been introduced, of which nine have successfully established wild-breeding populations. This introduction history produces a rate of introduction of one bird species every 14 years, and a successful establishment of one new bird species every 55 years. In contrast, the minimum geological age of St Helena is estimated to be

c. 7 million years, which is the date of its last volcanic eruption (Chaffey *et al.*, 1989). St. Helena is known to have had at least 22 native bird species at the time of discovery, of which 12 are now extinct (Blackburn *et al.*, 2004). Natural colonisation therefore produced a success rate of one new bird species every 320,000 years. Even if we assume that 99% of naturally colonising species have subsequently gone extinct, the rate of exotic introduction was still more than 50 times greater than that of natural colonisation (Blackburn *et al.*, 2009).

The exotic bird species that were moved by humans (transported and introduced) during the period of the Acclimatisation Societies were clustered into a relatively limited set of bird taxa and were not a randomly selected (or distributed) set of species from all possible extant taxa (Blackburn *et al.*, 2009). Analysis of global introduction data has revealed that some families contain significantly more introduced bird species than expected. These are: the Phasianidae (pheasants, partridges and quails); Anatidae (ducks and waterfowl); Columbidae (pigeons and doves); Psittacidae (parrots); and Passeridae (Old World sparrows) (Blackburn & Duncan, 2001; Lockwood, 1999). More than half of all introduced species came from just these five families, despite the families including less than 15% of all extant bird species. Two other families with relatively high representation of introduced taxa are the Odontophoridae (New World quails) and Fringillidae (true finches). This taxonomic bias strongly reflected the purposeful introduction of birds to new locations for the provision of hunting, game and food (pheasants, partridges, ducks and pigeons), as well as for their aesthetic and/or domestic qualities (sparrows, finches, pigeons and parrots).

Even within the era of Acclimatisation, however, the identities of the species introduced were subject to change (Blackburn *et al.*, 2009). A high proportion of early introductions concerned game birds (Galliformes), but this order's representation in introduction events has steadily declined over time. In contrast, the proportion of events that relate to parrots (Psittaciformes) has increased steadily since the 1850s. Passerine introductions appeared to peak in the second half of the nineteenth century, while the proportion of introductions concerning waterfowl (Anseriformes) varied little throughout this period (Figure 2.2). This temporal variation in the taxonomic composition of bird introductions within the Acclimatisation period most likely reflects the changes in reasons for introducing bird species. Early introductions consisted of species regarded as beneficial to colonists for their survival and livelihood, particularly game birds and waterfowl. Passerine introductions peaked towards the end of the popularity of Acclimatisation Societies, when many small-bodied songbird species were introduced for aesthetic reasons to supplement native avifaunas with familiar and conspicuous garden species, sometimes under the guise of insect and horticultural biocontrol (Pipek *et al.*, 2015).

Fig. 2.1: Biogeographic sources of introduction for 125 extant bird species introduced to New Zealand during the period 1773–1952. Native breeding ranges were obtained from the ADHoC (Avian Diversity Hotspots Consortium) database, first published by Orme *et al*. (2005). Native ranges were projected using ESRI ArcMap GIS software (version 9.3, 2008), and allocated to a biogeographic realm (Nearctic, Neotropical, Palaearctic, Afrotropical, Indo-Malay, Australasian and Oceanic). Each realm was awarded a score depending on the proportion of a species' range that fell within its boundaries. For example, if the entire range was enclosed then it received a score of 1, if 50% of the range then 0.5, and so on. This process was repeated for all 125 species with the resultant scores as follows: Palaearctic = 57.66, Australasian = 30.16, Nearctic = 19.16, Indo-Malay = 9.66, Afrotropical = 2.16, Neotropical = 2.16, Oceanic = 1. The size of the directional arrows were weighted to represent the number of species from each biogeographic realm that have been introduced to New Zealand. The darkness of arrow colours corresponds to the rate of successful establishment ranging from zero in the Neotropics to 38.7% in the Afrotropics. The rate of successful establishment for the Palearctic (Europe, including the United Kingdom) was 34.7%.

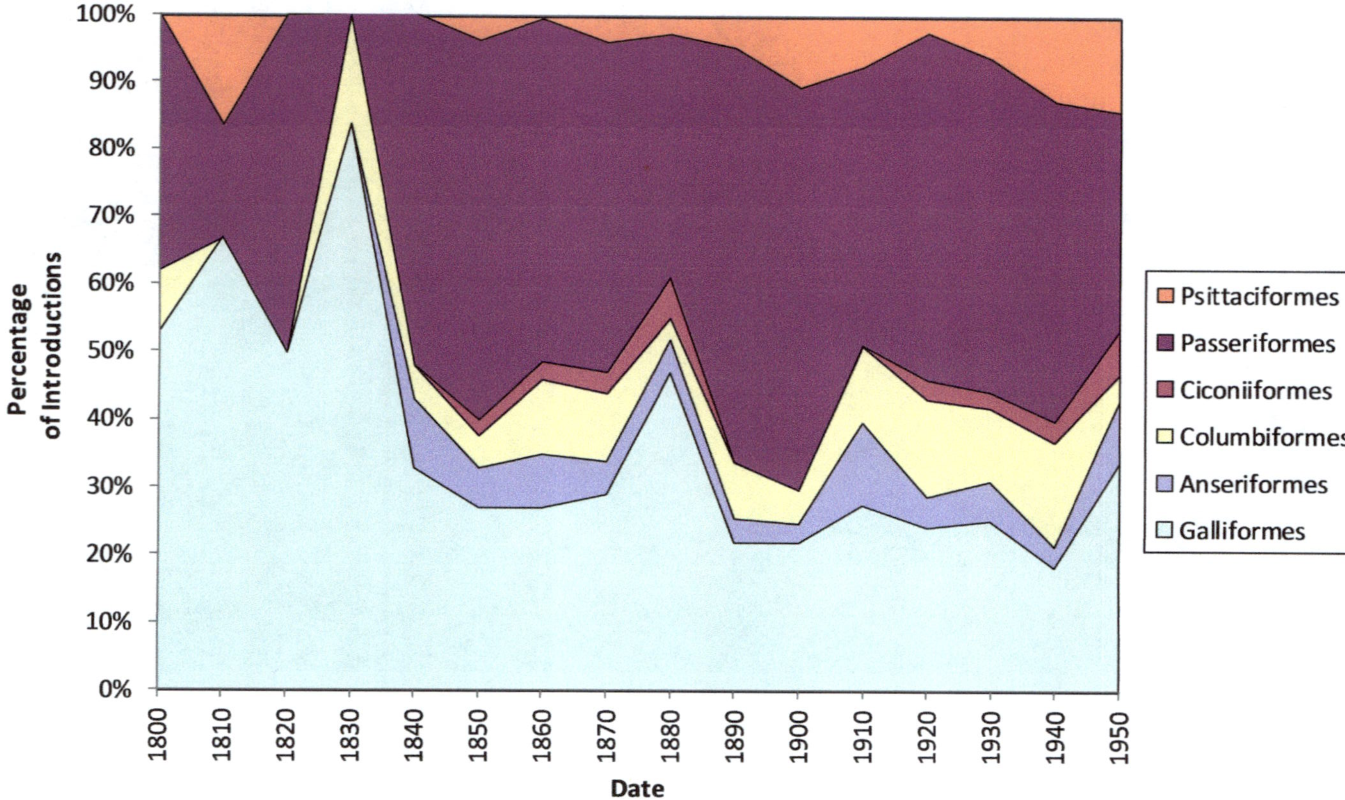

Fig. 2.2: The proportional taxonomic composition of avian introduction events in the period 1800-1950. Only the six bird orders with more than 30 introductions with an estimate of date of introduction are included. Reproduced with permission from Blackburn *et al.* (2009).

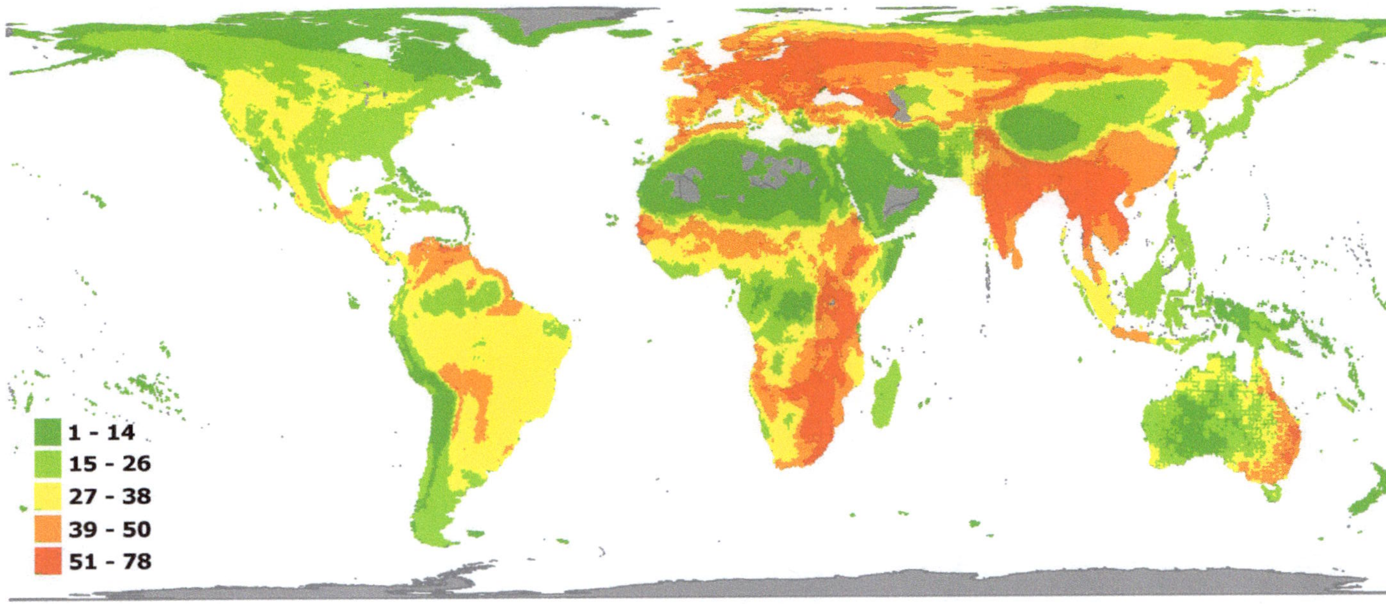

Fig. 2.3: Native breeding ranges for 639 extant bird species that have been introduced outside their native breeding ranges. Breeding ranges for species with three or more introduction records in the GAVIA (Global Avian Invasions Atlas) database (Dyer & Blackburn, unpublished data) were obtained from the ADHoC (Avian Diversity Hotspots Consortium) database, first published by Orme *et al.* (2005). All maps were created using ESRI ArcMap GIS software (version 9.3, 2008). In order to identify areas of high levels of introduction sources, a species richness map was created by projecting the native range maps onto a hexagonal grid of the world, resulting in a geodesic discrete global grid, defined on an icosahedron and projected onto the sphere using the inverse Icosahedral Snyder Equal Area projection. This resulted in a hexagonal grid composed of cells that retain their shape and area (2591.3 km^2) throughout the world. The colour then assigned to each hexagonal cell reflects the number of species' ranges that intersect that cell.

The era of the Acclimatisation Societies led to bird species being introduced to all major regions of the world, and to the majority of ice-free latitudes (Duncan *et al.*, 2003), but predominantly to islands (Cassey, 2003). Globally, more than two-thirds of all past avian introductions were to islands (Blackburn *et al.*, 2009), despite islands constituting only a small fraction of all land area around the globe. The location of these islands, primarily British colonies, largely dictates latitudinal variation in the distribution of exotic bird species moved in this period. For example, in the Northern Hemisphere, there were a relatively large number of introductions to low latitudes as a consequence of the geographic location of the Hawaiian and Caribbean Islands, whereas peaks of introduction in the Southern Hemisphere reflect the very large numbers of introductions to islands in the Pacific and Indian Oceans. Although species have been sourced from all major biogeographic regions, the Palaearctic region 'donated' the most species during this era and has one of the highest percentages of native breeding bird species introduced elsewhere. Other significant donor regions for exotic birds include sub-Saharan and southern Africa, the tropical regions of South America, and southern Asia, as well as Indonesia and eastern Australia (Figure 2.3). Eventually, however, it was recognised that the introduction of exotic species was ecologically unsound, and the practice slowly died out. For example, New Zealand Acclimatisation Societies had changed their role from introducing species to preventing further introductions of exotic species by the end of the Second World War (Simberloff & Rejmanek, 2011).

2.3 Sitting Around in Bars: the Influence of the Pet Trade on Current Avian Biogeography

Deliberate introductions were the main pathway of bird introductions during the era of Acclimatisation Societies, but are now in decline globally as scientific, conservation and political opinions have turned against them (Hulme *et al.*, 2008). However, bird invasions have not ceased; it is just that their primary cause has changed. Today, bird invasions are driven largely by the demand for caged birds. Most have accidental (e.g. pet birds escaping from cages) or recreational (e.g. for religious or festive reasons) origins, rather than resulting from deliberate establishment efforts (Eguchi & Amano, 2004; Leven & Corlett, 2004; Lever, 2005). In the current era of globalisation, the main driver of exotic introductions is therefore wildlife trade (Westphal *et al.*, 2008).

Throughout history, bird-keeping has predominantly been a pastime for the wealthy, and thus concerned relatively low numbers of birds. Only in the last century did the activity become generally popular, encouraged by accelerated economic growth in the developed world. The increase of the wealth of the middle classes, coupled with the improvement of international transport capacity, which permitted faster and cheaper trade between far distant regions, opened the door to the commercial trade of millions of birds (Hulme, 2009). The demand for pets increased in the Communist Block after the collapse of the Soviet Union (Chiron *et al.*, 2014), and has also grown in developing countries such as Brazil (Regueira & Bernard, 2012), Mexico (Cantu-Guzman *et al.*, 2007) and South Africa (Goss & Cumming, 2013), as these societies progressively achieved higher living standards. In these cases, bird-keeping is a symbol of higher socio-economic status among increasingly urban societies (Jepson & Ladle, 2005). Most recently, the growth of the internet has facilitated the sale and circulation of species worldwide (Derraik & Phillips, 2009; Kikillus *et al.*, 2012). The consequences of these trends are that more than one million birds, of over a thousand species, are legally traded around the world on an annual basis (Butchart, 2008; Gilardi, 2006; Karesh *et al.*, 2005). Bird-keeping is currently one of the most popular hobbies in the world (Carrete & Tella, 2008), and social workers and other healthcare professionals believe that such pets help many people to lead healthier, happier lives (Anderson, 2003).

The greatest effect that wildlife trade has had on bird transportation has been the reversal of the direction of introduction, which in the era of Acclimatisation was from the Old World to the colonies (e.g. birds from Europe sent to America, Australasia and Africa; Figure 2.1). Now, transport is primarily from the former colonies (birds from South America, Africa, southeast Asia and Oceania) to the population centres of Europe, continental Asia and North America (Jeschke & Strayer, 2005). In South America, passerines (Passeriformes), parrots (Psittaciformes), doves (Columbiformes) and toucans (Piciformes) are the most frequently traded species, whereas tanagers (Thraupidae), New World sparrows (Emberizidae) and troupials (Icteridae) are the most commonly traded families (Alves & Brooks, 2010; Dauphine, 2008; Regueira & Bernard, 2012). From southeast Asia, the most commonly traded birds are small softbill cagebirds, such as babblers (Timalidae) and mynas (Sturnidae) (Nijman, 2010). The Chinese market has acted as a major hub for the re-export of birds to other parts of the world: for example, the Netherlands imports birds from African countries and re-exports them to Hong Kong (Lau *et al.*, 1997). Live birds from China are mainly exported to Europe, other Asian countries and the USA. In the recent past, most native birds from these countries were exported to Europe; however, formal operations have been drastically reduced since the European Union (EU) imposed a ban on wild bird imports, resulting in the closure of this important market. This and other national and international regulations on wildlife trade (Cooper & Rosser, 2002) mean that in Western markets, such as the USA, Europe and Australia, the private demand for pet animals is usually satisfied by domestic trade (Anderson, 2003). Exotic animals legally traded in these markets are a combination of wild caught birds imported prior to the bans, birds imported with special permits, descendants of wild caught birds raised in captivity, and/or native birds exempt from trade regulations (Anderson, 2003). Legal international trade is directed mainly to Mexico, followed by Asia and Africa (FAO, 2011).

The consequence of modern bird translocations is the redistribution of species to 'new' biogeographical regions where these taxa were historically absent. One of the avian families that has particularly benefited from introductions is the parrots (Cassey, Blackburn, Russell, Jones, & Lockwood, 2004). The majority of birds currently in pet markets around the world are either parrots or passerines (Figure 2.4). These often wild caught birds are largely sourced from tropical areas of Africa and southeast Asia (Alves & Brooks, 2010; Dauphine, 2008; Li & Jiang, 2014; Nijman, 2010) with around 1,800 species. However, many species are now threatened by illegal capture and trade that is largely driven by the culture of keeping wild birds as pets. As a result of deliberate or accidental releases, the USA is now home to feral populations of several exotic species of parrots (Bull, 1973; Butler, 2005; Lever, 2005); The USA was originally home to two native parrot species, but these were driven to extinction during the 20th century (Butler, 2005). The best known examples of established exotic parrot species in the USA are the Monk Parakeet (*Myopsitta monachus*) (Russello *et al.*, 2008) and the Rose-ringed Parakeet (*Psittacula krameri*) (Butler, 2005), but there are presently at least seven other parrot species established there (Butler, 2005). Novel parrot species have also established successfully in Europe. Monk and Rose-ringed Parakeets have feral populations in the United Kingdom, Germany, France, Belgium, the Netherlands, Spain and Italy, among others (Sol *et al.*, 1997; Strubbe & Matthysen, 2009; Strubbe, 2009; Mori *et al.*, 2013). The Rose-ringed Parakeet is also established in several countries in the Middle East and Arabian Peninsula (Lever, 2005), and is the most common aviary escapee in Australia (Henderson *et al.*, 2011). Global transport and communication networks continue to increase. As a consequence, more goods are being traded and the type of species being transported has changed, as well as the associated risks.

Less is known about the extent of avian invasions in developing countries. In these emerging economies, mainly situated in tropical or subtropical climates, the number of recorded introduced species is still low compared to other regions. However, it is predicted that these areas should reach high exotic species richness, mainly due to their current increasing economic growth (Lin *et al.*, 2007) and their great ecological diversity (Levine, 2000). The number of introduced (and established) species is surely under-reported given the low research effort devoted to alien species compared with other global regions (Speziale *et al.*, 2012), and may be higher than is widely appreciated. For example, a recent study by Fontoura *et al.* (2013) identified 59 bird species that had been introduced to the large South American country of Brazil, of which 14 species have established or probably established non-native populations. The largely accidental establishment of new species in these regions is likely to increase, and may even exceed the rates of introduction during the era of the Acclimatisation Societies (Su *et al.*, 2014).

2.4 Future Trends

From a global perspective, the effect of the widespread and repeated introductions of certain bird species (particularly Old World sparrows and New World quails), combined with the endangerment and extinction of other species (particularly the rails, petrels and shearwaters) points to a general pattern of biotic homogenisation (Lockwood *et al.*, 2000). As this process continues we increasingly expect to find the same species, and very similar ecological communities, in localities thousands of kilometres apart (Blackburn *et al.*, 2009). In the most extreme situation, we could face a world where the zoogeographical realms are no longer identifiable (McKinney & Lockwood 1999; Rosenzweig, 2001).

A growing concern with the trade in live animals (including birds) is that, combined with the loss and degradation of habitats, the over-exploitation of wildlife can heavily deplete native populations and even bring some species close to extinction (Beissinger, 2001; Chapin *et al.*, 2000; Peres, 2001). This has motivated the creation of laws and international trade agreements to safeguard certain species from over-exploitation. The principal instrument for controlling international trade in wild species is the Convention on International Trade in Endangered Species (CITES), an agreement between governments with the aim of ensuring that international trade in specimens of wild animals and plants does not threaten their survival (Cooper & Rosser, 2002). Since its implementation, 180 member states have passed legislation to adhere to the Convention (http://www.cites.org/eng/disc/parties/index.php), effectively regulating wildlife trade in their country. Parallel to this, the strong evidence that invasive species cause declines in abundances of native species and undesirable changes in ecosystem function (Mack *et al.*, 2000; Sala *et al.*, 2000), as well as economic losses (Hulme *et al.*, 2010; Pimentel *et al.*, 2000), has promoted the implementation of legal tools to regulate invasive species.

The implementation of international restrictions on the global trade in wild birds reduced the trade from an estimated 7.5 million birds a year during the early 1970s to around 2.5 million in the 1990s (Inskipp, 1979; 1990; Leader-Williams & Tibanyenda, 1996). In the USA, the Lacey Act of 1900 provided jurisdiction to prohibit the importation and transport of wildlife included in a list considered 'injurious' (Anderson, 1995). Later, the legal trade of exotic birds into the USA was stopped by the Wild Bird Conservation Act of 1992, which banned the import of all CITES-listed birds (Pain *et al.*, 2006). In Australia, the Environment Protection and Biodiversity Conservation Act 1999 effectively prohibited the export of live wildlife and heavily regulated the import of all CITES-listed species (Alacs & Georges, 2008). In Europe, wildlife regulations were more strongly oriented towards preventing the spread of avian influenza and other pathogenic zoonotic diseases in human beings. The EU ban on live bird imports was adopted in 2007, motivated by the outbreak of Severe Acute Respiratory Syndrome (SARS) and avian influenza, both diseases being linked to wildlife trade (Chomel *et al.*, 2007; Karesh *et al.*, 2005). Laws protecting wildlife also appear in the countries of origin of the traded species (Alves & Brooks, 2010).

Fig. 2.4: Examples of (A) parrots and (B) passerines for sale in pet shops in Taiwan. The parrot species depicted are, from top to bottom and back to front, Military Macaw (*Ara militaris*), White Cockatoo (*Cacatua alba*), Blue-crowned Conure (*Aratinga acuticaudata*), Sun Conure (*Aratinga sostitialis*), Blue-crowned Lorikeet (*Vini australis*), Monk Parakeet (*Myopsitta monachus*) in green and blue colour-morphs, Yellow-bibbed Lory (*Lorius chlorocercus*), Red-shouldered Macaw (*Diopsittaca nobilis*), Australian King Parrot (*Alisterus scapularis*) and Blue-streaked Lory (*Eos bornea*); while the passerines are, from top to bottom and left to right, Japanese White-eye (*Zosterops japonicus*), Nutmeg Mannikin (*Lonchura punctulata)* and Taiwan Scimitar-babbler (*Pomatorhinus musicus*) (the bottom right cage contains two bantam chickens (*Gallus gallus*)). Photo credits: T.M. Blackburn.

Biological invasions by alien species are one of the consequences of human activities, and so it is important that we consider the economic and social dimensions of the problem when dealing with exotic species (Perrings *et al.*, 2005). In most cases, people are naïve about exotic birds, not even being able to recognize which species in the wild are non-native where they live. In the worst of cases, exotic species are highly valued by people for hunting, commercial or religious purposes, or for aesthetic reasons; for example, people enjoy seeing exotic parakeets at bird feeders (Butler, 2005). Popular exotic species can be so well integrated into public perception that in some cases even conservation organisations promote laws protecting them. This cultural attachment is detrimental for developing control and eradication programs. At the same time, people are often also not conscious of the impacts of exotic invasive species. This may become even more true in the future as the current younger generations are raised in a landscape filled with exotic species, unaware of the legacy of past biological invasions (Decocq, 2010; Papworth *et al.*, 2009).

Strategies to address the problems caused by biological invasions need to begin by educating people on the value of native biodiversity. At the same time, agencies can provide education on exotic species, not only on the problems that invasive species pose to native ecosystems, culture and economy (Pfeiffer & Voeks, 2008), but also on the positive impacts of a small number of exotic species. In a scenario of declining conservation budgets, efforts dedicated to environmental conservation are increasingly being redirected to actions that either produce clear (and immediate) benefits or that do not entail costs (Kareiva *et al.*, 2014). Conservation actions are also likely to be focussed increasingly on responses to climate change, for which suggested mitigation strategies include the controversial technique of assisted colonisation (Hoegh-Guldberg *et al.*, 2008; Loss *et al.*, 2011; Ricciardi & Simberloff, 2009; Wiegand *et al.*, 2005) – essentially, the introduction of exotic species for conservation purposes. Unfortunately, the consequence of these policies on exotic species management is that they are likely to enhance the chances of establishment and spread of new species. This will be particularly true for exotic bird species, for which we currently have little knowledge of the long-term impacts on environmental and human wellbeing (Baker *et al.*, 2014; Kumschick *et al.*, 2013; Shirley & Kark, 2009).

In a nutshell

- Since the earliest of times, civilisations have translocated bird species, both domesticated and non-domesticated, as sources of food, for ornamentation, for game hunting, or as pets.
- There are two distinct periods of major activity in the transportation of bird species: (1) the era of the Acclimatisation Societies, during the great European diaspora between the eighteenth and twentieth centuries; and (2) the era of the international trade in wild birds for bird-keeping from the late-twentieth century to the present.
- The direction of introductions has reversed between the two main periods of bird transportation: from the European 'motherland' to the colonies in the era of the Acclimatisation Societies (birds from Europe introduced to America, Australasia and Africa), to transport from the former colonies to the Old World (birds from South America, Africa, southeast Asia and Oceania introduced to Europe, continental Asia and North America) in the era of international trade.
- The legal international trade in birds has recently been curtailed in developed countries by legislation to restrict the distribution of avian infectious diseases.
- An enormous trade and economy still exists around keeping cage-birds, particularly in developing countries such as Brazil, Mexico and South Africa, and across China and southeast Asia.
- It is likely that the accidental establishment of new species in these regions will increase and may even exceed the rates of introduction during the era of the intentional introductions of the Acclimatisation Societies.

2.5 Bibliography

Alacs, E., Georges, A. (2008). Wildlife across our borders: a review of the illegal trade in Australia. Australian Journal of Forensic Sciences, 40 (2), 147–160.

Alves, R. R. N., Brooks, S. E. (2010). Bird-keeping in the Caatinga , NE Brazil. Human Ecology, 38 (1), 147–156.

Anderson, P. K. (2003). A Bird in the House: An Anthropological Perspective on Companion Parrots. Society & Animals, 11 (4), 393–418.

Anderson, R. S. (1995). The Lacey Act: America' s Premier Weapon in the Fight against Unlawful Wildlife Trafficking. Public Land and Resources Law Review, 16, 27–85.

Baker, J., Harvey, K. J., French, K. (2014). Threats from introduced birds to native birds. Emu, 114 (1), 1–12.

Beissinger, S. R. (2001). Trade in live wild birds: potentials, principles and practices of sustainable use. In J. D. Reynolds, G. M. Mace, K. H. Redford, J. G. Robinson (Eds.), Conservation of Exploited Species. Cambridge: Cambridge University Press.

Blackburn, T. M., Cassey, P., Duncan, R. P., Evans, K. L., Gaston, K. J. (2004). Avian extinction and mammalian introductions on oceanic islands. Science, 305 (5692), 1955–1958.

Blackburn, T. M., Duncan, R. P. (2001). Establishment patterns of exotic birds are constrained by non-random patterns in introduction. Journal of Biogeography, 28(7), 927–939.

Blackburn, T. M., Lockwood, J. L., Cassey, P. (2009). Avian Invasions. The Ecology & Evolution of Exotic Birds. Oxford Avian Biology Series (Vol. 1). Oxford: Oxford University Press.

Blackburn, T. M., Pyšek, P., Bacher, S., Carlton, J. T., Duncan, R. P., Jarošík, V. *et al.* (2011). A proposed unified framework for biological invasions. Trends in Ecology and Evolution, 26 (7), 333–339.

Bonaccorso, E., Peterson, A. T., Navarro-Sigüenza, A. G., Fleischer, R. C. (2010). Molecular phylogenetics and evolution molecular systematics and evolution of the Cyanocorax jays. Molecular Phylogenetics and Evolution, 54 (3), 897–909.

Bull, J. (1973). Exotic Birds in the New York City. The Wilson Bulletin, 85 (4), 501–505.

Butchart, S. H. M. (2008). Red List Indices to measure the sustainability of species use and impacts of invasive alien species. Bird Conservation International, 18 (Suplement 1), S245–S262.

Butler, C. J. (2005). Feral parrots in the continental United States and United Kingdom: past, present, and future. Journal of Avian Medicine and Surgery, 19 (2), 142–149.

Cantu-Guzman, J. C., Sanchez-Saldana, M. E., Grosselet, M., Silva-Gamez, J. (2007). The Illegal Parrot Trade in Mexico. A Comprehensive Assessment. Bosques de las Lomas, Mexico: Defenders of Wildlife.

Carrete, M., Tella, J. (2008). Wild-bird trade and exotic invasions: a new link of conservation concern? Frontiers in Ecology and the Environment, 6 (4), 207–211.

Cassey, P. (2003). A comparative analysis of the relative success of introduced land birds on islands. Evolutionary Ecology Research, 5 (7), 1011–1021.

Cassey, P., Blackburn, T. M., Russell, G. J., Jones, K. E., Lockwood, J. L. (2004). Influences on the transport and establishment of exotic bird species: an analysis of the parrots (Psittaciformes) of the world. Global Change Biology, 10 (4), 417–426.

Chaffey, D. J., Cliff, R. A., Wilson, B. M. (1989). Characterization of the St Helena magma source. Geological Society, London, Special Publications, 42 (1), 257–276.

Chapin, F. S., Zavaleta, E. S., Eviner, V. T., Naylor, R. L., Vitousek, P. M., Reynolds, H. L., *et al.* (2000). Consequences of changing biodiversity. Nature, 405 (6783), 234–242.

Chiron, F., Shirley, S. M., Kark, S. (2010). Behind the Iron Curtain: socio-economic and political factors shaped exotic bird introductions into Europe. Biological Conservation, 143 (2), 351–356.

Chomel, B. B., Belotto, A., Meslin, F.X. (2007). Wildlife, exotic pets, and emerging zoonoses. Emerging Infectious Diseases, 13 (1), 6–11.

Cooper, M. E., Rosser, A. M. (2002). International regulation of wildlife trade: relevant legislation and organisations. Revue Scientifique Et Technique De L Office International Des Epizooties, 21 (1), 103–123.

Crosby Jr, A. W. (1972). The Columbian Exchange: Biological and Cultural Consequences of 1492. Westport: Greenwood.

Dauphine, N. (2008). Notes on the Live Bird Trade in Northen Peru. Proceedings of the Fourth International Partners in Flight Conference: Tundra to Tropics, 220–222.

Decocq, G. (2010). Invisibility promotes invasibility. Frontiers in Ecology and the Environment, 8 (7), 346–347.

Derraik, J. G. B., Phillips, S. (2009). Online trade poses a threat to biosecurity in New Zealand. Biological Invasions, 12 (6), 1477–1480.

Duncan, R. P., Blackburn, T. M., Sol, D. (2003). The ecology of bird introductions. Annual Review of Ecology, Evolution, and Systematics, 34 (1), 71–98.

Eguchi, K., Amano, H. E. (2004). Spread of exotic birds in Japan. Ornithological Science, 3 (1), 3–11.

FAO (2011). International trade in wild birds, and related bird movements, in Latin America and the Caribbean. Animal Production and Health, 166.

Fontoura, P. M., Dyer, E., Blackburn, T. M., Orsi, M. L. (2013). Non-native bird species in Brazil. Neotropical Biology and Conservation, 8 (3), 165–175.

Gilardi, J. D. (2006). Captured for conservation: will cages save wild birds? A response to Cooney & Jepson. Oryx, 40 (1), 24–26.

Goss, J. R., Cumming, G. S. (2013). Networks of wildlife translocations in developing countries: an emerging conservation issue? Frontiers in Ecology and the Environment, 11 (5), 243–250.

Haemig, P. D. (1978). Aztec emperor Auitzotl and the Great-Tailed Grackle. Biotropica, 10 (1), 11–17.
Haemig, P. D. (1979). Secret of the Painted Jay. Biotropica, 11 (2), 81–87.
Haemig, P. D. (2012). Introduction of the Great-Tailed Grackle (*Quiscalus mexicanus*) by Aztec emperor Auitzotl: provenance of the historical account. The Auk, 129 (1), 70–75.
Henderson, W., Bomford, M., Cassey, P. (2011). Managing the risk of exotic vertebrate incursions in Australia. Wildlife Research, 38 (6), 501–508.
Hoegh-Guldberg, O., Hughes, L., Mclntyre, S., Lindenmayer, D. B., Parmesan, C., Thomas, C. D., *et al.* (2008). Assisted colonization and rapid climate change. Science, 321 (5887), 345–346.
Hughes, J. D. (2010). Europe as consumer of exotic biodiversity: greek and roman times. Landscape Research, 28, 37–41.
Hulme, P. E. (2009). Trade, transport and trouble: managing invasive species pathways in an era of globalization. Journal of Applied Ecology, 46 (1), 10–18.
Hulme, P. E., Bacher, S., Kenis, M., Klotz, S., Kuhn, I., Minchin, D., Vila, M. (2008). Grasping at the routes of biological invasions: a framework for integrating pathways into policy. Journal of Applied Ecology, 45 (2), 403–414.
Hulme, P. E., Pysek, P., Nentwig, W., Vila, M. (2010). Will threat of biological invasions unite the European Union? Science, 324 (5923), 4–5.
Inskipp, T. P. (1979). The extent of world trade and the mortality involved. In P. Barclay-Smith, R.D. Chancellor (Eds.), Thirteenth Bulletin of the International Council for Bird Preservation. London, UK: International Council for Bird Preservation.
Inskipp, T. P. (1990). Overview of the numbers and value of birds in trade. In J.B Thomsen, S.R. Edwards, T.A. Mulliken (Eds.), Symposium on Trade in Wild Birds, Twentieth World Conference of International Council for Bird Preservation. Hamilton, New Zealand.
Jennison, G. (1937). Animals for Show and Pleasure in Ancient Rome. Manchester: Manchester University Press.
Jepson, P., Ladle, R. J. (2005). Bird-keeping in Indonesia: conservation impacts and the potential for substitution-based conservation responses. Oryx, 39 (04), 442.
Jeschke, J. M., Strayer, D. L. (2005). Invasion success of vertebrates in Europe and North America. Proceedings of the National Academy of Sciences of the United States of America, 102 (20), 7198–7202.
Jetz, W., Thomas, G. H., Joy, J. B., Hartmann, K., Mooers, A. O. (2012). The global diversity of birds in space and time. Nature, 491 (7424), 444–448.
Kareiva, P., Watts, S., McDonald, R., Boucher, T., Uauy, C., Akhunov, E. *et al.* (2014). Domesticated nature: shaping and ecosystems landscapes for human welfare. Science, 316 (5833), 1866–1869.
Karesh, W. B., Cook, R. A., Bennett, E. L., Newcomb, J. (2005). Wildlife trade and global disease emergence. Emerging Infectious Diseases, 11 (7), 1000–1002.
Kikillus, K. H., Hare, K., Hartley, S. (2012). Online trading tools as a method of estimating propagule pressure via the pet-release pathway. Biological Invasions, 14 (12), 2657–2664.
Kumschick, S., Bacher, S., Blackburn, T. M. (2013). What determines the impact of alien birds and mammals in Europe? Biological Invasions, 15 (4), 785–797.
Lau, M. W.-N., Ades, G., Goodyer, N., Zou, F. (1997). Wildlife Trade in Southern China including Hong Kong and Macao. Hong Kong: Biodiversity Working Group of the China Council for International Cooperation on Environment and Development Project.
Leader-Williams, N., Tibanyenda, R. K. (1996). The Live Bird Trade in Tanzania. Gland, Switzerland and Cambridge, United Kingdom: IUCN.
Leven, M. R., Corlett, R. T. (2004). Invasive birds in Hong Kong, China. Ornithological Science, 3 (1), 43–55.
Lever, C. (1979). The Naturalized Animals of the British Isles. London: Granada.
Lever, C. (2005). Naturalised Birds of the World. London, UK: T & AD Poyser.

Levine, J. M. (2000). Species diversity and biological invasions: relating local process to community pattern. *Science, 852* (5467), 852–854.
Li, L., Jiang, Z. (2014). International trade of CITES listed bird species in China. PloS one, 9 (2), 1–8.
Lin, W., Zhou, G., Cheng, X., Xu, R. (2007). Fast economic development accelerates biological invasions in China. PloS one, 2 (11), e1208.
Lockwood, J. L. (1999). Using taxonomy to predict success among introduced avifauna: relative importance of transport and establishment. Conservation Biology, 13 (3), 560–567.
Lockwood, J. L., Brooks, T. M., McKinney, M. L. (2000). Taxonomic homogenization of the global avifauna. Animal Conservation, 3 (1), 27–35.
Loss, S. R., Terwilliger, L. A., Peterson, A. C. (2011). Assisted colonization: integrating conservation strategies in the face of climate change. Biological Conservation, 144 (1), 92–100.
Mack, R. N., Simberloff, D., Lonsdale, W. M., Evans, H., Clout, M., Bazzaz, F. A. (2000). Biotic invasions: causes, epidemiology, global consequences, and control. Ecological Applications, 10 (3), 689–710.
Mason, I. L. (1984). Evolution of Domesticated Animals. London: Longman.
McDowall, R. M. (1994). Gamekeepers for the nation: the story of New Zealand's acclimatisation societies, 1861–1990. Christchurch: Canterbury University Press.
McKinney, M., Lockwood, J. (1999). Biotic homogenization: a few winners replacing many losers in the next mass extinction. Trends in Ecology and Evolution, 14 (11), 450–453.
Miao, Y.W., Peng, M.-S., Wu, G.-S., Ouyang, Y.-N., Yang, Z.-Y., Yu, N., *et al.* (2013). Chicken domestication: an updated perspective based on mitochondrial genomes. Heredity, 110 (3), 277–282.
Mori, E., Di Febbraro, M., Foresta, M., Melis, P., Romanazzi, E., Notari, A. (2013). Assessment of the current distribution of free-living parrots and parakeets (Aves: Psittaciformes) in Italy: a synthesis of published data and new records. Italian Journal of Zoology, 80 (2), 37–41.
Nijman, V. (2010). An overview of international wildlife trade from Southeast Asia. Biodiversity and Conservation, 19 (4), 1101–1114.
Orme, C. D. L., Davies, R. G., Burgess, M., Eigenbrod, F., Pickup, N., Olson, V. A. *et al.* (2005). Global hotspots of species richness are not congruent with endemism or threat. Nature, 436 (7053), 1016–1019.
Pain, D. J., Martins, T. L. F., Boussekey, M., Diaz, S. H., Downs, C. T., Ekstrom, J. M. M. *et al.* (2006). Impact of protection on nest take and nesting success of parrots in Africa, Asia and Australasia. Animal Conservation, 9 (3), 322–330.
Papworth, S. K., Rist, J., Coad, L., Milner-Gulland, E. J. (2009). Evidence for shifting baseline syndrome in conservation. Conservation Letters, 2 (2), 93–100.
Peres, C. A. (2001). Synergistic effects of subsistence hunting and habitat fragmentation on Amazonian forest vertebrates. Conservation Biology, 15 (6), 1490–1505.
Perrings, C., Dehnen-Schmutz, K., Touza, J., Williamson, M. (2005). How to manage biological invasions under globalization. Trends in Ecology & Evolution, 20 (5), 212–215.
Pfeiffer, J. M., Voeks, R. A. (2008). Biological invasions and biocultural diversity : linking ecological and cultural systems. Environmental Conservation, 35 (4), 281–293.
Pimentel, D., Lach, L., Zuniga, R., Morrison, D. (2000). Environmental and economic costs of nonindigenous species in the United States. Bioscience, 50 (1), 53–65.
Pipek, P., Pysek, P., Blackburn, T. M. (2015). How the yellowhammer became a Kiwi: the history of an alien bird invasion revealed. NeoBiota, 24, 1-31.
Regueira, R. F. S., Bernard, E. (2012). Wildlife sinks: Quantifying the impact of illegal bird trade in street markets in Brazil. Biological Conservation, 149 (1), 16–22.
Ricciardi, A., Simberloff, D. (2009). Assisted colonization: good intentions and dubious risk assessment. Trends in Ecology and Evolution, 24 (9), 476–477.
Ricklefs, R. E., Bermingham, E. (2008). Likely human introduction of the Red-Legged Thrush (Turdus Plumbeus) to Dominica, West Indies. The Auk, 125 (2), 299–303.

Rosenzweig, M. (2001). The four questions: What does the introduction of exotic species do to diversity? Evolutionary Ecology Research, 3 (3), 361–367.
Russello, M. A., Avery, M. L., Wright, T. F. (2008). Genetic evidence links invasive monk parakeet populations in the United States to the international pet trade. BMC Evolutionary Biology, 8 (217), 1–11.
Sala, O., Chapin, F., Armesto, J., Berlow, E., Bloomfield, J., Dirzo, R. *et al.* (2000). Global biodiversity scenarios for the year 2100. Science, 287 (5459), 1770–1774.
Shirley, S. M., Kark, S. (2009). The role of species traits and taxonomic patterns in alien bird impacts. Global Ecology and Biogeography, 18 (4), 450–459.
Simberloff, D., Rejmanek, M. (2011). Encyclopedia of Biological Invasions. Berkeley: University of California Press.
Sol, D., Santos, D., Feria, E., Clavell, J. (1997). Habitat selection by the monk parakeet during colonization of a new area in Spain. Condor, 99 (1), 39–46.
Speziale, K. L., Lambertucci, S. A., Carrete, M., Tella, J. L. (2012). Dealing with non-native species: what makes the difference in South America? Biological Invasions, 14 (8), 1609–1621.
Steadman, D. W., Greiner, E. C., Wood, C. S. (1990). Absence of blood parasites in indigenous and introduced birds from the Cook Islands, Pacific South. Conservation Biology, 4 (4), 398–404.
Storey, A. A., Ramírez, J. M., Quiroz, D., Burley, D. V, David, J., Walter, R. *et al.* (2007). Radiocarbon introduction and DNA evidence for a pre-Columbian to Chile of Polynesian chickens. Proceedings of the National Academy of Sciences of the United States of America, 104 (25), 10335–10339.
Strubbe, D. (2009). Invasive ring-necked parakeets Psittacula krameri in Europe: invasion success, habitat selection and impact on native bird species. Universiteit Antwerpen.
Strubbe, D., Matthysen, E. (2009). Establishment success of invasive ring-necked and monk parakeets in Europe. Journal of Biogeography, 36, 2264–2278.
Su, S., Cassey, P., Blackburn, T. M. (2014). Patterns of non-randomness in the composition and characteristics of the Taiwanese bird trade. Biological Invasions, 16 (12), 2563-2575.
Sykes, N. (2012). A social perspective on the introduction of exotic animals: the case of the chicken. World Archaeology, 44 (1), 158–169.
Tella, J. L. (2011). The unknown extent of ancient bird introductions. Ardeola, 58 (2), 399–404.
West, B., Zhou, B. (1989). Did chickens go north? New evidence for domestication. World's Poultry Science Journal, 45 (3), 205–218.
Westphal, M. I., Browne, M., Mackinnon, K., Noble, I. (2008). The link between international trade and the global distribution of invasive alien species. Biological Invasions, 10 (4), 391–398.
Wiegand, T., Revilla, E., Moloney, K. A. (2005). Effects of habitat loss and fragmentation on population dynamics. Conservation Biology, 19 (1), 108–121.
Yalden, D. W. (1999). History of British Mammals. London: T. & A.D. Poyser.
Zhou, Z. (2004). The origin and early evolution of birds: discoveries, disputes, and perspectives from fossil evidence. Naturwissenschaften, 91 (10), 455–471.

Mark D. Sytsma, Toni Pennington

3 Vectors for Spread of Invasive Freshwater Vascular Plants with a North American Analysis

3.1 Introduction

Freshwater plants, or macrophytes, are important to the structure and function of lentic and lotic systems. They influence chemical and physical attributes of the aquatic environment while also providing habitat for invertebrates and fish as well as forage for waterfowl (Butcher, 1933; Spence, 1967; Westlake, 1982; Carpenter & Lodge, 1986; Sand-Jensen *et al.*, 1989). Growth form and characteristics of individual species are important determinants of aquatic plant function in aquatic ecosystems. Changes in community composition caused by introduction of alien invasive species of macrophytes can have cascading effects throughout aquatic food webs and alter ecosystem services provided by aquatic plants.

3.1.1 Growth Forms of Freshwater Plants

Growth form influences the relative importance of vectors of introduction and the type of system changes caused by invasive species. For example, emergent and floating-leaved plants with attractive foliage or flowers are popular ornamentals and are common in the horticultural trade, whereas submersed species with limp stems easily attach to boats. Invasive submersed plants can impede water flow and alter habitat structure for fish and invertebrates (Vermaat *et al.*, 2000; Toft *et al.*, 2003; Colon-Gaud *et al.*, 2004), whereas floating species can impede light penetration and gas exchange with the atmosphere (Frodge *et al.*, 1990; Goodwin *et al.*, 2008).

3.1.1.1 Reproduction and Dispersal in Aquatic Plants

The ability to reproduce vegetatively is ubiquitous among aquatic plants because it functions efficiently in aquatic environments (Philbrick & Les, 1996). Vegetative reproduction and clonal growth permits rapid expansion of favorable genotypes under conditions that are relatively favorable for plant growth, especially for submersed species, e.g., risk of desiccation is low and temperatures are moderate. Some highly invasive aquatic plants are dioecious (staminate and pistillate flowers borne on different individuals) and only one sex is invasive, illustrating the efficiency of vegetative reproduction. For example, in northern Europe, *Elodea canadensis* is widespread and nearly all the plants are female (Hutchinson, 1975). Similarly, in the western United States of America, *Egeria densa* are all male (Carter & Sytsma, 2001) and in New Zealand, South Africa, and North America all *Myriophyllum aquaticum* are female (Guillarmod, 1979; Orchard, 1979; Aiken, 1981).

Aquatic plants produce a variety of vegetative propagules (Grace, 1993). In many submersed macrophytes, plant fragments as small as a single node are capable of establishing new plants, so every node of the plant stem is a potential propagule. Specialized vegetative propagules, however, may convey advantages in survival and establishment. In *Hydrilla verticillata*, larger propagules (tubers) produce more competitive plants than smaller propagules (turions) (Spencer & Rejmánek, 1989) and have increased survival (Bowes *et al.*, 1979). Continued production of turions by *H. verticillata* may be advantageous if turions and tubers represent different survival and dispersal strategies. Spencer *et al.* (1987) suggested that turions, which are formed on the stem of the plant, are better suited for dispersal and occupation of open space where they are less likely to face competition, and that tubers, which are formed in the sediment and are not as easily dispersed, are more efficient at maintaining established stands where intraspecific competition would be more intense.

Flowering, pollination, and seed germination are problematic for many species of aquatic plants, especially for submersed species (Bornette & Puijalon, 2009). Flowers of aquatic and terrestrial plants in the same family are morphologically similar. For submersed species, the differences are primarily in their mechanisms to achieve an aerial position for wind or insect pollination or, more rarely, in the production of completely submersed, hydrophilous flowers. Sculthorpe (1967) provides numerous examples of reduced fertility and poor seed viability in aquatic plants as well as examples of pseudovivipary where vegetative propagules replace flowers, particularly in submersed species. Nevertheless, production of numerous hybrids in the *Potamogeton* (Kaplan *et al.*, 2009) and *Myriophyllum* (Moody & Les, 2002) genera suggests that sexual reproduction is common in submersed aquatic plants. Moody & Les (2002) proposed that hybridization between introduced and native *Myriophyllum* species may be responsible for the invasiveness of the species outside their native range.

Sexual reproduction is more common in emergent species than in submersed and floating-leaved species (Philbrick & Les, 1996), and is important to long-term survival of aquatic plant species in systems subject to fluctuating water levels (Van der Valk & Davis, 1978; Westcott *et al.*, 1997; Combroux & Bornette, 2004). Seeds of aquatic plants survive longer than vegetative propagules and may permit escape from adverse conditions for decades to centuries. *Nelumbo nucifera* seeds as old as 1,300 years have been germinated (Shen-Miller *et al.*, 1995). Vegetative propagules are shorter-lived (Van & Steward, 1990; Kunii, 1993) and function primarily in perennation and dispersal.

Vectors

Vectors for dispersal of invasive freshwater plants can be coarsely classified as primary (vectors for initial introduction to a new habitat) and secondary (natural processes for spread following establishment). Primary vectors are human-mediated and may function over intercontinental, inter-watershed, or intra-watershed scales. Secondary vectors can also result in long-distance dispersal, but typically function more locally within a watershed.

Primary Vectors

Primary vectors for freshwater plants include shipping, trailered boats, and intentional importation. Intentional importation can result in invasion of natural systems by escape from cultivation, contaminated shipments, or by direct introduction. Association of any individual species introduction with a specific vector is often impossible because the lag in population growth following introduction provides a temporal discontinuity (e.g., *Eichhornia crassipes* was first introduced into Italy in the first half of the 19th century, but became invasive in late 20th century (Brundu *et al.*, 2013)) and because multiple vectors may be acting simultaneously.

Intentional Importation

Freshwater plants are intentionally introduced for a variety of beneficial uses. Often the biological characteristics that make the plants attractive for their intended use, e.g., rapid growth, wide environmental tolerances, and few pests, also make them invasive if they escape from cultivation or are intentionally introduced into natural systems.

Aquarium and Water Gardening

Water gardening is a growing hobby and a major source of invasive freshwater plants. Invasive plants in the USA are introduced primarily through the ornamental plant trade (Lehan *et al.*, 2013). In 2003, there were 16 million households in the USA with water gardening retail sales totaling $1.56 billion (Crosson, n.d.). Les & Mehrhoff (1999) found that up to 88% of the invasive aquatic plants in the USA entered as cultivated plants. In Europe, about 7 million aquatic plants are imported each year, primarily for aquarium use, but only 10 of the 247 species are considered potentially invasive (Brunel, 2009). Champion (1998, cited in Champion & Clayton, 2000) reported that 75% of the naturalized aquatic plants in New Zealand were introduced as ornamentals. There is seasonality in the aquatic plant trade, with sales peaking in July in the Pacific northwest of the USA (Strecker *et al.*, 2011). Strecker *et al.* (2011) found that propagule pressure of plants and fish released by aquarists was the same (< 1 released per aquarist per year), but released aquarium plants had a higher probability of establishment. In a detailed study of aquatic plant propagule pressure from aquaria releases in Montreal, Cohen *et al.* (2007) estimated that 3,015 plant propagules are released each year into the St Lawrence Seaway, and found that *Cabomba caroliniana* and *Egeria densa*, two known invasive species, were among the top seven species released.

Invasive freshwater plants may be introduced as hitchhikers with otherwise innocuous plants or products. Maki and Galatowitsch (2004) found that 93% of the orders received from aquatic plant vendors in the USA included plant and animal species that were not ordered, and that 10% of the orders included federal noxious weeds or other nonindigenous species. They reported that the frequency of hitchhiking species varied with growth form of plants ordered. Orders of submersed and floating species included hitchhiking plants 100% of the time; emergent plant orders included hitchhiking plants 62% of the time. *Hydrilla verticillata*, *Potamogeton crispus*, and *Salvinia molesta*, all highly invasive species, were among the plants that Maki and Galatowitsch (2004) received as unordered hitchhikers. The *Hydrilla verticillata* invasion in the northeastern (Les, 1996 cited in Kay & Hoyle, 2001) and western USA (Boersma *et al.*, 2006; Akers, 2011) were the result of contamination of water lily shipments. The first known invasive submerged plant in New Zealand, *Elodea canadensis*, was introduced as a contaminant in a shipment of fish eggs, and *Hydrodictyon reticulatum* was introduced into New Zealand with a shipment of ornamental fish (Champion & Clayton, 2000).

Food

There is a long history of human use of freshwater plant seeds, fruits, and perennating organs for food (Sculthorpe, 1967). Human transport and introduction of species used for food may relate to human migration and reluctance to sample unfamiliar food (Mack, 1999 cited in Mack & Lonsdale, 2001). Several aquatic plants used for food have become invasive when introduced outside their native range. Champion & Clayton (2000) reported that *Rorippa nasturtium-aquaticum* was introduced to New Zealand for culinary purposes and became a major weed problem, and that *Alternanthera philoxeroides* is cultivated in New Zealand as a food crop. *Ipomoea aquatica* has been cultivated for centuries in Asia and was introduced and is cultivated in the southern USA even though it is a federally listed noxious weed (Van & Madeira, 1998). *Trapa natans* was a significant food source for prehistoric Europeans (Karg, 2006), and has been shipped throughout the world (Sculthorpe, 1967). It is invasive in lakes in Kashmir, where it is an important food resource (Masoodi, 2013), and in the northeastern USA (Orth & Moore, 1984; Nieder *et al.*, 2004).

Medicine

Sculthorpe (1967) reviewed freshwater plants that have had historical medicinal uses, several of which are considered invasive outside their native range. *Acorus calamus* was introduced into Europe for medicinal purposes in 1574 (Weber and Brändle, 1996), where it forms monospecific stands and displaces native species (Dykyjová, 1980). *Monochoria vaginalis*, a common folk remedy for a variety of ailments (Latha & Latha, 2014), has been spread throughout Asia and the Pacific islands and is a common weed in rice fields from Japan (Shibayama, 2001) to California (Barrett & Seaman, 1980). Although they were most likely dispersed as ornamentals rather than for medical uses, several invasive *Nymphaea* species contain compounds with potential medical applications (Rajagopal & Sasikala, 2008; Bose *et al.*, 2012; Jesurun *et al.*, 2013)

Forage

Use of freshwater plants as forage for livestock has received extensive attention (see review by Little, 1979). Nearly all of the most invasive aquatic plant species have been considered as livestock feed. For most species it is unlikely that they were introduced with the intention of increasing forage. Rather, their use as forage appears to be a utilitarian exploitation of rampant growth resulting from introduction via a different vector. Introduction for forage, however, appears to have led to creation of invasive *Phalaris arundinacea* populations. Repeated introductions of *P. arundinacea* into North America from Europe for forage increased genetic diversity of the species in its North American native range and allowed development of a highly invasive phenotype (Lavergne & Molosfky, 2007). More generally, Roman & Darling (2007) concluded that high propagule pressure through repeated introductions can add genetic diversity, decrease founder effects, and increase success of a variety of introduced aquatic taxa.

Industrial Use

Because of their rapid growth rate, several invasive aquatic plant species have been proposed for treatment of industrial and domestic wastewater (Brix & Schierup, 1989). Floating species, such as *Pistia stratiotes, Eichhornia crassipes*, and *Salvinia* spp., provide tertiary treatment of wastewater and the nutrients the plants sequester can be easily removed from the system by harvesting the plants. *M. spicatum* was suggested as an efficient species for phytoremediation of industrial processes (Hughes *et al.*, 1997; Lesage *et al.*, 2007; Brundu *et al.*, 2013).

Shipping

The role of shipping in dispersal of aquatic invasive species is well documented (National Research Council, 1996). There are multiple sub-vectors associated with shipping, including hull fouling, ballast, sea chests, and dunnage. Solid ballast was used on ships until the 1880s when water-ballasting technology was developed. Water ballast can be pumped much more quickly than solid ballast can be loaded or offloaded, which shortened ship time in port and made shipping more profitable. The role of ballast water in dispersal of plant propagules is not well documented; however, the importance of solid ballast in introducing nonindigenous species to port areas has been documented for over 100 years (Mack, 2003). Nelson (1917) reported 92 species unique to a solid ballast disposal site near Portland, Oregon, USA, many of which were first reports in the USA. *Schoenoplectus californicus*, *Zizania latifolia*, and *Alterniflora philoxeroides* were likely solid ballast introductions into New Zealand (de Lange *et al.*, 1998). Solid ballast was a vector for *Veronica beccabunga* introduction into New England (Les & Mehrhoff, 1999) and for *Butomus umbellatus* introduction into the St. Lawrence River (Countryman, 1970).

The importance of shipping as a vector for plants dropped substantially with the abandonment of solid ballast and the adoption of water-ballast technology (Mills *et al.*, 1996; Riccardi, 2006; Keller *et al.*, 2009). Furthermore, shipping and ballasting occurs primarily in marine and estuarine systems, and current ballast management strategies include mid-ocean ballast water exchange with movement toward ballast water treatment, all of which reduce the importance of shipping as a vector for freshwater macrophytes.

Boats and Trailers

Multiple environmental impacts of boating on freshwater systems have been long recognized (Liddle & Scorgie, 1980; Mosisch & Arthington, 1998). Given the speed and long distance that they travel over land, trailered boats have been a major focus of study as vectors for aquatic invasive species. Much of the research on recreational boats as vectors for invasive species has focused on *Dreissena polymorpha* and *D. rostriformis bugensis* (Rothlisberger *et al.*, 2011; Choi *et al.*, 2013; Dalton & Cottrell, 2013). Many models of mussel dispersal by boat vectors that were stimulated by the mussel invasion of North America (Buchan & Padilla, 1999; Bossenbroek *et al.*, 2001; Johnson *et al.*, 2001; Leung *et al.*, 2006) are equally useful for understanding dispersal of aquatic plants. Few models have been developed specifically for aquatic plant dispersal by boats (Jacobs & MacIsaac, 2009).

The importance of boats as dispersal vectors seems intuitively obvious; however, definitive documentation of its importance is not available. Biological characteristics of the invader and the source water body, environmental conditions in the receiving water body, and transit time interact to determine the effectiveness of the boat vector in dispersal of aquatic plants. Johnstone *et al.* (1985) found that the distribution of five invasive submerged plant species in New Zealand was associated with boating activity, and that boats leaving lakes carried plant fragments only when the haul-out area was near an invasive plant bed. They also reported that biotic factors, such as lateral bud frequency, internode length, and resistance to desiccation may influence the effectiveness of boats as vectors for aquatic plants. The length of *M. spicatum* fragments and whether they originated from the plant apex or near the bottom of the stem influenced desiccation rate and survival (Mcalarnen *et al.*, 2012) and by inference distance that a viable fragment could be transported between lakes. Jerde *et al.* (2012) and Barnes *et al.* (2013) also described differences in the desiccation rate of invasive freshwater plant species and related it to probability of survival during interlake transport by boats. In Minnesota, USA, distance to the nearest invaded lake, and by inference the importance of boats as vectors, was a predictor of presence of *M. spicatum*; however, lake size, alkalinity, Secchi depth, and lake depth were also significant predictors (Roley & Newman, 2008). Factors associated with boating activity, such as the number of boat ramps and proximity to roads, were less important than water quality factors, especially inorganic carbon concentration, in predicting *M. spicatum* presence in lakes in Wisconsin, USA (Buchan & Padilla, 2000). In addition to physically moving plant propagules, boat wakes and propeller-wash uproot plants and propellers produce plant fragments that can be dispersed via multiple secondary vectors (Mumma *et al.*, 1996; Owens *et al.*, 2001).

Secondary vectors

Secondary vectors are processes that can facilitate dispersal following introduction by primary vectors (Ridley, 1930 cited by Mack & Lonsdale, 2001). Secondary vectors for freshwater plants are not mediated by humans and typically function at a watershed scale, but can result in rapid, long-distance dispersal of established populations of invasive species.

Hydrochory (Water Currents)

Hydrochory is the dispersal of plant propagules by water and is the primary mode of dispersal for many aquatic plants (Sculthorpe, 1967; Sarneel, 2012). Much of the research on the importance of hydrochory has focused on dispersal in lotic systems and riparian plant species (Johansson *et al.*, 1996; Mills *et al.*, 1996; Nilsson *et al.*, 2010). The importance of hydrochory in lentic systems is less studied; however, Nilsson *et al.* (2010) hypothesized that zoochory or anemochory may be more important for lentic species than hydrochory.

The importance of hydrochory in dispersal of freshwater plants is a function of plant growth form, propagule buoyancy and type (vegetative versus generative), and the timing of hydrology and plant phenology. Boedeltje *et al.* (2003) found that that 100% of the propagules from free-floating plants, 98.9% of the propagules from submerged plants, and 23.7% of the propagules from emergent plants were vegetative in a lowland stream in The Netherlands. Seeds and vegetative propagules function differently in hydrochory. Vegetative propagules are usually larger than seeds and more buoyant, which allows them to disperse longer distances, but also makes them more likely to become trapped by obstacles. Vegetative propagules have a greater probability to establish but are more short-lived than seeds (Johansson & Nilsson, 1993). Buoyancy is an important determinant of dispersal distance and fate of dispersing propagules in flowing systems. Highly buoyant stem fragments tend to be deposited in shallow water or trapped in riparian vegetation, whereas less buoyant fragments that float beneath the water surface are more likely to be retained in deeper water with established submerged vegetation and obstacles in the streambed (Riis & Sand-Jensen, 2006). High flow and turbulence can overcome buoyancy effects (Andersson *et al.*, 2000) and non-buoyant seeds can be transported in the bed load of more turbulent, fast moving rivers (Markwith & Leigh, 2008); however, seed production must be timed to seasonal hydrology to ensure long-distance dispersal (Truscott *et al.*, 2006).

Dams and reservoirs can impede hydrochory by altering hydrology (timing, magnitude, and duration of high and low flows and the rate of change in flow), the timing of exposure of shorelines, and act as a sink for downstream movement of seeds (Nilsson *et al.*, 2010). Rood *et al.* (2010) found that dams impeded the downstream movement of riparian weeds in the Snake River, USA, which they attributed to repeated reservoir drawdown and refilling for hydropower production. These activities result in alternating periods of flood and drought that impede riparian plant establishment. The high disturbance regime may also facilitate establishment of other invaders. Non-indigenous species, including *M. spicatum*, are more likely to occur in reservoirs than in natural lakes in the Laurentian Great Lakes region (Johnson *et al.*, 2008). Johnson *et al.* (2008) attributed this to the young age of reservoirs, increased niche availability, and disturbance in most impoundments. They further argued that reservoirs may act as invasive "hubs" *sensu* (Muirhead & MacIsaac, 2005) that serve as a source of propagules for invasion of nearby natural water bodies.

Anemochory (Wind)
Although wind dispersal of free-floating plants is commonly observed, there are few published studies documenting its importance. Wind has moved even large mats of *Eichhornia crassipes* in Lake Victoria and resulted in markedly different daily and seasonal differences in distribution (Albright *et al.*, 2004). Wind was also cited as a vector for dispersal of water hyacinth in Bahia, Brazil (Fidelman, 2005), and of *Salvinia molesta* on Lake Kariba, Africa (Mitchell, 1973) and the Sepik River, Papua New Guinea (Mitchell *et al.*, 1980). The effectiveness of anemochory in dispersal of free-floating plants is a function of the "sail area" that the leaves provide, which is a plastic phenotypic trait in some plants. The morphology of free-floating plants is often density dependent: crowded stands of *E. crassipes* and *S. molesta* tend to form upright leaves (Agami & Reddy, 1990; Jacono *et al.*, 2001) that facilitate anemochory.

Fruits and seeds of aquatic plants lack structures known to facilitate aerial transport and aerial dispersal is typically limited to a few meters (Sculthorpe, 1967; Cook, 1985). Champion and Clayton (2000) suggested, however, that *Typha* species were introduced to New Zealand from Australia by windblown seeds. Sarneel *et al.* (2014) found that wind had an important role in dispersal of floating seeds in lentic systems. They reported that increasing wind speed increased dispersal speed but decreased dispersal distance.

Zoochory (Animals)
Over 150 years ago in the *Origin of Species*, Charles Darwin (1859) proposed that the wide range of some freshwater plant species was a result of dispersal on the plumage and muddy feet of waterfowl. Zoochory has been suggested as an important mechanism for maintaining plant population genetic diversity in rivers in the face of continuous downstream movement of propagules (Honnay *et al.*, 2010; Chen *et al.*, 2009). Waterfowl are perhaps the most-studied zoochorous vector for aquatic plants (Figuerola & Green, 2002; Charalambidou & Santamaría, 2002). Dispersal by fish (Agami & Waisel, 1988; Pollux *et al.*, 2007) and beavers (Medwecka-Kornaś & Hawro, 1993) has also been documented. Dispersal by waterfowl is thought to likely facilitate longer distance dispersal than mammals or fish and has been implicated in maintenance of continent-scale biodiversity and as an important vector for adjustment of aquatic plant communities in response to climate change (Raulings *et al.*, 2011; Viana *et al.*, 2013).

Zoochory can occur through transport of plant propagules internally in the digestive tract (endozoochory) or externally on the animal's body (exozoochory). It is clear that waterbirds and some fish consume aquatic plant propagules, but the importance of endozoochory as a dispersal agent for aquatic plants is unclear. Plant propagules differ in their tolerance to gut passage and waterfowl species differ in their gut metabolism and migratory patterns (Figuerola & Green, 2002). Clausen *et al.* (2002) suggested that long-distance endozoochory by waterfowl is likely to be rare because: 1) most long-distance movements of waterfowl are out of phase with the reproductive efforts of the plants, and if birds arrive at sites when plants still bear seeds then the birds are likely to leave well after the seed stock has been depleted; 2) most long-distance seed transport by birds is likely to be uni-directional, from north to south during autumn migration; 3) most gut contents are likely to be discharged within 300 km of departure; and 4) in many cases birds will arrive in habitats much different from the ones they left, reducing the probability of environmental match and establishment in the receiving environment. Although the possibility of an individual waterfowl transporting aquatic plant seeds may be rare, the large numbers of migrating waterfowl may, collectively, make them effective dispersal agents (Mueller & van der Valk, 2002). Endozoochory in fish and mammals is less well studied than in waterfowl. Agami & Waisel (1988) and Pollux *et al.* (2007) found that fish species and plant species were both important considerations in fish dispersal of aquatic plant seeds. Gottsberger (1978) documented fish dispersal of allochthonous seeds and fruits by fish in Amazonia.

Exozoochory is more difficult to quantify than endozoochory. There are surprisingly few rigorous studies of plant propagules transported attached to birds and mammals. Sculthorpe (1967, p. 357) cited "innumerable" reports of waterfowl carrying fragments of submersed aquatic plants and suggested that even the larger flying aquatic insects could transport small plant fragments. Johnstone *et al.* (1985) found that several highly invasive submerged plants were absent from lakes that were near lakes colonized by these species and concluded that waterfowl were ineffective vectors for plants, even over short distances. Cook (1990) examined transport of *Nymphoides peltata* seeds on ducks. He found that the marginal trichomes on the *N. peltata* seeds allowed the seeds to adhere to duck feathers even in a dry atmosphere. When the trichomes were removed, the seeds did not adhere after they dried. Brochet *et al.* (2010) found that endozoochory was much more important than exozoochory in dispersal of aquatic plant propagules by teal (*Anas crecca*) in the Carmargue in southern France. They found 21 plant taxa were transported internally and 10 were transported externally on the birds. Up to 171 endochorous propagules were found per bird, but no bird had carried more than one propagule externally. Exozoochory by mammals is poorly documented although aquatic mammals may be expected to be effective vectors for small floating plants such as *Lemna*, *Salvinia*, *Spirodela*, and *Wolffia* species. Manatees and turtles have been suggested as vectors for seagrass dispersal (Kendrick *et al.*, 2012), but we could find no published reports of these species as vectors for freshwater aquatic plants. MedweckaKornaś & Hawro (1993) reported that beavers transported plant propagules on the food, branches, stones, and mud used in construction of dams.

3.2 Relative Importance of Aquatic Plant Vectors

Keller *et al.* (2009) reported that 71% (22 of 31) of the nonindigenous freshwater plants in Great Britain were introduced as ornamentals, six percent were agricultural imports, 10% were introduced as contaminants, and the vector for 13% of the species could not be discerned. The relative importance of vectors of introduction of invasive freshwater plants in the USA was assessed using the United States Geological Survey Nuisance Aquatic Species (USGS NAS) database. This database includes 34,150 records of invasive freshwater plants (including plants that can survive in fresh-brackish water) in the USA. The database was populated through online reports and literature surveys. Twenty-one species account for 80% of the freshwater plant records in the database (Table 3.1). Approximately half of the database records are for emergent species (Figure 3.1). Free-floating and submersed plants each account for about one-quarter of the database entries. Rooted, floating-leaf plants comprise only 1% of the database records.

Tab. 3.1: Species of freshwater invasive plants recorded in the USGS NAS database that comprise 80% of the database records.

Growth Form	Species	% of Database Records
Emergent	*Eichhornia crassipes*	11.00
Submersed	*Myriophyllum spicatum*	8.14
Submersed	*Hydrilla verticillata*	8.11
Emergent	*Panicum repens*	5.58
Emergent	*Alternanthera philoxeroides*	5.53
Emergent	*Lythrum salicaria*	4.94
Floating	*Salvinia minima*	4.88
Floating	*Pistia stratiotes*	4.61
Submersed	*Potamogeton crispus*	4.00
Emergent	*Urochloa mutica*	3.48
Emergent	*Colocasia esculenta*	3.36
Emergent	*Nasturtium officinale*	2.46
Emergent	*Iris pseudacorus*	1.84
Emergent	*Myriophyllum aquaticum*	1.78
Submersed	*Egeria densa*	1.78
Emergent	*Agrostis gigantea*	1.77
Emergent	*Persicaria maculosa*	1.63

continued **Tab. 3.1:** Species of freshwater invasive plants recorded in the USGS NAS database that comprise 80% of the database records.

Growth Form	Species	% of Database Records
Emergent	*Lysimachia nummularia*	1.29
Emergent	*Echinochloa crus-galli*	1.25
Submersed	*Najas minor*	1.22
Emergent	*Typha angustifolia*	1.05

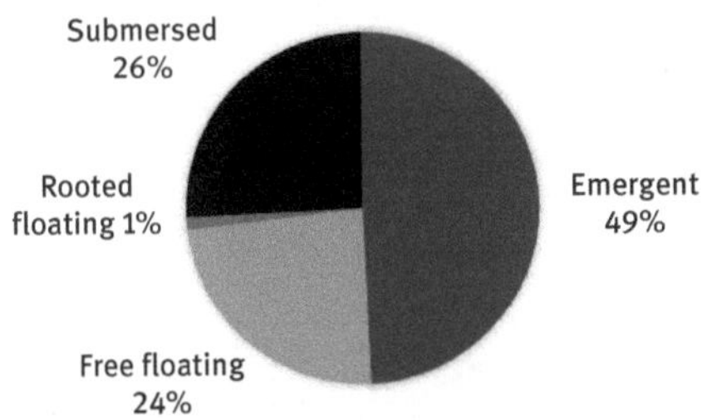

Fig. 3.1: Proportion of freshwater plant growth forms represented in the USGS NAS database.

The database contains records on 136 different species of freshwater aquatic plants that include assignment of a primary vector of introduction, although the vector descriptions differed from those described above (Table 3.2). Some plants that were entered into the database from multiple sites have more than one primary vector assigned to them. For example, *E. crassipes* has 3,657 records in the database, which include four different primary vector assignments (dispersed, hitchhiker, planted/escaped, and released). A total of 182 individual records had a primary vector assignment. Natural dispersal, hitchhiking, planted/escaped, and released vectors were each associated with 21 to 27% of the records. Shipping was associated with nine percent of the records (Figure 3.2).

Tab. 3.2: Categories of dispersal of freshwater aquatic plants in the USGS NAS database.

Dispersed	Natural dispersal by water, animals, etc.
Hitchhiker	Introduction via fishing/boating, aquaculture, or on other introduced plants
Planted/Escaped	Intentionally planted for wildlife habitat, erosion control, or as an ornamental or escaped from cultivation
Released	Aquarium or other unspecified release into the environment
Shipping	Solid or water ballast

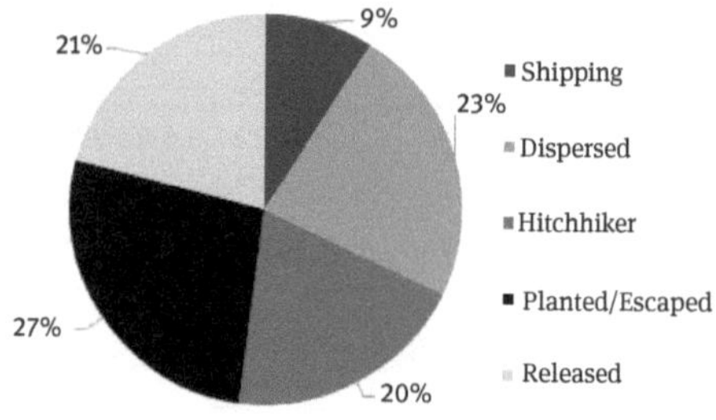

Fig. 3.2: Primary vector associated with freshwater plant records in the USGS NAS database.

The database included 3,573 records that included both growth form and vector fields. The relative importance of vectors differed with growth form. Natural dispersal was the most important vector for emergent plants, free-floating and submersed plants were primarily dispersed by hitchhiking, and planting was the dominant vector for rooted-floating plants (Figure 3.3). Emergent plants often produce abundant rhizomes, stems, or other specialized vegetative organs that are adapted to natural dispersal by hydrochory (Sarneel, 2013). Free-floating plants are located at the surface and submersed plants are typically flaccid, characteristics that facilitate attachment to boats and hitchhiking. Rooted, floating-leaved plants are popular ornamentals and are intentionally planted in water gardens (Nash & Thorpe, 1998).

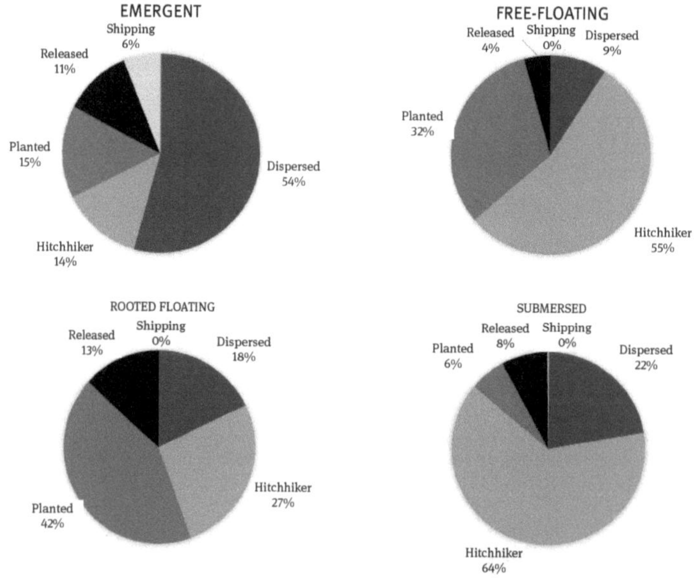

Fig. 3.3: Relative importance of vector by growth form of invasive freshwater aquatic plants in the USGS NAS database.

3.2.1 Conclusions

Vector management is the key to successful and economical management of alien invasive species. Development of the predictive models of alien invasive species dispersal that are required for effective management of aquatic plant invasions will require combining understanding of organism biology, probability of introduction, and site suitability (Buchan & Padilla, 2000; Vander Zanden & Olden, 2008; Jacobs & MacIsaac, 2009; Tamayo & Olden, 2013). Our understanding of all three of these key elements is inadequate for aquatic plants. In addition, several environmental and social factors can be expected to alter each of them to some degree, which will further complicate our ability to make useful predictions. For example, climate change that results in alteration of seasonal precipitation patterns or the timing of significant hydrological events could have multiple, interacting influences on aquatic plant vectors. Water scarcity could reduce popularity of water gardening, the most important primary vector, and increase the number of reservoirs, which would alter hydrology and hydrochory, the most important secondary vector. Other factors could also influence the relative importance of aquatic plant vectors. Increases in fuel cost could lead to a decrease in the number recreational boats and a reduction in the boating vector strength (National Marine Manufacturers Association Canada, 2012). Increased border security in response to terrorism will have the ancillary benefit of increasing inspections and strengthening of biosecurity enforcement, which will reduce instances of intentional introduction.

More detailed and in-depth analysis of primary vectors for aquatic plant introduction is required for cost-effective suppression of vector strength. Recent assessment of vectors for marine alien invasive species introduction into California provides an excellent model (California Ocean Science Trust, 2014) that could be applied to freshwater invasions in general and freshwater plants specifically. Better understanding of the underlying sociological factors that influence individual interest in aquatic plant culture would aid in development of effective outreach and education programs to minimize escape or release from cultivation. Finally, research on factors that control propagule establishment success, and how those factors interact in time and space with vectors, is needed for development of early detection programs that are necessary for cost-effective eradication of pioneering populations of invasive aquatic plants.

3.3 Acknowledgements

We thank Pam Fuller for providing the plant information from the USGS NAS database and Vanessa Morgan for helping with the analysis of the database information.

In a nutshell

- Invasive freshwater aquatic plants degrade fish and wildlife habitat and other ecosystem services provided by aquatic systems.
- A vector is the physical means or agent by which species are transported (see Chapter 1).
- Primary vectors of introduction, such as intentional introduction for ornamental use or accidental introduction on boats, are human-mediated and are subject to interdiction with appropriate vector management.
- Secondary vectors are natural dispersal mechanisms, such as waterfowl and water currents, which can function in local and long-distance dispersal of established species.
- Factors that determine success of propagule establishment and how it interacts with vector strength are poorly understood.
- Climate change may alter the importance of vectors in unpredictable ways and complicate development of the predictive models that are necessary for cost-effective early detection and rapid response to new introductions..

3.4 Bibliography

Agami, M., Reddy, K.R. (1990). Competition for space between *Eichhornia crassipes* (Mart.) Solms and *Pistia stratiotes* L. cultured in nutrient-rich water. Aquatic Botany, 38, 195-208.

Agami, M., Waisel, Y. (1988). The role of fish in distribution and germination of seeds of the submerged macrophytes *Najas marina* L. and *Ruppia maritima* L. Oecologia, 76, 83-88.

Aiken, S.G. (1981). A conspectus of *Myriophyllum* (Halagoraceae) in North America. Brittonia, 33, 57-89.

Akers, P. (2011). The California Department of Food and Agriculture Hydrilla Eradication Program Annual Progress Report 2011. Sacramento: California Department of Food and Agriculture.

Albright, T.P., Moorhouse, T.G., McNabb, T.J (2004). The rise and fall of water hyacinth in Lake Victoria and the Kagera River basin, 1989-2001. Journal of Aquatic Plant Management, 42, 73-84.

Andersson, E., Nilsson, C., Johansson, M.E. (2000). Plant dispersal in boreal rivers and its relation to the diversity of riparian flora. Journal of Biogeography, 27, 1095-1106.

Barnes, M.A., Jerde, C.L., Keller, D., Chadderton, W.L., Howeth, J.G., Lodge, D.M. (2013). Viability of aquatic plant fragments following desiccation. Invasive Plant Science and Management, 6, 320-325.

Barrett, S.C.H., Seaman, D.E. (1980). The weed flora of Californian rice fields. Aquatic Botany, 9, 351-376.

Boedeltje, G., Bakker, J.P., Bekker, R.M., Van Groenendael, J.M., Soesbergen, M. (2003). Plant dispersal in a lowland stream in relation to occurrence and three specific life-history traits of a species pool. Journal of Ecology, 91, 855-866.

Boersma, P.D., Reichard, S.H., Van Buren, A.N. (2006). Invasive Species in the Pacific Northwest. Seattle: University of Washington Press.

Bornette, G., Puijalon, S. (2009). Macrophytes: Ecology of Aquatic Plants. Encyclopedia of Life Sciences. Chichester: John Wiley & Sons, Ltd., DOI: 10.1002/9780470015902.a0020475.
Bose, A., Ray, S.D., Sahoo, M. (2012). Evaluation of analgesic and antioxidant potential of ethanolic extract of *Nymphaea alba* rhizome. Oxidants and Antioxidants in Medical Science, 1, 217-223.
Bossenbroek, J.M., Kraft, C.E., Nekola, J. C. (2001). Prediction of long-distance dispersal using gravity models: zebra mussel invasion of inland lakes. Ecological Applications, 11, 1778-1788.
Bowes, G., Holaday, A.S., Haller, W.T. (1979). Seasonal variation in the biomass, tuber density, and photosynthetic metabolism of hydrilla in three Florida lakes. Journal of Aquatic Plant Management, 1, 61-65.
Brochet, A.L., Guillemain, M., Fritz, H., Gauthier-Clerc, M., Green, A.J. (2010). Plant dispersal by teal (*Anas crecca*) in the Camargue: duck guts are more important than their feet. Freshwater Biology, 55, 1262-1273.
Brix, H., Schierup, H.H. (1989). The use of aquatic macrophytes in water-pollution control. Ambio, 18, 100-107.
Brundu, G., Azzella, M.M., Blasi, C., Camarda, I., Iberite, M., Celesti-Grapow, L. (2013). The silent invasion of *Eichhornia crassipes* (Mart.) Solms. in Italy. Plant Biosystems, 147, 1120-1127.
Brunel, S. (2009). Pathway analysis: aquatic plants imported in 10 EPPO countries. OEPP/EPPO Bulletin, 39, 201-213.
Buchan, L.A.J., Padilla, D.K. (1999). Estimating the probability of long-distance overland dispersal of invading aquatic species. Ecological Applications, 9, 254-265.
Buchan, L.A.J, Padilla, D.K. (2000). Predicting the likelihood of Eurasian watermilfoil presence in lakes, a macrophyte monitoring tool. Ecological Applications, 10, 1442-1455.
Butcher, R.W. (1933). Studies on the ecology of rivers: I. On the distribution of macrophytic vegetation in the rivers of Britain. Journal of Ecology, 21 (1), 58-91.
California Ocean Science Trust (2014). Aquatic Invasive Species Vector Risk Assessments. http://calost.org/science-initiatives/?page=aquatic-invasive-species, Accessed 8 April 2014.
Carpenter, S.R., Lodge, D.M. (1986). Effects of submersed macrophytes on ecosystem processes. Aquatic Botany, 26, 341-370.
Carter, M.C., Sytsma, M.D. (2001). Comparison of the genetic structure of North and South American populations of a clonal aquatic plant. Biological Invasions, 3, 113-118.
Champion, P.D., Clayton, J.S. (2000). Border Control for Potential Aquatic Weeds. Stage 1. Weed Risk Model. Science for Conservation 141. Wellington, New Zealand: Department of Conservation.
Charalambidou, I., Santamaría, L. (2002). Waterbirds as endozoochorus dispersers of aquatic organisms: a review of experimental evidence. Act Oecologia, 23, 165-176.
Chen, Y., Li, X., Yin, L., Cheng, Y., Li, W. (2009). Genetic diversity and migration patterns of the aquatic macrophyte *Potamogeton malainus* in a potamo-lacustrine system. Freshwater Biology, 54, 1178-1188.
Choi, W.J., Gerstenberger, S., McMahon, R. F., Wong, W.H. (2013). Estimating survival rates of quagga mussel (*Dreissena rostriformis bugensis*) veliger larvae under summer and autumn temperature regimes in water of trailered watercraft at Lake Mead, USA. Management of Biological Invasions, 4, 61-69.
Clausen, P., Nolet, B.A., Fox, A.D., Klaassen, M. (2002). Long-distance endozoochorous dispersal of submerged macrophyte seeds by migratory waterbirds in northern Europe – a critical review of possibilities and limitation. Acta Oecologia, 23, 191-2-3.
Cohen, J., Mirotchnick, N., Leung, B. (2007). Thousands introduced annually: the aquarium pathway for non-indigenous plants to the St Lawrence Seaway. Frontiers in Ecology and the Environment, 5, 528-532.
Colon-Gaud, J-C., Kelso, W.E., Rutherford, D.A. (2004). Spatial distribution of macroinvertebrates inhabiting hydrilla and coontail beds in the Atchafalaya Basin, Louisiana. Journal of Aquatic Plant Management, 42, 85-91.

Combroux, I.C.S., Bornette, G. (2004). Propagule banks and regeneration strategies of aquatic plants. Journal of Vegetation Science, 15, 13-20.
Cook, C.D.K. (1985). Range extensions of aquatic vascular plant species. Journal of Aquatic Plant Management, 23, 1-6.
Cook, C.D.K. (1990). Seed dispersal of *Nymphoides peltata* (S.G. Gmelin) O. Kuntze (Menyanthaceae). Aquatic Botany, 37, 325-340.
Countryman, W.D. (1970). The history, spread and present distribution of some immigrant aquatic weeds in New England. Hyacinth Control Journal, 8, 50-52.
Crosson, H. (n.d.). Keeping Aquatic Plants in Their Place: Common Sense Tips to Protect Lakes and Rivers. www.ridnis.ucdavis.edu/Docs/HCrossonArtcle.pdf, Accessed March 31, 2014.
Dalton, L.B., Cottrell, S. (2013). Quagga and zebra mussel rick via veliger transfer by overland hauled boats. Management of Biological Invasions, 4, 129-133.
Darwin, C. (1859). On the Origin of Species by Means of Natural Selection, or the Preservation of Favoured Races in the Struggle for Life. London: John Murry, Albemarble Street.
de Lange, P.J., Gardner, R.O., Champion, P., Tanner, C.C. (1998). *Schoenoplectus californicus* (Cyperaceae) in New Zealand. New Zealand Journal of Botany, 36, 319-327.
Dykyjová, D. (1980). Ecology of *Acorus calamus*. Folia Geobotanica and Phytotaxonomica, 15, 29-57.
Figerola, J., Green, A.J. (2002). Dispersal of aquatic organisms by waterbirds: a review of past research and priorities for future studies. Freshwater Biology, 47, 483-494.
Fidelman, P. (2005). A contribution to the mitigation of impacts from the aquatic macrophyte *Eichhornia crassipes* on the coastal zone of southern Bahia, Brazil. Journal of Integrated Coastal Zone Management, 4, 1-5.
Frodge, J.D., Thomas, G.L., Pauly, G.B. (1990). Effects of canopy formation by floating and submersed aquatic macrophytes on the water quality of two shallow Pacific Northwest lakes. Aquatic Botany, 38, 231-248.
Goodwin, K., Caraco, N., Cole J. (2008). Temporal dynamics of dissolved oxygen in a floating-leaved macrophyte bed. Freshwater Biology, 53, 1632-1641.
Gottsberger, G. (1978). Seed dispersal by fish in the inundated regions of Humaita, Amazonia. Biotropica, 10, 170-183.
Grace, J. (1993). The adaptive significance of clonal reproduction in angiosperms: an aquatic perspective. Aquatic Botany, 44, 159-180.
Guillarmod, A.J. (1979). Water weeds of South Africa. Aquatic Botany, 6, 377-391.
Honnay, O. Jacquemyn, H., Nackaerts, K., Breyne, P., Van Looy, K. (2010). Patterns of population genetic diversity in riparian and aquatic plant species along rivers. Journal of Biogeography, 37, 1730-1739.
Hughes, J.B., Shanks, J., Vanderford, M., Lauritzen, J., Bhadra, R. (1997). Transformation of TNT by aquatic plants and plant tissue cultures. Environmental Science and Technology, 31, 266-271.
Hutchinson, G. E. (1975). A Treatise on Limnology. Volume III. Limnological Botany. New York: John Wiley.
Jacobs, M.J., MacIsaac, H.J. (2009). Modelling spread of the invasive macrophyte *Cabomba caroliniana*. Freshwater Biology, 54, 296-305.
Jacono, C., Davern, T.R., Center, T.D. (2001). The adventive status of *Salvinia minima* and *S. molesta* in the southern United States and the related distribution of the weevil *Cyrtobagous salviniae*. Castanea, 66, 214-226.
Jerde, C.L., Barnes, M.A., DeBuysser, E.K., Noveroske, A., Chadderton, W.L., Lodge, D.M. (2012). Eurasian watermilfoil fitness loss and invasion potential following desiccation during simulated overland transport. Aquatic Invasions, 7, 135-142.
Jesurun, J. Jagadeesh, S., Ganesan, S., Rao, V., Eerike, M. (2013). Anti-inflammatory activity of ethanolic extract of *Nymphaea alba* flower in Swiss albino mice. International Journal of Medical Research and Health Sciences, 2, 474-478.

Johansson, M.E., Nilsson, C. (1993). Hydrochory, population dynamics and distribution of the clonal aquatic plant *Ranunculus lingua*. Journal of Ecology, 81, 81-91.
Johansson, M.E., Nilsson, C., Nilsson, E. (1996). Do rivers function as corridors for plant dispersal? Journal of Vegetation Science, 7, 593-598.
Johnson, L.E., Ricciardi, A., Carlton, J.T. (2001). Overland dispersal of aquatic invasive species: a risk assessment of transient recreational boating. Ecological Applications, 11, 1789-1799.
Johnson, P.T.L., Olden, J.D., Vander Zanden, M.J. (2008). Dam invaders: impoundments facilitate biological invasions of freshwaters. Frontiers in Ecology and the Environment, 6, 357-363, DOI: 10.1890/070156.
Johnstone, I.M., Coffey, B.T., Howard-Williams, C. (1985). The role of recreational boat traffic in interlake dispersal of macrophytes: a New Zealand case study. Journal of Environmental Management, 20, 263-279.
Kaplan, Z., Fehrer, J., Hellquist, B. (2009). New hybrid combinations revealed by molecular analysis: the unknown side of North American pondweed diversity (*Potamogeton*). Systematic Botany, 34, 625-642.
Karg, S. (2006). The water chestnut (*Trapa natans* L.) as a food resource during the 4th to 1st millennia BC at Lake Federsee, Bad Buchau (southern Germany). Environmental Archeology, 11, 125-130.
Kay, S.H., Hoyle, S.T. (2001). Mail order, the internet, and invasive aquatic weeds. Journal of Aquatic Plant Management, 39, 88-91.
Keller, R.P., Zu Ermgassen, P.S.E., Aldridge, D. V. (2009). Vectors and timing of freshwater invasions in Great Britain. Conservation Biology, 6, 1526-1534.
Kendrick, G.A., Waycott, M., Carruthers, T.J.B., Cambridge, M.L., Hovey, R., Krauss, S. L. *et al.* (2012). The central role of dispersal in the maintenance and persistence of seagrass populations. BioScience, 62, 56-65.
Kunii, H. (1993). Rhizome longevity in two floating-leaved aquatic macrophytes, *Nymphaea tetragona* and *Brasenia schreberi*. Journal of Aquatic Plant Management, 31, 94-98.
Latha, B.K., Latha, M.S. (2014). Curative efficacy of methanolic extract of *Monochoria vaginalis* against carbontetra chloride (CCl_4) induced liver fibrosis in male wistar rats. Journal of Pharmacy and Biological Sciences, 9, 71-76.
Lavergne, S., Molofsky, J. (2007). Increased genetic variation and evolutionally potential drive the success of an invasive grass. Proceeding of the National Academy of Sciences, 104, 3883-3888.
Lehan, N.E., Murphy, J.R., Thorburn, L.P., Bradley, B.A. (2013). Accidental introductions are an important source of invasive plants in the continental United States. American Journal of Botany, 100, 1287-1293.
Les, D.S. (1996). *Hydrilla verticillata* threatens New England. Aquatic Exotic News 3(1): 1-2. Northeast Sea Grant Network, Connecticut Sea Grant, Groton, CT.
Les, D.H., Mehrhoff, L.J. (1999). Introduction of nonindigenous aquatic vascular plants in southern New England: a historical perspective. Biological Invasions, 1, 281-300.
Lesage, E., Mundia, C., Rousseau, D.O.L., Van de Moortel, A.M.K., Du Laing, G., Meers, E., Tack, F.M.G., De Pauw, N., Verloo, M.G. (2007). Sorption of Co, Cu, Ni, and Zn from industrial effluents by the submerged aquatic macrophyte *Myriophyllum spicatum* L. Ecological Engineering, 30, 320-325.
Leung, B., Brossenbroek, J.M., Lodge, D.M. (2006). Boats, pathways, and aquatic biological invasions: estimating dispersal potential with gravity models. Biological Invasions, 8, 241-254.
Liddle, M.J., Scorgie, H.R.A. (1980). The effects of recreation on freshwater plants and animals: a review. Biological Conservation, 17, 183-206.
Little, E.C.S. (1979). Handbook of utilization of aquatic plants. FAO Fisheries Technical Paper (187). Rome: FAO.

Mack, R.N. (1999). The motivation for importing potentially invasive plant species: a primal urge? In D. Eldridge, D. Freudenberger (Eds.), Proceedings of the VI International Rangelands Congress 2 (p.557-562). Aitkenvale, Queensland, Australia: International Rangelands Congress, Inc.

Mack, R.N. (2003). Global plant dispersal, naturalization, and invasion: pathways, modes, and circumstances. In G.M. Ruiz, J.T. Carlton (Eds.), Invasive Species: Vectors and Management Strategies (p.3-30). Washington: Island Press.

Mack, R.N., Lonsdale, W.M. (2001). Humans as global plant dispersers: getting more than we bargained for. BioScience, 51, 95-102.

Maki, K., Galatowitsch, S. (2004). Movement of invasive aquatic plants into Minnesota (USA) through horticultural trade. Biological Conservation, 118, 389-396.

Markwith, S.H., Leigh, D.S. (2008). Subaqueous hydrochory: open-channel hydraulic modeling of non-buoyant seed movement. Freshwater Biology, 53, 2274-2286, DOI: 10.1111/j.1365-2427.2008.02054.x.

Masoodi, A. (2013). Ecology of Important Invasive Macrophytes and Emerging Management Scenarios for Wular Lake, Kashmir, India. Ph. D. Thesis, Department of Botany, Aligarh Muslim University, Aligarh (U.P.) India.

Mcalarnen, L.A., Barnes, M.A., Jerde, C.L., Lodge, D. M. (2012). Simulated overland transport of Eurasian watermilfoil: survival of desiccated plant fragments. Journal of Aquatic Plant Management, 50, 147-149.

Medwecka-Kornaś, A., Hawro, R. (1993). Vegetation on beaver dams in the Ojcow National Park (Southern Poland). Phytocoenologia, 23, 611-618.

Mills, E.L., Strayer, D.L., Scheuerell, M.D., Carlton, J.T. (1996). Exotic species in the Hudson River Basin: a history of invasions and introductions. Estuaries, 19, 814-823.

Mitchell, D.S. (1973). Supply of plant nutrient chemicals in Lake Kariba. Man-made Lakes: Their Problems and Environmental Effects, Geophysical Monograph Series, 17, 165-169.

Mitchell, D.S., Tomislav, P., Viner, A.B. (1980). The water-fern, *Salvinia molesta*, in the Sepik River, Papua New Guinea. Environmental Conservation, 7, 115-122.

Moody, M.L., Les, D.H. (2002). Evidence of hybridity in invasive watermilfoil (*Myriophyllum*) populations. Proceedings of the National Academy of Sciences, 99, 14867-14871.

Mosisch, T.D., Arthington, A. H. (1998). The impacts of power boating and water skiing on lakes and reservoirs. Lakes and Reservoirs: Research and Management, 3, 1-17.

Mueller, M.H., van der Valk, A. (2002). The potential role of ducks in wetland seed dispersal. Wetlands, 22, 170-178.

Muirhead, J.R., MacIsaac, H. (2005). Development of inland lakes as hubs in an invasive network. Journal of Applied Ecology, 42, 80-90.

Mumma, M.T., Cichra, C.E., Sowards, J.T. (1996). Effects of recreation on the submersed aquatic plant community of Rainbow River, Florida. Journal of Aquatic Plant Management, 34, 53-56.

Nash, H., Thorpe, S. (1998). Aquatic Plants and Their Cultivation. A Complete Guide for Water Gardeners. New York: Sterling Publishing Co.

National Marine Manufacturers Association Canada (2012). The Economic Impact of Recreational Boating in Canada. Ontario, Canada: NMMA Canada.

National Research Council (1996). Stemming the Tide: Controlling Introductions of Nonindigenous Species by Ships' Ballast Water. Washington, DC: The National Academies Press.

Nelson, T.C. (1917). The introduction of foreign weeds in ballast as illustrated by ballast-plants at Linton, Oregon. Torreya, 17, 151-160.

Nieder, W.C., Barnaba, E., Findlay, S.E.G., Haskins, S., Holochuck, N., Blair, E.A. (2004). Distribution and abundance of submerged aquatic vegetation and *Trapa natans* in the Hudson River estuary. Journal of Coastal Research, 45, 150-161.

Nilsson, C., Brown, R. L., Jansson, R., Merritt, D.M. (2010). The role of hydrochory in structuring riparian and wetland vegetation. Biological Reviews, 85, 837-858, DOI: 10.1111/j.1469-185X.2010.00129.x.

Orchard, A.E. (1979). *Myriophyllum* (Halagoraceae) in Australis. I. New Zealand: a revision of the genus and synopsis of the family. Brunonia, 2, 247-287.

Orth, R.J., Moore, K.A. (1984). Distribution and abundance of submerged aquatic vegetation in Chesapeake Bay: an historical perspective. Estuaries, 7, 531-540.

Owens, C.S., Madsen, J.D., Smart, R.M., Steward, R. M. (2001). Dispersal of native and nonnative aquatic plant species in the San Marcos River, Texas. Journal of Aquatic Plant Management, 39, 75-79.

Philbrick, C.T., Les, D.H. (1996). Evolution of aquatic angiosperm reproductive systems. BioScience, 46, 813-826.

Pollux, B.J.A., Ouborg, N.J., Van Groenendael, J.M., Klaassen, M. (2007). Consequences of intraspecific seed-size variation in *Sparganium emersum* for dispersal by fish. Functional Ecology, 21, 1084-1091.

Rajagopal, K., Sasikala, K. (2008). Antihyperglycaemic and antihyperlipidaemic effects of *Nymphaea stellata* in alloxan-induced diabetic rats. Singapore Medical Journal, 49, 137-141.

Raulings, E., Morris, K., Thompson, R., Mac Nally, R. (2011). Do birds of a feather disperse plants together? Freshwater Biology, 56, 1390-1402.

Ricciardi, A. (2006). Patterns of invasion in the Laurentian Great Lakes in relation to changes in vector activity. Diversity and Distributions, 12, 425-433.

Ridley, H.N. (1930). The Dispersal of Plant throughout the World. Kent (UK): L. Reeve.

Riis, T., Sand-Jenson, K. (2006). Dispersal of plant fragments in small streams. Freshwater Biology, 51, 274-286.

Roley, S.S., Newman, R. M. (2008). Predicting Eurasian watermilfoil invasions in Minnesota. Lake and Reservoir Management, 24, 361-369.

Roman, J., Darling, J.A. (2007). Paradox lost: genetic diversity and success of aquatic invasions. Trends in Ecology and Evolution, 22 (9), 454-464.

Rood, S.B., Braatne, J.H., Goater, L.A. (2010). Favorable fragmentation: river reservoirs can impede downstream expansion of riparian weeds. Ecological Applications, 20, 1664-1677.

Rothlisberger, J.D., Chadderton, W.L., McNulty, J., Lodge, D.M. (2011). Aquatic invasive species transport via trailered boats: what is being moved, who is moving it, and what can be done? Fisheries, 35, 121-132, DOI: 10.1577/1548-8446-35.3.121.

Sand-Jensen, K. (1989). Environmental variables and their effects on photosynthesis of aquatic plant communities. Aquatic Botany, 34, 5-25.

Sarneel, J.M. (2012). The dispersal capacity of vegetative propagules of riparian fen species. Hydrobiologia, 710, 219-225.

Sarneel, J.M., Beltman, B., Buijze, A., Groen, R., Soons, M.B. (2014). The role of wind in dispersal of floating seeds in slow-moving or stagnant water bodies. Journal of Vegetation Science, 25, 262-274.

Sculthorpe, C.D. (1967). The biology of aquatic vascular plants. London: Edward Arnold Ltd.

Shen-Miller, J., Mudgett, M.B., Schopf, J.W. Clarke, S., Berger R. (1995). Exceptional seed longevity and robust growth: ancient sacred lotus from China. American Journal of Botany, 82, 1367-1380.

Shibayama, H. (2001). Weeds and weed management in rice production in Japan. Weed Biology and Management, 1, 53-60.

Spence, D.H. (1967). Factors controlling the distribution of freshwater macrophytes with particular reference to the lochs of Scotland. Journal of Ecology, 55 (1), 147-170.

Spencer, D.F., Rejmánek, M. (1989). Propagule type influences competition between two submersed aquatic macrophytes. Oecologia, 81, 132-137.

Spencer, D.F., Anderson, L.W.J., Ames, M.D., Ryan, F.J. (1987). Variation in *Hydrilla verticillata* (L.f.) Royle propagule weight. Journal of Aquatic Plant Management, 25, 11-14.
Strecker, A.L., Campbell, P.M., Olden, J. D. (2011). The aquarium trade as an invasion pathway in the Pacific Northwest. Fisheries, 36, 74-85.
Tamayo, M., Olden, J.D. (2013). Forecasting the vulnerability of lakes to aquatic plant invasions. Invasive Plant Science and Management, 7 (1), 32-45.
Toft, J.D., Simenstad, C.A., Cordell, J.R., Grimaldo, L.F. (2003). The effects of introduced water hyacinth on habitat structure, invertebrate assemblages, and fish diets. Estuaries, 26 (3), 746-758.
Truscott, A-M., Soulsby, C., Palmer, S.C.F., Newell, L., Hulme, P.E. (2006). The dispersal characteristics of the invasive plant *Mimulus guttatus* and the ecological significance of increased occurrence of high-flow events. Journal of Ecology, 94, 1080-1091.
Van der Valk, A.G., Davis, C.B. (1978). The role of seed banks in the vegetation dynamics of prairie glacial marshes. Ecology, 59, 322-335.
Van, T., Madeira, P.T. (1998). Random amplified polymorphic DNA analysis of water spinach (*Ipomoea aquatic*) in Florida. Journal of Aquatic Plant Management, 36, 107-111.
Van, T.K., Steward, K.K. (1990). Longevity of monoecious Hydrilla propagules. Journal of Aquatic Plant Management, 28, 74-76.
Vander Zanden, M.J., Olden, J.D. (2008). A management framework for preventing the secondary spread of aquatic invasive species. Canadian Journal of Fisheries and Aquatic Sciences, 65, 1512-1522.
Vermaat, J.E., Santamaría, L., Roos. P.J. (2000). Water flow across and sediment trapping in submerged macrophyte beds of contrasting growth form. Archiv für Hydrobiologie, 148 (4), 549-562.
Viana, D. S., Santamaría, L. Michot, T. C., Figuerola, J. (2013). Migratory strategies of waterbirds shape the continental-scale dispersal of aquatic organisms. Ecography, 36, 430-438.
Weber, M., Brändle, R. (1996). Aspects of the extreme anoxia tolerance of the sweet flag, *Acorus calamus* L. Folia Gerbotanica and Phytotaxonomica, 31, 37-46.
Westcott, K., Whillans, T.H., Fox, M.G. (1997). Viability and abundance of seeds of submerged macrophytes in the sediment of disturbed and reference shoreline marshes in Lake Ontario. Canadian Journal of Botany, 75, 451-456.
Westlake, D.F. (1982). The primary productivity of water plants. In J.J. Symoens, S.S. Hooper, P. Compère (Eds.), Studies on Aquatic Vascular Plants (p.165-180). Brussels: Royal Botanical Society of Belgium.

George K. Roderick, Maria Navajas

4 Invasions of Terrestrial Arthropods: Mechanisms, Pathways, and Dynamics

4.1 Introduction

Insects and other terrestrial arthropods are particularly notorious as invasive species, both in terms of numbers of species (Hulme *et al.*, 2008) as well as in their ecological (Lodge, 1993; Wilcove *et al.*, 1998) and economic (Pimentel *et al.*, 2000; Pimentel *et al.*, 2005) impacts. Much of the success of terrestrial arthropods as invasive species stems from the same biological features thought to account for their diversity: insects and other terrestrial arthropods are small, often reproduce quickly, have diverse lifestyles, occupy many habitats, and are well protected from the external environment by an exoskeleton (Resh & Cardé, 2009; Gullan & Cranston, 2014). Invasive insects and other arthropods are linked to nearly every human activity. For example, invasive insects and mites include countless domesticated pests in agricultural systems, affecting both plant and animal products. Other arthropods are pests associated with the enterprises of forestry and horticulture. Still more impact urban systems and human health, while others are threats to biodiversity in natural ecosystems. Invasive arthropods play many ecological roles in natural and managed ecosystems, including functioning as all types of consumers (herbivores, predators, parasites, internal and external), but also as vectors of disease for plants and animals. It is this biological and functional diversity of invasive arthropods that makes it difficult to develop simple management strategies for their control, as well as to propose and enact legislation to limit their spread.

Efforts to categorize invasions have provided some structure with which to understand the biology of invasive insects and predict their spread and impact. Particularly useful in this regard has been the distinction between natural dispersal and dispersal associated with human activities (Falk-Petersen *et al.*, 2006), as well as identification of stages of the invasion process itself from dispersal of propagules to successful colonization, and eventual spread and impact, either ecological or economic (Facon *et al.*, 2006; Blackburn *et al.*, 2011). For management, understanding pathways of invasion is critical (Hulme *et al.*, 2008); knowledge of pathways also provides important insights into the interacting biological and socio-economic factors responsible for the spread and impact of invasive species.

In this chapter we use the framework of Hulme *et al.* (2008) to outline the main mechanisms and pathways by which insects and other terrestrial arthropods invade new habitats. We also review novel tools that have contributed to a better understanding of the invasion process and the general trends that these approaches have revealed about insect invasions. Finally, we emphasize the importance of policy and risk assessment in the management of invasive arthropods, and recognize the significant role that they play in both natural and managed ecosystems.

4.2 Mechanisms and Pathways

Hulme *et al.* (2008) identified three mechanisms through which non-indigenous species typically invade new habitats: (1) purposeful importation of a commodity; (2) arrival of a vector involved in transportation; and (3) natural dispersal from another region (Table 4.1). Each of these mechanisms is associated with one or more pathways of invasion: (1) live commodities (organisms for sale) may be released or escape, and other commodities may have associated contaminants; (2) vectors move invasive species as stowaways; and (3) dispersal by invasive organisms may follow corridors or may move unguided. Here, we adopt Hulme *et al.*'s (2008) framework to describe the means by which insects invade new habitats and to illustrate their associated ecological diversity.

Tab. 4.1: Mechanisms, pathways and activities associated with terrestrial arthropod invasions following the invasion classification of Hulme *et al.* (2008).

Mechanism	Pathway	Activity	Insect examples	References
Commodity	Release	Biological control	Ladybird beetles, Coccinellidae Africanized honeybee, *Apis mellifera scutellata*	(Simberloff & Stiling, 1996; Snyder *et al.*, 2004; Majerus *et al.*, 2006; Lombaert *et al.*, 2010; Roderick *et al.*, 2012)
	Escape	Biological control	Asian harlequin ladybird, *Harmonia axyridis*, especially in glass houses	(Majerus *et al.*, 2006)
	Contaminant	Plant trade Wood products Grain Soil	Glassywinged sharpshooter, *Homalodisca vitripennis* Whiteflies, *Bemisia spp.* Hemlock woolly adelgid, *Adelges tsugae* Tephritid fruit flies, *Ceratitus capitata, Bactrocera spp.* Pine beetles, *Dendroctonus spp.* Asian long-horned beetle, *Anoplophora glabripennis* Asian gypsy moth, *Lymantria dispar asiatica* Spider mites, e.g. tomato spider mite, *Tetranychus evansi*	(Clarke *et al.*, 2005; Malacrida *et al.*, 2007; Petit *et al.*, 2008; Petit *et al.*, 2009; Hadjistylli *et al.*, 2010; Nardi *et al.*, 2010; Boubou *et al.*, 2012)

continued **Tab. 4.1:** Mechanisms, pathways and activities associated with terrestrial arthropod invasions following the invasion classification of Hulme *et al.* (2008).

Mechanism	Pathway	Activity	Insect examples	References
Vector	Stowaway	Water transportation (voyaging, rubber tires) Cargo holds Vehicles Animals	Mosquitos, *Culicidae, Aedes spp.* *Gypsy moth, Lymantria dispar dispar* Horse-chestnut leafminer, *Cameraria ohridella*	(Gilbert *et al.*, 2005; Johnson *et al.*, 2006; Benedict *et al.*, 2007)
Dispersal	Corridors	Trails Roadsides	Argentine ant, *Linepithema humile*	(Holway, 1995)
	Unaided	Natural dispersal	Brown planthopper, *Nilaparvata lugens* Colorado potato beetle, *Leptinotarsa decemlineata* Ladybird beetles, Coccinellidae	(Snyder *et al.*, 2004; Majerus *et al.*, 2006; Lombaert *et al.*, 2010; Gillespie *et al.*, 2012)

4.2.1 Commodities: Release, Escape, Contaminants

Previous reviews of invasive species (Hulme *et al.*, 2008), including insects and other terrestrial arthropods (e.g., Howarth, 1996; Yano *et al.*, 1999; Kiritani & Yamamura, 2003; Sax *et al.*, 2005) have illustrated the importance of movement of commodities for invasions, a process that is associated with the three following pathways: release, escape, and as contaminants. Live commodities, such as horticultural plants, can become invasive if they escape as weeds to occupy new habitats (see Chapter 3). Similar pathways are possible for insect commodities, such as those intentionally released for biological control. While current programs of biological control using insects are tightly regulated, including pre-release quarantine and host-range testing, some earlier introductions for biological control have had severe negative impacts (Simberloff & Stiling, 1996; Roderick & Howarth, 1999; Snyder *et al.*, 2004). For insects, adaptation to novel conditions, particularly novel hosts or new physical environments, is possible in ways not predicted by pre-release testing (Roderick *et al.*, 2012). For example, following use in glass houses for biological control, the Asian harlequin ladybird, *Harmonia axyridis*, escaped and became invasive in Britain and elsewhere (Majerus *et al.*, 2006; Lombaert *et al.*, 2010). Another example of an insect invasion first introduced as a commodity is the release and spread of the more excitable Africanized honeybee, *Apis mellifera scutellata*, in the Americas (Hall & Muralidharan, 1989).

A vast diversity of invasive arthropods are contaminants of commodities, which spread through global trade or other human transport (Hulme *et al.*, 2008). Insect contaminants of horticultural, agricultural, and forestry products are particularly important in this regard. Noted examples include *Bemisia* whiteflies, which were moved around the US on ornamental poinsettia and world-wide on other plant species (Perring *et al.*, 1993; Hadjistylli *et al.*, 2010), and the glassy-winged sharp-shooter, *Homalodisca vitripennis,* which spread internationally through movement of citrus and vine hosts (Petit *et al.*, 2008). Contaminants of food products also include the recent invasions of Asian citrus psyllid, *Diaphorina citri*; numerous species of fruit flies, *Bactrocera spp.* (Clarke *et al.*, 2005); and the tomato spider mite, *Tetranychus evansi* (Boubou *et al.*, 2012). Movement of wood products, including wooden shipping pallets and lumber, is thought to have spread Asian gypsy moths, *Lymantria dispar asiatica*; pine beetles, *Dendroctonus spp.*; Formosan subterranean termites, *Coptotermes formosanus*; Asian long-horned beetles, *Anoplophora glabripennis*; and Hemlock woolly adelgids, *Adelges tsugae,* to name only a few. Insect contaminants are also common in grain supplies and feed, seeds, stored products, and soil (Hulme *et al.*, 2008). Unfortunately, with an increase in global trade, contaminants of commodities will continue to be a worldwide problem.

4.2.2 Vectors: Stowaways

A fourth pathway for terrestrial arthropods to arrive in new habitats is as stowaways associated with some vehicle or animal vector; this pathway is common for many invasive plants and animals (see Chapters 1, 2, 3, 5, 6). For example, the Polynesian tiger mosquito, *Aedes polynesiensis,* is thought to have stowed away in water containers transported by ancient Polynesians when voyaging across the Pacific. More recently, the Asian tiger mosquito, *Aedes albopictus,* travelled in water found inside discarded automobile tires (Benedict *et al.*, 2007). The glassy-winged sharpshooter, *Homalodisca vitripennis,* and many other species have been observed in the cargo holds of airplanes (Liebhold *et al.*, 2006), and bedbugs, *Cimex lectularius,* move with human belongings (Saenz *et al.*, 2012), presumably in luggage. Flightless gypsy moths, *Lymantria dispar dispar,* have been spread through transport on vehicles in North America (Johnson *et al.*, 2006), a vector which has also been proposed for spread of the horse-chestnut leafminer, *Cameraria ohridella,* in Europe (Gilbert *et al.*, 2005).

Many arthropods are transported to new habitats in association with other animals (Chapter 6). For example, the distribution of ticks carrying Lyme disease is associated with vertebrate hosts (Ostfeld *et al.*, 2006; Swei *et al.*, 2011). Insects can also be vectored by birds, sometimes over great distances, either as external stowaways or in bird guts inside seeds (Gillespie *et al.*, 2012). Finally, humans transport their own domesticated arthropod parasites, especially lice and mites.

4.2.3 Dispersal: Corridors or Unguided

Hulme *et al.* (2008) recognized two pathways of dispersal, either along dispersal corridors or unguided (unaided). While dispersal corridors are likely more important for freshwater aquatic species than for terrestrial species (see Chapters 3, 5), terrestrial arthropods can also disperse to new areas facilitated by ecological corridors, including disturbed roadsides, railways, or walking trails. For example, the invasive Argentine ant, *Linepithema humile*, moves along roads and trails into native forest in Hawaii where it is a serious ecological pest (Krushelnycky & Gillespie, 2008). Terrestrial arthropods can also spread unguided by corridors from one area to another. As one might expect, this pathway is common for insects with great aerial dispersal ability, such as the rice brown planthopper, *Nilaparvata lugens*, which is known to move seasonally between tropical and temperate regions in South East Asia (Denno & Roderick, 1990; Mun *et al.*, 1999; Zhu *et al.*, 2000). However, even less dispersive insects can move efficiently, locally and regionally: examples include the Colorado potato beetle, (Grapputo *et al.*, 2005); various ladybird coccinellid beetles (Snyder *et al.*, 2004; Majerus *et al.*, 2006; Lombaert *et al.*, 2010); and many species of ants (Holway *et al.*, 2002).

4.3 New Tools and Approaches

Invasive insects and other terrestrial arthropods are model systems for the study of invasive species, in that their effects are important economically and ecologically, their distribution is global, and many species are easily collected and monitored. Much of our current understanding of insect invasions comes from new research approaches. Active areas of research include the use of molecular population genetics facilitated by high-throughput DNA sequencing to infer the origins of colonization events and other features of demographic history (Davies *et al.*, 1999b; Estoup & Guillemaud, 2010; Boubou *et al.*, 2012). In this regard, insect collections are proving invaluable as sources of DNA for studies of origins in addition to providing documentation of historical ranges (Carey, 1991; Suarez & Tsutsui, 2004; Malacrida *et al.*, 2007; Marsico *et al.*, 2010). Insects in collections can also provide information on food webs and other ecological interactions, such as through examination of pollen or stable isotopes (Hobson *et al.*, 2012). Recent advances in making predictions of range expansion associated with global change, especially with changes in climate and land use, are possible through using collection data, online databases, niche modeling, and integral projection models (Suarez *et al.*, 2001; Migeon & Dorkeld, 2006-2015; Rapacciuolo *et al.*, 2012; Meynard *et al.*, 2013; Berkeley, 2014; Merow *et al.*, 2014; DAISIE, 2015). Finally, citizen science is allowing the public to participate in large scientific endeavors and at the same time benefit from new knowledge. Insect-related examples of citizen science include identification tools, including *Discover Life* (Pickering,

2009) and *iNaturalist* (Ueda & Loarie, 2013), as well as targeted research focusing on changing geographic distributions, such as the *Lost Ladybug Project* (Cornell University, 2014).

4.4 The Invasion Process

Because of their worldwide economic importance, particularly as pests of agriculture, as urban associates, and as vectors of human disease, the invasion process of many terrestrial arthropods has been studied in great detail. Several common themes emerge that often characterize arthropod invasions:

- Invading populations can be small in size, and often with invasive genotypes unrepresentative of the species as a whole; examples include tephritid fruit flies; mites; mosquitoes (*Culicidae* spp.); and ants (Davies *et al.*, 1999a; Fonseca *et al.*, 2000; Holway *et al.*, 2002; Navajas & Boursot, 2003; Roderick & Navajas, 2003; Navajas *et al.*, 2009). That invasions of terrestrial arthropods can be successful despite small initial population sizes, and thus low genetic diversity that should limit potential adaptation, is a paradox (but see below).
- Many insect invasions involve cryptic invasions of more than one colonization event. For example, molecular genetic studies of medflies, *Ceratitis capitata*, olive flies, *Bactrocera oleae*, oriental fruit flies, *Bactrocera dorsalis*, and other tephritid fruit flies show that multiple, often cryptic invasions are common (Davies *et al.*, 1999a; Clarke *et al.*, 2005; Nardi *et al.*, 2005; Malacrida *et al.*, 2007; Nardi *et al.*, 2010). Similar results have been found in Bemesia whiteflies (Hadjistylli *et al.*, 2010), glassy-winged sharpshooters, *Homalodisca vitripennis* (Petit *et al.*, 2008), Colorado potato beetles, *Leptinotarsa decemlineata* (Grapputo *et al.*, 2005), and other terrestrial arthropods, particularly mites (Boubou *et al.*, 2012). Mixing or hybridization associated with multiple colonization events may increase the genetic variation in colonizing populations, which in theory should contribute to the ability of invasive populations to adapt to novel conditions.
- In many invasive arthropod species, invading populations originate from populations that were invasive elsewhere. An example is the serial invasion history of the medfly, *Ceratitis capitata*, which originated in sub-Saharan Africa and spread to the Mediterranean, and then world-wide (Davies *et al.*, 1999a). Another example is the spread of the glassy-winged sharpshooter from one island to the next within and between island archipelagoes in French Polynesia (Petit *et al.*, 2008). Such species may have overcome limitations of small populations and low genetic diversity through selection in previous colonization episodes.
- A lag period of many generations may occur before the populations reach a size large enough to be noticed or to cause economic damage. Examples include the gypsy moth, *Lymantria dispar dispar* (Johnson *et al.*, 2006), and light brown apple moth *Epiphyas postvittana* (Suckling & Brockerhoff, 2010), among others.

- Species interactions, or lack thereof, are critical in many insect invasions, especially escape from competitors and enemies (Torchin *et al.*, 2003). Other interactions among species may facilitate arthropod invasions, such the presence of host plants or other resources (Leong *et al.*, 2014). For example, the Colorado potato beetle is thought to have moved from native solanaceous species to potatoes with the arrival of European settlers in the American West (Grapputo *et al.*, 2005). For sap-feeding insects, microbe symbionts may provide essential amino acids necessary to switch to new plant host species (McFall-Ngai *et al.*, 2013).

4.5 Policy and Management

Understanding the mechanisms of invasion and associated pathways is critical for management, including monitoring, interception, and policies to restrict trade (Hulme, 2006; Petit *et al.*, 2009). Knowledge of pathways is also necessary to predict future spread and impact. Where invasions involve mechanisms associated with commodities or human activity, such information can aid in understanding the process of invasion. For example, when invasive arthropods are contaminants of commodities, the occurrence and traits of contaminants can be at least partially understood by the commodity itself (Hulme *et al.*, 2008). Likewise, understanding vectors of transportation provides testable hypotheses for the spread of species associated with those vectors (Carey, 1991). For example, airplane and shipping routes, coupled with climate matching, predict aspects of insect invasions (Liebhold *et al.*, 2006; Tatum & Hay, 2007).

4.6 Risk Assessment

Risk assessment associated with predicting the spread and impact of invasive species is difficult in general and particularly so for insects and other arthropods (Shogren, 2000). In part, this is because a changing environment and novel sets of species interactions create uncertainties, but also because invasive propagules can be rare (Drake & Lodge, 2006; Simberloff, 2009; Gillespie *et al.*, 2012). Also, there are few incentives for studying the risk of invasions for non-economic species. For example, national and international initiatives such as the European Food Safety Authority (2015) and USDA APHIS (2015) focus on species of commercial interest but are less concerned about risks to natural environments (but see Gilioli *et al.*, 2014). Risk assessment is a major research gap in the study of invasive species and a topic that will become even more critical in the context of global change in climate and land use (Barnosky *et al.*, 2012).

4.7 Living with Invasive Arthropods

Established insect populations are difficult to eradicate, prompting a reanalysis of both their new roles in ecological communities and our public perception of non-native species (Davis *et al.*, 2011; Richardson & Ricciardi, 2013). In most cases, management must necessarily turn from eradication to mitigation and adaptation and, where control is still an option, policy makers must prioritize efforts on the early stages of invasion, and particularly exclusion (Simberloff, 2009; Simberloff *et al.*, 2012). Exclusion is a difficult task and one that is a prime focus for national and international plant health agencies (e.g., EFSA, 2015; USDA APHIS, 2015). A key question is how invasive species will respond to global climate change and what the resulting impacts will be (Barnosky *et al.*, 2012; Biermann *et al.*, 2012). Will invasive insect species move to more suitable climates, adapt to new climates, or die out? Clearly, invasive arthropods are a global problem, and any solutions will necessarily be multidisciplinary and require international collaboration.

4.8 Conclusions

Terrestrial arthropod invaders are diverse and are associated with many pathways of introduction. New tools, including collection science, molecular population genetics, computational modeling, climate/niche modeling, and integral population modeling are providing novel insights into the dynamics of invasions of terrestrial insects and other arthropods. Many invasive populations of terrestrial arthropods are the result of multiple, often cryptic, colonization events, which adds genetic diversity to founding populations but also complicates management. Terrestrial arthropod populations are difficult to eradicate once established, so the best strategy for management is limiting propagules and establishment.

4.9 Acknowledgements

We thank the editors João Canning-Clode and Anssi Vainikka for encouragement and suggestions. This work was supported by the Institut National de la Recherche Agronomique, France, the French Agence Nationale de la Recherche (ID grant ANR-14-JFAC-0006-01), the US National Science Foundation, the US Department of Agriculture, and the France-Berkeley Fund.

In a nutshell

- Terrestrial arthropods are important as invasive species in terms of species numbers as well as ecological and economic impacts.
- Characteristics of terrestrial arthropods, their life histories, and their diversity of lifestyles, contribute both to their success and to difficulty in their management.
- Insects and other terrestrial arthropods illustrate the full range of invasion mechanisms and associated pathways; many are contaminants of commodities.
- New tools, including molecular population genetics, high throughput sequencing, computational methods, climate modeling, and collection science are providing novel insights into the dynamics of invasions.
- Invasive terrestrial arthropods often show a pattern of serial invasions, in which an invasive population gives rise to new invasive populations.
- Many invasive arthropod populations are the result of multiple, often cryptic, colonization events, which adds genetic diversity to founding populations but also complicates management. Invasive populations often stem from other invasive populations.
- Population lags are common, in which invasive populations grow undetected for many generations before reaching a threshold where they become abundant enough to be noticed or cause economic damage.
- While knowledge of mechanisms and pathways is critical to management of invasive arthropods, management is most effective at the early stages of the invasion process, particularly by preventing the initial colonization.
- As a result of their success, invasive terrestrial arthropods are now important elements of biological communities in both managed and natural ecosystems.

4.10 Bibliography

Barnosky, A. D., Hadly, E. A., Bascompte, J., Berlow, E. L., Brown, J. H., Fortelius, M., Getz, W. M. *et al.* (2012). Approaching a state shift in Earth's biosphere. Nature, 486, 52-58.

Benedict, M. Q., Levine, R. S., Hawley, W. A., Lounibos, L. P. (2007). Spread of the tiger: global risk of invasion by the mosquito Aedes albopictus. Vector-Borne and Zoonotic Diseases, 7, 76-85.

Berkeley (2014). Holos, Berkeley EcoEngine. Retrieved 1 February 2015, from https://ecoengine.berkeley.edu/

Biermann, F., Abbott, K., Andresen, S., Bäckstrand, K., Bernstein, S., Betsill, M. M. *et al.* (2012). Navigating the anthropocene: Improving earth system governance. Science, 335, 1306-1307.

Blackburn, T. M., Pysek, P., Bacher, S., Carlton, J. T., Duncan, R. P., Jarosik, V., Wilson, J. R., Richardson, D. M. (2011). A proposed unified framework for biological invasions. Trends in Ecology and Evolution, 26, 333-339.

Boubou, A., Migeon, A., Roderick, G. K., Auger, P., Cornuet, J.-M., Magalhães, S., Navajas, M. (2012). Test of colonisation scenarios reveals complex invasion history of the red tomato spider mite *Tetranychus evansi*. PLoS One, 7, e35601.

Carey, J. R. (1991). Establishment of the Mediterranean fruit fly in California. Science, 253, 1369-1373.

Clarke, A. R., Armstrong, K. F., Carmichael, A. E., Milne, J. R., Raghu, S., Roderick, G. K., Yeates, D. K. (2005). Invasive phytophagous pests arising through a recent tropical evolutionary radiation: The *Bactrocera dorsalis* complex of tropical fruit flies. Annual Review of Entomology, 50, 293-319.

Cornell University (2014). Lost Ladybug Project. Retrieved 1 February 2015, from http://www.lostladybug.org/

DAISIE (2015). Delivering Alien Invasive Species Inventories for Europe. Retrieved 1 February 2015, from http://www.europe-aliens.org

Davies, N., Villablanca, F. X., Roderick, G. K. (1999a). Bioinvasions of the medfly, *Ceratitis capitata*: source estimation using DNA sequences at multiple intron loci. Genetics, 153, 351-360.

Davies, N., Villablanca, F. X., Roderick, G. K. (1999b). Determining the sources of individuals in recently founded populations: multilocus genotyping in non-equilibrium genetics. Trends in Ecology and Evolution, 14, 17-21.

Davis, M. A., Chew, M. K., Hobbs, R. J., Lugo, A. E., Ewel, J. J., Vermeij, G. J. *et al.* (2011). Don't judge species on their origins. Nature, 474, 153-154.

Denno, R. F., Roderick, G. K. (1990). Population biology of planthoppers. Annual Review of Entomology, 35, 489-520.

Drake, J. M., Lodge, D. M. (2006). Allee effects, propagule pressure and the probability of establishment: risk analysis for biological invasions. Biological Invasions, 8, 365-375.

EFSA (2015). European Food Safety Authority. Retrieved 1 February 2015, from http://www.efsa.europa.eu

Estoup, A., Guillemaud, T. (2010). Reconstructing routes of invasion using genetic data: why, how and so what? Molecular Ecology, 19, 4113-4130.

Facon, B., Genton, B. J., Shykoff, J., Jarne, P., Estoup, A., David, P. (2006). A general eco-evolutionary framework for understanding invasions. Trends in Ecology and Evolution, 21, 130-135.

Falk-Petersen, J., Bøhn, T., Sandlund, O. T. (2006). On the Numerous Concepts in Invasion Biology. Biological Invasions, 8, 1409-1424.

Fonseca, D. M., LaPointe, D. A., Fleischer, R. C. (2000). Bottlenecks and multiple introductions: population genetics of the vector of avian malaria in Hawaii. Molecular Ecology, 9, 1803-1814.

Gilbert, M., Guichard, S., Freise, J., Gregoire, J.-C., Heitland, W., Straw, N., Tilbury, C., Augustin, S. (2005). Forecasting *Cameraria ohridella* invasion dynamics in recently invaded countries: from validation to prediction. Journal of Applied Ecology, 42, 805-813.

Gilioli, G., Schrader, G., Baker, R., Ceglarska, E., Kertész, V., Lövei, G., Navajas, M., Rossi, V., Tramontini, S., van Lenteren, J. (2014). Environmental risk assessment for plant pests: A procedure to evaluate their impacts on ecosystem services. Science of the Total Environment, 468, 475-486.

Gillespie, R. G., Baldwin, B. G., Waters, J. M., Fraser, C. I., Nikula, R., Roderick, G. K. (2012). Long-distance dispersal: a framework for hypothesis testing. Trends in Ecology and Evolution, 27, 52-61.

Grapputo, A., Boman, S., Lindstroem, L., Lyytinen, A., Mappes, J. (2005). The voyage of an invasive species across continents: genetic diversity of North American and European Colorado potato beetle populations. Molecular Ecology, 14, 4207-4219.

Gullan, P. J., Cranston, P. S. (2014). The Insects: An Outline of Entomology (5th ed.). Chichester, West Sussex: John Wiley and Sons, Ltd.

Hadjistylli, M., Brown, J. K., Roderick, G. K. (2010). Tools and recent progress in studying gene flow and population genetics of the *Bemisia tabaci* sibling species group. In P. A. Stansly, S. E. Naranjo (Eds.), Bemisia: Bionomics and Management of a Global Pest (pp.69-103). Dordrecht: Springer Science+Business Media B.V.

Hall, H. G., Muralidharan, K. (1989). Evidence from mitochondrial DNA that African honey bees spread as continuous maternal lineages. Nature, 339, 211-213.

Hobson, K. A., Soto, D. X., Paulson, D. R., Wassenaar, L. I., Matthews, J. H. (2012). A dragonfly (δ2H) isoscape for North America: a new tool for determining natal origins of migratory aquatic emergent insects. Methods in Ecology and Evolution, 3, 766-772.
Holway, D. A. (1995). Distribution of the Argentine ant (*Linepithema humile*) in northern California. Conservation Biology, 9 (6), 1634-1637.
Holway, D. A., Lach, L., Suarez, A. V., Tsutsui, N. D., Case, T. J. (2002). The causes and consequences of ant invasions. Annual Review of Ecology and Systematics, 33, 181-233.
Howarth, F. G. (1996). The major taxonomic groups that become invasive alien pests in Hawaii and the characteristics that make them pestiferous. Technical Report No. 1996.022. Honolulu, Hawaii: Hawaii Biological Survey.
Hulme, P. E. (2006). Beyond control: wider implications for the management of biological invasions. Journal of Applied Ecology, 43, 835-847.
Hulme, P. E., Bacher, S., Kenis, M., Klotz, S., Kühn, I., Minchin, D. *et al.* (2008). Grasping at the routes of biological invasions: a framework for integrating pathways into policy. Journal of Applied Ecology, 45, 403-414.
Johnson, D. M., Liebhold, A. M., Tobin, P. C., Bjørnstad, O. N. (2006). Allee effects and pulsed invasion by the gypsy moth. Nature, 444, 361-363.
Kiritani, K., & Yamamura, K. (2003). Exotic insects and their pathways for invasion. In G. M. Ruiz & J. T. Carlton (Eds.), Invasive Species: vectors and management strategies (pp.44-67). Washington, DC: Island Press.
Krushelnycky, P. D., Gillespie, R. G. (2008). Compositional and functional stability of arthropod communities in the face of ant invasions. Ecological Applications, 18, 1547-1562.
Leong, M., Kremen, C., Roderick, G. K. (2014). Pollinator Interactions with yellow starthistle (*Centaurea solstitialis*) across urban, agricultural, and natural landscapes. PLoS ONE 9(1): e86357. doi:10.1371/journal.pone.0086357.
Liebhold, A. M., Work, T. T., McCullough, D. G., Cavey, J. F. (2006). Airline Baggage as a Pathway for Alien Insect Species Invading the United States. American Entomologist, 52, 48-54.
Lodge, D. M. (1993). Biological invasions: lessons for ecology. Trends in Ecology and Evolution, 8, 133-137.
Lombaert, E., Guillemaud, T., Cornuet, J.-M., Malausa, T., Facon, B., Estoup, A. (2010). Bridgehead effect in the worldwide invasion of the biocontrol harlequin ladybird. PLoS One, 5, e9743.
Majerus, M., Strawson, V., Roy, H. (2006). The potential impacts of the arrival of the harlequin ladybird, *Harmonia axyridis* (Pallas)(Coleoptera: Coccinellidae), in Britain. Ecological Entomology, 31, 207-215.
Malacrida, A. R., Gomulski, L. M., Bonizzoni, M., Bertin, S., Gasperi, G., Guglielmino, C. R. (2007). Globalization and fruitfly invasion and expansion: the medfly paradigm. Genetica, 131, 1-9.
Marsico, T. D., Burt, J. W., Espeland, E. K., Gilchrist, G. W., Jamieson, M. A., Lindström, L., Roderick, G. K., Swope, S., Szücs, M., Tstusui, N. D. (2010). Underutilized resources for studying the evolution of invasive species during their introduction, establishment, and lag phases. Evolutionary Applications, 3, 203-219.
McFall-Ngai, M., Hadfield, M. G., Bosch, T. C. G., Carey, H. V., Domazet-Lošo, T., Douglas, A. E. *et al.* (2013). Animals in a bacterial world, a new imperative for the life sciences. Proceedings of the National Academy of Sciences, 110, 3229-3236.
Merow, C., Dahlgren, J. P., Metcalf, C. J. E., Childs, D. Z., Evans, M. E. K., Jongejans, E., Record, S., Rees, M., Salguero-Gómez, R., McMahon, S. M. (2014). Advancing population ecology with integral projection models: a practical guide. Methods in Ecology and Evolution, 5, 99-110.
Meynard, C. N., Migeon, A., Navajas, M. (2013). Uncertainties in predicting species distributions under climate change: A case study using Tetranychus evansi (Acari: Tetranychidae), a widespread agricultural pest. PLoS One, 8, e66445.

Migeon, A., & Dorkeld, F. (2006-2015). Spider Mites Web: a comprehensive database for the Tetranychidae. Retrieved 1 February 2015, from http://www.montpellier.inra.fr/CBGP/spmweb

Mun, J. H., Song, Y. H., Heong, K. L., Roderick, G. K. (1999). Genetic variation among Asian populations of rice planthoppers, *Nilaparvata lugens* and *Sogatella furcifera* (Hemiptera: Delphacidae): mitochondrial DNA sequences. Bulletin of Entomological Research, 89, 245-253.

Nardi, F., Carapelli, A., Boore, J. L., Roderick, G. K., Dallai, R., Frati, F. (2010). Domestication of olive fly through a multi-regional host shift to cultivated olives: Comparative dating using complete mitochondrial genomes. Molecular Phylogenetics and Evolution, 57, 678-686.

Nardi, F., Crapelli, A., Dallai, R., Roderick, G. K., Frati, F. (2005). Population structure and colonization history of the olive fly, *Bactrocera oleae* (Diptera, Tephritidae). Molecular Ecology, 14, 2729-2738.

Navajas, M., Anderson, D., de Guzman, L., Huang, Z., Clement, J., Zhou, T., le Conte, Y. (2009). New Asian types of *Varroa destructor*: a potential new threat for world apiculture. Apidologie, 41, 181-193.

Navajas, M., Boursot, P. (2003). Nuclear ribosomal differentiation between two closely mite species polyphyletic for mitochondrial DNA. Proceedings of the Royal Society of London B, Biological Sciences, 270, S124-S127.

Ostfeld, R. S., Canham, C. D., Oggenfuss, K., Winchcombe, R. J., Keesing, F. (2006). Climate, deer, rodents, and acorns as determinants of variation in Lyme-disease risk. PLoS Biology, 4, e145.

Perring, T. M., Cooper, A. D., Rodriquez, R. J., Farrar, C. A., Bellows, T. S., Jr. (1993). Identification of a whitefly species by genomic and behavioral studies. Science, 259, 74-77.

Petit, J. N., Hoddle, M. S., Grandgirard, J., Roderick, G. K., Davies, N. (2008). Invasion dynamics of the glassy-winged sharpshooter *Homalodisca vitripennis (*Germar) (Hemiptera: Cicadellidae) in French Polynesia. Biological Invasions, 10, 955-967.

Petit, J. N., Hoddle, M. S., Grandgirard, J., Roderick, G. K., Davies, N. (2009). Successful spread of a biocontrol agent reveals a biosecurity failure: Elucidating long distance invasion pathways for *Gonatocerus ashmeadi* in French Polynesia. BioControl, 54, 485-495.

Pickering, J. (2009). Discover life. Retrieved 1 February 2015, from http://www.discoverlife.org

Pimentel, D., Lack, L., Suniga, R., Morrison, D. (2000). Environmental and economic costs of nonindigenous species in the United States. Bioscience, 50, 53-65.

Pimentel, D., Zuniga, R., Morrison, D. (2005). Update on the environmental and economic costs associated with alien-invasive species in the United States. Ecological Economics, 52, 273-288.

Rapacciuolo, G., Roy, D. B., Gillings, S., Fox, R., Walker, K., Purvis, A. (2012). Climatic associations of British species distributions show good transferability in time but low predictive accuracy for range change. PLoS One, 7, e40212.

Resh, V. H., Cardé, R. T. (2009). Encyclopedia of insects. San Diego, London: Elsevier.

Richardson, D. M., Ricciardi, A. (2013). Misleading criticisms of invasion science: a field guide. Diversity and Distributions, 19, 1461-1467.

Roderick, G. K., Howarth, F. G. (1999). Invasion genetics: natural colonizations, non-indigenous species, and classical biological control. In E. Yano, K. Matsuo, M. Shiyomi, D. Andow (Eds.), Biological Invasions of Pests and Beneficial Organisms (Vol. NIAES Series 3, pp. 98-108). Tsukuba, Japan: National Institute of Agro-Environmental Sciences.

Roderick, G. K., Hufbauer, R. A., Navajas, M. (2012). Evolution and biological control. Evolutionary Applications, 5, 419-423.

Roderick, G. K., Navajas, M. (2003). Genes in novel environments: Genetics and evolution in biological control. Nature Reviews Genetics, 4, 889-899.

Saenz, V. L., Booth, W., Schal, C., Vargo, E. L. (2012). Genetic analysis of bed bug populations reveals small propagule size within individual infestations but high genetic diversity across infestations from the eastern United States. Journal of Medical Entomology, 49, 865-875.

Sax, D. F., Stchowicz, J. J., Gaines, S. D. (2005). Species Invasions: Insights into Ecology, Evolution, and Biogeography. Sunderland: Sinauer Associates.

Shogren, J. F. (2000). Risk reduction strategies against the 'explosive Invader'. In C. Perrings, M. Williamson, D. Dalmazzone (Eds.), The Economics of Biological Invasions. Cheltenham, UK: Edward Elgar.

Simberloff, D. (2009). The Role of Propagule Pressure in Biological Invasions. Annual Review of Ecology, Evolution, and Systematics, 40, 81-102.

Simberloff, D., Martin, J. L., Genovesi, P., Maris, V., Wardle, D. A., Aronson, J. *et al.* (2012). Impacts of biological invasions: what's what and the way forward. Trends in Ecology and Evolution, 28, 59-66.

Simberloff, D., Stiling, P. (1996). Risks of species introduced for biological control. Biological Conservation, 78, 185-192.

Snyder, W. E., Clevenger, G. M., Eigenbrode, S. D. (2004). Intraguild predation and successful invasion by introduced ladybird beetles. Oecologia, 140, 559-565.

Suarez, A. V., Holway, D. A., Case, T. J. (2001). Predicting patterns of spread in biological invasions dominated by long-distance jump dispersal: Insights from Argentine ants. Proceedings of the National Academy of Sciences (USA), 98, 1095-1100.

Suarez, A. V., Tsutsui, N. D. (2004). The value of museum collections for research and society. Bioscience, 66-74.

Suckling, D., Brockerhoff, E. (2010). Invasion biology, ecology, and management of the light brown apple moth (Tortricidae). Annual Review of Entomology, 55, 285-306.

Swei, A., Ostfeld, R. S., Lane, R. S., Briggs, C. J. (2011). Impact of the experimental removal of lizards on Lyme disease risk. Proceedings of the Royal Society B, Biological Sciences, 278, 2970-2978.

Tatum, A. J., Hay, S. I. (2007). Climatic similarity and biological exchange in the worldwide airline transportation network. Proceedings of the Royal Society of London B, 274, 1489-1496.

Torchin, M. E., Lafferty, K. D., Dobson, A. P., McKenzie, V. J., Kuris, A. M. (2003). Introduced species and their missing parasites. Nature, 421, 628-630.

Ueda, K., Loarie, S. (2013). iNaturalist. Retrieved 1 February 2015, from http://www.inaturalist.org

USDA APHIS (2015). United States Department of Agricutlure, Animal and Plant Health Inspection Service. Retrieved 1 February 2015, from http://www.aphis.usda.gov/wps/portal/aphis/home/

Wilcove, D. S., Rothstein, D., Dubow, J., Phillips, A., Losos, E. (1998). Quantifying threats to imperiled species in the United States. BioScience, 48, 607-615.

Yano, E., Matsuo, K., Shiyomi, M., Andow, D. (Eds.) (1999). Biological Invasions of Pests and Beneficial Organisms (Vol. NIAES Series 3). Tsukuba, Japan: National Institute of Agro-Environmental Sciences.

Zhu, M., Song, Y.-H., Uhm, K.-B., Turner, R. W., Lee, J.-H., Roderick, G. K. (2000). Simulation of the long range migration of brown planthopper, *Nilaparvata lugens* (Stål), by using a boundary layer atmospheric model and geographic information system. Journal of Asia-Pacific Entomology, 3, 25-32.

Pam L. Fuller

5 Vectors of Invasions in Freshwater Invertebrates and Fishes

5.1 Introduction

Without human assistance, the terrestrial environment and oceans represent barriers to the dispersal of freshwater aquatic organisms. The ability to overcome such barriers depends on the existence of anthropogenic vectors that can transport live organisms to new areas, and the species' biology to survive the transportation and transplantation into the new environment (Johnson *et al.*, 2006).

Fishes and invertebrates are intentionally introduced for sport, food, forage, and biocontrol, and may be unintentionally introduced by accidental escapes, man-made water connections, hitchhikers, stock contamination, escapes from captivity such as aquaculture, zoos or personal pets, and releases by individuals not cognizant of the consequences. Once a species is introduced to a drainage, it is likely to be introduced and expand into nearby drainages (Marr *et al.*, 2010). These introductions result in a homogenization of fauna between drainages and even zoogeographic regions (Leprieur *et al.*, 2008; Lever, 1996; Rahel, 2000). There are now at least seven fish species [Rainbow Trout (*Oncorhynchus mykiss*), Brown Trout (*Salmo trutta*), Brook Trout (*Salvelinus fontinalis*), Western Mosquitofish (*Gambusia affinis*), Guppy (*Poecilia reticulata*), Common Carp (*Cyprinus carpio*), and Goldfish (*Carassius auratus*)] that are present in all six of the Earth's zoogeographic regions; before human introductions there were none (Lever, 1996; Rahel, 2007).

Simberloff and Rejmánek (2010) define a vector as a mechanism or pathway (e.g., shipping or canal) that is responsible for the introduction of nonindigenous species. Vectors may change in size and frequency (e.g., Azevedo-Santos *et al.*, 2001; Bij de Vaate *et al.*, 2002; Marr *et al.*, 2010; Rahel, 2004). Donor regions, species transported, and receiving areas can also change over time with changes in trade, hobby interests, and habitat. Vectors come into and out of play over time. The number of fish, mollusk and crustacean species introduced in the United States increased dramatically after 1950 due to improvements in transportation, an increased interest in the ornamental fish trade, and improved methods of shipping (Fuller *et al.*, 1999; USGS, 2014; Welcomme, 1988) (Figure 5.1). Other causes for changing vectors include the opening and closing of canals and the growth of e-commerce. Different areas may have very different strengths of the same vector. Aquarium releases account for more than 50% of the introduced fishes in Australia (Koehn & MacKenzie, 2004), yet reports of aquarium dumping or collection of aquarium species are rare in Brazil (Magalhães & Jacobi, 2013) and only 2% of fishes are introduced via the ornamental trade in Colombia (Álvarez-Leon *et al.*, 2013). Donor regions may differ for different areas. For example, most (63%) of the fish introductions in the United States are the result of transplants, which may be as close as an adjacent non-native drainage, or as far as the other side of the country (USGS, 2014). However, in Colombia, only 1% of the fishes are native transplants (species native to the country but moved to non-native drainages) (Álvarez-Leon *et al.*, 2013). The Great Lakes began seeing invaders from the Ponto-Caspian region when European shipping ports were invaded as a result of the opening of the Rhine-Main-Danube Canal in 1992 (Ricciardi & MacIsaac, 2000).

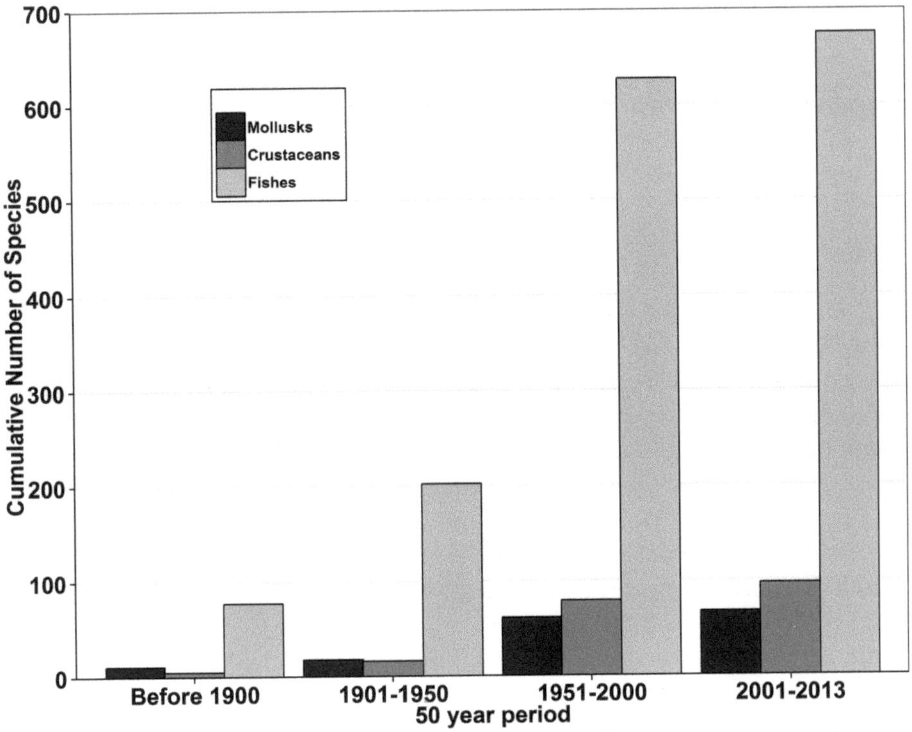

Fig. 5.1: Cumulative number of species of freshwater molluscs, crustaceans, and fishes introduced to non-native waters of the United States in 50 year periods. Source: USGS, 2014.

The focus of this chapter is on the pathways and vectors involved in the spread of biota to the natural environment. Information concerning the impacts of these introduced species will not be included in this chapter but are addressed in Chapters 6 and 8. Information on risk assessment, vector management options or policy is not presented. Policies are presented in Chapters 15 and 17, and Chapter 16 relating to ballast water management options.

5.2 Fisheries/Stocking

Freshwater fishes and invertebrates are stocked for a variety of reasons, including to create a food resource, for sport, for forage, as biocontrol agents, and for conservation purposes (to create additional protected populations). These species can also be inadvertently stocked as a result of stock contamination or stock misidentification. Stockings may be legally conducted by natural resource agencies or permitted individuals, or they may be conducted illegally by unauthorized individuals (Fuller *et al.* 1999; Rahel, 2004; Vashro, 1990; 1995). For most countries, stocking accounts for a large percentage of the introduced fishes, for example 30% in Europe (Keller *et al.*, 2011); 57% in North America (Crossman & Cudmore, 1999); 61% of species that are introduced to a new state in the US (Rahel (2000); 52% in Colombia (Álvarez-Leon *et al.*, 2013); most species in Africa (Ogutu-Ohwayo & Hecky, 1991); but only 17% in Singapore (Ng *et al.* 1993).

5.3 Stocking for Sport and Food

The intent of stocking is to "improve" a fauna perceived as lacking (Dextrase & Coscarelli, 1999). Gozlan (2010) attributed 12% of 624 successful fish introductions worldwide to angling and 7% to commercial fisheries. Stocking in the United States began in earnest in the late 1800s. The first major effort to transport species from one side of the US to the other took place in 1871 when Dr. Livingstone Stone collected several freshwater fish species from the Hudson River and transported them in milk cans via train across the country to Stockton, California (Dill & Cordone, 1997; Moyle, 2002; Smith, 1896). In early days they were also moved around by means of coffee pots and horseback (Moyle, 2002). One hundred and eight fish species have been stocked in the US for sport and/or food (USGS, 2014). Additional species have been added at a fairly constant rate (Figure 5.2).

Initially, many states stocked species because the native fauna was declining as a result of human development. Subsequent reasons included immigrants desiring familiar food items, the need to feed new immigrants, and the need to cater to a growing sport-fishing industry (Whelan, 2004). However, philosophies about stocking non-native species have evolved over time in the United States (Rahel, 1997). For example, Idaho's policy states that "nonnative species of fish will be introduced only in waters where they are not expected to adversely impact stocks of wild native fish" (LaBar & Few, 2004). Some states have ceased stocking certain non-native species (LaBar & Frew, 2004; Jackson *et al.*, 2004; Whelan, 2004). Fewer new species are being stocked, partly because most sport fish demand has been met and partly because of the greater awareness of introduced species (Rahel, 1997, 2004). However, public pressure is an important influence on agencies to stock cultured sport fishes (Jackson *et al.*, 2004).

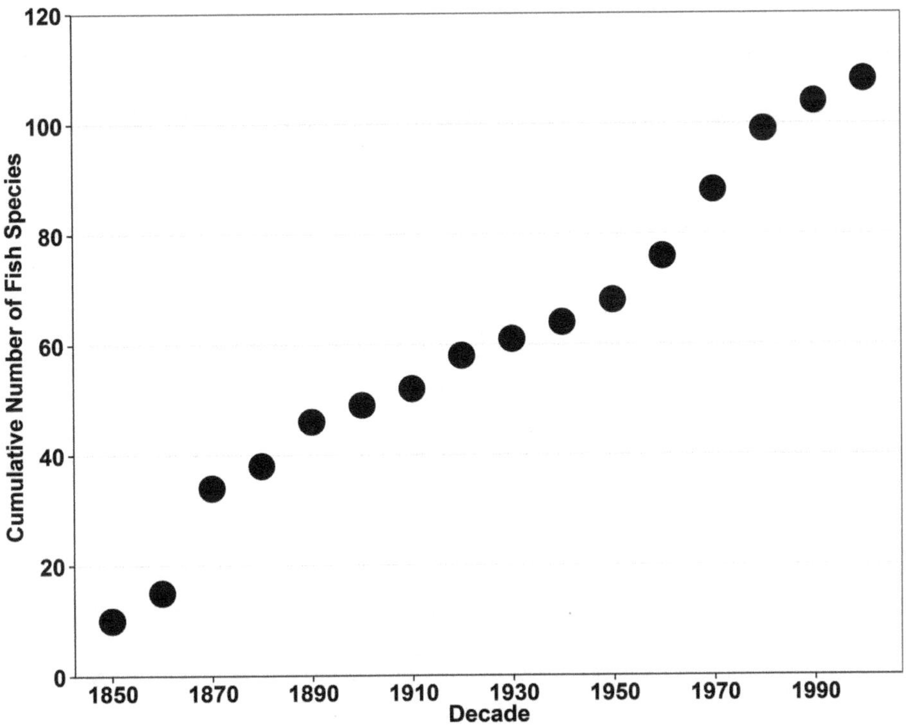

Fig. 5.2: Cumulative total number of fish species stocked for food or sport in freshwaters of the US by decade.

Although agency-sanctioned stocking has been the major reason for fish introductions in the past (Fuller *et al.*, 1999; Rahel, 2004), unauthorized stockings are now the major driver (Rahel, 2004; Vashro, 1990, 1995). These illegal stockings often interfere with the intended management of the water body and can result in environmental and economic impacts (Johnson *et al.*, 2009). Millions of dollars have been spent to control illegally introduced Northern Pike (*Esox lucius*) in Lake Davis, California; Lake Trout (*Salvelinus namaycush*) in Yellowstone Lake, Wyoming; and several species in the Upper Colorado Basin (Johnson *et al.*, 2009).

Species Examples

Stocking is especially common in reservoirs as a fisheries management technique (Kohler & Hubert, 1999; Nickum, 2004). These reservoirs are often near population centers and provide recreational activity. Most species stocked for sport are large, piscivorous top predators (Fuller *et al.*, 1999; USGS, 2014). Examples of species stocked for sport or food include centrarchids (black basses *Micropterus* spp.; sunfishes *Lepomis* spp.; crappies *Pomoxis* spp.); pikes (*Esox* spp.); perch (*Perca* spp.); walleyes

(*Sander* spp.); ictalurid catfish (*Ameiurus* spp., *Ictalurus* spp.); salmonids (*Salmo* spp., *Salvelinus* spp., and *Oncorhynchus* spp.); moronid bass (*Morone* spp.); tilapia (*Oreochromis* spp.); Peacock Bass (*Cichla* spp.); and Tench (*Tinca tinca*) (Álvarez-Leon *et al.*, 2013; Britton & Orsi, 2012; Ogutu-Ohwayo & Hecky, 1991; Radonski *et al.*, 1984; Fuller *et al.*, 1999). As a result of these stocking efforts, introduced sport fishes are now the dominant fauna in many North American waters (Dextrase & Coscarelli, 1999).

Crossman and Cudmore (1999) reported that 48 species of North American fish have been introduced as stocking contaminants. Examples include Rainwater Killifish (*Lucania parva*) and some sunfishes (*Lepomis* spp.) (Dextrase & Coscarelli, 1999); the Naked Goby (*Gobiosoma bosc*) stocked into inland Texas reservoirs along with marine species (Fuller *et al.*, 1999); and madtoms (*Noturus* spp.) stocked with bullheads (*Ameiurus* spp.) (Fuller *et al.*, 1999). Ng *et al.* (1993) reported that the Chinese Barb (*Puntius semifasciolatus*) and *Rasborinus lineatus*, now *Metzia lineata*, probably entered Singapore as contaminants in food and sport fish stockings.

Some invertebrates have also been introduced through stocking. Crayfish are sometimes illegally stocked for the purpose of creating a new population to harvest for food (Johnson *et al.*, 2009; Larson & Olden, 2011). The Red Swamp Crayfish (*Procambarus clarkii*), which is native to the southern United States and northeastern Mexico, has been introduced to Africa, Asia, Europe, and non-native areas of North America primarily through stocking to establish harvestable wild populations and aquaculture escapes (Hobbs *et al.*, 1989; Larson & Olden, 2011).

The Chinese Mystery-Snail (*Cipangopaludina chinensis*) is native to East Asia (Karatayev *et al.*, 2009). These snails were recorded in Chinese markets in San Francisco as early as 1892, where they were initially brought in for food (Karatayev *et al.*, 2009). They became established on the West Coast by the turn of the 20th century (Clench & Fuller, 1965; Karatayev *et al.*, 2009). Later this species was used in the ornamental trade and continued its spread via that pathway (Howells, 2001).

The cladoceran *Daphnia lumholtzi* is believed to have been a hitchhiker with African-sourced Nile Perch (*Lates niloticus*) that were stocked in Texas reservoirs (Havel & Hebert, 1993). Likewise, bivalve species such as *Pseudodon vondembuschianus* probably entered Singapore as parasitic glochidia larvae on the gills of stocked sport fish. An undetermined number of Unionid mussels have been introduced outside their range in the US, as a result of hitchhiking on stocked fish. Aquaculture facilities have been found to contain mussels in their ponds, which would parasitize the fish. Subsequently, these fish are stocked in drainages that are not native to the mussels (J.D. Williams, USGS retired, personal communication; USGS, 2014).

5.3.1 Stocking for Forage

Stocking of reservoirs is a common practice. As the habitat changes from lotic to lentic a new suite of fishes can be introduced, usually piscivorous predators. Often prey (forage) species are also stocked to feed the predator species. At least 40 fish species (6%) have been stocked as forage in the US, with 50% of those being cyprinids (USGS, 2014); whereas in Colombia, only 2% of introduced fishes were stocked for forage (Ålvarez-Leon *et al.*, 2013).

Species Examples

Species stocked for forage in the US include shads (*Alosa* spp.; *Dorosoma* spp.); chubs (*Siphateles/Gila* spp.); dace (*Rhinichthys* spp.); Fathead Minnow (*Pimephales promelas*); Golden Shiner (*Notemigonus crysoleucas*); ciscoes (*Coregonus* and *Prosopium* spp.); shiners (*Cyprinella* and *Notropis* spp.); silversides (*Menidia* spp.); and Bluegill (*Lepomis macrochirus*), which are often stocked in conjunction with Largemouth Bass (*Micropterus salmoides*) (Fuller *et al.*, 1999; USGS, 2014). Species stocked for forage in Amazonian Brazil are the South American Jeju (*Erythrinus erythrinus*) and the knifefish, *Brachyhypopomus gauderio* (Britton & Orsi, 2012).

Although fish may be the most common type of forage stocked, crayfish and other crustaceans are also stocked as forage. The Virile Crayfish (*Orconectes virilis*) (Johnson, 1986; Larson & Olden, 2011) and Red Swamp crayfish (*Procambarus clarkii*) have been stocked in the western US (Larson & Olden, 2011; Riegel, 1959). *Procambarus clarkii* was also used as bullfrog food in the American West, Hawaii and Japan (Hobbs *et al.*, 1989; Riegel, 1959).

The Opossum shrimp, traditionally known as *Mysis relicta* (Audzijonytė & Väinölä, 2005; Lasenby *et al.*, 1986), has been introduced to more than 130 western and northwestern lakes in the US (Audzijonytė & Väinölä, 2005; Dextrase & Coscarelli, 1999); 21 locations in British Columbia; 9 in Norway; one in Finland; and 61 in Sweden (Lasenby *et al.*, 1986). *Mysis relicta* has been stocked as forage for salmonids such as Kokanee, (*Oncorhynchus nerka*), Rainbow Trout (*Oncorhynchus mykiss*), Brown Trout (*Salmo trutta*) and char (*Salvelinus stagnalis/alpinus*) (Lasenby *et al.*, 1986; Nesler & Bergersen, 1991).

5.3.2 Stocking for Biocontrol

Aquatic organisms have been used as a control mechanism against a variety of pest species such as aquatic weeds, mosquitoes, snails, and other fishes. Stocking species has been used as an alternative to pesticides and herbicides (chemical control). They have the benefit of not being toxic and remaining effective as long as they persist. They are able to cover large or otherwise inaccessible areas, and the effects continue without a reapplication of chemicals. The use of biocontrol organisms is especially useful when the cost of chemical treatment is prohibitive, undesirable, or impractical. On the reverse side, biocontrol agents may be worse than chemicals because they are unpredictable and irreversible (Cowie, 2001). Many biocontrols have become pests themselves.

Species Examples

Israeli aquaculturists have been using fish as biological control agents since the late 1960s (Shireman, 1984). They use Silver Carp (*Hypophthalmichthys molitrix*) for phytoplankton, Bighead Carp (*Hypophthalmichthys nobilis*) for zooplankton, Grass Carp (*Ctenopharyngodon idella*) for submerged plants, Black Carp (*Mylopharyngodon piceus*) for snails, Common Carp (*Cyprinus carpio*) for filamentous algae, Sea Bass (species not identified) for larval fishes, Blue Tilapia (*Oreochromis aureus*) for bottom sediments and mullet (*Mugil* spp.) for detritus (Shireman, 1984).

Both fishes and snails have been introduced in attempts to control aquatic vegetation, snails, and mosquitoes. Predatory fish can be stocked to control prey species populations. Species that have been used to control aquatic plants and algae include Mozambique Tilapia (*Oreochromis mossambicus*), Redbreast Tilapia (*Tilapia rendalli*), Grass Carp (*Ctenopharyngodon idella*), and Silver Carp (*Hypophthalmichthys molitrix*) (Shireman, 1984; Welcomme, 1988, Chilton & Muoneke, 1992), Giant Ramshorn Snail (*Marisa cornuarietis*) (Cowie, 2001) and Golden Apple Snail (*Pomacea canaliculata*) (Seaman & Porterfield, 1964; Carlsson & Lacoursiere, 2005).

Snail control is important for controlling the intermediate snail hosts of schistosomiasis (*Schistosoma mansoni*) and trematodes. Species used for snail control include the Red-rimmed Melania (*Melanoides tuberculatus*), Giant Ramshorn Snail (*Marisa cornuarietis*) (Cowie, 2001; Rocha-Miranda & Martins-Silva, 2006; Pointier, 2001; Pointier & Augustin, 1999; Pointier & David, 2004; Seaman & Porterfield, 1964), Black Carp (*Mylopharyngodon piceus*) (Ben-Ami & Heller, 2001; Nico *et al.*, 2005; Welcomme, 1988), Alluaud's Haplo (*Astatoreochromis alluaudi*), the Snaileater (*Sargochromis* [=*Haplochromis*] *mellandi*) (Welcomme, 1988), and the Common Carp (*Cyprinus carpio*) (Su Sin, 2006). It has been noted that fish are not effective biocontrol agents for gastropods in large reservoirs; however, they are more effective in small ponds (Welcomme, 1988).

Mosquito-borne diseases (such as malaria, yellow fever, dengue fever, chikungunya, filariasis, encephalitis, etc.) continue to be a major problem in almost all tropical and subtropical countries. They are responsible for the transmission of the pathogens causing some of the most life-threatening human diseases (Chandra *et al.*, 2008; Stauffer *et al.*, 1997). Larvivorous fishes from at least 216 fish species drawn from 30 families have been used in biocontrol (Stauffer *et al.*, 1997) since the early 1900s (Chandra *et al.*, 2008). One of the earliest and most widespread species to be used for this purpose is the Western Mosquitofish (Froese & Pauly, 2014; Fuller *et al.*, 1999). However, Courtenay & Meffe (1989) point out that the native invertebrates and fishes are often more effective at mosquito control.

In some cases, the introduction of one exotic fish to control another can be a useful fisheries management tool (Shafland, 1999). The Peacock Bass (*Cichla ocellaris*) (Shafland, 1999), Northern Pike (*Esox lucius*), Muskellunge (*Esox masquinongy*) (Pflieger, 1997), and Pacific salmon (*Oncorhynchus kisutch* and *O. tshawytscha*) (Emery, 1985) have all been used to reduced populations of smaller prey species.

5.4 Bait Release

Bait release or bait bucket transfer involves the transport and release of aquatic organisms used for sport-fishing activities to a basin where they were previously absent (Ludwig & Leitch, 1996). Sometimes the bait is able to escape the hook while still alive, other times it is dumped by the angler at the end of the day (Litvak & Mandrak, 1993). In the US, baitfish providers often move these fish long distances, including across drainages, states and even the country (Goodwin *et al.*, 2004). Also the anglers themselves may move the bait across drainage divides by buying in one watershed and fishing in another (Litvak & Mandrak, 1993; Ludwig & Leitch, 1996). Many species have been introduced into new watersheds, or even countries via this vector (e.g. Heard, 1956; Fuller *et al.*, 1999; Fuller, 2003; Thompson & Alam, 2005; Thaulow *et al.*, 2013).

A study conducted by Litvak & Mandrak (1993) found that 41% of anglers admitted to releasing live bait after use. The same survey found nearly all the anglers who released their bait thought they were doing a good thing for the environment (Litvak & Mandrak, 1993). When the authors examined the purchase location and the angling destination, they concluded that 18 of the 28 species found in the dealers' bait tanks may have been used outside their native range. Therefore, it is not surprising that so many species are introduced in this manner; in Canada, Ontario alone has more than 65 legal baitfish species (Cudmore & Mandrak, 2005). Ludwig and Leitch (1996) concluded with certainty that at least 1,000 bait transfers from the Mississippi Basin to the Hudson Bay Basin would take place in one year.

Bait buckets can transfer a variety of contaminant species, including undetected or misidentified non-target fishes, and invertebrates not visible to the naked eye. Examples of these are Round Gobies (*Neogobius melanostomus*), Zebra Mussel (*Dreissena polymorpha*) veligers, and Spiny Waterfleas (*Bythotrephes cederstroemi*) (Litvak & Mandrak, 1993).

Ludwig and Leitch (1996) purchased baitfish from 21 retailers and found that 28.5% of the samples contained unexpected fish species, indicating that the dealers do not maintain "clean" stock. Similarly, Litvak and Mandrak (1993) found 6 out of 28 species (21% of species) in surveyed bait tanks to be illegal. LoVullo and Stauffer (1993) found 7 illegal species out of 16 (44%) in a state-wide survey of Pennsylvania bait shops. Kercheis (1998) found 10 illegal species and contamination in 4% to 16% of the bait shops inspected in Maine. A study of bait shops in Missouri found illegal crayfish species. Furthermore, the shop owners did not know how to identify the various species (DiStefano *et al.*, 2009).

Species Examples

The primary baitfish species sold in the US are Fathead Minnows (*Pimephales promelas*), Golden Shiners (*Notemigonus crysoleucas*), shiners (*Notropis* spp.), chubs of various taxa, suckers (*Catostomus* spp.) and Goldfish (*Carassius auratus*) (Nielsen, 1982; Goodwin *et al.*, 2004). All of these species have been introduced to areas outside their native range (USGS, 2014). In Brazil, Gaudério (*Brachyhypopomus gauderio*) and Jeju (*Erythrinus erythrinus*) have been introduced through bait release to the upper Parana River basin (Britton & Orsi, 2012).

It is not just fish that are used for bait. Crayfish, frogs, mussels, snails, worms, and salamanders are also used (e.g. Jensen & Watters, 1999; Keller *et al.*, 2007; DiStefano *et al.*, 2009; Larson & Olden, 2011). Rusty Crayfish (*Orconectes rusticus*) (Hobbs *et al.*, 1989; Larson & Olden, 2011; USGS, 2014), Virile Crayfish (*Orconectes virilis*) (Larson & Olden, 2011; USGS, 2014), and Signal Crayfish (*Pacifastacus leniusculus*) (Riegel, 1959) have all been transplanted to new areas in the US as a result of bait release.

5.5 Aquarium Release

Aquariums have become very popular all over the world, particularly after World War II (Courtenay, 1999). The result is that there are billions of fish kept in captivity (Krishnakumar *et al.*, 2009) with the potential to be released. Studies have estimated that 2,000 to more than 5,000 freshwater fish species are traded globally each year (McDowall, 2004; Livengood & Chapman, 2007; Krishnakumar *et al.*, 2009). Pet store surveys have recorded 308 (Rixon *et al.*, 2005) to 730 (Chapman *et al.*, 1997) species available for sale in the US, and more than 700 species found in stores in southwestern Europe (Maceda-Veiga *et al.*, 2013). In addition to pet stores, hobbyists can also order a vast number of animals on-line (Weigle *et al.*, 2005; Keller & Lodge, 2007; Magalhães & Jacobi, 2010). The growth of e-commerce has contributed to the increase of ornamental introductions in many ecosystems (Magalhães & Jacobi, 2010).

The aquarium trade has long been documented as a vector of fish release (Courtenay *et al.*, 1984; Lachner *et al.*, 1970; Welcomme, 1984). Gozlan (2008) reported that 21% of global fish introductions were related to the ornamental trade. Strecker *et al.* (2011) suggest that between 2,500 and 21,000 individual fish are released each year in the Puget Sound region alone. Gertzen *et al.* (2008) calculated that more than 10,000 fishes were released annually from Montreal (Quebec, Canada). Not surprisingly, release sites are often situated near population centers or public areas, such as fairgrounds or parks (Copp *et al.*, 2005; Courtenay & Taylor, 1984; USGS, 2014). In fact, Copp *et al.* (2005) found that numbers of species introductions decreased as distance from the nearest road, footpath, or inhabitation increased. As a result of these releases, approximately 175 species of exotic fish have been introduced into the United States, including Hawaii; 94 of these have become established, or at least reproduced and overwintered for some period of time (Crossman & Cudmore, 1999; USGS, 2014). Because aquariums hold more than fish, the aquarium trade is responsible for over 150 species of fish, invertebrates, plants and microorganisms becoming established in the wild (Fuller *et al.*, 1999; Fuller, 2003; Padilla & Williams, 2004; Chang *et al.*, 2009). The number of introduced mollusks is far smaller than introduced fishes. Snails are often inadvertently imported on aquatic aquarium plants (Cowie & Robinson, 2003; Smith, 1989). Rates of introductions of snails have increased since 1970s and 1980s because of an increase in the aquarium trade (Pontier, 1999; Cowie & Robinson, 2003). Eight of the 22 mollusk species introduced into North America have been linked to the aquarium trade (Mackie, 1999). That number is now 30 species, including two records of inland octopus introductions (USGS, 2014). Padilla and Williams (2004) include a table that lists introduced mollusks worldwide.

The reasons commonly cited for discard of aquarium fish are large size, aggressiveness, high reproductive rates, fish illness, and believing that releasing the animal is the most humane option (Courtenay, 1999; Duggan *et al.*, 2006; Gertzen *et al.*, 2008; Krishnakumar *et al.*, 2009; Padilla & Williams, 2004; Severinghaus & Chi, 1999). Other reasons cited were growing bored with the aquarium and moving (Gertzen *et al.*, 2008).

Species Examples

In North America, the fish species most often introduced include cichlids, characins, exotic cyprinids, exotic catfishes (from several families), and livebearers (Courtenay & Hensley, 1984; Crossman & Cudmore, 1999; Fuller *et al.*, 1999; USGS, 2014). Large fish species, such as Redtail Catfish (*Phractocephalus hemioliopterus*) and Pacu species (*Colossoma* and *Piaractus* spp.), are almost always found as single individuals (USGS, 2014), which decreases their chance of reproduction. Sailfin catfish/Plecos/Hypostomus (Loricariidae; primarily *Pterygoplichthys* spp.) have been released in large enough numbers that they are established and are becoming a growing problem around the world (ISSG, 2013; Krishnakumar *et al.*, 2009; CEC, 2009; Ng *et al.*, 1993; USGS, 2014). Smaller species commonly released include the Goldfish (*Carassius auratus*) (Courtenay & Stauffer, 1990; Nico & Fuller, 1999; Welcomme, 1988), Convict Cichlid (*Cichlasoma nigrofasciatum*), and livebearers (family Poeciliidae) like the Sailfin Molly (*Poecilia velifera*) and Swordtail (*Xiphophorus hellerii*) (Contreras-MacBeath *et al.*, 1998; Fuller *et al.*, 1999; Golani & Mires, 2000; Morgan & Gill, 2001).

Gastropods are the dominant group of mollusks that are introduced via the aquarium trade (Padilla & Williams, 2004; USGS, 2014). Many different species are introduced to the wild via aquarium release (Cowie & Robinson, 2003; Howells, 2001; Mackie, 1999). Sometimes these are intentionally purchased as algae-cleaners; however, they are more commonly hitchhikers on aquatic plants that hobbyists purchase (Madsen & Frandsen, 1989).

5.6 Aquaculture Escapes

The food benefits of aquaculture practice also come with risks. Unfortunately, many countries rely on introduced, rather than native, species for their production (Casal, 2006). In 2008, non-native fish comprised 63% of the total value of freshwater fish cultured (Britton & Orsi, 2012). In fact, most aquatic species cultured in the US are not native to their culture sites (Naylor *et al.*, 2001). The Food and Agriculture Organization of the United Nations (1996 [in Casal, 2006]) states, “In relation to aquaculture, experience has shown that animals will usually escape the confines of a facility. As a consequence, the introduction of aquatic organisms for aquaculture should be considered as a purposeful introduction into the wild, even though the quarantine/ hatchery facility may be a closed system". Aquaculture has become a leading vector in fish transplants worldwide (Naylor *et al.*, 2001) and is responsible for 51% of all fish introductions world-wide (Gozlan, 2010); 74% in Mexico (Contreras-Balderas, 1999); 42% in Colombia (Álvarez-Leon *et al.*, 2013) and Germany (Welcomme, 1988); 27% of non-native freshwater animals species in Europe (Keller *et al.*, 2011); and 16% of freshwater fish species in the US (Fuller *et al.*, 1999; USGS, 2014). Organisms are likely to escape culture facilities during flood events, cleaning of ponds, or from improper outflow barriers (Fuller *et al.*, 1999; Padilla & Williams, 2004).

When escapes do occur, they can result in large numbers of individuals being released (Arndt *et al.*, 2002; Azevedo-Santos *et al.*, 2001; Goldburg & Triplett, 1997). Several different containment methods are used in aquaculture including cages, net pens, raceways, and earthen berms (Goldburg & Triplett, 1997). Net pens and cages are placed directly in the body of water and are inherently more risky. A study conducted in Furnas Reservoir, Brazil, where there are more than 5,000 cages of Nile Tilapia (*Oreochromis niloticus*), found just over a quarter of the aquaculturists reported escape due to cage damage. But containment failure is only one reason for introductions. The same study conducted in Furnas Reservoir found that there is a high incidence of accidental release during size sorting, fish capture and juvenile stocking, and a smaller proportion that are intentionally released. Often the farmers were unaware of this being an issue (Azevedo-Santos *et al.*, 2001). Nile Tilapia is now established in this South American reservoir as a result of these releases and escapes (Azevedo-Santos *et al.*, 2001).

Species Examples

As of 1988, Welcomme found that although there are many species that had been introduced sparingly around the world, only ten had been introduced to more than ten countries, and only four had been introduced to more than 30 countries. Those four were the Mozambique Tilapia (*Oreochromis mossambicus*) (31), Grass Carp (35), Common Carp (41), and Rainbow Trout (44). Since that time, the numbers of introductions for those four species have grown substantially: Mozambique Tilapia (102 countries and islands); Grass Carp (91); Common Carp (122); and Rainbow Trout (93). The next species on Welcomme's list was Nile Tilapia (*Oreochromis niloticus*), which was in 26 countries in 1988 and is now in 82 countries as of 2014 (Froese & Pauly, 2014).

Just over 29% of world production of crustaceans takes place in fresh water (FAO, 2012). Shrimp, even when raised in fresh water, are considered a marine crop. Freshwater species grown include the Red Swamp Crayfish (*Procambarus clarkii*), Chinese Mitten Crab (*Eriocheir sinensis*), Oriental Shrimp (*Palaemon macrodactylus*), and Giant River Prawn (*Macrobrachium rosenbergii*) (FAO, 2012:37).

The Golden Apple Snail (*Pomacea canaliculata*) was introduced to Taiwan from South America in 1980 for aquaculture (Naylor, 1996; Carlsson & Lacoursiere, 2005; Cowie & Robinson, 2003). Unfortunately, the local people did not accept them as a food item. They were discarded and spread throughout rice fields, wetlands, and irrigation ditches (Hayes *et al.*, 2008; Carlsson & Lacoursiere, 2005). This species was also introduced into Hawaii for the same reason (Cowie, 1998; Cowie & Robinson, 2003). It is now a serious agricultural pest in both areas (Lach & Cowie, 1999; Cowie & Robinson, 2003) in addition to having spread to the wild in much of Southeast Asia and Australia (FDACS, 2002).

As with all introductions, when one species is introduced there is a risk of other species being introduced along with it. These can be diseases, parasites, or hitchhikers such as snails or parasitic larval Unionid mussels. Examples of aquaculture hitchhikers that have become established include the New Zealand Mud Snail (*Potamopyrgus antipodarum*) (U.S. Congress, 1993; Cowie & Robinson, 2003), Topmouth Gudgeon (*Pseudorasbora parva*) (Copp *et al.*, 2005), swimbladder nematode (*Anguillicoloides crassus*) (Barse & Secor, 1999; Fries *et al.*, 1996), Asian tapeworm (*Bothriocephalus opsarichthydis*) (Ganzhorn *et al.*, 1992; Hoffman & Schubert, 1984), and the mussel *Sinanodonta woodiana* (Beran, 2008; Kraszewski & Zdanowski, 2007; Paunovic *et al.*, 2006; Pou-Rovira *et al.*, 2009; Popa & Popa, 2006; Watters, 1997).

5.7 Water Conveyance Structures

Canals have served as corridors for the spread of non-native fishes and invertebrates since the early part of the nineteenth century (Daniels, 2001). Marsh (1874) was one of the early authors to suggest transfer of fishes via canal connections as a biological corridor (Daniels, 2001). Early settlers to North America had constructed a vast array of canals in the Northeast by the late 1700s (Mills *et al.*, 1993). In the case of New York State (US), all five major drainages have been connected by canals in the past 200 years, resulting in the homogenization of fish communities across all the drainages (Daniels, 2001).

Although canals have indeed resulted in migrations of species from one watershed to another that was previously inaccessible, they are most often used in secondary expansion of species introduced into a non-native watershed by another vector (Mills *et al.*, 1999). For example, zebra mussels were first introduced to the Great Lakes via shipping, but expanded into the Mississippi Basin via canal (Hebert *et al.*, 1989; MacIsaac *et al.*, 1992; USGS, 2014). Canals have been implicated as either the primary or secondary vector for at least 140 species introduced into the Great Lakes (Mills *et al.*, 1999).

Canal Examples

The Erie Canal stretches from Albany to Buffalo and connects the Atlantic Ocean (via the Hudson River) to Lake Erie (and Lake Ontario via a branch of the canal) (Figure 5.3). At the opening ceremony in 1825, a keg of Lake Erie water was symbolically poured into the Atlantic Ocean (Finch, 1925; Mills *et al.*, 1993). Fourteen fish species, three gastropods, seven bivalves, and two crayfish have invaded the Hudson River via the Erie Canal (Mills *et al.*, 1997). Fewer species have invaded the Great Lakes (Mills *et al.*, 1999).

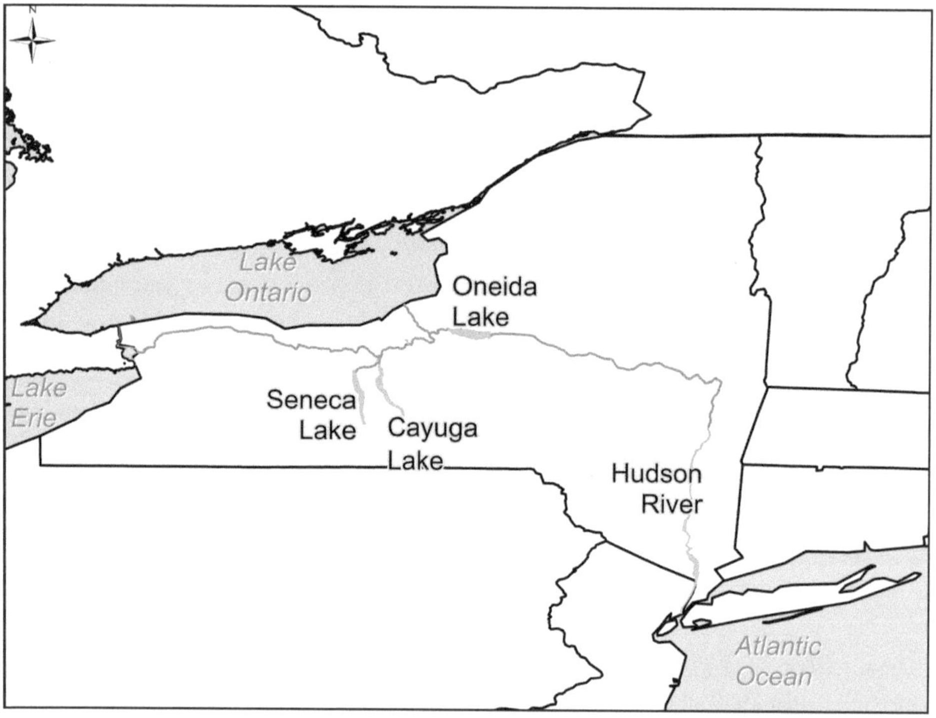

Fig. 5.3: Map of the Erie Canal showing the linkage between the Atlantic Ocean and Lakes Erie and Ontario.

In the late 1800s, the city of Chicago reversed the flow of the Chicago River in order to flush sewage downstream, away from their drinking water in Lake Michigan (Rasmussen, 2002; Moy *et al.*, 2011). Once species move from the Great Lakes southward through the canal, they have access to two thirds of the country via the Mississippi River and its major tributaries (Figure 5.4, Figure 5.5). Several species introduced to the Great Lakes via shipping have moved out of the Great Lakes into the Mississippi Basin (Rasmussen, 2002; Veraldi *et al.*, 2011; USGS, 2014). But the main concern recently has been the three species of Asian carp [Bighead Carp (*H. nobilis*), Silver Carp (*H. molitrix*), and Black Carp (*Mylopharyngodon piceus*)] that escaped aquaculture facilities decades ago, established in the Mississippi River, and have expanded northward towards the Great Lakes (Conover *et al.*, 2007; USGS, 2014). Because of the great concern about the potential impacts of these species in the Great Lakes, millions of dollars have been spent on electric barriers, monitoring, and researching ways to prevent these carp species from gaining access (Conover *et al.*, 2007; Moy *et al.*, 2011).

Man-made canals have played a large role in the dispersal of Ponto-Caspian macroinvertebrates in Europe (Bij de Vaate *et al.*, 2002). Of the three main European canal corridors (Figure 5.6), the Rhine-Main-Danube Canal is the southernmost. The Main-Danube Canal connects the North Sea to the Black Sea by connecting the south-flowing Danube River to the Main, which then flows into the north-flowing Rhine River in Germany (Figure 5.7) (Ricciardi & MacIsaac, 2000). The canal, which opened in 1992, has been a major invasion corridor to the Ponto-Caspian/Black Sea since it opened (Bij de Vaate *et al.*, 2002; Leuven *et al.*, 2009; Ricciardi & MacIsaac, 2000). Mobile species are capable of migrating on their own; sessile organisms expand via the ships that use the canals (Bij de Vaate *et al.*, 2002). Because many of the ships in that area are bound for the Great Lakes, this area serves as an invasion source for the Great Lakes (US) (Ricciardi & MacIsaac, 2000).

Aqueducts are another water conveyance that move animals in addition to water. The California Aqueduct extends more than 440 miles from Northern California to Southern California (US) (Figure 5.8). There are many such aqueducts in the American West that transport water between watersheds or basins. The Sacramento Pikeminnow (*Ptychocheilus grandis*), Hitch (*Lavinia exilicauda*), Splittail (*Pogonichthys microlepidotus*), Sacramento Sucker (*Catostomus occidentalis*), Inland Silverside (*Menidia beryllina*), Prickly Sculpin (*Cottus asper*), Striped Bass (*Morone saxatilis*), Tule Perch (*Hysterocarpus traski*), Shimofuri Goby (*Tridentiger bifasciatus*) (Daniels, 2001; Moyle 2002; Swift *et al.*, 1993) and Asian Clam (*Corbicula fluminea*) (Counts, 1986) have been transported in water from the California Aqueduct and Central Valley Project to Southern California reservoirs.

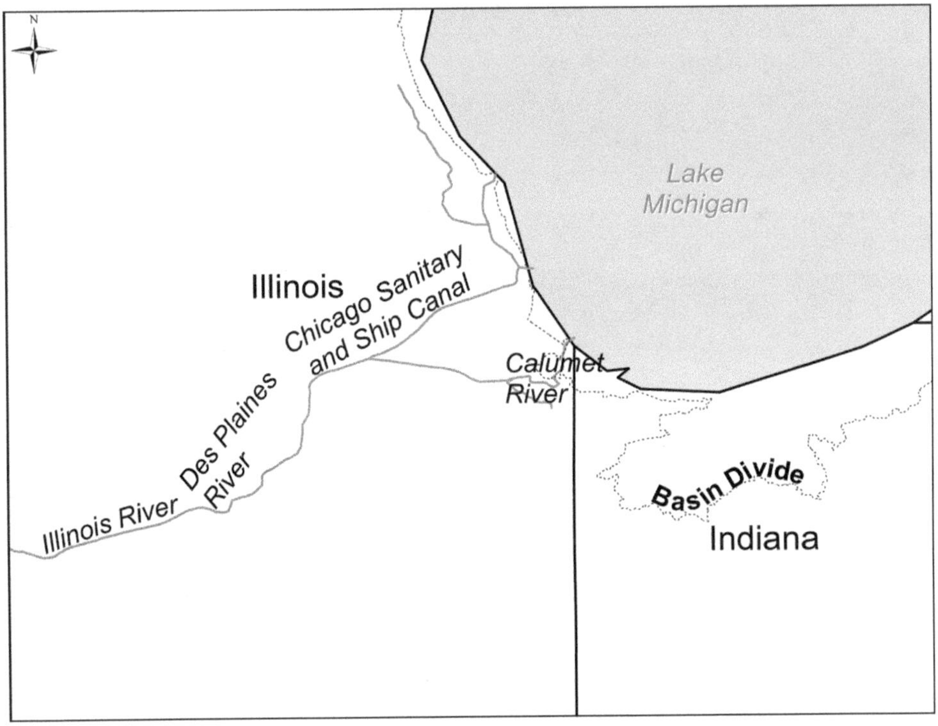

Fig. 5.4: Chicago Shipping and Sanitary Canal.

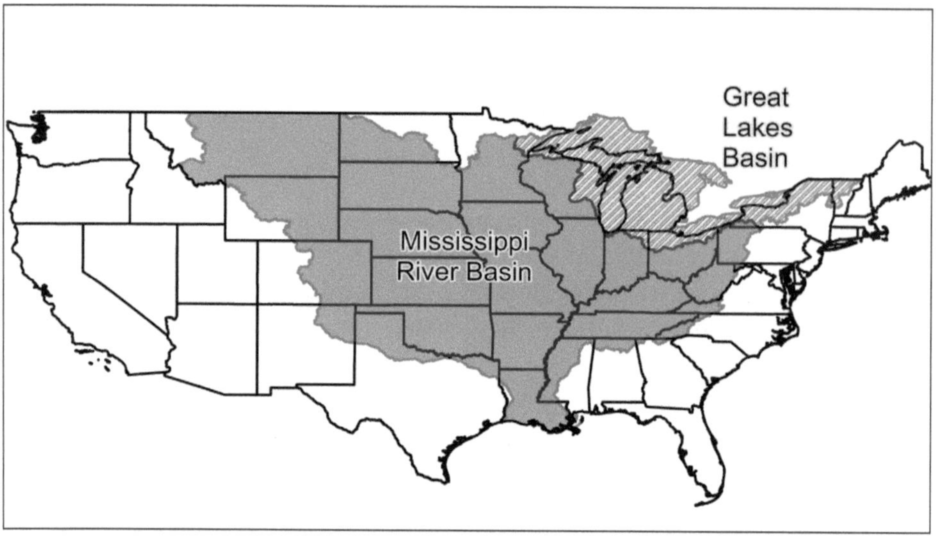

Fig. 5.5: Map showing the extent of the Mississippi Basin. Species that leave the Great Lakes basin via the Chicago Shipping and Sanitary Canal have access to this entire basin.

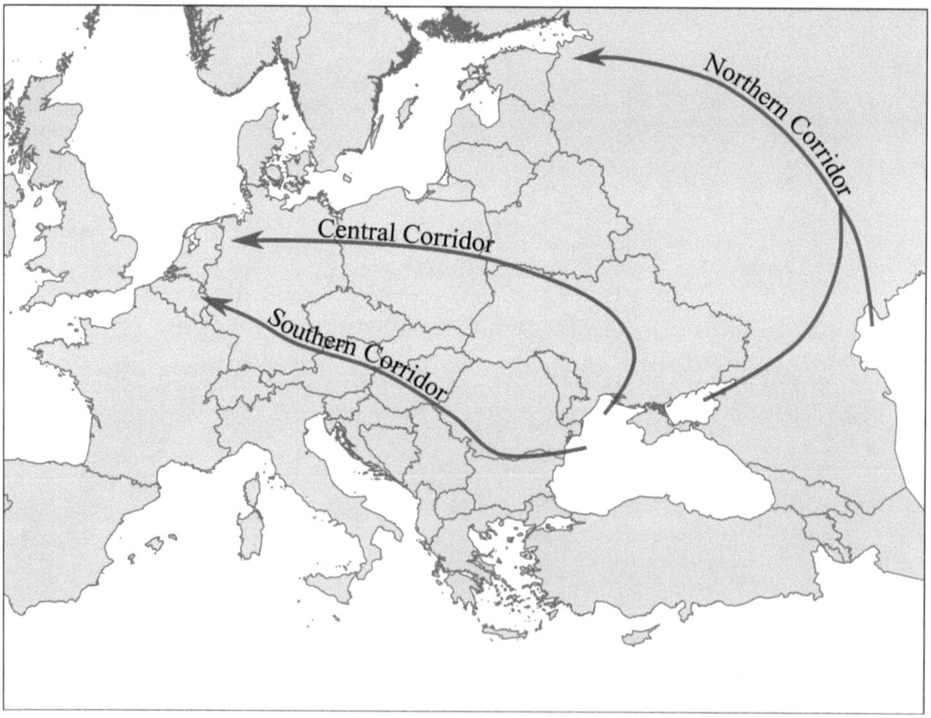

Fig. 5.6: Map indicating the three major canal corridors in Europe. Source: Bij de Vaate *et al.*, 2002. © Canadian Science Publishing or its licensors.

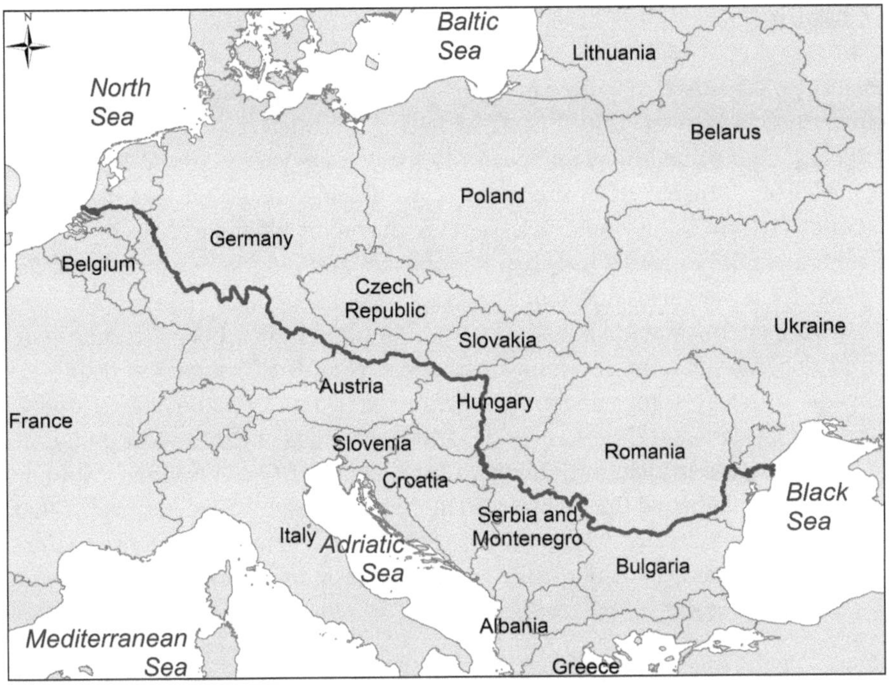

Fig. 5.7: Map showing the Rhine- Main-Danube Canal that connects the Black Sea to the North Sea.

Fig. 5.8: Map showing the California Aqueduct.

5.8 Shipping

The majority of species transported in ballast are invertebrates (Carlton, 1985; Carlton & Geller, 1993; Ruiz *et al.*, 2000). And, although the interior tanks of a ship may present challenging conditions for species to survive a voyage (Carlton, 1985), some fishes have been transported by this method (Wonham *et al.*, 2000). The majority of fishes collected from ballast tanks in a study conducted by Wonham *et al.* (2000) were post-larval juveniles or adults less than 90 mm standard length. However, some fish as large as 39 cm were also collected.

Fishes are not uncommon in ballast tanks (Carlton & Geller, 1993; Wonham *et al.*, 2000); 31 fish species from 21 families have been recorded (Wonham *et al.*, 2000). World-wide, at least 40 introductions of 32 species from 11 families are attributed to ballast water transport (Wonham *et al.*, 2000). Families of fishes most frequently found in ballast water include Gobiidae, Clupeidae, and Gasterosteidae (Wonham *et al.*, 2000). It is theorized that gobies and blennies are transported because of their crevicolous nature (Carlton, 1985; Wonham *et al.*, 2000). Fishes with extensive lateral line systems, such as gobies and ruffe, may also be more prone, as are schooling species such as Clupeids (Wonham *et al.*, 2000).

Species Examples

Species that have been transferred from Europe to the Great Lakes include the Round Goby, the Tubenose Goby (*Proterorhinus semilunaris* (formerly *P. marmoratus*)), the Ruffe (*Gymnocephalus cernua* (formerly *G. cernuus*)) (Busiahn, 1993; Leigh 1998; Pratt *et al.* 1992; Ricciardi & MacIsaac, 2000; Simon & Vondruska, 1991); and the European Flounder (*Platichthys flesus*) (Crossman 1984; Cudmore-Vokey & Crossman 1999; Emery & Teleki 1978). The flounder is the only species that failed to establish.

Forty-one freshwater invertebrates have been introduced into the Great Lakes via this vector (see USGS, 2014 and NOAA, 2014), including the Zebra Mussel (Hebert *et al.*, 1989), Quagga Mussel (*D. bugensis*) (Ricciardi & MacIsaac, 2000), *Cercopagis pengoi* Waterflea (Ricciardi & MacIsaac, 2000), *Echinogammarus ischnus* (Ricciardi & MacIsaac, 2000), and Bloody Shrimp (*Hemimysis anomala*) (Pothoven *et al.*, 2007). The primary source of invading organisms in the Great Lakes is the Ponto-Caspian region (Ricciardi & MacIsaac, 2000).

5.9 Recreational Boats and Equipment

Aquatic organisms can be carried on devices used for recreation including boating, fishing, personal watercraft, and SCUBA. Various species of bait and Zebra Mussel larvae can be transported in boat live wells, Spiny Waterflea (*Cercopagis pengoi*) on fishing line, New Zealand Mudsnails (*Potamopyrgus antipodarum*) on felt-soled waders, and Quagga Mussels in personal watercraft. Trailered recreational boats have been shown to be of primary importance for transporting zebra and quagga mussels to unconnected bodies of water as either veligers or attached adults (Johnson & Padilla, 1996; Pollux *et al.*, 2003; Wong, 2011; Choi *et al.*, 21013). Voluntary guidelines have been established in the US to prevent such introductions (Aquatic Nuisance Species Task Force, 2012).

5.10 Prayer Releases/Released Food Items

In East Asia, some people believe they will accrue religious merits by freeing captive animals into the wild as prayers to the gods (Severinghouse & Chi, 1999). They believe they are benefitting the animal, improving their karma, and may also believe that these acts can extend benefits to their loved ones (Liu *et al.*, 2013). These releases may be organized by temples and involve large numbers of animals, or they may involve a single person (Severinghaus & Chi, 1999). Liu *et al.* (2012) found they are predominantly released by Buddhists and Taoists. Both local and exotic species may be involved (Severinghaus & Chi, 1999). In Asia, these animals may be purchased from pet stores, but in the United States, they are often purchased from live food markets (Liu *et al.*, 2012; Severinghaus & Chi, 1999). In Taipei, Taiwan, approximately 30% of those surveyed had released prayer animals, with 64% of them doing it on an individual basis rather than a temple event (Severinghaus & Chi, 1999). When organized as an event, many people are involved and thousands of animals are released (Liu, 1996 in Severinghaus & Chi, 1999). In 2004, it was estimated that the people of Taiwan spent nearly $6 million annually to release 200 million animals (Agoramoorthy & Hsu, 2005).

Species Examples
In the United States, it is believed that releases of Northern Snakehead (*Channa argus*), swamp eels (*Monopterus* spp.), and some Bighead and Silver Carp have been the result of prayer release (Courtenay & Williams, 2004; Kolar *et al.*, 2007; Nico *et al.*, 2011). In 2007, a large ceremonial release into the Passaic River, New Jersey, obtained press coverage (Henry, 2007). Many of the species released were not native.

5.11 Conclusions

Vectors, species, donor regions, recipient regions, and risk are constantly changing with changes in economies, culture, and transportation. The world is losing the natural composition of communities and unique faunal regions to non-native introductions. We are learning more about the consequences of these introductions and are more thoughtful now before conducting them intentionally. However, unintentional introductions and those conducted by unknowing individuals are much more difficult to control. Solutions will need to be multifaceted. Various actions can be taken in the realms of public outreach and education, policy (importation, possession, release; both Federal and state), regulations (state fishing regulations), risk assessments before introductions, technology (e.g., canal barriers), and aquaculture (e.g., sterile or mono-sex animals).

In a nutshell

- Globally, most fish introductions are caused by aquaculture; second is aquarium release.
- In the US, most fish introductions are caused by stocking; second is ornamental.
- Crustaceans and mollusks are moved in the aquarium trade, for aquaculture, and stocked for food, forage and biocontrol.
- Illegal stocking is a growing problem that affects managed fisheries.
- Releases by individuals (e.g., bait and aquarium release) are a growing problem in the US.
- Areas with more human activity have more fish introductions.
- Philosophies about introducing new species are changing. There is less authorized stocking of new non-natives and more precautions taken with aquaculture.
- The primary source of ballast water introductions to the freshwater Great Lakes is the Ponto-Caspian region of Europe and Asia.
- Solutions to reducing introductions need to be multi-faceted.
- Public education about invasive species and transport methods is important!
- The Food and Agriculture Organization Database on Introductions of Aquatic Species (DIAS) and FishBase (Froese & Pauly, 2014) both document fish introductions at a global level (Casal, 2006).
- The United State Geological Survey maintains a database of aquatic introductions to the US (USGS, 2014).

5.12 Bibliography

Agoramoorthy, G., Hsu, M. J. (2005). Religious freeing of wildlife promotes alien species invasion. BioScience, 55(1), 6-6.

Álvarez-Leon, R., Orozco-Rey, R. H., Paramo-Fonseca, M. E., Restrepo-Santamaria, D. (2013). Lista de los Peces Fósiles y Actuales de Colombia: Nombres Científicos Válido, Distribución, Geográfica, Diagnosis & Nombres Comunes e Indígenas. Primera Edición. Bogota D.C., Colombia: Eco Prints Diseño Gráfico y Audiovisual Ltda.

Aquatic Nuisance Species Task Force. (2012). Voluntary guidelines to prevent the introduction and spread of aquatic invasive species: recreational activities. Arlington, VA: Aquatic Nuisance Species Task Force, US Fish and Wildlife Service. Retrieved from http://anstaskforce.gov/Documents/AIS_Recreation_Guidelines_Final_Draft.pdf. Accessed 6 February 2014.

Arndt, G. M., Gessner, J., Raymakers, C. (2002). Trends in farming, trade and occurrence of native and exotic sturgeons in natural habitats in Central and Western Europe. Journal of Applied Ichthyology, 18 (4-6), 444-448.

Audzijonytė, A., Väinölä, R. (2005). Diversity and distributions of circumpolar fresh-and brackish-water *Mysis* (Crustacea: Mysida): descriptions of *M. relicta* Lovén, 1862, *M. salemaai* n. sp., *M. segerstralei* n. sp. and *M. diluviana* n. sp., based on molecular and morphological characters. Hydrobiologia, 544 (1), 89-141.

Azevedo-Santos, V. M. D., Rigolin-Sá, O., Pelicice, F. M. (2011). Growing, losing or introducing? Cage aquaculture as a vector for the introduction of non-native fish in Furnas Reservoir, Minas Gerais, Brazil. Neotropical Ichthyology, 9 (4), 915-919.

Barse, A. M., Secor, D. H. (1999). An exotic nematode parasite of the American eel. Fisheries, 24 (2), 6-10.

Ben-Ami, F., Heller, J. (2001). Biological control of aquatic pest snails by the black carp *Mylopharyngodon piceus*. Biological Control, 22 (2), 131-138.

Beran, L. (2008). Expansion of *Sinanodonta woodiana* (Lea, 1834) (Bivalvia: Unionidae) in the Czech Republic. Aquatic Invasions, 3 (1), 91-94.

Bij de Vaate, A., Jazdzewski, K., Ketelaars, H. A., Gollasch, S., Van der Velde, G. (2002). Geographical patterns in range extension of Ponto-Caspian macroinvertebrate species in Europe. Canadian Journal of Fisheries and Aquatic Sciences, 59 (7), 1159-1174.

Britton, J. R., Orsi, M. L. (2012). Non-native fish in aquaculture and sport fishing in Brazil: Economic benefits versus risks to fish diversity in the upper River Paraná Basin. Reviews in Fish Biology and Fisheries, 22 (3), 555-565.

Busiahn, T. R. (1993). Can the ruffe be contained before it becomes your problem. Fisheries, 18 (8), 22-23.

Carlsson, N. O., Lacoursiere, J. O. (2005). Herbivory on aquatic vascular plants by the introduced golden apple snail (*Pomacea canaliculata*) in Lao PDR. Biological Invasions, 7 (2), 233-241.

Carlton, J. T. (1985). Transoceanic and interoceanic dispersal of coastal marine organisms: the biology of ballast water. Oceanography and Marine Biology: An Annual Review, 23.

Carlton, J. T., Geller, J. B. (1993). Ecological roulette: The global transport of nonindigenous marine organisms. Science, 261 (5117), 78-82.

Casal, C. M. V. (2006). Global documentation of fish introductions: the growing crisis and recommendations for action. Biological Invasions, 8 (1), 3-11.

CEC (Commission on Environmental Cooperation) (2009). Proposed tri-national risk assessment guidelines for invasive alien species: test cases for the snakeheads (Channidae) and armored catfishes (Loricariidae) in North American waters. Montreal: Commission for Environmental Cooperation.

Chandra, G., Bhattacharjee, I., Chatterjee, S. N., Ghosh, A. (2008). Mosquito control by larvivorous fish. Indian Journal of Medical Research, 127 (1), 13.

Chang, A. L., Grossman, J. D., Spezio, T. S., *et al.* (2009). Tackling aquatic invasions: risks and opportunities for the aquarium fish industry. Biological Invasions, 11 (4), 773-785.

Chapman, F. A., Fitz-Coy, S. A., Thunberg, E. M., Adams, C. M. (1997). United States of America trade in ornamental fish. Journal of the World Aquaculture Society, 28 (1), 1-10.

Chilton II, E. W., & Muoneke, M. I. (1992). Biology and management of grass carp (*Ctenopharyngodon idella*, Cyprinidae) for vegetation control: a North American perspective. Reviews in Fish Biology and Fisheries, 2 (4), 283-320.

Choi, W. J.; Gerstenberger, S., McMahon, R. F.; Wong, W. H. (2013). Estimating survival rates of quagga mussel (*Dreissena rostriformis bugensis*) veliger larvae under summer and autumn temperature regimes in residual water of trailered watercraft at Lake Mead, USA. Management of Biological Invasions, 2013 (4), 61-69.

Clench, W. J., Fuller, S. L. H. (1965). The genus Viviparus (Viviparidae) in North America. Occasional Papers on Mollusks, 2 (32), 385-412.

Conover, G., Simmonds, R., Whalen, M. (Eds.) (2007). Management and control plan for bighead, black, grass, and silver carps in the United States. Washington, D.C.: Asian Carp Working

Group, Aquatic Nuisance Species Task Force. Retrieved from http://asiancarp.org/Documents/Carps_Management_Plan.pdf

Contreras-Balderas, S. (1999). Annotated checklist of introduced invasive fishes in Mexico, with examples of some recent introductions. In R. Claudi, J.H. Leach (Eds), Nonindigenous Freshwater Organisms: Vectors, biology, and impacts (pp.33-54). Boca Raton, FL: Lewis Publishers.

Contreras-MacBeath, T., Mojica, H. M., Wilson, R. C. (1998). Negative impact on the aquatic ecosystems of the state of Morelos, Mexico from introduced aquarium and other commercial fish. Aquarium Sciences and Conservation, 2 (2), 67-78.

Copp, G. H., Bianco, P. G., Bogutskaya, N. G., *et al.* (2005). To be, or not to be, a non-native freshwater fish? Journal of Applied Ichthyology, 21 (4), 242-262.

Counts III, C. L. (1986). The Zoogeography and History of the Invasion of the United States by Corbicula fluminea (Bivalvia: Corbiculidae). In Proceedings of the Second International Corbicula Symposium, American Malacological Bulletin, Special Edition (No. 2, pp. 7-39).

Courtenay, W.R., Jr. (1999). Aquariums and water gardens as vectors of introduction. In Nonindigenous freshwater organisms: vectors, biology and impacts. In R. Claudi, J.H. Leach (Eds). Nonindigenous Freshwater Organisms: Vectors, biology, and impacts (pp.127–128). Boca Raton, FL: Lewis Publishers.

Courtenay Jr, W. R., Hensley, D. A., Taylor, J. N., McCann, J. A. (1984). Distribution of exotic fishes in the continental United States. In W. R. Courtenay, Jr., J. R. Stauffer, Jr., (Eds.), Distribution, biology and management of exotic fishes (pp.41-77). Baltimore, MD: Johns Hopkins University Press.

Courtenay, W.R., Jr., Meffe,G.K. (1989). Small fishes in strange places: a review of introduced poeciliids. In G.K. Meffe, F.F. Snelson, Jr, (Eds), Ecology and evolution of livebearing fishes (Poeciliidae) (pp.319-331). Englewood Cliffs, NJ: Prentice Hall.

Courtenay, W. R., Stauffer, J. R. (1990). The introduced fish problem and the aquarium fish industry. Journal of the World Aquaculture Society, 21 (3), 145-159.

Courtenay, W.R., Jr., Taylor, J. N. (1984). The exotic ichthyofauna of the contiguous United States with preliminary observations on intra-national transplants. EIFAC Technology Paper, 42 (2), 466-487.

Courtenay, Jr., W.R., Williams, J.D. (2004). Snakeheads (Pisces: Channidae): A biological synopsis and risk assessment. Circular 1251. United States Geological Survey, Gainesville, FL. Retrieved from: http://fl.biology.usgs.gov/Snakehead_circ_1251/html/flyer.html, Accessed 29 December 2014.

Cowie, R. H. (1998). Patterns of introduction of non-indigenous non-marine snails and slugs in the Hawaiian Islands. Biodiversity and Conservation, 7 (3), 349-368.

Cowie, R. H. (2001). Can snails ever be effective and safe biocontrol agents? International Journal of Pest Management, 47 (1), 23-40.

Cowie, R. H., Robinson, D. G. (2003). Pathways of introduction of nonindigenous land and freshwater snails and slugs. In G. M. Ruiz, J. T. Carlton (Eds.), Invasive species: vectors and management strategies (pp.93-122). Washington: Island Press.

Crossman, E. J. (1984). Introductions of Exotic Fishes into Canada. In W. R. Courtenay, Jr., J. R. Stauffer, Jr., (Eds.), Distribution, biology, and management of exotic fishes (pp.78-101). Baltimore, MD: Johns Hopkins University Press.

Crossman, E. J., Cudmore, B. C. (1999). Summary of North American fish introductions through the aquarium/horticulture trade. In R. Claudi, J.H. Leach (Eds.), Nonindigenous freshwater organisms: Vectors, biology, and impacts (pp.129–33). Boca Raton, FL: Lewis Publishers.

Cudmore, B., Mandrak, N.E. (2005). The Baitfish Primer. Burlington, Ontario: Fisheries and Oceans Canada, Great Lakes Laboratory for Fisheries and Aquatic Sciences.

Cudmore-Vokey, B., Crossman, E. J. (1999). Checklists of the fish fauna of the Laurentian Great Lakes and their connecting channels. Canadian Manuscript Report of Fisheries and Aquatic Sciences, 2500. Burlington: Fisheries and Oceans Canada.

Daniels, R. A. (2001). Untested assumptions: the role of canals in the dispersal of sea lamprey, alewife, and other fishes in the eastern United States. Environmental Biology of Fishes, 60 (4), 309-329.

Dextrase, A. J. Coscarelli, M. A. (1999). Intentional introductions of nonindigenous freshwater organisms in North America. In R. Claudi, J.H. Leach (Eds.), Nonindigenous freshwater organisms: Vectors, biology, and impacts (pp.61-98). Boca Raton, FL: Lewis Publishers.

Dill, W. A., Cordone, A. J. (1997). History and status of introduced fishes in California, 1871-1996. Fish Bulletin 178. Sacramento, CA: California Department of Fish and Game. Retrieved from http://content.cdlib.org/view?docId=kt8p30069f&brand=calisphere, Accessed 30 December 2013.

DiStefano, R. J., Litvan, M. E., Horner, P. T. (2009). The bait industry as a potential vector for alien crayfish introductions: problem recognition by fisheries agencies and a Missouri evaluation. Fisheries, 34 (12), 586-597.

Duggan, I. C., Rixon, C. A., MacIsaac, H. J. (2006). Popularity and propagule pressure: determinants of introduction and establishment of aquarium fish. Biological invasions, 8 (2), 377-382.

Emery, L. (1985). Review of fish introduced into the Great Lakes, 1819-1974. Technical Report No. 45. Ann Arbor, MI: Great Lakes Fishery Commission.

Emery, A. R., Teleki, G. (1978). European founder (*Platyichthys flesus*) captured in Lake Erie, Ontario. Canadian Field-Naturalist, 92 (1), 89-91.

(FAO) Food and Agriculture Organization of the United Nations (1996) Precautionary approach to capture fisheries and species introductions. Elaborated by the Technical Consultation on the Precautionary Approach to Capture Fisheries (Including Species Introductions). Lysekil, Sweden, 6–13 June 1995. FAO Technical Guidelines for Responsible Fisheries. No. 2. Rome: FAO.

(FAO) Food and Agriculture Organization of the United Nations (2012).The State of World Fisheries and Aquaculture - 2012. Rome: Food and Agriculture Organization of the United Nations. Retrieved from http://www.fao.org/docrep/016/i2727e/i2727e00.htm, Accessed 6 November 2013.

(FAO) Food and Agriculture Organization of the United Nations (2013a). Database on Introductions of Aquatic Species (DIAS). http://www.fao.org/fishery/dias/en

(FAO) Food and Agriculture Organization of the United Nations (2013b). Global Aquaculture Production 1950-2011. http://www.fao.org/figis/servlet/TabSelector?tb_ds=Aquaculture&tb_mode=TABLE&tb_act=SELECT&tb_grp=COUNTRY, Accessed 6 November 2013.

FDACS (Florida Department of Agriculture and Consumer Services) (2002). Apple Snails. Technical Bulletin Number 3. Tallahassee, FL: Florida Department of Agriculture and Consumer Services.

Finch, R.G. (1925). The Story of the New York State Canals: historical and commercial information. NY, Albany: State of New York, State Engineer and Surveyor. (Copyright renewed NYS Canal Corporation 1998). Retrieved from http://www.canals.ny.gov/history/finch_history.pdf

Fries, L. T., Williams, D. J., Johnson, S. K. (1996). Notes: Occurrence of *Anguillicola crassus*, an Exotic Parasitic Swim Bladder Nematode of Eels, in the Southeastern United States. Transactions of the American Fisheries Society, 125 (5), 794-797.

Froese, R., Pauly, D. (Eds.) (2014). FishBase. World Wide Web electronic publication. Retrieved from http://www.fishbase.org, Accessed 7 January 2014.

Fuller, P. L. (2003). Freshwater Aquatic Vertebrate Introductions in the United States: Patterns and Pathways. In G. Ruiz, J. T. Carlton (Eds.), Invasive species: Vectors and management strategies (pp.123-151). Washington, DC.: Island Press.

Fuller, P. L., Nico, L. G., Williams, J. D. (1999). Nonindigenous Fishes Introduced to Inland Waters of the United States, Special Publication 27. Bethesda, MD: American Fisheries Society.

Ganzhorn, J., Rohovec, J. S., Fryer, J. L. (1992). Dissemination of microbial pathogens through introductions and transfers of finfish. In A. Rosenfield, R. Mann (Eds.), Dispersal of living organisms into aquatic ecosystems (pp.175-192). College Park, MD: Maryland Sea Grant.

Gertzen, E., Familiar, O., Leung, B. (2008). Quantifying invasion pathways: fish introductions from the aquarium trade. Canadian Journal of Fisheries and Aquatic Sciences, 65 (7), 1265-1273.

Golani, D., Mires, D. (2000). Introduction of fishes to the freshwater system of Israel. Israeli Journal of Aquaculture/Bamidgeh, 52 (2), 47-60.

Goldburg, R. Triplett, T. (1997). Murky Waters: Environmental Effects of Aquaculture in the US. New York: The Environmental Defense Fund.

Goodwin, A. E., Peterson, J. E., Meyers, T. R., Money, D. J. (2004). Transmission of exotic fish viruses: the relative risks of wild and cultured bait. Fisheries, 29 (5), 19-23.

Gozlan, R. E. (2008). Introduction of non-native freshwater fish: is it all bad? Fish and Fisheries, 9 (1), 106-115.

Gozlan, R. E., Britton, J. R., Cowx, I., Copp, G. H. (2010). Current knowledge on non-native freshwater fish introductions. Journal of Fish Biology, 76 (4), 751-786.

Havel, J. E., Hebert, P. D. (1993). *Daphnia lumholtzi* in North America: another exotic zooplankter. Limnology and Oceanography, 38 (8), 1823-1827.

Hayes, K. A., Joshi, R. C., Thiengo, S. C., Cowie, R. H. (2008). Out of South America: multiple origins of non-native apple snails in Asia. Diversity and Distributions, 14 (4), 701-712.

Heard, W. (1956). Live bait imports: *Chrosomus eos* and *Eucalia inconstans* as potential additions to Oklahoma's fish fauna. Oklahoma Academy of Science, 37, 47-48.

Hebert, P. D., Muncaster, B. W., Mackie, G. L. (1989). Ecological and genetic studies on *Dreissena polymorpha* (Pallas): a new mollusc in the Great Lakes. Canadian Journal of Fisheries and Aquatic Sciences, 46 (9), 1587-1591.

Henry, S. (2007). Buddhists release creatures into Passaic. Herald News. Retrieved from NorthJersey.com. Accessed 15 August 2007.

Hobbs III, H. H., Jass, J. P., Huner, J. V. (1989). A review of global crayfish introductions with particular emphasis on two North American species (Decapoda, Cambaridae). Crustaceana, 299-316.

Hoffman, F. L., Schubert, G. (1984). Some parasites of exotic fishes. In W.R. Courtenay Jr., J.R. Stauffer, Jr. (Eds), Distribution, biology, and management of exotic fishes (pp.233-261). Baltimore, MD: The Johns Hopkins University Press.

Howells, R. G. (2001). History and status of applesnail (*Pomacea* spp.) introductions in Texas. Texas Parks and Wildlife Department, Management Data Series No. 183. Austin, TX: Texas Parks and Wildlife Department, Inland Fisheries Division.

Jackson, J. R., Boxrucker, J. C., Willis, D. W. (2004). Trends in agency use of propagated fishes as a management tool in inland fisheries. American Fisheries Society Symposium, 44, 121-138.

Jensen, J. B., Waters, C. (1999). The spring lizard bait industry in the state of Georgia, USA. Herpetological Review, 30 (1), 50-51.

Johnson, B. M., Arlinghaus, R., Martinez, P. J. (2009). Are we doing all we can to stem the tide of illegal fish stocking? Fisheries, 34 (8), 389-394.

Johnson, J. E. (1986). Inventory of Utah crayfish with notes on current distribution. Western North American Naturalist, 46 (4), 625-631. Retrieved from http://www.biodiversitylibrary.org/item/33077#page/651/mode/1up, Accessed 17 January 2014

Johnson, L. E., Bossenbroek, J. M., Kraft, C. E. (2006). Patterns and pathways in the post-establishment spread of non-indigenous aquatic species: the slowing invasion of North American inland lakes by the zebra mussel. Biological Invasions, 8 (3), 475-489.

Johnson, L. E., Padilla, D. K. (1996). Geographic spread of exotic species: Ecological lessons and opportunities from the invasion of the zebra mussel *Dreissena polymorpha*. Biological Conservation, 78 (1), 23-33.

Karatayev, A. Y., Burlakova, L. E., Karatayev, V. A., Padilla, D. K. (2009). Introduction, distribution, spread, and impacts of exotic freshwater gastropods in Texas. Hydrobiologia, 619 (1), 181-194.

Keller, R. P., Cox, A. N., Van Loon, C., Lodge, D. M., Herborg, L. M., Rothlisberger, J. (2007). From bait shops to the forest floor: earthworm use and disposal by anglers. The American Midland Naturalist, 158 (2), 321-328.

Keller, R. P., Geist, J., Jeschke, J. M., Kühn, I. (2011). Invasive species in Europe: ecology, status, and policy. Environmental Sciences Europe, 23 (1), 1-17

Keller, R. P., Lodge, D. M. (2007). Species invasions from commerce in live aquatic organisms: problems and possible solutions. Bioscience, 57 (5), 428-436.

Kircheis, F. W. (1998). Species composition and economic value of Maine's winter baitfish industry. North American Journal of Fisheries Management, 18 (1), 175-180.

Koehn, J. D., MacKenzie, R. F. (2004). Priority management actions for alien freshwater fish species in Australia. New Zealand Journal of Marine and Freshwater Research, 38 (3), 457-472, DOI: 10.1080/00288330.2004.9517253

Kohler, C. C., Hubert, W. A. (1999). Inland fisheries management in North America, Bethesda. Maryland: American Fisheries Society.

Kolar, C. S., Chapman, D. C., Courtenay, W. R., Housel, C. M., Williams, J. D., Jennings, D. P. (2007). Bigheaded carps: a biological synopsis and risk assessment. American Fisheries Society, Special Publication 33, Bethesda, Maryland: American Fisheries Society.

Kraszewski, A., Zdanowski, B. (2007). *Sinanodonta woodiana* (Lea, 1834) (Mollusca)-a new mussel species in Poland: occurrence and habitat preferences in a heated lake system. Polish Journal of Ecology, 55 (2), 337-356.

Krishnakumar, K., Raghavan, R., Prasad, G., Bijukumar, A., Sekharan, M., Pereira, B., Ali, A. (2009). When pets become pests– exotic aquarium fishes and biological invasions in Kerala, India. Current Science, 97 (4), 474-476.

LaBar, G. W., Frew, T. (2004). Using a fish-stocking database to demonstrate temporal changes in stocking patterns. In M. Nickum, Mazik, P., Nickum, J., D. MacKinlay (Eds.), Propagated fish in resource management (pp.113-118). American Fisheries Society Symposium 44. Bethesda, MD: American Fisheries Society.Lach, L. O. R. I., Cowie, R. H. (1999). The spread of the introduced freshwater apple snail *Pomacea canaliculata* (Lamarck) (Gastropoda: Ampullariidae) on Oahu, Hawaii. Bishop Museum Occasional Papers, 58, 66-71.

Lachner, E. A., Robins, C. R., Courtenay, Jr., W. R. (1970). Exotic fishes and other aquatic organisms introduced into North America. Smithsonian Contributions to Zoology, 59, 1-29.

Lasenby, D. C., Northcote, T. G., Fürst, M. (1986). Theory, practice, and effects of *Mysis relicta* introductions to North American and Scandinavian lakes. Canadian Journal of Fisheries and Aquatic Sciences, 43 (6), 1277-1284.

Larson, E. R., Olden, J. D. (2011). The state of crayfish in the Pacific Northwest. Fisheries, 36 (2), 60-73.

Leigh, P. (1998). Benefits and costs of the ruffe control program for the Great Lakes fishery. Journal of Great Lakes Research, 24 (2), 351-360.

Leprieur, F., Beauchard, O., Blanchet, S., Oberdorff, T., Brosse, S. (2008). Fish invasions in the world's river systems: when natural processes are blurred by human activities. PLoS Biology, 6 (2), e28.

Leuven, R. S., van der Velde, G., Baijens, I., Snijders, J., van der Zwart, C., Lenders, H. R., bij de Vaate, A. (2009). The river Rhine: a global highway for dispersal of aquatic invasive species. Biological Invasions, 11 (9), 1989-2008.

Lever, C. (1996). Naturalized fishes of the world. Bath, Great Britain: Academic Press.

Litvak, M. K., Mandrak, N. E. (1993). Ecology of freshwater baitfish use in Canada and the United States. Fisheries, 18 (12), 6-13.

Liu, L. S. (1996). The recapture and sale of animals released for prayers. Animals' Voice of Taiwan 11, 15. (In Chinese).

Liu, X., McGarrity, M. E., Li, Y. (2012). The influence of traditional Buddhist wildlife release on biological invasions. Conservation Letters, 5 (2), 107-114.

Livengood, E. J., Chapman, F. A. (2007). The Ornamental Fish Trade: An Introduction with Perspectives for Responsible Aquarium Fish Ownership - FA124. Gainesville, FL: University of Florida, Florida Cooperative Extension Service, Institute of Food and Agricultural Sciences, Department of Fisheries and Aquatic Sciences.

LoVullo, T. J., Stauffer Jr, J. R. (1993). The retail bait-fish industry in Pennsylvania: source of introduced species. Journal of the Pennsylvania Academy of Science, 67 (1), 13-15.

Ludwig Jr, H. R., Leitch, J. A. (1996). Interbasin transfer of aquatic biota via anglers' bait buckets. Fisheries, 21 (7), 14-18.

Maceda-Veiga, A., Escribano-Alacid, J., de Sostoa, A., García-Berthou, E. (2013). The aquarium trade as a potential source of fish introductions in southwestern Europe. Biological Invasions, 15, 2707-2716.

MacIsaac, H. J., Sprules, G., Johannson, O. E., Leach, J. H. (1992). Filtering impacts of larval and sessile zebra mussels (*Dreissena polymorpha*) in western Lake Erie. Oecologia, 92 (1), 30-39.

Mackie, G. L. (1999). Mollusc Introductions through aquarium trade. In R. Claudi, J.H. Leach (Eds.), Nonindigenous freshwater organisms: Vectors, biology, and Impacts (pp.135-149). Boca Raton, FL: Lewis Publishers.

Madsen, H., Frandsen, F. (1989). The spread of freshwater snails including those of medical and veterinary importance. Acta Tropica, 46 (3), 139-146.

Marr, S. M., Marchetti, M. P., Olden, J. D., *et al.* (2010). Freshwater fish introductions in Mediterranean-climate regions: are there commonalities in the conservation problem? Diversity and Distributions, 16 (4), 606-619.

Marsh, G.P. (1874). The Earth as modified by human action. New York: Charles Scribner's Sons.

Magalhães, A. L. B., Jacobi, C. M. (2013). Invasion risks posed by ornamental freshwater fish trade to southeastern Brazilian rivers. Neotropical Ichthyology, 11 (2), 433-441.

Magalhães, A. L. B. Jacobi, C. M. (2010). E-commerce of freshwater aquarium fishes: potential disseminator of exotic species in Brazil. Acta Scientiarum. Biological Sciences, 32 (3), 243-248.

McDowall, R. M. (2004). Shoot first, and then ask questions: A look at aquarium fish imports and invasiveness in New Zealand. New Zealand Journal of Marine and Freshwater Research, 38 (3), 503-510, DOI: 10.1080/00288330.2004.9517256.

Mills, E. L., Leach, J. H., Carlton, J. T., Secor, C. L. (1993). Exotic species in the Great Lakes: a history of biotic crises and anthropogenic introductions. Journal of Great Lakes Research, 19 (1), 1-54.

Mills, E. L., Scheuerell, M. D., Carlton, J. T., Strayer, D. L. (1997). Biological invasions in the Hudson River basin: an inventory and historical analysis. New York State Museum Circular 57, 1-51, University of the State of New York, State Education Department.

Mills, E. L., Chrisman, J. R., Holeck, K. T. (1999). The role of canals in the spread of non-indigenous species in North America. In R. Claudi, J.H. Leach (Eds), Nonindigenous Freshwater Organisms: Vectors, biology, and impacts (pp.347-380). Boca Raton, FL: Lewis Publishers.

Morgan, D. L., Gill, H. S. (2001). The green swordtail *Xiphophorus helleri* Heckel (Poeciliidae): another aquarium fish established in the wild in Western Australia. Records of the Western Australian Museum, 20 (3), 349-352.

Moy, P. B., Polls, I. R. W. I. N., Dettmers, J. M. (2011). The Chicago sanitary and ship canal aquatic nuisance species dispersal barrier. In D. C. Chapman, M. H. Hoff (Eds.), Invasive Asian carps in North America (pp.121-137). Bethesda, MD: American Fisheries Society Special Publication 74.

Moyle, P. B. (2002). Inland Fishes of California. Berkeley, CA: University of California Press.

Naylor, R. (1996). Invasions in agriculture: assessing the cost of the golden apple snail in Asia. Ambio, 25 (70), 443-448.

Naylor, R. L., Williams, S. L., Strong, D. R. (2001). Aquaculture--A gateway for exotic species. Science, 294 (5547), 1655-1656.

Nesler, T. P., Bergersen, E. P. (1991). Mysids in fisheries: hard lessons from headlong introductions Symposium (No. 9). Bethesda: American Fisheries Society.

Ng, P. K., Chou, L. M., Lam, T. J. (1993). The status and impact of introduced freshwater animals in Singapore. Biological Conservation, 64 (1), 19-24.
Nico, L. G., Fuller, P. L. (1999). Spatial and temporal patterns of nonindigenous fish introductions in the United States. Fisheries, 24 (1), 16-27.
Nico, L. G., Sharp, P., Collins, T. M. (2011). Imported Asian swamp eels (Synbranchidae: *Monopterus*) in North American live food markets: Potential vectors of non-native parasites. Aquatic Invasions, 6 (1), 69-76.
Nico, L. G., Williams, J. D., Jelks, H. L. (2005). Black carp: biological synopsis and risk assessment of an introduced fish. Bethesda: American Fisheries Society.
Nickum, M. J. (Ed.) (2004). Propagated fish in resource management (Vol. 44). Bethesda: American Fisheries Society.
Nielsen, L. A. (1982). The bait-fish industry in Ohio and West Virginia, with special reference to the Ohio River sport fishery. North American Journal of Fisheries Management, 2 (3), 232-238.
NOAA (2014). Great Lakes Aquatic Nonindigenous Species Information System. Retrieved from http://www.glerl.noaa.gov/res/Programs/glansis/nas_database.html Accessed 9 January 2014.
Ogutu-Ohwayo, R., Hecky, R. E. (1991). Fish introductions in Africa and some of their implications. Canadian Journal of Fisheries and Aquatic Sciences, 48 (S1), 8-12.
Padilla, D. K., Williams, S. L. (2004). Beyond ballast water: aquarium and ornamental trades as sources of invasive species in aquatic ecosystems. Frontiers in Ecology and the Environment, 2 (3), 131-138.
Paunovic, M., Csányi, B., Simic, V., Stojanovic, B., Cakic, P. (2006). Distribution of *Anodonta (Sinanodonta) woodiana* (Lea, 1834) in inland waters of Serbia. Aquatic Invasions, 1 (3), 154-160.
Pflieger, W. L. (1975). The Fishes of Missouri. Missouri Department of Conservation. Jefferson City, Missouri, USA.
Pointier, J. P. (2001). Invading freshwater snails and biological control in Martinique Island, French West Indies. Memórias do Instituto Oswaldo Cruz, 96 Suppl., 67-74.
Pointier, J. P., Augustin, D. (1999). Biological control and invading freshwater snails. A case study. Comptes Rendus de l'Académie des Sciences-Series III-Sciences de la Vie, 322 (12), 1093-1098.
Pointier, J. P., David, P. (2004). Biological control of *Biomphalaria glabrata*, the intermediate host of schistosomes, by *Marisa cornuarietis* in ponds of Guadeloupe: long-term impact on the local snail fauna and aquatic flora. Biological Control, 29 (1), 81-89.
Popa, O. Popa, L. O. (2006). *Sinanodonta woodiana* (Lea, 1834), *Corbicula fluminea* (O. F. Muller, 1774), *Dreissena bugensis* (Andrusov, 1897) (Mollusca: Bivalvia): alien invasive species in Romanian fauna. Travaux du Museum National d'Histoire Naturalle, 49, 7-12.
Pothoven, S. A., Grigorovich, I. A., Fahnenstiel, G. L., Balcer, M. D. (2007). Introduction of the Ponto-Caspian bloody-red mysid *Hemimysis anomala* into the Lake Michigan basin. Journal of Great Lakes Research, 33 (1), 285-292.
Pou-Rovira, Q., Araujo, R., Boix, D., Clavero, M., Feo, C., Ordeix, M., Zamora, L. (2009). Presence of the alien chinese pond mussel *Anodonta woodiana* (Lea, 1834)(Bivalvia, Unionidae) in the Iberian Peninsula. Graellsia, 65 (1), 67-70.
Pratt, D. M., Blust, W. H., Selgeby, J. H. (1992). Ruffe, *Gymnocephalus cernuus*: newly introduced in North America. Canadian Journal of Fisheries and Aquatic Sciences, 49 (8), 1616-1618.
Radonski, G.C., N.S. Prosser, N. S., Martin, R. G., Stroud, R. H. (1984). Exotic fishes and sport fishing. In W.R., Jr., Courtenay, J.R. Stauffer, Jr, (Eds.), Distribution, biology, and management of exotic fishes (pp.313-321). Baltimore, MD: John Hopkins University Press.
Rahel F.J. (1997) From Johnny Appleseed to Dr. Frankenstein: changing values and the legacy of fisheries management. Fisheries, 22 (8), 8–9.

Rahel, F. J. (2000). Homogenization of fish faunas across the United States. Science, 288 (5467), 854-856.

Rahel, F. J. (2004). Unauthorized fish introductions: Fisheries management of the people, for the people, or by the people? American Fisheries Society Symposium, 44, 431-443.

Rahel, F. J. (2007). Biogeographic barriers, connectivity and homogenization of freshwater faunas: it's a small world after all. Freshwater Biology, 52 (4), 696-710.

Rasmussen, J. L. (2002). The Cal-Sag and Chicago Sanitary and Ship Canal: A Perspective on the Spread and Control of Selected Aquatic Nuisance Fish Species. Rock Island, IL: U.S. Fish and Wildlife Service.

Riegel, J. A. (1959). The systematics and distribution of crayfishes in California. California Fish and Game, 45 (1), 29-50.

Ricciardi, A., MacIsaac, H. J. (2000). Recent mass invasion of the North American Great Lakes by Ponto–Caspian species. Trends in Ecology and Evolution, 15 (2), 62-65.

Rixon, C. A., Duggan, I. C., Bergeron, N. M., Ricciardi, A., Macisaac, H. J. (2005). Invasion risks posed by the aquarium trade and live fish markets on the Laurentian Great Lakes. Biodiversity and Conservation, 14 (6), 1365-1381.

Rocha-Miranda, F., Martins-Silva, M. J. (2006). First record of the invasive snail *Melanoides tuberculatus* (Gastropoda: Prosobranchia: Thiaridae) in the Paranã River basin, GO, Brazil. Brazilian Journal of Biology, 66 (4), 1109-1115.

Ruiz, G. M., Fofonoff, P. W., Carlton, J. T., Wonham, M. J., Hines, A. H. (2000). Invasion of coastal marine communities in North America: apparent patterns, processes, and biases. Annual Review of Ecology and Systematics, 31, 481-531.

Seaman, D. E., Porterfield, W. A. (1964). Control of aquatic weeds by the snail *Marisa cornuarietis*. Weeds, 87-92.

Severinghaus, L. L., Chi, L. (1999) Prayer animal release in Taiwan. Biological Conservation, 89, 301–304.

Shafland, P. L. (1999). The introduced butterfly peacock (*Cichla ocellaris*) in Florida. I. Fish community analyses. Reviews in Fisheries Science, 7 (2), 71-94.

Shireman, J. V. (1984). Control of aquatic weeds with exotic fishes. In W.R. Courtenay, J.D. Stauffer, Distribution, biology and management of exotic fishes (pp.302-312). Baltimore, MD: The John Hopkins University. Press.

Simberloff, D., Rejmánek, M. (Eds.) (2010). Encyclopedia of biological invasions (Vol. 3). Oakland, CA: University of California Press.

Simon, T. P., Vondruska, J. T. (1991). Larval identification of the ruffe, *Gymnocephalus cernuus* (Linnaeus) (Percidae: Percini), in the St. Louis River Estuary, Lake Superior drainage basin, Minnesota. Canadian Journal of Zoology, 69 (2), 436-442.

Smith, B.J. (1989) Travelling Snails. Journal of Medical and Applied Malacology, 1, 195–204.

Stauffer, J. R., Arnegard, M. E., Cetron, M., *et al.* (1997). Controlling vectors and hosts of parasitic diseases using fishes. BioScience, 47 (1), 41-49.

Strecker, A. L., Campbell, P. M., Olden, J. D. (2011). The aquarium trade as an invasion pathway in the Pacific Northwest. Fisheries, 36 (2), 74-85.

Su Sin, T. (2006). Evaluation of different species of fish for biological control of golden apple snail *Pomacea canaliculata* (Lamarck) in rice. Crop Protection, 25 (9), 1004-1012.

Swift, C. C., Haglund, T. R., Ruiz, M., Fisher, R. N. (1993). The status and distribution of the freshwater fishes of southern California. Bulletin of the Southern California Academy of Sciences, 92 (3), 101-167.

Thaulow, J., Borgstrøm, R., Heun, M. (2014). Genetic analyses of European minnow, *Phoxinus phoxinus*, in a river system outside its native range indicate multiple invasions from different sources. Fisheries Management and Ecology, 21 (1), 75-81.

Thompson, J. A., Alam, S. K. (2005). Analysis of customs trade data to characterize importation of live bait. Fisheries, 30 (6), 36-39.
U.S. Congress. (1993). Office of Technology Assessment. (1993). Harmful non-indigenous species in the United States. OTA-F-565. Washington, DC: US Government Printing Office.
USGS (US. Geological Survey) (2014). Nonindigenous Aquatic Species Database. Retrieved from http://nas.er.usgs.gov.
Vashro, J. (1990). Illegal aliens. Montana Outdoors, 21 (4), 34-37.
Vashro, J. (1995). The bucket brigade. Montana Outdoors, 26 (5), 34-35.
Veraldi, F.M., Baerwaldt, K., Herman, B., Herleth-King, S., Shanks, M., Kring,L., Hannes, A. (2011). Non-Native Species of Concern and Dispersal Risk for the Great Lakes and Mississippi River Interbasin Study. U.S. Army Corps of Engineers. Retrieved from http://www.glmris.anl.gov/documents/docs/Non-Native_Species.pdf Accessed 5 December 2013.
Watters, G. T. (1997). A synthesis and review of the expanding range of the Asian freshwater mussel *Anodonta woodiana* (Lea, 1834)(Bivalvia, Unionidae). The Veliger, 40 (2), 152-156.
Weigle, S. M., Smith, L. D., Carlton, J. T., Pederson, J. (2005). Assessing the risk of introducing exotic species via the live marine species trade. Conservation Biology, 19 (1), 213-223.
Welcomme R.L. (1984). International transfers of Inland fish species. In W.R. Courtenay Jr., J.R. Stauffer Jr. (Eds.), Distribution, biology, and management of exotic fishes (pp.22-40). Baltimore, MD: Johns Hopkins University Press.
Welcomme, R. L. (Ed.). (1988). International Introductions of Inland Aquatic Species (Vol. 294). Food and Agriculture Organization of the United Nations. Available on-line: http://www.fao.org/docrep/X5628E/x5628e04.htm. Accessed 7 February 2014
Whelan, G. (2004). A historical perspective on the philosophy behind the use of propagated fish in fisheries management: Michigan's 130-year experience. Propagated Fish in Resource Management, 44, 307-315.
Wong, W. H., Gerstenberger, S. (2011). Quagga mussels in the western United States: Monitoring and Management. Aquatic Invasions, 6 (2), 125.
Wonham, M. J., Carlton, J. T., Ruiz, G. M., Smith, L. D. (2000). Fish and ships: relating dispersal frequency to success in biological invasions. Marine Biology, 136 (6), 1111-1121.

Christina M. Romagosa

6 Contribution of the Live Animal Trade to Biological Invasions

6.1 Introduction

Live animals are an important biological commodity, and their trade causes the global movement of millions of individuals annually (Fernandez & Luxmoore, 1997). Although the trade in live animals has a long history, dating back to at least Greek and Roman times, the emphasis was most often on novelty and exotic foods for the noble classes (Lever, 1992; Cassey *et al.*, see Chapter 2). Today, the ability to obtain live animals is no longer restricted to nobility, and species are traded widely for a variety of purposes. Advances in air transport and shipping techniques have facilitated the movement of more species and individuals used for the live animal trade than in the past (Roe *et al.*, 2002; Fuller, 2003). The trade in live animals is largely unregulated (Smith *et al.*, 2009), and the number of non-native introductions through this pathway continues to rise; therefore, future introductions and their associated impacts will continue to be of concern (Kraus, 2003; Goss & Cumming, 2013).

This chapter describes the live animal trade, the specific pathways through which animals are transported, and how each pathway has contributed to species introductions worldwide. The term pathway is considered here as the specific purpose for which a non-native species was transported to an area beyond its native range. I have attempted to further break down the live animal trade pathway into sectors based on the purpose for trade, but the boundaries that separate each sector can be fuzzy, and cannot be considered mutually exclusive. Many species are traded for more than one purpose, and can also be introduced through multiple pathways. I primarily focus on the pathways for tetrapods; a summary of the pathways of invertebrate and fish invasions is covered by Fuller (see Chapter 5). I have provided several examples from the United States, in large part because these events are well-documented, and also because it reflects my experience. However, I do include additional information with examples drawn from around the world.

Introduction is defined here as the occurrence of a species outside of its native range, regardless of whether the species is of foreign or domestic origin (i.e., individuals of a species within a country that are transplanted outside their native range). I include here discussion of introductions of species that have become established (breeding population) as well as those that did not do so. While introduction of one individual is typically not enough to establish a viable population, it is still important to document that species' presence outside of captivity because it records a trend of escaping or being intentionally released.

6.2 Captive Animals

This section describes the reasons for which animals are traded specifically to be kept in captivity for non-scientific purposes. The patterns by which animals are kept in captivity differ by geography and culture. For example, from animal ownership surveys conducted in Indonesia, Costa Rica, and the United States, the percentage of households that owned birds were 35.7% (Jepson & Ladle, 2005), 23.3% (Drews, 2001), and 3.9% (American Veterinary Medical Association, 2007), respectively. As far as scale, it is important to consider that 3.9% of households in the United States translates to about 11 million birds in captivity, whereas 35.7% of households in Indonesia refers to about 2 million birds in captivity. Some animals described in this section can be loosely categorized as "pets", but I also include here animals that are kept in captivity for a variety of purposes. Introductions for most vertebrates in trade occur through the pathways described here; therefore, this section is particularly thorough.

6.2.1 Pets

The definition of pet can vary widely: from companion animals to animals kept as part of a more impersonal menagerie or collection. The types of pets can range from domesticated animals, such as dogs and cats, to those that are collected directly from the wild. Dogs and cats are excluded from this chapter, as their trade and introduction history is a topic too expansive to be suitably covered here.

Birds have the longest history as captive animals. Records of birds in captivity, such as mynahs and parrots, date back to at least 1000 BC in Asia (Law, 1923). There are various motivators for keeping birds: for their aesthetic value, their song, their ability to mimic speech, their ability to bond to their owners, and for the social status they can bring (Beissinger, 2001; Shepherd *et al.*, 2004). The importance of each of these motivators differs among cultures. For example, in Taiwan, passerines are valued for their song (Su *et al.*, 2014), whereas in the United States parrots are more popular for their ability to interact with their owners (Anderson, 2003).

The captive bird trade was active in Europe by the 1800s, and the United States followed suit by the mid-1800s. By the late 1800s, Oldys (1907) estimated that approximately 700 species and 800,000 individuals were imported into Europe, and more than 200 species and 300,000 individuals were imported into the United States. The global trade in wild birds peaked in the mid-1970s, and it was estimated that about 7.5 million birds were traded annually. In the 1990s, this number declined to about 3 million birds annually. More recently, these figures have dropped further, owing to the enforcement of international regulations related to bird conservation and human health, such as the United States Wild Bird Conservation Act of 1992, and the European Union's ban on wild bird imports in 2007 (FAO, 2011). While the decline in these numbers seems encouraging, birds are still removed from the wild at unsustainable numbers, and trade in some regions is, in fact, increasing.

For most countries in Latin America, Africa, and Asia, the keeping of birds is centered around native species, rather than the importation of non-native species. For example, it is estimated that before 2008, most of Mexico's parrots caught in the wild were sold at markets within Mexico's borders (Cantú *et al.*, 2007). However, after Mexico banned the trade in native parrots, the country quickly surpassed all others in the importing of CITES-listed species, mostly Monk Parakeets (*Myiopsitta monachus*) (CITES trade database, http://trade.cites.org/).

In the last 30 years, there has been a surge of interest in amphibians and reptiles as pets, both in the United States and abroad (Hoover, 1998; Auliya, 2003; Tapley *et al.*, 2011). The United States and the European Union are the primary consumers of reptiles and amphibians as pets. In the United States, the quantity of individuals imported from some of these groups has more than doubled since the early 1970s. For example, approximately 320,000 lizards and snakes were imported per year in the early 1970s, and over a million individuals were imported per year in the early 2000s (Romagosa, 2015). Many individuals imported in recent years belong to just a few species that are popular pet animals, such as green iguanas (*Iguana iguana*), Chinese water dragons (*Physignathus cocincinus*), and ball pythons (*Python regius*).

Interest in a more diverse selection of amphibian and reptile species for pets has been accompanied by an increase in the number of species imported from these groups. This phenomenon is particularly apparent in the diversity of lizard species; the number of lizard species imported in 1970–71 (176 species) versus 2009–10 (287 species) increased by 63% (Romagosa, 2015). The trade in these groups for pets is also increasing in Asia, where the most commonly kept reptiles are turtles (Zhang & Yin, 2014). Many consumers view reptiles and amphibians as an easy-care pet, as these pets often do not require as much attention as birds and mammals. Unfortunately, many consumers either receive erroneous or no information regarding their care and find themselves unprepared for the adult size and longevity of some species, leading to the introduction of unwanted animals to the wild.

The trade in live mammals as pets has never exhibited the species diversity and scale seen in other taxonomic groups. Non-human primates are popular as pets around the world, and have been so for thousands of years (Mack & Mittermeier, 1984; Hughes, 2003). In wealthy regions, such as the United States, European Union, and Japan, the interest in small mammals as pets is increasing. Gerbils, hamsters, guinea pigs and rabbits have long been common, but unique animals such as sugar gliders, degus, and prairie dogs are increasing (Grier, 2006; Lennox, 2007; Jekl *et al.*, 2011; Moutou & Pastoret, 2010). Trade in captive-bred desert hamsters (*Phodopus* spp.) in the United States has increased at an alarming rate, with approximately a million individuals imported between 2004-2013, and with no record of their importation before 1998 (Romagosa, 2015). Given the penchant for small mammals to escape their enclosures (Long, 2003), the sudden appearance in trade of such a quantity of animals is of concern.

When it comes to releases of individuals into the wild, the pet trade is cited as the primary pathway for vertebrate introductions (Kraus, 2003). Some of these introductions are accidental, and some are intentional. Any owner of these animals can attest that some of these groups are particularly prone to escaping (Owre, 1973; Pyšek *et al.*, 2010). Unfortunately, these animals can also be intentionally released by irresponsible or naïve owners. At a somewhat larger scale, there can be releases by breeders or dealers. For example, James (1997) reported that in addition to accidental escapes, dealers of exotic pet birds often release large numbers of recently imported birds to avoid quarantine restrictions, and aviculturists purposely release other unwanted birds. Unauthorized release of pets and surplus commercial stocks are not limited to birds; the release of non-native fish, mammals, amphibians, and reptiles also occurs with increasing regularity around the world (Kraus, 2009; Krysko *et al.*, 2011; Witmer & Hall, 2011; Fuller, see Chapter 5). As an additional source of introductions, in Brazil it was noted that the authorities will often release birds and reptiles after they have seized illegal animals (Destro *et al.*, 2012). Issues of genetic introgression, where non-native species interbreed with native species, abound with animals released through the pet trade.

To further complicate introductions, some species are released intentionally to create a wild breeding stock, to then funnel back into the pet trade (Kraus *et al.*, 2012). Other non-native populations, perhaps not intentionally released for re-sale, are also harvested for the pet trade (Krysko *et al.*, 2004). Once an established population is well known among the public, people may also release their unwanted animals into that population. For example, people have been caught illegally releasing their unwanted pet iguanas in a well-established population on Key Biscayne, Florida (Krysko *et al.*, 2010). Finally, additional populations can also be seeded across a region, creating challenging management scenarios, as seen in Florida and Hawaii with chameleons (Kraus *et al.*, 2012; Rochford *et al.*, 2013).

6.2.2 Competition

Birds are also kept for singing competitions in parts of Europe, Latin America, and Asia (Nash, 1993; Anderson, 2005; Hanks, 2005; Birkhead & Van Balen, 2008; Gama & Sassi, 2008; Alves *et al.*, 2010; Su *et al.*, 2014). While song has always been a large component of the appeal of keeping birds, the species selected for this purpose can differ from those species simply favored for song. In each region, specific bird species are prized for their singing abilities for competition: Europe, Chaffinch (*Fringilla coelebs*; Birkhead and van Balen, 2008); South America, seedeaters (*Oryzoborus angolensis and Oryzoborus crassirostris;* Hanks, 2005); Asia, the Zebra Dove (*Geopelia striata*), Canary (*Serinus canaria*), Black-throated Laughingthrush (*Garrulax chinensis*), Chinese Hwamei (*Garrulax canorus*), Orange-headed Thrush (*Zoothera citrina),*White-rumped Shama (*Copsychus malabaricus),* and Oriental Magpie-robin (*Copsychus saularis)* and various white-eye species *(Zosterops* spp.*)* (Shepherd *et al.*, 2004; Jepson & Ladle,

2009; Shepherd, 2011; Jepson *et al.*, 2011). An additional bird-keeping subculture includes birds kept for fighting competitions. For example, the Saffron Finch (*Sicalis flaveola*) is used for these types of competitions in Brazil (Gama & Sassi, 2008).

Many of the species described above have established wild populations. Exemplary singers can fetch high prices, and the release of inferior individuals has been documented (Nash, 1993; Li *et al.*, 2010). Additionally, there is evidence that interbreeding between subspecies and/or races of birds has occurred at a regional scale. Examples of this admixture include the Oriental Magpie-robins in Singapore (Sheldon *et al.*, 2009) and the Chinese and Taiwan Hwamei (*Garrulax canorus* and *G. taewanus*) (Li *et al.*, 2010). As people emigrate from their respective countries, so do their pastimes. The Guyanese pastime of bird racing has been documented in New York (Sell, 2009), and finches used for fighting contests by Brazilian immigrants were seized by officials in Connecticut and Massachusetts in 2009–2010 (Peters, 2010). At this point in time, the species associated with these pastimes have not been documented in the wild at these locations.

6.2.3 Religious Purposes

The release of animals for religious purposes, known as merit release, is prevalent in many parts of Asia, but has also been documented in North America and Australia (Liu *et al.*, 2012). The modern-day ritual of merit release has changed considerably from traditional practices, and quantity of animals necessary to supply the demand is enormous (Shiu & Stokes, 2008; Agoramoorthy & Hsu, 2005). Birds tend to be the most common taxonomic group used for this practice (Severinghaus & Chi, 1999), but individuals from all taxonomic groups, from invertebrates to mammals, are captured from the wild solely for this purpose (Liu *et al.*, 2012). An estimated annual sale of 680,000–1,050,000 birds for release in Hong Kong alone exemplifies the potential scale of the practice across the region (Chan, 2006). The birds selected for this trade tend to be small, inexpensive, and typically lack the attractive plumage and song that are in demand for pets or singing contests (Severinghaus & Chi, 1999; Edmunds, 2011). In much of Asia, birds sold for this purpose are typically wild-caught native species, though as wild populations decline, more non-native birds are being imported (Nash, 1993). Many of these imported birds have been documented with avian flu (H5N1), and therefore represent a risk of disease transmission to other birds as well as humans (Chan, 2006). Turtles and amphibians are also common merit release animals, but these individuals tend to be captive-bred or farmed non-native species, such as the slider turtle (*Trachemys scripta*) and bullfrog (*Lithobates catesbeianus*) (Severinghaus & Chi, 1999; Liu *et al.*, 2013).

Several species of birds have established populations through merit releases in Hong Kong, Singapore, and Taiwan (Gilbert *et al.*, 2012). As in the case for birds kept for song, interbreeding has resulted in genetic introgression among native and non-native bulbuls in Taiwan (Gilbert *et al.*, 2012). Merit releases of bullfrogs and sliders, both among the 100 worst invaders in the world (Lowe *et al.*, 2000), add to the numerous pathways by which these species have established wild populations (Liu *et al.*, 2012).

6.2.4 Private Zoos/Tourist Attractions

All across the world, there have been zoos or menageries, travelling circuses, and tourist attractions that have displayed animals. These types of collections have been in existence since Ancient Egyptian times, and persisted through Ancient Greece and Rome until today (Hoage *et al.*, 1996). Animals from all taxonomic groups were found in these collections, and became the source of various established populations (Hughes, 2003). This sector is not as important a source for vertebrate introductions, but introductions have occurred. Hulme *et al.* (2008) reports the escape of 48 bird and 20 mammal species from zoos in Europe, and Kraus (2009) reports the release of 17 reptile species from zoos worldwide. Some of these introductions are notable, such as the escape and establishment of Sacred Ibis (*Threskiornis aethiopicus*) populations throughout Europe (Clergeau & Yesou, 2006). Releases from private menageries have also occurred; among the oddest is the establishment of the hippopotamus in Colombia (*Hippopotamus amphibious)* (Valderrama-Vâsquez, 2012).

In the United States, after the Federal Highway Act of 1921 encouraged the building of national highways (Drumm, 2009), roadside animal attractions began to appear across the landscape (Mays, 2009). Animals were released to populate some of these attractions, and then many of the attractions closed after they were circumvented by the routes taken by larger interstates, and the style of American travel and tourism began to change (Drumm, 2009; Mays, 2009). In some cases, their closure led to the additional release of animals once housed there. Most primate introductions in the state of Florida are reported to be due to previous tourist attractions (Layne, 1997).

6.3 Exploitative Trade

This section describes the trade in vertebrates specifically for their exploitation. These pathways are ones where the end destination of the animals traded is for direct consumption, such as for food or traditional medicine, or for scientific research.

6.3.1 Food - Wild Sources

Consumption of animals sourced from the wild has long existed, traditionally as a protein source for subsistence hunters (Wilkie & Carpenter, 1999). Often these animals will also fulfill a medicinal as well as a nutritional role (Secretariat of the Convention on Biological Diversity, 2011). The increase in demand for wild meat in urban and international markets drives the movement of these animals from the wild into different regions (Lindsey *et al.*, 2013). Depending on cultural preferences, wild meat is supplied in markets either alive or in parts (MacDonald *et al.*, 2011). In some regions, live animals can be sold for twice the amount commanded for carcasses (Mendelson *et al.*, 2003; Hennessey & Rogers, 2008).

Amphibians and reptiles are important components of this trade. Millions of amphibians are harvested from the wild to fulfill the global demand for frog legs (Warkentin *et al.*, 2009). Commercial frog farming is performing an increasing role in meeting this demand (see Commercial Wildlife Farming), but there is still a large trade in wild-caught specimens (Altherr *et al.*, 2011). Currently, the United States and European Union are the leading importers; Indonesia and China are the primary exporters, although there is also a large domestic market within these countries (Warkentin *et al.*, 2009).

Almost all species of turtles have been used for food (Klemens & Thorbjarnarson, 1995), and the oldest records of the movement of animals for food may be for this group. By the 1600s, live giant tortoises served as a fresh food source for sailors travelling in the Indian and Pacific Oceans. The tortoises transported well, and were able to survive long periods without food or water (Bowman, 1966). These same species were also introduced to several islands (Kraus, 2009).

Currently, countries in Asia are the leading consumers of turtles in the food trade. As the trade in wild-caught turtles for this purpose began to deplete domestic supply, importation of turtles from other regions increased (Ades *et al.*, 2000; Lee *et al.*, 2004). Most turtles in the international trade are intended for Asian food markets. This demand has also encouraged the large-scale farming of turtles in several countries (see section on Commercial Wildlife Farming). A staggering 13 million turtles, both wild-caught and farmed, were exported annually (2002–2006) from the United States, primarily to Asian countries (Romagosa, 2015). The consumption of snakes in China has increased over time, and the corresponding increase in imports suggests that, like turtles, the domestic harvest may be exhausted (Zhou & Jiang, 2004).

After pets, more bird species are used for food than any other purpose (Butchart, 2008). These species are primarily from the Orders Galliformes, Anseriformes and Columbiformes, but there is also a market for small birds. Passerines are considered to be delicacies in some Mediterranean countries (Italy, Cyprus, and Lebanon), and are traded—mostly illegally—for this purpose (BIO Intelligence Service, 2011). Additional groups consumed are Psittacids in South America (Duplaix, 2001), owls in India (Ahmed, 2010), and herons and egrets in China (Lee *et al.*, 2004).

Among mammals, ungulates and rodents make up the largest proportion of biomass collected for food, but primates and carnivores are also common (Fa *et al.*, 2002). Additional groups that dominate the food markets in China include two mustelid species (*Melogale moschata*, *Arctonyx collaris*) and the masked palm civet (*Paguma larvata*) (Lee *et al.*, 2004).

The opportunity for cross-species and zoonotic disease transmission through the consumption of wild animals is high (Karesh *et al.*, 2005; Subramanian, 2012). The SARS and avian flu outbreaks were tied to contact with wild animals in China's live food markets (Karesh *et al.*, 2005; Karesh *et al.*, 2007). The consumption of wild animals slowed after each outbreak, but the tradition has once again gained in popularity, which should lead to resumption in trade (Xu *et al.*, 2007).

6.3.2 Traditional Medicine

Throughout human history, people have used wildlife resources to treat various ailments and improve their overall health (Alves & Rosa, 2013). Live animals are collected from the wild and sold for this purpose throughout the world, as their use in traditional medicine is still prevalent. As mentioned in the previous section, the animals used for food also often serve a medicinal use; therefore trade trends in some of these species are similar. Mammals (primates and carnivores) and reptiles (snakes, lizards and turtles) are most prevalent in this trade, and the largest consumer of these animals and their products is China. The trade is susceptible to shifts due to novel medicinal claims, such as the increased trade in Tokay geckos (*Gekko gecko*) into Malaysia after an unfounded claim was made that the species can serve as a cure for HIV/AIDS (Caillabet, 2013). Tokay geckos already have an important role in the medicinal and pet trades throughout the world. Although Tokay geckos have several established populations outside of their native range, they are protected or harvest-restricted throughout parts of their native range in Asia.

No introductions of vertebrate species have been documented for the medicinal trade, but some do exist for invertebrates. The giant African snail (*Achatina fulica*) was introduced to Hawaii for unspecified medicinal purposes, as well as to Reunion Island (Cowie & Robinson, 2003).

6.3.3 Commercial Wildlife Farming

With increasing human populations, and changing dynamics of demand, the need for a stable supply of live wildlife for various purposes has encouraged the farming of non-livestock animals. The commercial farming and/or ranching of animals now exists mainly to source the food, skin/fur, and pet trades. These operations can produce billions of animals a year, but the legitimacy of some of these operations, and their reliance on wild-caught individuals, has been questioned (Shi *et al.*, 2007; Brooks *et al.*, 2010; Lyons & Natusch, 2011). Wildlife farms have been documented for a variety of animals, including but not limited to: frogs, turtles, crocodilians, iguanas, pythons, rodents, ungulates, and porcupines (Wildlife Conservation Society, 2008).

As an early example, the importance of the frog harvesting industry in the state of California in the late 1800s, and subsequent declines of the native frog (*Rana draytonii*), led to the creation of frog farms and the introduction of the bullfrog to the state (Jennings & Hayes, 1985). With the realization that the demand for frog meat outweighed the supply in wild populations, commercial frog farms popped up all over the world, particularly in Asia and South America. Increased demand for turtles for food and medicinal uses has also led to large-scale farming efforts in various countries such as the United States and parts of Asia. A recent publication estimated that there are over 1000 turtle farms in China with a combined value of more than one billion US dollars (Shi *et al.*, 2007).

Wildlife farming at a large scale almost inevitably results in animals eventually escaping (Mockrin *et al.*, 2005). Populations of ungulates, frogs, and turtles have established in many parts of the world as a result of commercial wildlife farming. Nutria (*Myocastor coypu*), muskrats (*Ondatra zibethicus*), and American mink (*Neovison vison*), escapees from fur farms around the world, are among the best examples of detrimental introductions through this medium (Long, 2003). Evidence also exists for the genetic introgression of turtles from some of these farms in native populations (Shi *et al.*, 2005; Fong & Chen, 2010l). The potential for disease transmission within the farms and to native species is also of concern; the farming of bullfrogs has contributed to the spread of chytrid fungus (Mazzoni *et al.*, 2003).

6.3.4 Bait

In the United States, use of salamanders as live bait for freshwater fishing is common. Salamanders from several genera (*Ambystoma*, *Desmognathus*, *Gyrinophilus*, and *Pseudotriton*) are collected from the wild either directly by fisherman, or collected by the thousands and then sold in bait shops (Picco & Collins, 2008; Bonett *et al.*, 2007). These salamanders are transported and then often released into regions different from where they were collected (Fuller, 2003). The successful establishment of tiger salamander (*Ambystoma tigrinum*) and seal salamander (*Desmognathus monticola*) populations are attributed to this pathway (Mabry & Verrell, 2003; Bonett *et al.*, 2007). A few releases of other amphibians through the bait trade have been documented (*Hyla wrightorum, Lithobates blairi*), but these species are not known to have established wild populations (Kraus, 2009). It seems likely that many other amphibian species have been moved through this pathway, not just in the United States, but documentation is difficult to locate. The bait trade has been implicated in the spread of ranavirus and chytrid fungus (Picco & Collins, 2008), and is also an important pathway for earthworms, fish and crayfish (Keller *et al.*, 2007; Fuller, see Chapter 5).

6.3.5 Scientific Research/Education

The use of animals for dissection and research dates back to Aristotle's times (384–322 BC) (Nutton, 1995). Since that time, the transport of live animals for this purpose has expanded in both scope and scale. The primary tetrapod groups used for and introduced through scientific research and education purposes are amphibians and mammals.

Amphibians were commonly used for scientific research in the 1880s, and a trade developed in several species specifically for laboratory research and as educational dissection tools (Holmes, 1993). In the late 1930s, a diagnostic pregnancy testing tool was developed using African clawed-frogs (*Xenopus laevis)*, which led to a boom in their trade. Even after the development of more modern methods for pregnancy testing, the trade in clawed-frogs and related species still continues, as they are widely used in scientific research as well as the pet trade. *X. laevis* is among the most widely distributed captive amphibian species in the world, and many populations have become established in the wild due to intentional and unintentional introductions by research laboratories and pet owners/suppliers (Measey *et al.*, 2012).

Rodents and lagomorphs are the most common mammalian groups used for scientific research and education purposes. Because many of these species are so ubiquitous, having been introduced all over the world, it is difficult to pinpoint the specific pathway for all wild populations. Non-human primates are another commonly transported group for scientific research purposes. The primate trade for the biomedical and pharmaceutical research markets peaked in the late 1960s and early 1970s (Nijman *et al.*, 2011); several hundred thousand primates were traded for this purpose. The United States was then by far the largest importer of all primates (LeCornu & Rowan, 1978), and remains so today.

Some established populations of introduced primates originate from scientific research animals. For example, some colonies of Rhesus monkeys (United States-Florida, Caribbean, and South America), and chimpanzees (United States-Georgia) were introduced with the specific purpose of establishing breeding colonies from which individuals could be collected for scientific research (Layne, 1997; Long, 2003). Interestingly, populations of *Chlorocebus* monkeys that were established in Barbados and Nevis/St. Kitts in the 1700s are now the primary source for these species in the scientific research trade (Kavanagh, 1984).

6.4 Trade for Direct Intentional Introduction

This section describes the reasons why live vertebrates were traded to intentionally introduce them to the wild. This process has occurred for thousands of years, but experienced an increase during and after the colonial period. In the 1800s, various Acclimatization Societies were formed across several continents, whose main interest was to introduce non-native plants and animals. These groups did so for various purposes: to improve the aesthetic quality of the environment, to provide animals for sport, and to help control pests. They did so for at least a hundred years until they lost favor among the public and the scientific community (Lever, 1992). Their lofty goals were not abandoned, however. Many individuals and government agencies took matters into their own hands and continued the work started by these societies, albeit in a more rational manner (Lewin, 1971). This section has the most overlap with the previous sections, but a few subtleties deem it worthy of its own discussion. There are many published descriptions about these introductions, as well as a brief treatise of Acclimatization Societies and bird introductions in a previous chapter (Cassey *et al.*, see Chapter 2), therefore I will only discuss them briefly.

6.4.1 Aesthetics

Many birds and, to a lesser extent, mammals, have been introduced to improve the aesthetics of the landscape by Acclimatization Societies as well as other individuals seeking the same goal (Long, 1981; 2003). One of the many bird species introduced was the Mute Swan (*Cygnus olor*) (Phillips, 1928), which is now considered to be a pest in the northeastern United States (Tatu *et al.*, 2007). The majority of the birds released for their aesthetic value around the world include species from the families Psittacidae, Fringillidae, Estrildidae, Anseridae, Anatidae and Phasiandae (Long, 1981). The release of animals for aesthetic reasons still occurs today, and species continue to establish self-sustained breeding populations.

6.4.2 Sport

Game birds and mammals are the primary groups that have been transported and introduced for this purpose. While some species were introduced with the intention of keeping them contained within the boundaries of a preserve, others were fully liberated with the intention of creating wild, free-living populations as a method of game enhancement.

Game birds have been introduced worldwide (Long, 1981), but the documentation of the importation and introduction of game birds is reasonably well documented for the United States. The introduction of game birds in the United States began at least by the mid-1700s (Phillips, 1928; Bump, 1963). Importation of game birds and introductions continued thereafter, some by private individuals and others by various levels of government. More than 1.6 million game birds were imported into the United

States between 1901–1942, most of which were intended for release to the wild (Banks, 1976). The game bird releases by the United States government were formalized as the Foreign Game Importation Program of the United States Fish and Wildlife Service, created in the mid-1940s, and discontinued in 1970. Toward the end of this period (1966–1968) this program released 94,486 game birds (Bohl & Bump, 1970).

The introduction of small and large game mammals has also occurred worldwide, most notably in New Zealand and the United States. Many of these animals have been introduced to game ranches, where the maintenance of wild animals occurs in defined areas delineated by fences. This pathway is the most important in relation to successful mammal introductions (Kraus, 2003). Presnall (1958) reports at least 10 species of ungulates that were introduced to the wild in the United States, and at least 5 of these species still have established populations (Long, 2003). The most successful ungulates introduced worldwide for sport include the wild pig (*Sus scrofa*), axis deer (*Axis axis*), fallow deer (*Dama dama*), red deer (*Cervus elaphus*) and Sika deer (*Cervus nippon*) (Long, 2003). Animals are still transported to stock game ranches, and additional species and populations will undoubtedly establish in the future.

6.4.3 Biological Control

In the late 1800s, several vertebrate species were introduced to serve as biological control agents. These introductions were later identified as ill-conceived because of the impacts the control agents would have on non-target species (Hoddle, 2004). Terrestrial vertebrates that were introduced for biological control of other vertebrates or arthropod pests include: several mustelids, such as the mongoose (*Herpestes javanicus*) and stoats (*Mustela erminea*); the cane toad (*Rhinella marina*); and several species of mynas (*Acridotheres* spp.) (Hoddle, 2004). The trade and introduction of vertebrates specifically for biological control rarely happens today.

6.5 Conclusions

Global trade has increased substantially since the 1950s (Nordstrom & Vaughn, 1999) and, with it, the risk of additional introductions has increased (Levine & D'Antonio, 2003). This phenomenon has created an increase in all biological commodities, including the live animal trade. Characterizing the contribution of trade to biological invasions begins with identifying the various pathways by which animals are transported to new locations. Efforts to understand how species are entrained into the various pathways will lead to better understanding of biological invasions as a whole (Su *et al.*, 2014).

In a nutshell

- The live animal trade is an important commodity market that drives the annual transport of millions of animals worldwide.
- This trade in live animals is the primary pathway for species introductions.
- To help best describe the contribution of the live animal trade to biological invasions in this chapter, it is separated into several pathway sectors, focusing on tetrapods.
- Live animals are traded to be kept in captivity as pets, for competitions, for religious purposes, and for zoos and tourist attractions.
- Pets and merit releases are the most important pathways for the introduction of most vertebrate groups.
- Live animals are also traded specifically for exploitation for food, fur and skins, traditional medicine, bait, and scientific research purposes.
- The demand for animals for exploitation exceeds supply, and commercial wildlife farming has been implemented to meet this demand.
- Animals are also transported to new regions for intentional introductions for aesthetic and sport (game) enhancement, and for biocontrol.
- The introduction of game animals is the most important pathway for mammal introductions through trade.
- Introductions through the live animal trade can be expected to continue with the expansion of global trade in all commodities.

6.6 Bibliography

Ades G., Banks C. B., Buhlmann, K. A., *et al.* (2000). Turtle trade in Northeast Asia: regional summary (China, Hong Kong and Taiwan). In P. P. van Dijk, B. L. Stuart, A. G. J. Rhodin (Eds.), Asian turtle trade: Proceedings of a workshop on conservation and trade of freshwater turtles and tortoises in Asia (pp.52–54). Lunenburg, MA: Chelonian Research Foundation.

Agoramoorthy, G., Hsu, M. J. (2005). Religious freeing of wildlife promotes alien species invasion. BioScience, 55 (1), 6-6.

Ahmed, A. (2010). Imperilled custodians of the night: A study of the illegal trade, trapping, and utilization of owls in India. India: TRAFFIC India/WWF-India.

Altherr, S., Goyenechea, A., Schubert, D. (2011). Canapés to extinction—the international trade in frogs' legs and its ecological impact. Munich, Germany & Washington, D.C.: Defenders of Wildlife and Animal Welfare Institute.

Alves, R. R. N, Rosa, I. L. (Eds.). (2013). Animals in traditional folk medicine: Implications for conservation. Berlin: Springer-Verlag.

Alves, R. R. N., Nogueira, E. E., Araujo, H. F., Brooks, S. E. (2010). Bird-keeping in the Caatinga, NE Brazil. Human Ecology, 38 (1), 147-156.

American Veterinary Medical Association. (2007). US pet ownership and demographics sourcebook. Schaumburg, Ill: American Veterinary Medical Association.

Anderson, P. K. (2003). A bird in the house: An anthropological perspective on companion parrots. Society and Animals, 11 (4), 393-418.

Anderson, W. W. (2005). Beyond the cockfight: masculinity and the Thai dove-cooing contest. Manusya Journal of Humanities, 9, 80-91.
Auliya, M. (2003). Hot trade in cool creatures: a review of the live reptile trade in the European Union in the 1990s with a focus on Germany. Brussels: TRAFFIC Europe.
Banks, R. C. (1976). Wildlife importation into the United States, 1900-1972. Washington, D.C.: US Department of the Interior, Fish and Wildlife Service.
Beissinger, S. R. (2001). Trade of live wild birds: potentials, principles, and practices of sustainable use. In J. D. Reynolds, G. M. Mace, K. H. Redford, J. G. Robinson (Eds.), Conservation of exploited species (pp.182–202). Cambridge, United Kingdom: Cambridge University Press.
Birkhead, T. R., Van Balen, S. (2008). Bird-keeping and the development of ornithological science. Archives of Natural History, 35, 281-305.
BIO Intelligence Service. (2011). Stocktaking of the main problems and review of national enforcement mechanisms for tackling illegal killing, trapping and trade of birds in the EU. Final report prepared for European Commission, DG Environment.
Bohl, W. H., Bump, G. (1970). Summary of foreign game bird liberations 1960 to 1968 and propagation 1966 to 1968. Washington, D.C.: US Department of the Interior, Bureau of Sport Fisheries and Wildlife.
Bonett, R. M., Kozak, K. H., Vieites, D. R., *et al.* (2007). The importance of comparative phylogeography in diagnosing introduced species: a lesson from the seal salamander, *Desmognathus monticola*. BMC Ecology, 7, 7, doi:10.1186/1472-6785-7-7.
Bowman, R. I. (Ed.). (1966). The Galápagos: Proceedings of the symposia of the Galápagos International Scientific Project. Berkeley & Los Angeles, California: University of California Press.
Brooks, E. G., Roberton, S. I., Bell, D. J. (2010). The conservation impact of commercial wildlife farming of porcupines in Vietnam. Biological Conservation, 143 (11), 2808-2814.
Bump, G. (1963). History and analysis of tetraonid introductions into North America. Journal of Wildlife Management, 27 (4), 855-867.
Butchart, S. H. (2008). Red List Indices to measure the sustainability of species use and impacts of invasive alien species. Bird Conservation International, 18 (1), S245-S262.
Caillabet, O. S. (2013). The trade in tokay geckos (*Gekko gecko*) in South-East Asia: with a case study on novel medicinal claims in Peninsular Malaysia. Selangor, Malaysia: TRAFFIC.
Cantú, J. C., Sánchez, M. E., Groselet, M., Silva, J. (2007). The illegal parrot trade in Mexico: a comprehensive assessment. Washington, DC: Defenders of Wildlife.
Chan, S. (2006). Religious release of birds in Hong Kong. Master of Philosophy Thesis. University of Hong Kong, Hong Kong.
Clergeau, P., Yesou, P. (2006). Behavioural flexibility and numerous potential sources of introduction for the sacred ibis: causes of concern in western Europe? Biological Invasions, 8 (6), 1381-1388.
Cowie, R. H., Robinson, D. G. (2003). Pathways of introduction of nonindigenous land and freshwater snails and slugs. In G. Ruiz, J. T. Carlton (Eds.), Invasive Species: Vectors and Management Strategies (pp.93-122). Washington, DC: Island Press.
Destro, G. F. G., Pimentel, T. L., Sabaini, R. M., Borges, R. C., Barreto R. M. (2012). Efforts to combat wild animals trafficking in Brazil. In G. A. Lameed (Ed.), Biodiversity enrichment in a diverse world (pp.421-436). Brazil: InTech.
Drews, C. (2001). Wild animals and other pets kept in Costa Rican households: Incidence, species, and numbers. Society and Animals, 9 (2), 107-126.
Drumm, P. (2009). Applied animal psychology at an American roadside attraction: animal behavior enterprises and the IQ zoo of hot springs, Arkansas. The American Journal of Psychology, 122 (4), 537-545.
Duplaix, N. (2001). Evaluation of the animal and plant trade in the Guianas: Preliminary findings. Paramaribo, Suriname: WWF-Guianas.

Edmunds, K. (2011). Avian influenza, the wild bird trade and local livelihoods: an interdisciplinary and mixed-methods approach. Doctoral dissertation, University of East Anglia.
Fa, J. E., Peres, C. A., Meeuwig, J. (2002). Bushmeat exploitation in tropical forests: an intercontinental comparison. Conservation Biology, 16 (1), 232-237.
FAO (2011). International trade in wild birds, and related bird movements, in Latin America and the Caribbean. Rome, Italy: Animal Production and Health Paper, FAO.
Fernandez, C., and R. Luxmoore. (1997). The value of the wildlife trade. In T. M. Swanson & R. A. Luxmoore (Eds.), Industrial reliance on biodiversity (pp.13-55). Cambridge, United Kingdom: World Conservation Press, WCMC.
Fong, J. J., Chen, T. H. (2010). DNA evidence for the hybridization of wild turtles in Taiwan: possible genetic pollution from trade animals. Conservation Genetics, 11 (5), 2061-2066.
Fuller, P.L. (2003). Freshwater aquatic vertebrate introductions in the United States: patterns and pathways. In G. M. Ruiz, J. T. Carlton, (Eds.), Invasive species: vectors and management strategies (pp.123-151). Washington, D.C.: Island Press.
Gama, T. F., Sassi, R. (2008). Aspectos do comércio ilegal de pássaros silvestres na cidade de João Pessoa, Paraíba, Brasil. Gaia Scientia, 2, 21-20.
Gilbert, M., Sokha, C., Joyner, P. H., Thomson, R. L., Poole, C. (2012). Characterizing the trade of wild birds for merit release in Phnom Penh, Cambodia and associated risks to health and ecology. Biological Conservation, 153, 10-16.
Goss, J. R., Cumming, G. S. (2013). Networks of wildlife translocations in developing countries: an emerging conservation issue? Frontiers in Ecology and the Environment, 11 (5), 243-250.
Grier, K. C. (2006). Pets in America: A history. *Chapel Hill: University of North Carolina University Press.*
Hanks, C. K. (2005). Spatial patterns in Guyana's wild bird trade. Master of Science Thesis, The University of Texas at Austin.
Hennessey, A. B., Rogers, J. (2008). A study of the bushmeat trade in Ouesso, Republic of Congo. Conservation and Society, 6 (2), 179–184.
Hoage, R. J., Roskell, A., Mansour, J. (1996). Menageries and zoos to 1900. In R. J. Hoage, W. A. Deiss (Eds.), New worlds, new animals: From menagerie to zoological park in the Nineteenth Century (pp.8-18). Baltimore, Maryland: John Hopkins University Press.
Hoddle, M. S. (2004). Restoring balance: using exotic species to control invasive exotic species. Conservation Biology, 18 (1), 38-49.
Holmes, F. L. (1993). The old martyr of science: The frog in experimental physiology. Journal of the History of Biology, 26 (2), 311-328.
Hoover, C. (1998). The U.S. role in the international live reptile trade: Amazon tree boas to Zululand dwarf chameleons. Washington, D.C.: Traffic North America.
Hughes, J. D. (2003). Europe as consumer of exotic biodiversity: Greek and Roman times. Landscape Research, 28 (1), 21-31.
Hulme, P. E., Bacher, S., Kenis, M., *et al.* (2008). Grasping at the routes of biological invasions: a framework for integrating pathways into policy. Journal of Applied Ecology, 45 (2), 403-414.
James, F. C. (1997). Nonindigenous birds. In D. Simberloff, D. Schmitz, T. Brown (Eds.), Strangers in Paradise: Impact and Management of Nonindigenous Species in Florida (pp.139-156). Washington, D.C.: Island Press.
Jekl, V., Hauptman, K., Knotek, Z. (2011). Diseases in pet degus: A retrospective study in 300 animals. Journal of Small Animal Practice, 52 (2), 107-112.
Jennings, M. R., Hayes, M. P. (1985). Pre-1900 overharvest of California red-legged frogs (*Rana aurora draytonii*): The inducement for bullfrog (*Rana catesbeiana*) introduction. Herpetologica, 94-103.
Jepson, P., Ladle, R. J. (2005). Bird-keeping in Indonesia: Conservation impacts and the potential for substitution-based conservation responses. Oryx, 39 (4), 442-448.

Jepson, P., Ladle, R. J. (2009). Governing bird-keeping in Java and Bali: evidence from a household survey. Oryx, 43 (3), 364-374.

Jepson, P., Ladle, R. J., Sujatnika. (2011). Assessing market-based conservation governance approaches: a socio-economic profile of Indonesian markets for wild birds. Oryx, 45 (4), 482-491.

Karesh, W. B., Cook, R. A., Bennett, E. L., Newcomb, J. (2005). Wildlife trade and global disease emergence. Emerging Infectious Diseases, 11 (7), 1000-1002.

Karesh, W. B., Cook, R. A., Gilbert, M., Newcomb, J. (2007). Implications of wildlife trade on the movement of avian influenza and other infectious diseases. Journal of Wildlife Diseases, 43 (3), 55-59.

Kavanagh, M. (1984). A review of the international primate trade. In D. Mack, R. A. Mittermeier (Eds.), The international primate trade: Vol.1, Legislation, trade and captive breeding (pp.49-90). Washington, D.C.: Traffic USA.

Keller, R. P., Cox, A. N., Van Loon, C., *et al.* (2007). From bait shops to the forest floor: earthworm use and disposal by anglers. American Midland Naturalist, 158 (2), 321-328.

Klemens, M. W., Thorbjarnarson, J. B. (1995). Reptiles as a food resource. Biodiversity and Conservation, 4 (3), 281-298.

Kraus, F. (2003). Invasion pathways for terrestrial vertebrates. In G. M. Ruiz and J. T. Carlton (Eds.), Invasive species: vectors and management strategies (pp.68-92). Washington, D.C.: Island Press.

Kraus, F. (2009). Alien Reptiles and Amphibians: a scientific compendium and analysis. United Kingdom: Springer-Verlag.

Kraus, F., Medeiros, A., Preston, D., *et al.* (2012). Diet and conservation implications of an invasive chameleon, *Chamaeleo jacksonii* (Squamata: Chamaeleonidae) in Hawaii. Biological Invasions, 14 (3), 579-593.

Krysko, K. L., Enge, K. M., King, F. W. (2004). The veiled chameleon, *Chamaeleo calyptratus*: A new exotic lizard species in Florida. Florida Scientist, 67, 249-253.

Krysko, K. L., Enge, K. M., Donlan, E. M., *et al.* (2010). The non-marine herpetofauna of Key Biscayne, Florida. Herpetological Conservation and Biology, 5 (1), 132-142.

Krysko K.L., Burgess J.P., Rochford M.R., *et al.* (2011). Verified non-indigenous amphibians and reptiles in Florida from 1863 through 2010: Outlining the invasion process and identifying invasion pathways and stages. Zootaxa, 3028, 1-64.

Law, S. C. (1923). Pet birds of Bengal (Volume 1). Calcutta, India: Thacker, Spink & Co.

Layne, J. (1997). Nonindigenous mammals. In D. Simberloff, D. Schmitz, T. Brown (Eds.), Strangers in Paradise: Impact and Management of Nonindigenous Species in Florida (pp.157-186). Washington, D.C.: Island Press.

LeCornu, A., Rowan, A. N. (1978). Trends in the use of nonhuman primates in biomedical programmes. Laboratory animals, 12 (4), 235-242.

Lee K. S., Lau M. W. N, & Chan, B. P. L. (2004). Wild animal trade monitoring at selected markets in Guangzhou and Shenzhen, South China, 2000–2003. Kadoorie Farm and Botanic Garden Technical Report no. 2, Hong Kong SAR.

Lennox, A. M. (2007). Emergency and critical care procedures in sugar gliders (*Petaurus breviceps*), African hedgehogs (*Atelerix albiventris*) and prairie dogs (*Cynomys* spp). Veterinary Clinics of North America: Exotic Animal Practice, 10 (2), 533-555.

Lever, C. (1992). They dined on eland: the story of acclimatization societies. London: Quiller Press.

Levine, J. M., D'Antonio, C. M. (2003). Forecasting biological invasions with increasing international trade. Conservation Biology, 17, 322-326.

Lewin, V. (1971). Exotic game birds of the Puu Waawaa Ranch, Hawaii. Journal of Wildlife Management, 35 (1), 141-155.

Li, S. H., Yeung, C. K. L., Han, L., Le, M. H., Wang, C. X., Ding, P., Yao, C. T. (2010). Genetic introgression between an introduced babbler, the Chinese hwamei *Leucodioptron c. canorum*, and the endemic Taiwan hwamei *L. taewanus*: a multiple marker systems analysis. Journal of Avian Biology, 41 (1), 64-73.

Lindsey, P. A., Balme, G., Becker, M., *et al.* (2013). The bushmeat trade in African savannas: Impacts, drivers, and possible solutions. Biological Conservation, 160, 80-96.

Liu, X., McGarrity, M. E., Li, Y. (2012). The influence of traditional Buddhist wildlife release on biological invasions. Conservation Letters, 5 (2), 107-114.

Liu, X., McGarrity, M. E., Bai, C., Ke, Z., Li, Y. (2013). Ecological knowledge reduces religious release of invasive species. Ecosphere, 4 (2), art21.

Long, J. L. 1981. Introduced birds of the world. New York, NY: Universe Books.

Long, J. L. 2003. Introduced mammals of the world. Collingwood, Australia: CSIRO Publishing.

Lowe, S., Browne, M., Boudjelas, S., *et al.* (2000). 100 of the world's worst invasive alien species: a selection from the global invasive species database. Auckland, New Zealand: Invasive Species Specialist Group.

Lyons, J. A., Natusch, D. J. (2011). Wildlife laundering through breeding farms: Illegal harvest, population declines and a means of regulating the trade of green pythons (*Morelia viridis*) from Indonesia. Biological Conservation, 144 (12), 3073-3081.

Mabry, M., Verrell, P. (2003). All are one and one is all: sexual uniformity among widely separated populations of the North American seal salamander, *Desmognathus monticola*. Biological Journal of the Linnean Society, 78 (1), 1-10.

Macdonald, D. W., Johnson, P. J., Albrechtsen, L., *et al.* (2011). Association of body mass with price of bushmeat in Nigeria and Cameroon. Conservation Biology, 25 (6), 1220-1228.

Mack, D., Mittermeier, R. A. (Eds.). (1984). The international primate trade: Vol.1, Legislation, trade and captive breeding. Washington D.C.: Traffic USA.

Mays, D. (2009). Gatorland: Survival of the fittest among Florida's mid-tier tourist attractions. Florida Historical Quarterly, 87 (4), 509-539.

Mazzoni, R., Cunningham, A. A., Daszak, P., *et al.* (2003). Emerging pathogen in wild amphibians and frogs (*Rana catesbeiana*) farmed for international trade. Emerging Infectious Diseases, 9 (8), 995-998.

Measey, G. J., Rödder, D., Green, S. L., *et al.* (2012). Ongoing invasions of the African clawed frog, *Xenopus laevis*: a global review. Biological Invasions, 14 (11), 2255-2270.

Mendelson, S., Cowlishaw, G., Rowcliffe, J. M. (2003). Anatomy of a bushmeat commodity chain in Takoradi, Ghana. Journal of Peasant Studies, 31 (1), 73-100.

Mockrin, M. H., Bennett, E. L., LaBruna, D. T. (2005). Wildlife farming: A viable alternative to hunting in tropical forests? WCS Working Paper No. 23. Bronx, NY: Wildlife Conservation Society.

Moutou, F., Pastoret, P. P. (2010). Why own an exotic pet? Revue Scientifique et Technique (International Office of Epizootics), 29 (2), 359-365.

Nash, S. V. (1993). Sold for a song: the trade in Southeast Asian non-CITES birds. Cambridge, United Kingdom: Traffic International.

Nijman, V., Nekaris, K. A. I., Donati, G., *et al.* (2011). Primate conservation: measuring and mitigating trade in primates. Endangered Species Research, 13, 159-161.

Nordstrom, H., Vaughan, S. (1999). Trade and the environment. Special Studies 4.Geneva: World Trade Organization.

Nutton, V. (1995). Medicine in the Greek world, 800-50 B.C. In L. I. Conrad, M. Neve, V. Nutton, *et al.*, The western medical tradition: 800 B.C.-1800 A.D. (pp.11-38). New York: Cambridge University Press.

Oldys, H. 1907. Cage-bird traffic of the United States. In Yearbook of the United States Department Agriculture, 1906 (pp.165-180). Washington, D.C.: United States Government Printing Office.

Owre, O. T. (1973). A consideration of the exotic avifauna of southeastern Florida. Wilson Bulletin, 85, 491–500.

Peters, S. L. (2010). Authorities crack down on finch-fighting ring. USA Today, accessed 10 May 2014 from http://usatoday30.usatoday.com/life/lifestyle/pets/2010-03-10-finchfights10_ST_N.htm?csp=earthlink-ent.

Phillips, J. C. (1928). Wild birds introduced or transplanted in North America. USDA Technical Bulletin No. 61. Washington, D.C.: US Department of Agriculture.

Picco, A. M., Collins, J. P. (2008). Amphibian commerce as a likely source of pathogen pollution. Conservation Biology, 22 (6), 1582-1589.

Presnall, C. C. (1958). The present status of exotic mammals in the United States. Journal of Wildlife Management, 22, 45-50.

Pyšek, P., Jarošík, V., Hulme, P. E., *et al.* (2010). Disentangling the role of environmental and human pressures on biological invasions across Europe. Proceedings of the National Academy of Sciences, 107 (27), 12157-12162.

Rochford, M. R., Edwards, J. R., Howell, P. L., *et al.* (2013). The Panther Chameleon, *Furcifer pardalis* (Cuvier 1829) (Chamaeleonidae), another introduced chameleon species in Florida. IRCF Reptiles and Amphibians, 20, 205-207.

Roe, D., Mulliken, T., Milledge, S., *et al.* (2002). Making a killing or making a living? Wildlife trade, trade controls and rural livelihoods. Biodiversity and livelihoods issues No. 6. London: IIED

Romagosa, C. M. (2015). Global trade in live vertebrates and the contribution to biological invasions. In R.P. Keller, M. Cadotte, G. Sandiford (Eds.), Invasive Species in a Globalized World (pp.115-146). Chicago, IL: University of Chicago Press.

Secretariat of the Convention on Biological Diversity (2011). Livelihood alternatives for the unsustainable use of bushmeat. Report prepared for the CBD Bushmeat Liaison Group. Technical Series No. 60, Montreal, SCBD.

Sell, S. (2009). Guyanese in Queens listen for a different kind of tweet. The Columbia Journalist, Accessed 10 May 2014 from http://www.columbiajournalist.org/www/15-guyanese-in-queens-listen-for-a-different-kind-of-tweet/story.

Severinghaus, L. L., Chi, L. (1999). Prayer animal release in Taiwan. Biological Conservation, 89 (3), 301-304.

Sheldon, F. H., Lohman, D. J., Lim, H. C., *et al.* (2009). Phylogeography of the magpie-robin species complex (Aves: Turdidae: *Copsychus*) reveals a Philippine species, an interesting isolating barrier and unusual dispersal patterns in the Indian Ocean and Southeast Asia. Journal of Biogeography, 36 (6), 1070-1083.

Shepherd, C. R., Sukumaran, J., Wich, S. A. (2004). Open season: An analysis of the pet trade in Medan, Sumatra 1997-2001. Selangor, Malaysia: TRAFFIC Southeast Asia.

Shepherd, C. R. (2011). Observations on trade in laughingthrushes (*Garrulax* spp.) in North Sumatra, Indonesia. Bird Conservation International, 21 (1), 86-91.

Shi, H., Parham, J.F., Lau, M. *et al.* (2007). Farming endangered turtles to extinction in China. Conservation Biology, 21 (1), 5-6.

Shi, H., Parham, J. F., Simison, W. B., *et al.* (2005). A report on the hybridization between two species of threatened Asian box turtles (Testudines: *Cuora*) in the wild on Hainan Island (China) with comments on the origin of '*serrata*'-like turtles. Amphibia-Reptilia, 26, 377-381.

Shiu, H., Stokes, L. (2008). Buddhist animal release practices: Historic, environmental, public health and economic concerns. Contemporary Buddhism, 9 (2), 181-196.

Smith, K., Behrens, M., Schloegel, L., *et al.* (2009). Reducing the risks of the wildlife trade. Science, 324, 594-595.

Su, S., Cassey, P., Blackburn, T. M. (2014). Patterns of non-randomness in the composition and characteristics of the Taiwanese bird trade. Biological Invasions, 16, 1-13.

Subramanian, M. (2012). Zoonotic disease risk and the bushmeat trade: Assessing awareness among hunters and traders in Sierra Leone. EcoHealth, 9 (4), 471-482.
Tapley, B., Griffiths, R. A., Bride, I. (2011). Dynamics of the trade in reptiles and amphibians within the United Kingdom over a ten-year period. Herpetological Journal, 21, 27-34.
Tatu, K. S., Anderson, J. T., Hindman, L. J., *et al.* (2007). Mute swans' impact on submerged aquatic vegetation in Chesapeake Bay. The Journal of Wildlife Management, 71 (5), 1431-1439.
Valderrama-Vásquez, C. A. (2012). Wild hippos in Colombia. Aliens: The Invasive Species Bulletin, 32, 8-12.
Warkentin, I. G., Bickford, D., Sodhi, N. S., Bradshaw, C. J. (2009). Eating frogs to extinction. Conservation Biology, 23 (4), 1056-1059.
Wildlife Conservation Society. (2008). Commercial wildlife farms in Vietnam: A problem or solution for conservation? Hanoi, Vietnam: Wildlife Conservation Society.
Wilkie, D. S., Carpenter, J. F. (1999). Bushmeat hunting in the Congo Basin: an assessment of impacts and options for mitigation. Biodiversity and Conservation, 8 (7), 927-955.
Witmer, G.W., Hall, P. (2011). Attempting to eradicate invasive Gambian giant pouched rats (*Cricetomys gambianus*) in the United States: lessons learned. In C. R. Veitch, M. N. Clout, D. R. Towns (Eds.), Island invasives: eradication and management (pp.131-134). Gland, Switzerland: IUCN.
Xu, L., Meng, M., Yin, F. (2007). The wild meat trade in southern China. In H. Xu, C. Kirkpatrick (Eds.), The state of wildlife trade in China: Information on the trade in wild animals and plants in China 2007 (pp.12-14). Beijing, China: TRAFFIC East Asia.
Zhang, L., Yin, F. (2014). Wildlife consumption and conservation awareness in China: a long way to go. Biodiversity and Conservation, 23 (9), 371-381.
Zhou, Z., Jiang, Z. (2004). International trade status and crisis for snake species in China. Conservation Biology, 18 (5), 1386-1394.

Part II. Biological Invasions in Aquatic Ecosystems and in Host Parasite Systems

João Canning-Clode, Filipa Paiva

Summary of Part II

Part of the challenge of biological invasions arises from its history, ecology, evolutionary knowledge and the potential effects of non-indigenous species. Globalization and anthropogenic activities have contributed to increasing numbers of invasive species, their diseases, and the evolution of parasites that can affect a whole ecosystem. Parasites and diseases can spread to distant locations when their hosts move or are transported to a new area. Some parasite species are also capable of colonizing new host species in the new location. The second part of this book is mostly dedicated to aquatic ecosystem invasions, but also to the significance of parasites in the context of biological invasions in both aquatic and terrestrial systems.

The opening chapter of this section (**Chapter 7**), solely authored by April Blakeslee, provides the results of an interesting meta-analysis on the effect of invasions of hosts and parasites on genetic diversity in marine systems. Typically, there are two possible scenarios that occur when species are transferred to a new location: i) a significant genetic bottleneck or ii) a significant reduction in parasite diversity in founding populations, i.e. parasite escape. This chapter uses a comprehensive meta-analysis of several studies on marine invertebrate introductions worldwide where one or both signatures were investigated. The main findings of this meta-analysis suggest that haplotype and parasite richness seem to be significantly lower in non-native versus native regions at large scales.

In addition, the crash of the native freshwater crayfish population in Europe due to an accidental introduction of the crayfish plague *Aphanomyces astaci* along with North American crayfishes urged fisheries departments and researchers to adopt a new attitude towards this ecological and economical disaster. The introduction of this parasite in Europe and the repeated introductions of its North American host species are illustrated in **Chapter 8** by Japo Jussila and co-authors as a classic example of a man-made ecological disaster. The arrival of the crayfish plague in Europe has resulted in a massive loss of native crayfishes and a high availability of new host species for the pathogen. Based on the crayfish scenario, authors discuss how consequences associated with the introduction of non-native species and their diseases affect the evolution of host-parasite interactions and alter the entire ecosystem.

Furthermore, in **Chapter 9** Laura Härkönen and Arja Kaitala review the main factors enabling invasion of the blood-feeding ectoparasitic deer ked (*Lipoptena cervi*) and the importance of host dynamics and life history attributes contributing to the differences in invasiveness among different host populations and geographical areas. This well known ectoparasite of the European Moose (*Alces alces*) originally had a wide distribution across the Old World, but has recently greatly expanded towards higher latitudes. The northward range expansion in Fennoscandia has resulted in detrimental effects in the moose's health. This chapter examines the differences in the invasion capability of the parasite among different host populations and geographical areas in relation to life history variation.

Also in the marine system, the wide distribution of the Pacific oyster *Crassostrea gigas* occurred not only due to deliberate introductions for farming, but also through vessel transport where oysters could attach easily and travel long distances. The species has recently arrived on Scandinavian coasts where it now exists in large numbers and has been intensively investigated to predict its future distribution based on climate change scenarios. **Chapter 10** by Ane Laugen and co-authors outlines the current understanding of *Crassostrea gigas*'s origin and dispersal routes in Scandinavia, predicted future distribution in relation to climate change, and observed alterations to habitat structure and ecosystem function.

Finally, the introduction of non-native seaweeds in different marine ecosystems has been increasing in past years mostly due to shipping traffic, and several of these non-native seaweeds are responsible for significant ecological and economic impacts worldwide. Closing this book section, **Chapter 11** by Antonella Petrocelli and Ester Cecere review the distribution and impact of three of the most spread invasive seaweeds all around the world: the chlorophycean *Codium fragile*, the rhodophycean *Gracilaria vermiculophylla*, and the phaeophycean *Undaria pinnatifida*. The authors further discuss the most significant vectors of introduction of non-native seaweeds, management actions, and legislation.

April Blakeslee

7 Parasites and Genetics in Marine Invertebrate Introductions: Signatures of Diversity Declines across Systems

7.1 Introduction

Over the last several decades, non-indigenous species (NIS) have become a global concern due to intentional and unintentional translocation of species around the world across large geographic distances and natural migratory barriers; this in turn has resulted in considerable evolutionary, ecological, and environmental impacts on native communities and habitats (Ruiz *et al.*, 2000; Simberloff *et al.*, 2013). In fact, species invasions have been ranked second only to habitat loss as a major force of ecological disturbance worldwide (Crowl *et al.*, 2008; Vitousek *et al.*, 1997). Across the biosphere, numerous species have become successful in non-native regions, including notorious aquatic examples like the zebra mussel (*Dreissena polymorpha*) in the Great Lakes (Carlton, 2008); the Eurasian reed (*Phragmites australis*) in freshwater and estuarine systems of North America (Saltonstall, 2002); the Asian carp (several cyprinid fish species) in river systems of the South and Midwest United States (Kolar *et al.*, 2005); the lionfish (*Pterois volitans*) in Western Atlantic and Caribbean waters (Albins & Hixon, 2013); and the European green crab (*Carcinus maenas*) in numerous coastal populations across North America, Asia, Australia, and South America (Carlton & Cohen, 2003).

Some alien species, like the ones listed above, may demonstrate massive population explosions and strong negative interactions with native organisms in their new environments even without prior co-evolutionary history with their novel communities and habitats (Simberloff *et al.*, 2013). While there are numerous hypotheses as to why this seemingly counter-intuitive pattern has been shown to occur over and over again across ecosystems, two prominent ecological and evolutionary explanations for a non-native species' success in unfamiliar territory include: 1) *a loss of natural enemies* (e.g., predators, competitors, parasites, disease) in non-native versus native populations (e.g., Keane and Crawley, 2002; Torchin *et al.*, 2001, 2002, 2003; Liu & Stiling, 2006; Blakeslee *et al.*, 2013) and 2) *a genetic bottleneck* in non-native versus native populations. While this latter consequence could result in deleterious effects, especially in small founding populations, the invasion process may actually enhance success in some species by selecting for the "hardiest" individuals in the founding population and eliminating the more sensitive ones—similar to the phenomenon observed in bacterial communities that show antibiotic resistance as a result of inadvertent selection for resistant alleles that become dominant in gene pools (Lavergne and Molofsky, 2007; Saltonstall, 2002; Simberloff, 2009). Interestingly, these two commonly observed signatures of biological invasion—enemy release and genetic bottlenecks—can also serve as lines of evidence or tools to better understand invasions or even resolve uncertain invasion histories (e.g., Blakeslee *et al.*, 2008).

7.1.1 Multiple Lines of Evidence in Marine NIS

Regardless of the reasons for NIS success, the reality is that questions surrounding biological invasions are typically the norm, making it challenging to predict their impacts or determine how to manage them. Because NIS may go undetected for years or even decades before they are recognized or pose any environmental, ecological, or economic impacts (Carlton, 1996a), understanding the how, when, and what of an invasion can be difficult, considering little historical, ecological, or evolutionary knowledge may exist, leading to numerous uncertainties and questions. Though not all-inclusive, these questions (Figure 7.1) may include: (1) *Introduction Vector*: the human-mediated mechanism of NIS establishment in new regions (e.g., shipping, agriculture/aquaculture, bait, biocontrol, canals, etc.), which is strongly associated with 'propagule pressure' (the number of individuals entrained within a vector and the number of introduction events; Kolar & Lodge, 2001); (2) *Source*: the region from which the NIS originated and specific source populations; (3) *Timing of Introduction*: when a NIS first became established in a non-native region; (4) *Genetic diversity of founding populations*: NIS may only introduce a subset of their source genetic diversity, which will be strongly influenced by propagule pressure (Roman & Darling, 2007; see Fig. 2-A); (5) *Associated biota*: NIS may carry with them free-living and/or symbiotic hitchhiking organisms (Torchin & Mitchell, 2004; see Figure 7.2-B); (6) *Influence on natives*: NIS may influence native biota and habitats, e.g., competitive and predatory interactions (e.g., Byers, 2009; Rilov, 2009); (7) *Geographic spread*: the ability of NIS to expand ranges beyond original sites of introduction may be aided or impeded by natural dispersal processes and/or multiple introductions; (8) *Cryptogenic species*: species that cannot be demonstrably classified as native or non-native (Carlton, 1996b), a particularly troublesome issue from a management perspective.

As a result of these many uncertainties, multiple lines of evidence may be required. Moreover, these uncertainties may be more pronounced in marine versus terrestrial or freshwater systems since marine biota are inherently more difficult to study and track, and historical information regarding their movements may be non-existent or poor. Therefore, piecing together the evidence necessary to resolve invasion histories in marine systems may require innovative tools (e.g., parasites and genetics—see below). Fortunately, NIS can demonstrate discernible ecological, geographical, and evolutionary signatures, and these can be used as "clues" to resolve uncertainties, such as distinguishing among the native and non-native species in a marine community. For example, Chapman & Carlton (1991) compiled a list of ten criteria that could be used to resolve ambiguous invasion histories and tested their criteria using a cryptogenic (=origin uncertain; Carlton, 1996b) species of isopod crustacean, *Synidotea laticauda*, in San Francisco Bay, California. Assembling substantial local and global data on the species for these ten criteria, Chapman & Carlton (1991) determined that the isopod was a non-native species, probably having arrived a century earlier on the hulls of ships from Pacific Asia to Pacific North America. Over the years, Chapman & Carlton's criteria (1991) have been cited and used to help resolve questionable invasion histories for numerous other marine species (e.g., Coles *et al.*, 1999; Willis *et al.*, 2004; Glasby *et al.*, 2006) and for other species considered cryptogenic (Carlton, 1996b).

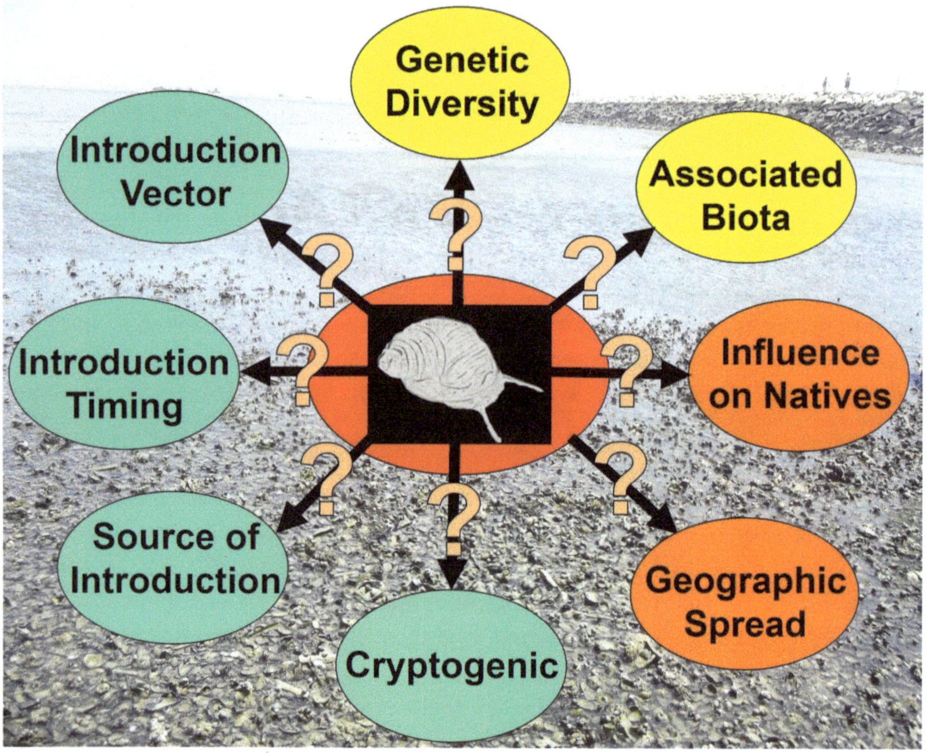

Fig. 7.1: Some of the numerous questions that may surround a newly discovered species in a marine population.

7.1.2 Genetics and Parasites in Marine NIS

Chapman & Carlton's study (1991) provides a very valuable set of criteria for helping resolve uncertain invasion histories, like cryptogenic status; yet, in some cases, additional lines of evidence are needed when available information is conflicting, vague, or poor (Blakeslee, 2007). Over the past couple of decades, molecular genetics has been a valuable tool for resolving ecological and evolutionary questions across biological disciplines, including marine invasions (e.g., Geller *et al.*, 2010). Additionally, parasites can impart important evidence of a host's ecological history and may sometimes provide more information regarding a host's distribution than the host itself (e.g., Criscione *et al.*, 2006). Thus, studies of parasite and genetic diversity can be important synergistic tools for understanding invasion histories; yet few studies have explored them together to look for emerging patterns.

A recent example of a problematic species where both parasites and genetics were used to resolve an ambiguous invasion history is the common periwinkle, *Littorina littorea*, a highly abundant marine snail found on both Atlantic coasts. A known native of Europe, the snail's status as native or non-native in northeastern North America was debated for over 100 years as a result of conflicting historical, genetic, ecological, and paleontological evidence, even though *L. littorea* is one of the most well-studied marine intertidal snails globally, and its native or non-native status had been examined in over a dozen publications from the late 1800s to 2000s (Blakeslee, 2007). Using novel parasite and genetic evidence, Blakeslee & Byers (2008) and Blakeslee *et al.* (2008) explored two common signatures of an invasion (parasite escape and genetic founder effects in the host and its most common parasite) and found significant reductions in diversity in snail and parasite populations in eastern North America compared to Europe. A further study by Brawley *et al.* (2009) found congruent molecular and shipping evidence for connections between North American populations and the British Isles, representing a potential source region for the snail's introduction.

Thus, when used together, parasite and genetic data can help resolve long-term ambiguities in a species' ecological history and can also provide powerful evidence for numerous other questions in biological invasion studies. Below, I further explore these two signatures (genetic bottlenecks and parasite escape) and the work that has been done independently on each. I then examine these two signatures *together* to look for emerging patterns, as well as what they can offer to our understanding of marine invasions.

7.1.3 Genetic Diversity and Founder Effects

Genetic data has been used in numerous studies to reveal species' invasive tracks, including source populations, introduction timing, and likely vectors (see Table 7.1 for numerous citations). Moreover, species introductions are often associated with 'founder effects', whereby founding populations demonstrate significant genetic bottlenecks compared to source populations (Grossberg & Cunningham, 2000). While this is a strong signature in many marine invasions, there remain some successful invaders that do not conform to this expectation, instead exhibiting little indication of a bottleneck, possibly due to multiple introductions and/or high propagule pressure (Roman & Darling, 2007). Depending on the type of introduction vector and invasion pathway, there could be multiple abiotic and/or biotic factors affecting NIS during the invasion process, and these divergent vectors and pathways may impact resulting genetic diversity in non-native populations (Figure 7.2A). While such a "genetic paradox" has been demonstrated in many free-living organisms, how these signatures manifest in parasites is much less clear. In fact, parasites may be more prone to genetic founder effects and genetically depauperate founding populations than hosts because of inherently smaller founding populations, lower genetic diversity, and complex life cycles (Figure 7.2B).

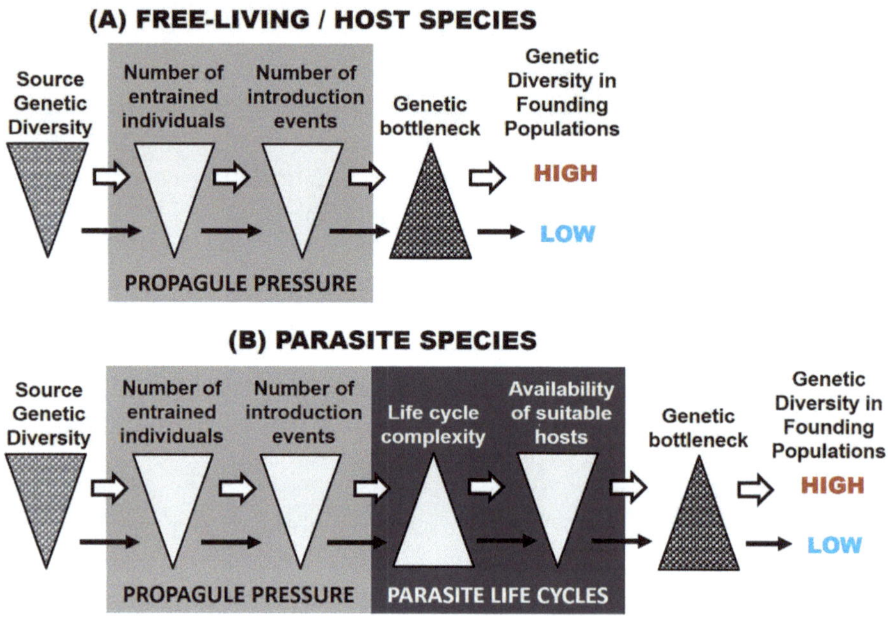

Fig. 7.2: Theoretical schematic for how source diversity and propagule pressure may influence genetic bottlenecks in free-living or host species (A) and parasite species (B) in non-native regions. For hosts (A), if source genetic diversity and propagule pressure are high, the extent of a genetic bottleneck is expected to be low and genetic diversity may then be high. For parasites (B), source genetic diversity and propagule pressure are still major factors influencing genetic bottlenecks in founding regions. However, parasites are dependent on hosts for life cycle completion; thus greater life cycle complexity (e.g., multi-host parasites) and lower host availability to complete life cycles could result in stronger genetic bottlenecks in parasite species. This figure has been adapted from Figure 1 in Roman & Darling (2007) with permission from the authors.

7.1.4 Enemy Release Hypothesis: Parasite Escape

Another well-studied hypothesis explaining why NIS may succeed in novel habitats is the enemy release hypothesis, which can occur when NIS leave behind natural enemies (competitors, predators, parasites) during the invasion process. This results in fewer enemies in the non-native range compared to the native range (Keane & Crawley, 2002; Torchin *et al.*, 2001, 2002, 2003). Specific to parasites, introductions can serve as a screening-out process, leading to lower parasite burdens (i.e., *parasite escape*) in the non-native region and the potential for ecological and physiological benefits in non-indigenous host populations (Torchin *et al.*, 2003). For example, a recent review of parasite escape in marine and estuarine systems spanning the last 2+ decades and 31 host-parasite systems (= 24 unique host species, 6 host *Classes*, and 20 parasite taxa) found parasite escape to continue to be a significant signature of marine invasions worldwide (Blakeslee *et al.*, 2013). On average, invading hosts carried with them approximately half the number of parasites in their native ranges, although some parasite groups contributed to that escape more than others. Thus, parasite escape has consistently been shown (via seminal works by Torchin *et al.* 2002, 2003, and the update by Blakeslee *et al.*, 2013) to be a strong signature of marine invasions worldwide. Similarly, a recent investigation (Jeschke *et al.*, 2012) exploring six prominent theories in invasion biology, including enemy release, found some invasion theories to demonstrate a “decline effect” with time and evidence, but enemy release has continued to show strong support with time, especially in marine systems. As such, parasite escape is a well-supported signature of invasion, and when compared to evidence from native systems, it could be a helpful line of evidence for resolving questionable invasion histories.

7.1.5 Study Questions

In this review, I searched the literature for marine investigations that included genetic diversity, parasite diversity, or both in native and non-native locations around the world. I explored parasite escape and genetic bottlenecks across studies and compared them to look for emerging patterns in both signatures of invasion, and I also focused on a subset of studies where more precise source areas were known. Finally, I explored a subset of data where hosts and parasites have both been investigated in native and non-native regions to determine if dissimilarities in propagule pressure and life history may have differentially influenced host versus parasite genetic diversities. Specifically, I addressed the following questions:

1. Do parasite escape and genetic bottlenecks show convergent patterns when averaged across marine NIS?
2. Is there a ‘source effect’, whereby parasite escape and genetic bottleneck signatures are less pronounced than in regional comparisons?
3. Are parasite escape and genetic bottlenecks influenced by vector type, NIS taxa, geography, and/or time since introduction?
4. Do parasites demonstrate more pronounced genetic bottlenecks in non-native regions than their hosts?

Tab. 7.1: The 61 study systems (40 unique species) included in this review of genetic and parasite diversity data in marine invertebrate species from native and non-native regions worldwide.

Species are listed alphabetically by Latin name. Class = the Class of the marine invertebrate species; F/P = free-living versus parasitic species; Vector = vector (mechanism) of introduction; Timing = timing of introduction; Native range (region) = the region where the species is native; Non-native range = the region where the species is non-native; Distance = the direct line distance (km) between the native and non-native region; Marker = the molecular marker the genetic data are based on; Genetic Bottleneck Index = the extent of the bottleneck; Non-native/Native Haplotype Diversity = the ratio of the non-native to the native haplotype diversity; Parasite Taxa = the parasite taxa in native and/or non-native regions; Parasite Escape Index = the extent of the parasite escape; Citations = the references for the genetic and parasite data.

Species: *Latin name* (Common name)	***Class***	**F/P**	**Vector**	**Timing**	**Native range (region)**	**Non-native range (region)**	**Distance**	**Marker**	**Genetic Bottleneck Index** $[(N_h - I_h)/N_h]$	**Non-native / Native Haplotype Diversity**	**Parasite Taxa**	**Parasite Escape Index** $[(N_p - I_p)/N_p]$	**Citations**	**Notes**
Asterias amurensis (Northern Pacific seastar)	Asteroidea	F	BWF	1990	Asia (Japan; Russia; North China; Korea: 60N-31N); Lat median: 46N; Lon @ median: 137E	Southern coast of Australia (37S-43S); Lat median: 40S; Lon @ median: 146E	3598	n/a	n/a	n/a	Ciliophora (N), Copepoda (N)	1.000	Torchin *et al.* (2002)	
Austrobilharzia variglandis	Trematoda	P	HOST (*Ilyanasaa obsoleta*)	1900	Eastern N. America (Canada to Georgia) (48N to 29N); Lat median: 39N; Lon @ median: 74W	Western N. America (SFB, WB and BB) (49N to 37N); Lat median: 46N; Lon @ median: 123W	4024	COI	0.652	0.717	n/a	n/a	Blakeslee & Fowler (2011); Blakeslee *et al.* (in prep)	Haplotype richness in native and non-native regions represent rarefied values
Batillaria attramentaria (=*cumingi*) (Asian hornsnail)	Gastropoda	F	OYS	1920	Asia (Japan, Hong Kong, other areas of Asia from 40N to equator); Lat median: 20N; Lon @ median: 110E	Western N. America (British Columbia, Washington, Elkhorn Slough, CA: 50N to 36N); Lat median: 43N; Lon @ median: 124W	11093	COI	0.667	0.122	Trematoda (N, I)	0.800	Byers (2000); Torchin *et al.* (2002); Torchin *et al.* (2005); Miura *et al.* (2006); Hechinger (2007); Lafferty & Kuris (2009)	Haplotype diversity represents the rarefied regional values in Figure 7.3

Tab. 7.1: The 61 study systems (40 unique species) included in this review of genetic and parasite diversity data in marine invertebrate species from native and non-native regions worldwide. (continued)

Species: ***Latin name*** (Common name)	***Class***	**F/P**	**Vector**	**Timing**	**Native range (region)**	**Non-native range (region)**	**Distan-ce**	**Marker**	**Genetic Bottleneck Index** $[(N_h - I_h)/N_h]$	**Non-native / Native Haplotype Diversity**	**Parasite Taxa**	**Parasite Escape Index** $[(N_p - I_p)/N_p]$	**Citations**	**Notes**
Batillaria australis	Gastropoda	F	BWF	1950	Southeastern Australia (Whitsunday Islands, Queensland; southwards to Victoria and Tasmania: 20S-43S); Lat median: 32S; Lon @ median: 152E	West Australia (Swan River estuary and Cockburn Sound: 31S-32S); Lat median: 32S; Lon @ median: 115E	3469	n/a	n/a	n/a	Trematoda (N, I)	0.625	Thomsen *et al.* (2010)	
Botrylloides violaceus (violet tunicate)	Tunicata	F	OYS/ BWF	1973	Northwest Pacific (Japan, southern China, Korea) (45N to 21N); Lat median: 33N; Lon @ median: 129E	Eastern N. America (Newfoundland to Chesapeake Bay) (51N to 36N); Lat median: 43N; Lon @ median: 69W	11360	COI	0.800	0.000	n/a	n/a	Lejeusne *et al.* (2011)	
Botryllus schlosseri (golden star tunicate)	Tunicata	F	AQC/ OYS/ BWF	1947	Mediterranean Sea? (45N to 30N; 5W to 36E); Lat median: 37N; Lon median: 16E	Western N. America (Alaska to Mexico) (57N to 30N); Lat median: 44N; Lon @ median: 123W	10100	COI	0.750	0.462	n/a	n/a	Lejeusne *et al.* (2011)	
Botryllus schlosseri (golden star tunicate)	Tunicata	F	AQC/ OYS/ BWF/ DBF	1838	Mediterranean Sea? (45N to 30N; 5W to 36E); Lat median: 37N; Lon median: 16E	Eastern N. America (Newfoundland to Florida) (51N to 28N); Lat median: 40N; Lon @ median: 74W	7472	COI	0.625	0.573	n/a	n/a	Lejeusne *et al.* (2011)	
Caprella mutica (Japanese skeleton shrimp)	Crustacea	F	BWF	1995	Asia (Russia, Japan) (53N to 30N); Lat median: 42N; Lon @ median: 139E	Europe (Scandinavia to British Isles) (71N to 47N); Lat median: 59N; Lon @ median: 5E	8010	COI	0.871	0.264	n/a	n/a	Ashton *et al.* (2008)	

continued **Tab. 7.1:** The 61 study systems (40 unique species) included in this review of genetic and parasite diversity data in marine invertebrate species from native and non-native regions worldwide.

Species: ***Latin name*** (Common name)	***Class***	**F/P**	**Vector**	**Timing**	**Native range (region)**	**Non-native range (region)**	**Distan-ce**	**Marker**	**Genetic Bottleneck Index $[(N_h - I_h)/N_h]$**	**Non-native / Native Haplotype Diversity**	**Parasite Taxa**	**Parasite Escape Index $[(N_p - I_p)/N_p]$**	**Citations**	**Notes**
Carcinus maenas (European green crab)	Crusta-cea	F	DBF/ BWF	1810	Europe (Norway to Portugal: 70N-37N); Lat median: 54N; Lon @ median: 5E	Eastern North America (New-foundland to North Carolina: 49N-34N); Lat median: 42N; Lon @ median: 70W	5459	COI	0.800	0.332	Acanthocepha (N, I), Cestoda (N), Copepoda (N), Fecamp-ida (N), Isopoda (N), Nematoda (N, I), Nemertea (N), Rhizo-cephala (N), Trematoda (N, I)	0.700	Torchin *et al.* (2001); Torchin *et al.* (2002); Ro-man (2006); Darling *et al.* (2008); Blakeslee *et al.* (2009); Pringle *et al.* (2011)	
Carcinus maenas (European green crab)	Crusta-cea	F	APM	1990	Europe (Norway to Portugal: 70N-37N); Lat median: 54N; Lon @ median: 5E	Western North America (British Columbia to San Francisco Bay, California: 50N-37N); Lat median: 44N; Lon @ median: 123W	8050	COI	0.980	0.000	Acantho-cepha (N), Cestoda (N, I), Copepoda (N), Fecampida (N), Isopoda (N), Nematoda (N), Nemertea (N, I), Rhizo-cephala (N), Trematoda (N)	0.800	Torchin *et al.* (2001); Torchin *et al.* (2002); Darling *et al.* (2008)	
Carcinus maenas (European green crab)	Crusta-cea	F	BWF	1980	Europe (Norway to Mediterranean Sea: 70N-35N); Lat median: 54N; Lon @ median: 5E	South Africa (Cape Peninsula, Cape Town, S. Africa: 33S-34S); Lat median: 34S; Lon @ median: 18E	9749	COI	0.800	1.064	Acanthocepha (N), Cestoda (N), Copepoda (N), Fecampida (N), Isopoda (N), Nematoda (N), Nemertea (N), Rhizo-cephala (N), Trematoda (N)	1.000	Torchin *et al.* (2001); Darling *et al.* (2008); Zetl-meisl *et al.* (2010);	

continued **Tab. 7.1:** The 61 study systems (40 unique species) included in this review of genetic and parasite diversity data in marine invertebrate species from native and non-native regions worldwide.

Species: *Latin name* (Common name)	*Class*	F/P	Vector	Timing	Native range (region)	Non-native range (region)	Distance	Marker	Genetic Bottleneck Index $[(N_h - I_h)/N_h]$	Non-native / Native Haplotype Diversity	Parasite Taxa	Parasite Escape Index $[(N_p - I_p)/N_p]$	Citations	Notes
Carcinus maenas (European green crab)	Crustacea	F	DBF/ BWF	1900	Europe (Norway to Portugal: 70N-37N); Lat median: 54N; Lon @ median: 5E	Australia (Victoria; Tasmania: 43S-33S); Lat median: 38S; Lon @ median: 145E	16504	COI	0.900	0.720	Acanthocepha (N), Cestoda (N, I), Copepoda (N), Fecampida (N), Isopoda (N), Nematoda (N, I), Nemertea (N), Rhizocephala (N), Trematoda (N)	0.800	Torchin *et al.* (2001); Darling *et al.* (2008); Zetlmeisl *et al.* (2010);	
Carcinus maenas (European green crab)	Crustacea	F	BWF	1980	Europe (Norway to Mediterranean Sea: 70N-35N); Lat median: 54N; Lon @ median: 5E	Japan (45N-30N); Tokyo Lat/Lon: 35N/139E	9263	COI	0.960	0.438	Acanthocepha (N), Cestoda (N), Copepoda (N), Fecampida (N), Isopoda (N), Nematoda (N), Nemertea (N), Rhizocephala (N), Trematoda (N)	1.000	Torchin *et al.* (2002); Darling *et al.* (2008)	
Carcinus maenas (European green crab)	Crustacea	F	BWF	1999	Europe (Norway to Mediterranean Sea: 70N-35N); Lat median: 54N; Lon @ median: 5E	Eastern Argentina (46S to 43S)	12830	COI	0.980	0.000	n/a	n/a	Darling *et al.* (2008)	
Ceratostoma inornatum (=Ocinebrellus inornatus) (Japanese oyster drill)	Gastropoda	F	OYS	1924	Asia (Japan, Korea) (45N to 31N); Lat median: 38N; Lon @ median: 139E	Western N. America (British Columbia to California (49N to 37N); Lat median: 43N; Lon @ median: 124W	7727	COI	0.667	n/a	n/a	n/a	Martel *et al.* (2004)	

continued **Tab. 7.1:** The 61 study systems (40 unique species) included in this review of genetic and parasite diversity data in marine invertebrate species from native and non-native regions worldwide.

Species: ***Latin name*** (Common name)	***Class***	**F/P**	**Vector**	**Timing**	**Native range (region)**	**Non-native range (region)**	**Distance**	**Marker**	**Genetic Bottleneck Index** $[(N_h - I_h)/N_h]$	**Non-native / Native Haplotype Diversity**	**Parasite Taxa**	**Parasite Escape Index** $[(N_p - I_p)/N_p]$	**Citations**	**Notes**
Ceratostoma inornatum (=Ocinebrellus inornatus) (Japanese oyster drill)	Gastropoda	F	OYS	1995	Asia (Japan, Korea) (45N to 31N); Lat median: 38N; Lon @ median: 139E	Europe (France) (50N to 43N); Lat median: 47N; Lon @ median: 2W	9793	COI	0.583	n/a	n/a	n/a	Martel *et al.* (2004)	
Cercaria batillaria (HL1)	Trematoda	P	HOST (*Batillaria attramentaria*)	1920	Asia (Japan, Hong Kong, other areas of Asia from 40N to equator); Lat median: 20N; Lon @ median: 110E	Western N. America (British Columbia, Washington, Elkhorn Slough, CA: 50N to 36N); Lat median: 43N; Lon @ median: 124W	11093	COI	0.125	0.989	n/a	n/a	Miura *et al.* (2006)	Haplotype diversity represents the rarefied regional values in Figure 3
Cercaria batillaria (HL6)	Trematoda	P	HOST (*Batillaria attramentaria*)	1920	Asia (Japan, Hong Kong, other areas of Asia from 40N to equator); Lat median: 20N; Lon @ median: 110E	Western N. America (British Columbia, Washington, Elkhorn Slough, CA: 50N to 36N); Lat median: 43N; Lon @ median: 124W	11093	COI	0.727	0.632	n/a	n/a	Miura *et al.* (2006)	Haplotype diversity represents the rarefied regional values in Figure 3
Cercaria parvicaudata	Trematoda	P	HOST (*Littorina littorea*)	1840	Europe (White Sea to Portgual) (70N to 40N); Lat median: 55N; Lon @ median: 8E	Northeastern N. America (Labrador to Delaware Bay) (51N to 38N); Lat median: 45N; Lon @ median: 66W	5144	COI	0.375	0.470	n/a	n/a	Blakeslee & Fowler (2011); Blakeslee (unpublished)	Haplotype richness in native and non-native regions represent rarefied values

continued **Tab. 7.1:** The 61 study systems (40 unique species) included in this review of genetic and parasite diversity data in marine invertebrate species from native and non-native regions worldwide.

Species: ***Latin name*** (Common name)	***Class***	**F/P**	**Vector**	**Timing**	**Native range (region)**	**Non-native range (region)**	**Distan-ce**	**Marker**	**Genetic Bottleneck Index $[(N_h - I_h)/N_h]$**	**Non-native / Native Haplotype Diversity**	**Parasite Taxa**	**Parasite Escape Index $[(N_p - I_p)/N_p]$**	**Citations**	**Notes**
Cercopagis pengoi (Fishook waterflea)	Crusta-cea	F	BWF	1198	Ponto-Caspian (Black, Azov, Caspian, and Aral Seas): Lat median: 41N; Lon median: 41E	Great Lakes; St. Lawrence River/Seaway (48N to 41N; 92W to 76W); Lat median: 44N; Lon @ median: 82W	8977	COI	0.857	n/a	n/a	n/a	Critescu *et al.* (2001)	
Chtlalamus proteus Caribbean barnacle)	Crusta-cea	F	BWF	1995	Gulf of Mexico; Caribbean Sea; South Atlantic (29N to 23S); Lat median: 3N; Lon @ median: 51W	Hawaiian islands (21N to 18N); Lat median: 20N; Lon @ median: 156W	11446	COI	0.111	n/a	n/a	n/a	Zardus & Hadfield (2005)	
Crassostrea angulata (Portuguese oyster)	Bivalvia	F	DEL	1500s	Portugal, Spain, France (48N to 36N); Lat median: 42N; Lon @ median: 8W	Taiwan (25N to 21N); Lat median: 23N; Lon @ median: 120E	11022	COI	0.556	n/a	n/a	n/a	Boudry *et al.* (1998); Foighil *et al.* (1998); Huvet *et al.* (2000)	
Crassostrea gigas (Pacific oyster)	Bivalvia	F	OYS	1900	Asia (Russia; east coast of China; Korea; Japan: 59N-22N); Lat median: 41N; Lon @ median: 129E	Western North America (southern Alaska to Humboldt Bay, California: 40N-60N); Lat median: 50N; Lon @ median: 125W	7596	ALLO-ZYME	-0.056	1.197	Copepoda (N, I), Nematoda (N), Trematoda (N), Turbellaria (N)	0.750	Mann *et al.* (1991); English *et al.* (2000)	**Mitochondrial data not available, so also including allozyme data here; i.e., data based upon mean alleles and observed heterozygosities from table 6 in English *et al.* (2000)

continued **Tab. 7.1:** The 61 study systems (40 unique species) included in this review of genetic and parasite diversity data in marine invertebrate species from native and non-native regions worldwide.

Species: ***Latin name*** (Common name)	*Class*	F/P	Vector	Timing	Native range (region)	Non-native range (region)	Distance	Marker	Genetic Bottleneck Index $[(N_h - I_h)/N_h]$	Non-native / Native Haplotype Diversity	Parasite Taxa	Parasite Escape Index $[(N_p - I_p)/N_p]$	Citations	Notes
Crassostrea gigas (Pacific oyster)	Bivalvia	F	OYS	1960	Asia (Russia; east coast of China; Korea; Japan: 59N-22N); Lat median: 41N; Lon @ median: 129E	Western Europe (Exe Estuary, Great Britain; Wadden Sea; France; Netherlands; Belgium; Germany; Denmark; Sweden; Norway: 65N-44N); Lat median: 55N; Lon @ median: 8E	7964	MNR	0.094	0.769	Copepoda (N, I), Nematoda (N), Polychaeta (I), Trematoda (N, I), Turbellaria (N, I)	-0.500	Mann *et al.* (1991); Aguierre-Macedo and Kennedy (1999); English *et al.* (2000); Krakau *et al.* (2006); Troost (2010); Elsner *et al.* (2011); Moeller *et al.* (2011); Thieltges *et al.* (2013)	**Only mitochondrial data in the non-native region (Moeller *et al.*, 2011), so also including allozyme data here; i.e., data based upon mean alleles and observed heterozygosities from table 6 in English *et al.* (2000)
Crassostrea gigas (Pacific oyster)	Bivalvia	F	BWF	1950	Asia (Russia; east coast of China; Korea; Japan: 59N-22N); Lat median: 41N; Lon @ median: 129E	New Zealand (46S-34S); Lat median: 40S; Lon @ median: 175E	10129	ALLOZYME	-0.007	0.875	Copepoda (N, I), Nematoda (N, I), Trematoda (N), Turbellaria (N, I)	0.000	Dinamami (1986); English *et al.* (2000)	**Mitochondrial data not available, so also including allozyme data here; i.e., data based upon mean alleles and observed heterozygosities from table 6 in English *et al.* (2000)
Cryptocotyle lingua	Trematoda	P	HOST (*Littorina littorea*)	1840	Europe (White Sea to Portugal) (70N to 40N); Lat median: 55N; Lon @ median: 8E	Northeastern N. America (Labrador to Delaware Bay) (51N to 38N); Lat median: 45N; Lon	5144	COI	0.851	0.902	n/a	n/a	Blakeslee *et al.* (2008)	Haplotype richness in native and non-native regions represent rarefied

continued **Tab. 7.1:** The 61 study systems (40 unique species) included in this review of genetic and parasite diversity data in marine invertebrate species from native and non-native regions worldwide.

Species: ***Latin name*** (Common name)	***Class***	**F/P**	**Vector**	**Timing**	**Native range (region)**	**Non-native range (region)**	**Distance**	**Marker**	**Genetic Bottleneck Index $[(N_h - I_h)/N_h]$**	**Non-native / Native Haplotype Diversity**	**Parasite Taxa**	**Parasite Escape Index $[(N_p - I_p)/N_p]$**	**Citations**	**Notes**
Cyclope nerite	Gastropoda	F	OYS	1976	Mediterranean Sea; Black Sea; Morocco; southern Portugal; southern Spain (46N to 27N); Lat median: 37N; Lon @ median: 16E	France (48N to 43N); Lat median: 46N; Lon @ median: 1W	1727	COI	-3.167	1.809	Trematoda (N, I)	0.500	Simon-Bouhet *et al.* (2006); Couceiro *et al.* (2012)	
Didemnum vexillum / sp. A (Didemnum tunicate)	Tunicata	F	BWF	1993	Asia (Japan) (45N to 30N); Lat median: 38N; Lon @ median: 139E	Western N. Amerca (Alaska to Mexico) (57N to 30N); Lat median: 44N; Lon @ median: 123W	7733	COI	0.375	n/a	n/a	n/a	Stefaniak *et al.* (2009)	
Didemnum vexillum / sp. A (Didemnum tunicate)	Tunicata	F	BWF	1982	Asia (Japan) (45N to 30N); Lat median: 38N; Lon @ median: 139E	Eastern N. America (Maine to Long Island) (45N to 40N); Lat median: 43N; Lon @ median: 70W	10537	COI	0.625	n/a	n/a	n/a	Stefaniak *et al.* (2009)	
Didemnum vexillum / sp. A (Didemnum tunicate)	Tunicata	F	BWF	1991	Asia (Japan) (45N to 30N); Lat median: 38N; Lon @ median: 139E	Europe (North Sea to Spain) (58N to 36N); Lat median: 43N; Lon @ median: 3W	10219	COI	0.625	n/a	n/a	n/a	Stefaniak *et al.* (2009)	
Echinogammarus ischnus (Scud)	Crustacea	F	BWF	1995	Ponto-Caspian (Black, Azov, Caspian, and Aral Seas): Lat median: 41N; Lon median: 41E	Great Lakes; St. Lawrence River/ Seaway (50N to 41N; 92W to 64W); Lat median: 45N; Lon @ median: 81W	8841	COI	0.818	0.000	n/a	n/a	Critescu *et al.* (2004)	

continued **Tab. 7.1:** The 61 study systems (40 unique species) included in this review of genetic and parasite diversity data in marine invertebrate species from native and non-native regions worldwide.

Species: *Latin name* (Common name)	*Class*	F/P	Vector	Timing	Native range (region)	Non-native range (region)	Distance	Marker	Genetic Bottleneck Index $[(N_h - I_h)/N_h]$	Non-native / Native Haplotype Diversity	Parasite Taxa	Parasite Escape Index $[(N_p - I_p)/N_p]$	Citations	Notes
***Ensis directus* (=*americanus*)** (razor shell)	Bivalvia	F	BWF	1978	Eastern N. America (Labrador to Florida) (59N to 25N); Lat median: 42N; Lon median: 70W	Europe (Southern Norway to Normandy; North Sea; British Isles) (59 N to 48N); Lat median: 54N; Lon median: 8E	5644	COI	-0.385	1.086	Trematoda (N, I)	n/a	Armonies and Reise (1999); Krakau *et al.* (2006); Thieltges *et al.* (2006); Vierna *et al.* (2012)	4 parasites in non-native range but native range is undersampled - suggestion that more exist but only one (*Himasthla quissitensis*) detected.
Gammarus tigrinus	Crustacea	F	OTHER	1931	Eastern N. America (Gulf of St Lawrence, Quebec to N. Florida) (49N to 29N); Lat median: 39N; Lon @ median: 74W	Europe (Baltic to Rhine; also British Isles) (60N to 49N); Lat median: 55N; Lon @ median: 12E	6319	COI	0.903	0.615	n/a	n/a	Kelly *et al.* (2006)	Averaged from Table 3
Gammarus tigrinus	Crustacea	F	OTHER	1931	Eastern N. America (Gulf of St Lawrence, Quebec to N. Florida) (49N to 29N); Lat median: 39N; Lon @ median: 74W	Great Lakes (48N to 41N; 91W to 76W); Lat median: 45N; Lon @ median: 83W	997	COI	0.968	n/a	n/a	n/a	Kelly *et al.* (2006)	
Gemma gemma (Gem clam)	Bivalvia	F	OYS	1880	Eastern N. America (Labrador to Florida; Gulf of Mexico) (53N to 25N); Lat median: 39N; Lon median: 74W	Western N. America (Puget Sound, WA to San Diego, CA) (47N to 32N); Lat median: 40N; Lon median: 124W	4231	COI	0.476	0.925	n/a	n/a	Hoos *et al.* (2010)	

continued **Tab. 7.1:** The 61 study systems (40 unique species) included in this review of genetic and parasite diversity data in marine invertebrate species from native and non-native regions worldwide.

Species: ***Latin name*** (Common name)	***Class***	**F/P**	**Vector**	**Timing**	**Native range (region)**	**Non-native range (region)**	**Distance**	**Marker**	**Genetic Bottleneck Index $[(N_h - I_h)/N_h]$**	**Non-native / Native Haplotype Diversity**	**Parasite Taxa**	**Parasite Escape Index $[(N_p - I_p)/N_p]$**	**Citations**	**Notes**
Haminoea japonica (Japanese bubble snail)	Gastropoda	F	OYS	1966	Asia (Japan; Korea; Thailand: 45N-13N); Lat median: 29N; Lon @ median: 122E	Wesern N. America (Boundary Bay, Washington, San Francisco Bay, Tomales Bay, California) (49N to 37N); Lat median: 43N; Lon @ median: 124W	9552	COI	0.909	0.544	Trematoda (I)	n/a	Brant *et al.* (2010); Hanson *et al.* (2013)	1 schistosome trematode found in San Francisco Bay, but not found in other locations on the west coast. No schistosome parasites observed in Japanese snails.
Haminoea japonica (Japanese bubble snail)	Gastropoda	F	OYS	1992	Asia (Japan; Korea; Thailand: 45N-13N); Lat median: 29N; Lon @ median: 122E	Europe (France, Italy, Spain, Mediterranean) (48N to 38N); Lat median: 43N; Lon @ median: 7E	9617	COI	0.909	0.333	n/a	n/a	Hanson *et al.* (2013)	No schistosome parasites observed in Japanese snails, nor European snails.
Hemigrapsus sanguineus (Asian shorecrab)	Crustacea	F	BWF	1988	Asia (Russia, Japan, Hong Kong) (52N to 22N); Lat median: 37N; Lon @ median: 141E	Eastern N. America (southern Maine to North Carolina) (43N to 35N); Lat median: 39N; Lon @ median: 74W	10829	COI	0.885	0.866	Acanthocephala (I), Nematoda (I), Microspora (N), Rhizocephala (N), Trematoda (N)	0.750	Blakeslee *et al.* (2009); Epifanio (2013); McDermott (2011); Blakeslee *et al.* (in prep)	Haplotype richness in native and non-native regions represent rarefied values
Himasthla quissitensis	Trematoda	P	HOST (*Ilyanasaa obsoleta*)	1900	Eastern N. America (Canada to Georgia) (48N to 29N); Lat median: 39N; Lon @ median: 74W	Western N. America (SFB, WB and BB) (49N to 37N); Lat median: 46N; Lon @ median: 123W	4024	COI	0.839	0.805	n/a	n/a	Blakeslee & Fowler (2011); Blakeslee *et al.* (in prep)	Haplotype richness in native and non-native regions represent rarefied values

continued **Tab. 7.1:** The 61 study systems (40 unique species) included in this review of genetic and parasite diversity data in marine invertebrate species from native and non-native regions worldwide.

Species: ***Latin name*** **(Common name)**	***Class***	**F/P**	**Vector**	**Timing**	**Native range (region)**	**Non-native range (region)**	**Distance**	**Marker**	**Genetic Bottleneck Index $[(N_h - I_h)/N_h]$**	**Non-native / Native Haplotype Diversity**	**Parasite Taxa**	**Parasite Escape Index $[(N_p - I_p)/N_p]$**	**Citations**	**Notes**
Ilyanassa obsoleta (eastern mud snail)	Gastropoda	F	OYS	1900	Eastern N. America (Canada to Georgia) (48N to 29N); Lat median: 39N; Lon @ median: 74W	Western N. America (SFB, WB and BB) (49N to 37N); Lat median: 46N; Lon @ median: 123W	4024	COI	0.481	1.025	Trematoda (N, I)	0.444	Curtis (1997); Blakeslee & Fowler (2011); Blakeslee *et al.* (2012); Blakeslee *et al.* (in prep)	Haplotype richness in native and non-native regions represent rarefied values
Lepocreadium setiferoides	Trematoda	P	HOST (*Ilyanasaa obsoleta*)	1900	Eastern N. America (Canada to Georgia) (48N to 29N); Lat median: 39N; Lon @ median: 74W	Western N. America (SFB, WB and BB) (49N to 37N); Lat median: 46N; Lon @ median: 123W	4024	COI	0.793	1.146	n/a	n/a	Blakeslee & Fowler (2011); Blakeslee *et al.* (in prep)	Haplotype richness in native and non-native regions represent rarefied values
Littorina littorea (common periwinkle snail)	Gastropoda	F	DBF	1840	Europe (White Sea, Russia to Portgual: 70N to 40N); Lat median: 55N; Lon @ median: 8E	Eastern North America (Labrador to Delaware Bay: 51N to 38N); Lat median: 45N; Lon @ median: 66W	5144	CYTB	0.873	1.096	Trematoda (N, I)	0.545	Blakeslee *et al.* (2008); Blakeslee & Byers (2008)	Haplotype richness in native and non-native regions represent rarefied values
Littorina littorea (common periwinkle snail)	Gastropoda	F	DEL	1960	Europe (White Sea, Russia to Portgual: 70N to 40N); Lat median: 55N; Lon @ median: 8E	Western North America (San Francisco Bay, California); Lat median: 37N; Lon @ median: 122W	8728	CYTB	0.938	1.092	Trematoda (N, I)	0.909	Chang *et al.* (2011); Blakeslee (unpublished)	Haplotype richness in native and non-native regions represent rarefied values

continued **Tab. 7.1:** The 61 study systems (40 unique species) included in this review of genetic and parasite diversity data in marine invertebrate species from native and non-native regions worldwide.

Species: ***Latin name*** (Common name)	***Class***	**F/P**	**Vector**	**Timing**	**Native range (region)**	**Non-native range (region)**	**Distance**	**Marker**	**Genetic Bottleneck Index** $[(N_h - I_h)/N_h]$	**Non-native / Native Haplotype Diversity**	**Parasite Taxa**	**Parasite Escape Index** $[(N_p - I_p)/N_p]$	**Citations**	**Notes**
Littorina saxatilis (rough periwinkle snail)	Gastropoda	F	APM	1990	Northeastern N. America (Labrador to Long Island, New York; 64N-40N); Lat median: 52N; Lon @ median: 65W	Western North America (San Francisco Bay, California: 37N); Lat median: 37N; Lon @ median: 122W	4678	COI	0.746	0.791	Trematoda (N, I)	0.786	Blakeslee & Fowler (2011); Blakeslee *et al.* (2012); Blakeslee *et al.* (unpublished)	Haplotype richness in native and non-native regions represent rarefied values
Littorina saxatilis (rough periwinkle snail)	Gastropoda	F	DBF	1792	Europe (Barents Sea to Portugal) (71N to36N); Lat median: 54N; Lon @ median: 8E	Mediterranean (Venice: 45N; 12E); Lat median: 45N; Lon @ median: 12E	1040	COI	0.987	0.000	n/a	n/a	Panova *et al.* (2011)	Haplotype richness in native and non-native regions represent rarefied values
Loxothylacus panopaei	Crustacea	P	HOST (*Rhithropanopeus harrisii*)	1964	Gulf of Mexico to southeast Florida; Caribbean (29N to 8N); Lat median: 19N; Lon @ median: 96W	Eastern N. America (Long Island Sound to central Florida) (40N to 28N); Lat median: 34N; Lon @ median: 77W	2511	COI	0.700	n/a	n/a	n/a	Kruse *et al.* (2007); Kruse *et al.* (2011)	
Mnemiopsis leidyi (Atlantic ctenophore)	Tentaculata	F	BWF	1980	Eastern North America (New York to Florida: 40N to 26N); Lat median: 33N; Lon @ median: 78W	Black Sea; Mediterranean Sea; Baltic Sea; North Sea (57N-31N); Lat median: 44N; Lon @ median: 34E	9027	ITS	-0.222	1.245	Amphipoda (N), Cnidaria (N, I), Nematoda (N, I), Trematoda (N)	0.500	Torchin *et al.* (2002); Selandar *et al.* (2010); Ghabooli *et al.* (2010)	*represents allelic diversity
Musculista senhousia (Asian date mussel)	Bivalvia	F	OYS	1980	Asia (Russia; Korea; Japan; China; Singapore: 60N to 1N); Lat median: 31N; Lon @ median: 121E	New Zealand (46S-34S); Lat median: 40S; Lon @ median: 175E	9651	COI	0.538	0.942	Copepoda (N, I)	0.667	Miller *et al.* (2008); Asif & Krug (2012)	

continued **Tab. 7.1:** The 61 study systems (40 unique species) included in this review of genetic and parasite diversity data in marine invertebrate species from native and non-native regions worldwide.

Species: *Latin name* (Common name)	*Class*	F/P	Vector	Timing	Native range (region)	Non-native range (region)	Distan-ce	Marker	Genetic Bottleneck Index $[(N_h - I_h)/N_h]$	Non-native / Native Haplotype Diversity	Parasite Taxa	Parasite Escape Index $[(N_p - I_p)/N_p]$	Citations	Notes
Musculista senhousia (Asian date mussel)	Bivalvia	F	OYS	1924	Asia (Russia; Korea; Japan; China; Singapore: 60N to 1N); Lat median: 31N; Lon @ median: 121E	Western N. America (Puget Sound, WA to San Diego, CA) (47N to 32N); Lat median: 40N; Lon median: 124W	9660	COI	0.590	0.675	n/a	n/a	Asif & Krug (2012)	
Mya arenaria (Soft shell clam)	Bivalvia	F	DEL	1600s	Eastern North America (Labrador to South Carolina: 54N-32N); Lat median: 43N; Lon @ median: 70W	Europe (northern Wadden Sea: 57N-53N); Lat median: 55N; Lon @ median: 8E	5530	COI	0.400	1.361	Copepoda (N), Nematoda (I), Trematoda (N, I), Turbellaria (I)	0.000	Thieltges *et al.* (2006); Petersen *et al.* (1992)	
Mytella charruana (Charru mussel)	Bivalvia	F	BWF	1986	Southeastern Pacific and southernwestern Atlantic (Mexico to Ecuador, Columbia to Argentina) (25N to 38S); Lat median: 7S; Lon @ median: 34W	Southeast N. America (South Carolina to Florida) (33N to 25N); Lat median: 29N; Lon @ median: 81W	6422	COI	0.400	0.836	n/a	n/a	Gillis *et al.* (2009)	
Mytilus galloprovincialis (Mediterranean mussel)	Bivalvia	F	BWF	1970	Mediterranean Sea, Black Sea, and Adriatic Sea (47N to 30N; 5W to 41E); Lat median: 39N; Lon median: 18E	Cape of Good Hope in South Africa to Lüderitz in southern Namibia (26S to 34S); Lat median: 30S; Lon @ median: 17E	7668	COI	0.400	n/a	Copepoda (N), Microspora (N), Trematoda (N), Turbellaria (N)	1.000	Villalba *et al.* (2007); Gerard *et al.* (2008)	

continued **Tab. 7.1:** The 61 study systems (40 unique species) included in this review of genetic and parasite diversity data in marine invertebrate species from native and non-native regions worldwide.

Species: ***Latin name*** (Common name)	***Class***	**F/P**	**Vector**	**Timing**	**Native range (region)**	**Non-native range (region)**	**Distance**	**Marker**	**Genetic Bottleneck Index** $[(N_h - I_h)/N_h]$	**Non-native / Native Haplotype Diversity**	**Parasite Taxa**	**Parasite Escape Index** $[(N_p - I_p)/N_p]$	**Citations**	**Notes**
Paralithoides camtschaticus (red king crab)	Crustacea	F	DEL	1960	North Pacific; Alaska (71N-34N); Lat median: 53N; Lon @ median: 160E	Barents Sea (76N-67N); Lat median: 72N; Lon @ median: 24E	5696	COI	0.400	n/a	Acanthocephala (N, I), Bivalvia (I), Copepoda (N, I), Nemertea (N, I), Isopoda (N), Rhizocephala (N), Turbellaria (I)	0.375	Hawkes *et al.* (1986); Sparks (1987); Kuris *et al.* (1991); Jansen *et al.* (1998); Hemmingsen *et al.* (2005)	
Potamopyrgus antipodarum (New Zealand mud snail)	Gastropoda	F	BWF	1859	New Zealand (36S to 46S); Lat median: 41S; Lon @ median: 173E	Europe (British Isles to Russia to Mediterranean) (70N to 40N); Lat median: 50N; Lon @ median: 11E	18287	16S	0.882	0.570	Trematoda (N, I)	0.714	Jokela & Livey (1995); Morley (2008)	
Rhithropanopeus harrisii (Harris mud crab)	Crustacea	F	OYS	1937	Eastern N. America (Gulf of St Lawrence to Florida) and Gulf of Mexico (Florida to Veracruz, Mexico) (49 N to 19N); Lat median: 34N; Lon @ median: 77W	Western N. America (Oregon and California to San Francisco Bay) (45N to 37N); Lat median: 42N; Lon @ median: 124W	11289	COI	0.864	n/a	n/a	n/a	Petersen (2006)	

continued **Tab. 7.1:** The 61 study systems (40 unique species) included in this review of genetic and parasite diversity data in marine invertebrate species from native and non-native regions worldwide.

Species: *Latin name* (Common name)	*Class*	F/P	Vector	Timing	Native range (region)	Non-native range (region)	Distance	Marker	Genetic Bottleneck Index $[(N_h - I_h)/N_h]$	Non-native / Native Haplotype Diversity	Parasite Taxa	Parasite Escape Index $[(N_p - I_p)/N_p]$	Citations	Notes
Rhithropanopeus harrisii (Harris mud crab)	Crustacea	F	DBF/ BWF	1874	Eastern N. America (Gulf of St Lawrence to Florida) and Gulf of Mexico (Florida to Veracruz, Mexico) (49 N to 19N); Lat median: 34N; Lon @ median: 77W	Europe (Baltic Sea to Portugal; Caspian and Black Seas; the Mediterranean Sea) (60N to 37N); Lat median: 49N; Lon @ median: 1E	6406	COI	0.727	0.690	Rhizocephala (N)	1.000	Projecto-Garcia *et al.* (2010); Fowler *et al.* (2013)	No parasites in the Baltic (Fowler *et al.*, 2013). *Loxothylacus panopaei* infects the crab in its native range, but no studies could be found that have looked for other parasites in the crab aside from *L. panopaei*.
Rhithropanopeus harrisii (Harris mud crab)	Crustacea	F	AQC	1988	Eastern N. America (Gulf of St Lawrence to Florida) and Gulf of Mexico (Florida to Veracruz, Mexico) (49 N to 19N); Lat median: 34N; Lon @ median: 77W	Texas Lakes (33N; 98W)	1946	COI	0.773	n/a	Rhizocephala (N)	1.000	Boyle *et al.* (2010)	No parasites have been found in inland lakes, and there is just one described parasite in the native range, Loxothylacus panopaei, but no studies have looked for other parasites in the crab.

continued **Tab. 7.1:** The 61 study systems (40 unique species) included in this review of genetic and parasite diversity data in marine invertebrate species from native and non-native regions worldwide.

Species: ***Latin name*** (Common name)	***Class***	**F/P**	**Vector**	**Timing**	**Native range (region)**	**Non-native range (region)**	**Distan-ce**	**Marker**	**Genetic Bottleneck Index $[(N_h - I_h)/N_h]$**	**Non-native / Native Haplotype Diversity**	**Parasite Taxa**	**Parasite Escape Index $[(N_p - I_p)/N_p]$**	**Citations**	**Notes**
Ruditapes philippinarum *(Manila clam)*	Bivalvia	F	OYS	1970	Indo-Pacific (40N-1N); Lat median: 21N; Lon @ median: 110E	Europe (France: 48N-43N); Lat median: 45N; Lon @ median: 1W	9894	n/a	n/a	n/a	Trematodes (N, I)	0.600	Rybakov *et al.* (1983); Rybakov (1987); Hua (1989); Mei (1994); Lee *et al.* (2001); Lasalle *et al.* (2007); Park *et al.* (2008); Dang *et al.* (2009); Yanagida *et al.* (2009)	
Styela clava (club sea squirt)	Tunicata	F	OYS / BWF	1933	Northwest Pacific (Shanghai to the Sea of Okhotsk and southeast-ern Bering Sea) (61N to 31N); Lat median: 46N; Lon @ median: 138E	Western N. America (British Columbia to San Diego Bay, California) (50N to 32N); Lat median: 41N; Lon @ median: 124W	7388	COI	0.375	n/a	n/a	n/a	Goldstein *et al.* (2011)	
Styela clava (club sea squirt)	Tunicata	F	OYS / BWF	1970	Northwest Pacific (Shanghai to the Sea of Okhotsk and southeast-ern Bering Sea) (61N to 31N); Lat median: 46N; Lon @ median: 138E	Eastern N. America (Prince Edward Island to Virginia) (46N to 37N); Lat me-dian: 42N; Lon @ median: 70W	9838	COI	0.375	n/a	n/a	n/a	Goldstein *et al.* (2011)	

continued **Tab. 7.1:** The 61 study systems (40 unique species) included in this review of genetic and parasite diversity data in marine invertebrate species from native and non-native regions worldwide.

Species: ***Latin name*** (Common name)	***Class***	**F/P**	**Vector**	**Timing**	**Native range (region)**	**Non-native range (region)**	**Distan-ce**	**Marker**	**Genetic Bottleneck Index** $[(N_h - I_h)/N_h]$	**Non-native / Native Haplotype Diversity**	**Parasite Taxa**	**Parasite Escape Index** $[(N_p - I_p)/N_p]$	**Citations**	**Notes**
Styela clava (club sea squirt)	Tunicata	F	OYS / BWF	1953	Northwest Pacific (Shanghai to the Sea of Okhotsk and southeast-ern Bering Sea) (61N to 31N); Lat median: 46N; Lon @ median: 138E	Europe (Denmark to Portugal) (57N to 36N); Lat me-dian: 46N; Lon @ median: 2W	9056	COI	0.438	n/a	n/a	n/a	Goldstein *et al.* (2011)	
Zoogonus rubellus	Tremat-oda	P	HOST (*Ily-anasaa obsol-eta*)	1900	Eastern N. America (Canada to Georgia) (48N to 29N); Lat median: 39N; Lon @ median: 74W	Western N. America (SFB, WB and BB) (49N to 37N); Lat median: 46N; Lon @ median: 123W	4024	COI	0.936	0.933	n/a	n/a	Blakeslee & Fowler (2011); Blakeslee *et al.* (in prep)	Haplotype richness in native and non-native regions rep-resent rarefied values

7.2 Methods

7.2.1 Data Sources

I gathered genetic and/or parasite data from studies of NIS marine or estuarine invertebrate species across populations in native and non-native regions. I focused on marine invertebrates because they are some of the most commonly introduced species globally (Cohen & Carlton, 1995; Ruiz *et al.*, 2000) and, additionally, they often serve as hosts to marine parasites (e.g. Lauckner, 1987a,b; Marcogliese, 2002; Torchin *et al.*, 2002). Although I attempted to include as many studies as possible, the data presented here are likely not exhaustive.

7.2.1.1 Genetic Diversity

To assess genetic diversity in marine and estuarine hosts in native and non-native regions, I used the list of species in Table 1 of Blakeslee *et al.* (2013) and in Table 1 of Roman & Darling (2007) as a first filter, and from there, I searched the literature for additional studies of NIS marine and estuarine invertebrates, concentrating on studies with mitochondrial (mt) DNA markers. I focused on mtDNA because mitochondrial markers (e.g., cytochrome oxidase I) have been used in numerous population genetics and bar-coding studies over the past couple of decades, resulting in ample available data for comparison and also allowing for the inclusion of introductions investigated in the recent past (Ratnasingham & Hebert, 2007). In a couple cases, mtDNA data was not available, and I instead reported nuclear markers.

7.2.1.2 Parasite Diversity

To assess parasite diversity and subsequent parasite escape in non-native versus native estuarine and marine regions, I primarily used the studies from Table 7.1 in Blakeslee *et al.* (2013) but also searched for any additional studies including parasite species richness in native and non-native regions in marine systems worldwide.

7.2.2 Data Extraction

For both genetic and parasite diversity, I extracted the following data from publications, online databases (e.g. Encyclopedia of Life, the USGS Nonindigenous Aquatic Species database, the Global Invasive Species Database, the National Exotic Marine and Estuarine Species Information System), regional websites reporting biogeographic information, or information in Table 7.1 of Blakeslee *et al.* (2013):

- *NIS Identification* to lowest taxonomic level as identified in publications.
- *NIS Taxa (including parasites)*: this included larger taxonomic groups. For free-living NIS, I used the *Class* level of classification; for parasites, I used the classi-

fication (often to *Class* or *Order* level) provided in published works. I ensured that this taxonomic classification was consistent across comparisons.

- *Free-living (F) or Parasitic (P)*: whether the species is free-living or parasitic. Many free-living species in this investigation also serve as hosts to marine parasites.
- *Introduction Vector*: as in Blakeslee *et al.* (2013), marine and estuarine hosts were categorized into the following bins based on their vector type:
  - APM—association with algal packing materials for live bait and trade
  - AQC—introductions associated with non-oyster aquaculture
  - BWF—ballast water and/or hull fouling associated with ballast-water-carrying vessels
  - CANAL—introductions following the creation of a canal, connecting two previously unconnected bodies of water
  - DEL—deliberate introductions not associated with aquaculture (e.g., research or bio-control).
  - DBF—dry ballast and/or fouling associated with solid-ballast-carrying vessels
  - HOST—a parasite that has been introduced with its host
  - OTHER—other accidental introductions
  - OYS—introductions associated with oyster transplantation
- *Timing of Introduction*: If multiple dates were listed, we used the earliest recorded date for timing of introduction; in addition, I assumed that the host introduction date was equivalent to the parasite introduction date. While in many cases this may not be correct (i.e., in cases of multiple introductions), introduction dates for parasites are typically not available or known; thus the host's introduction date was used as the best possible understanding of introduction timing for the parasite.
- *Native and Non-native Regions, and Native and Non-native Latitude and Longitude*: median whole number latitude was calculated from the most northern and southern extents of the host's native and invasive ranges, and longitude was classified at the median latitude point or, if within an enclosed sea, the median longitude point within that sea. Ranges for latitude and longitude were based upon reports from various databases and/or the literature and represented an approximation in order to calculate a relative direct line distance between the native and non-native regions (see below).
- *Distance (km) between Native and Non-native Regions*: using median whole number latitude and longitude values, I calculated distance between source and recipient ranges using NOAA's latitude/longitude distance calculator (http://www.nhc.noaa.gov/gccalc.shtml).
- *Molecular Marker*: the molecular marker used in the studies (with a focus on mitochondrial markers).
- *Genetic Bottleneck Index*: in order to directly compare with parasite escape, the genetic bottleneck index employed here uses the same formula as for parasite escape (Torchin *et al.*, 2003, see below) taking into account genetic richness

(i.e., haplotype richness for mitochondrial markers). It includes the total haplotype richness of a species' native range and the total haplotype richness of its non-native range as reported in publications and was calculated using the following formula: $[(N_h - I_h) / N_h]$, where N_h is haplotype richness in the native region and I_h is the haplotype richness in the introduced region. The index ranges from 0 to 1, where 0 would signify no bottleneck and values close to 1 would be a very strong bottleneck. Because these totals are influenced by sampling effort, rarefaction techniques were used in as many cases as possible to predict haplotype richness in native and non-native regions.

- *Ratio of Non-native to Native Genetic Diversity*: this is the ratio of the averaged haplotype diversity for all reported populations in the non-native region to the averaged haplotype diversity for all reported populations in the non-native region. The higher the ratio value (i.e., closer to 1.0), the more similar the two regions are in their haplotype diversity.
- *Parasite Escape Index*: this index includes the total taxonomic richness of parasites in a host's native range and the total taxonomic richness of parasites in its introduced range. The index is calculated as in Torchin *et al.* (2003): $[(N_p - I_p) / N_p]$, where N_p is the parasite taxonomic richness in the native region and I_p is the parasite taxonomic richness in the introduced region. The index ranges from 0 to 1, where 0 would signify no parasite escape and 1 would signify a complete escape from parasites. In non-native regions, parasite taxonomic richness in a host can include parasites introduced with the host, or those the host has newly acquired in its non-native range (Torchin & Mitchell, 2004).

7.2.3 Data Analysis

Three measures were used as response variables in analyses exploring patterns across the global dataset: the genetic bottleneck index, the ratio of non-native to native genetic diversity, and the parasite escape index. These indexes were compared across the various species represented in Table 1, and they were also analyzed for possible influences of vector type, NIS taxa, distance, and time since introduction using ANOVAs and post-hoc Tukey's tests. Pearson's correlations of the response variables were also performed for some analyses.

Where possible, I also explored these data using the known source area of an introduction rather than the whole native range; i.e., I calculated the genetic bottleneck and parasite escape indexes using data from the source area and the non-native region, and then compared it to the regional analysis (Table 7.1). This was to determine whether there would be differences in source versus whole region analyses since parasite escape might be overstated if the entire native range is included rather than the specific source populations from which the introduction originated (Colautti *et al.*, 2004; Colautti *et al.*, 2006).

Finally, I compared genetic data in hosts versus parasites to determine whether parasites are more likely to demonstrate stronger genetic bottleneck signatures and lower genetic diversity ratios than their hosts (Figure 7.2B). To date, few studies have investigated the genetic diversities of both host and parasite; thus, this analysis represents a preliminary exploration.

7.3 Results and Discussion

7.3.1 Trends in NIS Species Classification and Source/Recipient Regions

In this global review, I found 61 systems (Table 7.1) that included genetic diversity (focused on mtDNA), parasite diversity, or both in native and non-native regions. This yielded 40 unique marine invertebrate species, 31 of which were free-living and 9 of which were parasite species. Altogether, these species represented 7 invertebrate *Classes: Asteroidea*, *Bivalvia*, *Crustacea*, *Gastropoda*, *Tentaculata*, *Trematoda*, and *Tunicata*. Bivalves, crustaceans, and gastropods made up the majority of the species represented in the study systems (Figure 7.3), similar to several prior investigations (e.g., Cohen & Carlton, 1995; Ruiz *et al.*, 2000; Blakeslee *et al.*, 2013) demonstrating the dominance of these three *Classes* in marine invasions worldwide.

The native (source) regions *from which* the 61 systems originated included five continents: Asia, Australia, Europe, North America, and South America. Non-native (founding) regions where species were *introduced to* included the same five continents and additionally Africa (Figure 7.4). However, proportions of species introductions differed between founding and source regions. For example, Asia was the continent *from which* most introductions originated, followed by North America and Europe (Figure 7.4A), but

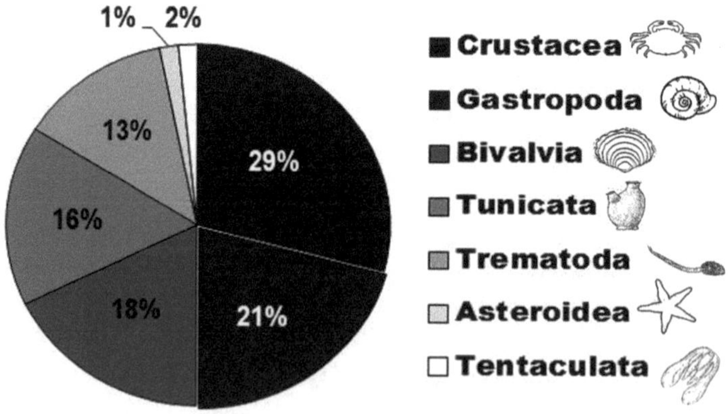

Fig. 7.3: The proportion of species in Table 7.1 that were the following seven Classes: Asteroidea, Bivalvia, Crustacea, Gastropoda, Tentaculata, Trematoda, and Tunicata. Crustaceans, gastropods, and bivalves made up two-thirds of all species introduced to new locations worldwide based on the studies in this investigation.

Asia had one of lowest proportions of species *introduced to* it (i.e., Asia was an important source but not recipient region). Instead, North America had the largest proportion of species introduced to it (>50% of all introductions; Figure 7.4B), while Europe was second highest, and collectively, North America and Europe made up over two-thirds of all the introductions *to* and *from* these regions. Several mechanisms may explain these patterns, including: *Global shipping*—in recent years, shipping has been dominated by Asian, North American and European ports, enhancing the likelihood of species transfer among these regions (Carlton, 1992; Ruiz *et al.*, 2000); *Oyster translocations*—this prominent global vector has been responsible for the accidental introduction of numerous hitchhiking species associated with oysters (e.g., bivalves, gastropods, crustaceans, and tunicates), and oysters in this vector primarily originate from two major regions: eastern North America, where *Crassostrea virginica* is native, and Asia, where *Crassostrea gigas* (Pacific oyster) is native (Ruesink *et al.*, 2005); and *Sampling bias*—many more studies published in English have been conducted in North America and Europe, and as such, reports of non-native species may be biased towards these two regions; for example,

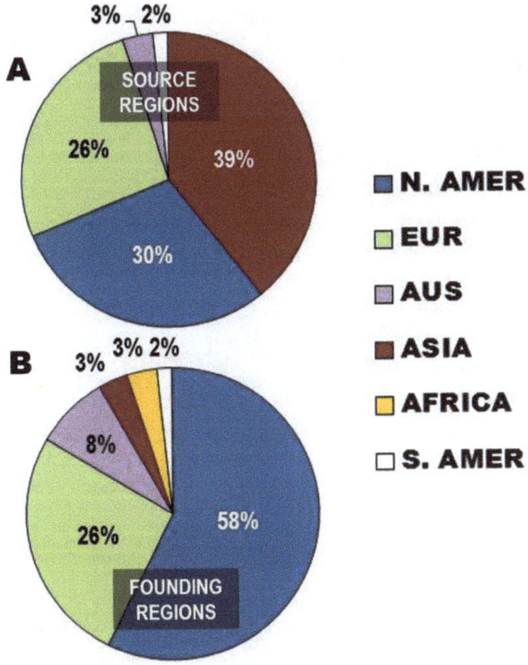

Fig. 7.4: The proportion of species in Table 7.1 that (A) came from (source regions) and were (B) introduced to (founding regions) the six continents listed above. Collectively, Asia, North America (N. AMER) and Europe (EUR) made up the largest proportion (95%) of the source regions, while Australia (AUS) and South America (S. AMER) made up the remaining 5%. In contrast, North America had the largest proportion (57%) of founding species introduced to the region, representing more than half the number of introductions for the species in Table 7.1, while Asia had one of the lowest proportions (3%).

Pysek *et al.* (2008) found clear sampling biases in invasion ecology research across continents, whereby the regions that were the most well studied were: North America > Europe > Australia > South America > Asia > Africa. With the exception of South America, these trends mirror our own data in terms of the founding/recipient regions (Figure 7.4B), where North America > Europe > Australia > Asia > Africa > South America.

7.3.2 Comparisons of Parasite Escape, Genetic Bottlenecks, and Haplotype Diversity across Studies

In general, the genetic bottleneck and parasite escape indexes showed fairly similar patterns, demonstrating about a 50% loss of haplotypes compared to a 66% loss of parasites in non-native regions compared to native regions (Figure 7.5). In fact, there was a significant positive correlation between parasite and genetic diversity losses in non-native versus native regions (Figure 7.6). This may suggest that the invasion process operates in a similar fashion for both parasite escape and genetic bottlenecks in influencing the number (and potentially types) of alleles and parasites that "survive" the process and are introduced (or not) to the new region.

While the two indexes appear to congruently support signatures of enemy release and genetic founder effects, the non-native to native haplotype diversity analysis provides seemingly contradictory results. In particular, non-native haplotype diversity represented about 75% of native haplotype diversity (Figure 7.5), suggesting less diversity loss at the population level in non-native regions rather than collectively across the region. Such a pattern for limited reductions in genetic diversity in

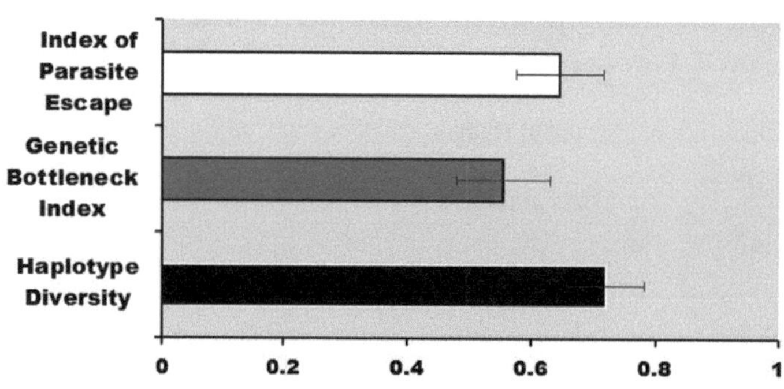

Fig. 7.5: The two indexes based on native and non-native diversity (see Methods and Table 7.1 for formulae) and the ratio of non-native to native haplotype diversity. Both indexes and the ratio represent averages (± SE) across all studies in the investigation. Both indexes demonstrate substantial levels of parasite escape and genetic bottlenecks (i.e., there has been a loss of more than 50% of the parasites and haplotypes in non-native regions). In contrast, haplotype diversity demonstrates a less substantial decline in average population-level genetic diversity in non-native versus native regions.

non-native versus native populations was the subject of the Roman & Darling (2007) paper, "Paradox lost: genetic diversity and the success of aquatic invasions". Roman & Darling (2007) hypothesized that this genetic 'paradox' – higher levels of genetic diversity than might be expected in recent founding events – could be due to multiple introductions and high levels of propagule pressure, which would result in a lessened bottleneck (see adapted Figure 7.2A). While this 'paradox' is likely playing a role here, especially for some groups (see below), another possible reason for this pattern could be because introduced populations may have been better sampled for genetic diversity than native populations, possibly limiting the ability to detect differences in average population diversity between native and non-native regions.

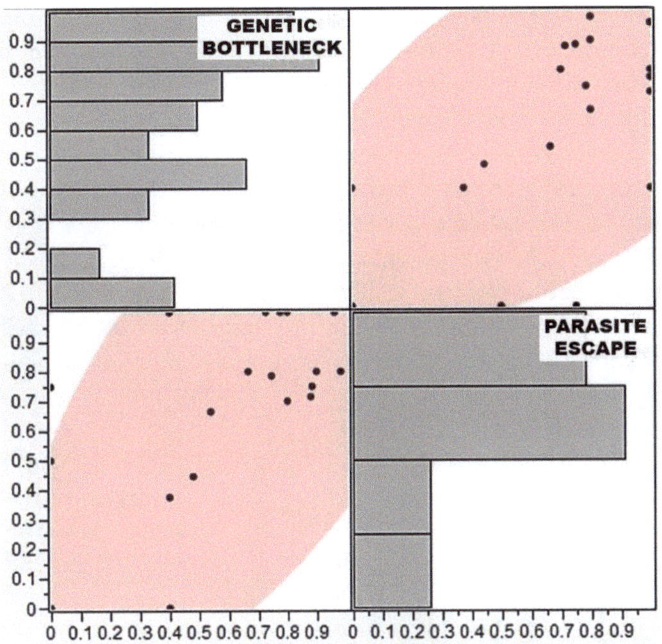

Fig. 7.6: Scatterplot Matrix of genetic bottleneck and parasite escape index correlations. This figure demonstrates correlations between the two indexes and histograms representing frequencies of proportion bins for each variable. Using a Pearson's pairwise correlation analysis, a significant positive correlation was found between the two indexes (Pearson's $r = 0.637$; $p = 0.0025$).

7.3.2.1 Source Area Analysis

While the results above suggest a substantial loss in both parasite and genetic diversity in non-native regions, they were based upon multiple native populations averaged across a larger regional exploration, which could overstate diversity losses if more precise source areas for introductions are not used for native versus non-native comparisons (Colautti *et al.*, 2004; Colautti *et al.*, 2005). The reality is that in many cases, a precise source area is unknown or difficult to pinpoint. Here, I attempted to explore whether I might find a 'source effect' in my data; however, source subregions could only be ascertained in six study species that also had available parasite and genetic diversity data. These six species included: *Batillaria attramentaria* (Asian hornsnail)–introduced from source populations in Japan to Pacific North America; *Carcinus maenas* (European green crab)–originally introduced from source populations in central/southern Europe to northeastern North America; *Ilyanassa obsoleta* (eastern mudsnail)–introduced from source populations in the mid-Atlantic USA to Pacific North America; *Littorina littorea* (common periwinkle)–introduced from the British Isles to northeastern North America; *Littorina saxatilis* (rough periwinkle)–introduced from northeastern USA to San Francisco Bay in Pacific North America; and *Rhithropanopeus harrisii* (Harris mud crab)–introduced from the Gulf of Mexico to inland Texas lakes in the USA. *Littorina littorea*, *L. saxatilis*, and *R. harrisii* have also been introduced to other locations around the world, but I only focused on the introductions described above for this analysis. Moreover, *C. maenas* has had two introduction events to northeastern North America (Roman, 2006), but this analysis focuses on the original 1800s introduction.

Altogether, there was some evidence for a 'source effect' on resulting patterns of parasite escape and genetic bottlenecks, but this was primarily for genetic diversity, where two species (*I. obsoleta* and *R. harrisii*) demonstrated substantial drops in genetic bottlenecks for source versus native regions, while another species demonstrated a modest decline (*B. attramentaria*) (Figure 7.7). In contrast, parasite escape showed much less of an effect, and only two species (*B. attramentaria* and *C. maenas*) demon-

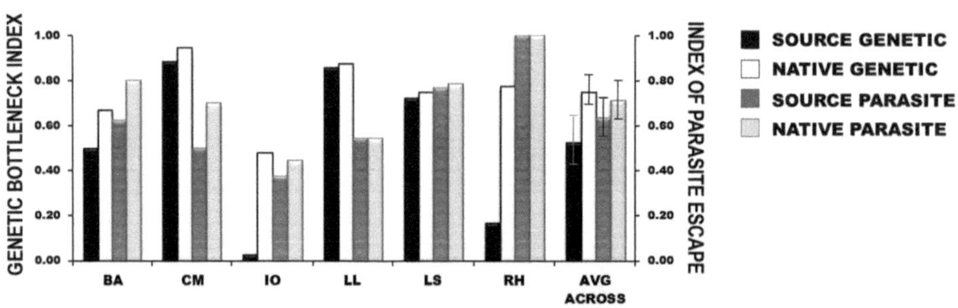

Fig. 7.7: An exploration of 6 study species where parasite and genetic diversity were reported for the native range as a whole and also for a more precise source area. The figure demonstrates the extent of the genetic bottleneck for the source area (black) and the larger native region (white), and the extent of parasite escape for the source area (dark gray) and the larger native region (light gray). Also calculated is the average across all six species. While the focus on source data can provide a more precise understanding of genetic and parasite diversity losses in non-native regions for some individuals, the analysis here of a small subset of the studies in Table 7.1 found no significant difference in parasite escape ($p = 0.64$) and genetic bottlenecks ($p = 0.478$) averaged across the six study species for source versus regional analyses. BA = *Batillaria attramentaria*, CM = *Carcinus maenas*, IO = *Ilyanassa obsoleta*, LL = *Littorina littorea*, LS = *Littorina saxatilis*, RH =*Rhithropanopeus harrisii*.

strated modest reductions in source versus native regions. When averaged across the six species, there was no evidence of a 'source effect' in one-way ANOVAs for genetic bottlenecks ($p = 0.478$) and parasite escape ($p = 0.649$), nor in a two-way ANOVA for both indexes ($p = 0.783$).

On the whole, these data suggest (albeit based on a small sample) that pinpointing precise source regions may help better understand effects on parasite and genetic diversity in non-native regions for some species. However, more data are needed to determine if these results are representative across systems, or if it is species- and/or invasion pathway-dependent. For example, in Figure 7.7, *I. obsoleta* showed evidence of a 'source effect' for genetic diversity; this may be due to its invasion vector—oysters—which are commonly associated with strong entrainment and transfer of propagules to non-native regions. *Ilyanassa obsoleta* was introduced to the west coast as a hitchhiking species with commercial shipments of the eastern oyster (*Crassostrea virginica*) (Carlton, 1992), and these shipments occurred on a massive scale sustained over many years (Miller, 2000); in addition, oysters were packaged for shipping in a manner ensuring their survival, and also enhancing the survival of hitchhiking organisms (Carlton, 1979). Thus, the intentional movement of oysters and associated individuals (including parasitized ones) has likely strongly influenced parasite escape and genetic bottlenecks in this species. In contrast, another intertidal snail species, *L. saxatilis,* demonstrates much greater levels of both parasite escape and genetic bottlenecks in its introduced region on the USA west coast, where there is little difference between source and regional analyses (Figure 7.7). Its introduction vector is much different: *L. saxatilis* was transferred to the west coast as an associate of packing algae in the live baitworm trade (Carlton & Cohen, 1998). In general, the magnitude of algal packing materials and associated individuals transferred with the live bait vector is far less than for commercial oysters, and the vector itself is accidental, which would promote fewer associated individuals than the intentional oyster vector (Blakeslee *et al.*, 2012).

7.3.3 Introduction Vector, NIS Taxa, Distance between Source and Recipient Region, and Time since Introduction

7.3.3.1 Introduction Vector

Because some vectors are associated with higher levels of propagule pressure than others, these vectors may be more likely to introduce parasites and alleles. When I explored this possibility, I found no significant differences ($p = 0.364$) in a two-way ANOVA for the two indexes with vector; however, when vectors were lumped into intentional (e.g., aquaculture/oysters) versus accidental (e.g., wet and dry ship ballast, hull fouling, and hosts) vectors, I found a significant difference between intentional and accidental vectors for both genetic bottlenecks ($p = 0.041$) and parasite escape ($p = 0.049$) in individual one-way ANOVAS and also in a two-way ANOVA (p = 0.042) (Fig. 8). As discussed in Blakeslee *et al.* (2013), accidental vectors like ballast water may be less likely to introduce parasites than intentional vectors like oysters because ballast water primarily transfers larvae from native to non-native locations and larvae are not typically the infective stages of most marine parasites; moreover, propagule pressure (especially related to introduction of parasites) would be expected to be higher for intentional introductions, like oysters, versus accidental vectors, like ballast water (Torchin & Mitchell, 2004). In fact, Torchin & Lafferty (2009) suggested that "ballast water introduction may be a particularly potent means for marine species to escape parasites." While some of these expectations might also hold true for genetic bottlenecks, introduced larvae via shipping vectors could still contribute to the genetic diversity of a NIS' non-native region, and this may help explain why there was a trend for the ballast water vector to have a higher index of parasite escape than a genetic bottleneck (Figure 7.8). Altogether, these results may also support some of the expectations presented in Figure 7.2A, where certain vectors would be more likely to lead to strong bottlenecks, while others show little difference between native and non-native regions as a result of high propagule pressure and multiple introductions.

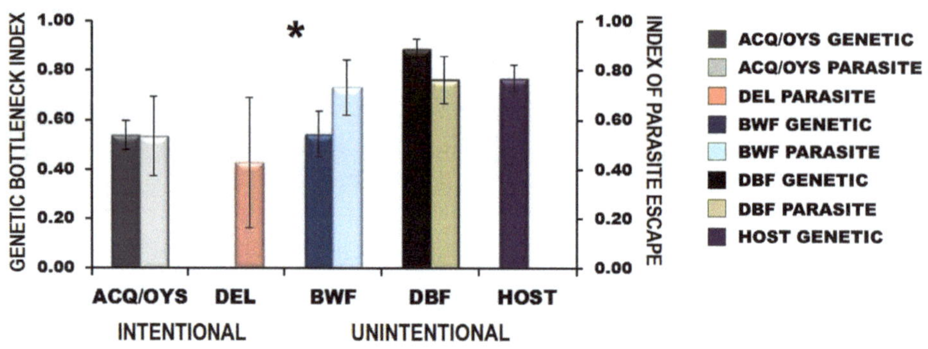

Fig. 7.8: Analysis of vector as a factor influencing the genetic bottleneck index (dark shades) and the parasite escape index (light shades). While there were no significant differences for vector, there was a significant difference between intentional and accidental vectors, whereby intentional vectors demonstrate lower levels of genetic bottlenecks and parasite escape than unintentional vectors (represented by a *). AQC/OYS = aquaculture/oysters; DEL = deliberate, non-aquaculture introduction; DBF = dry ballast & fouling; HOST = parasite introductions with their host.

7.3.3.2 NIS Taxonomy

When I explored the effect of NIS taxonomy (at the *Class* level) in native versus non-native regions in individual one-way ANOVAs, I found a significant effect of *Class* on the genetic bottleneck index ($p < 0.0001$) and the parasite escape index ($p = 0.043$), and also in a two-way ANOVA for both indexes ($p = 0.010$). For parasite escape, there was only sufficient data for bivalves, crustaceans, and gastropods. Post-hoc analyses demonstrated similar results for bivalves and crustaceans for both indexes, where crustaceans had significantly ($p < 0.05$) greater genetic and parasite diversity losses compared to bivalves. Moreover, for the genetic bottleneck index, two other groups—gastropods and trematodes—also demonstrated significantly ($p < 0.05$) greater diversity losses than bivalves (Figure 7.9). These data suggest the importance of taxonomic groups in influencing the introductions of alleles and associated parasites; e.g., congruent signatures between the two indexes for two *Classes*: crustaceans and bivalves. In both cases, crustaceans demonstrated much greater parasite escape and loss of haplotypes in non-native regions than did bivalves. Another interesting finding was how much lower the two indexes were for bivalves compared to the other investigated *Classes*. A similar result was observed just for parasite escape in the global review by Blakeslee *et al.* (2013). A possible explanation is that one of the most prominent bivalves in our analysis was the Pacific oyster, *Crassostrea gigas*, which is not only a vector for movement of other free-living organisms (including other bivalve species) but also for the transmission of hitchhiking parasites (Ruesink *et al.*, 2005). As a result, propagule pressure is likely much higher for this vector, and correspondingly, greater numbers of associated alleles and parasites could be transferred with the bivalve to introduced regions. Moreover, the results of this analysis further exemplify the expectations in Figure 7.2A; however, in this case, some taxonomic groups (e.g., bivalves) are more associated with higher propagule pressure and introduction of alleles and parasites than others (e.g., crustaceans) that demonstrate strong bottlenecks and parasite escape.

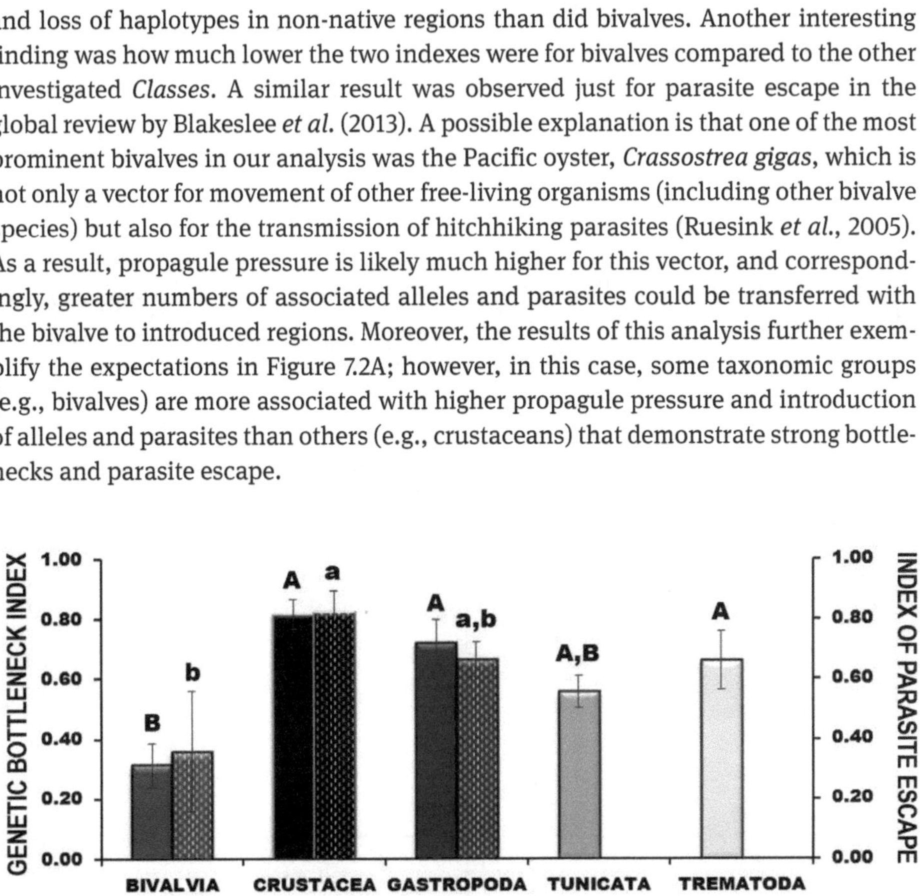

Fig. 7.9: Analysis of taxonomy as a factor influencing the genetic bottleneck index (dark shades) and the parasite escape index (patterned shades) by Class. There was a significant effect of Class on the genetic bottleneck index ($p < 0.001$). Post-hoc Tukey's tests revealed significant differences between crustaceans and bivalves ($p < 0.001$), gastropods and bivalves ($p = 0.007$), and trematodes and bivalves ($p = 0.021$); crustaceans and tunicates showed a nearly significant difference ($p = 0.063$). For parasite escape, the overall analysis was also significant ($p = 0.043$), and there was a significant difference between crustaceans and bivalves ($p = 0.034$). Significance is represented by upper case letters for genetic bottlenecks and lower case letters for parasite escape.

7.3.3.3 Distance between Source and Recipient Regions

Distance between source and recipient regions might be expected to influence parasite and genetic diversity in non-native regions because distance can serve as a proxy for transit time and stress on hitchhiking organisms (Miller & Ruiz, 2009). In other words, if the distance between the source and recipient regions is short, the native and non-native ranges are likely to experience more frequent connectivity, to share more phylogenetically similar taxa, and to allow for a greater proportion of entrained species (and parasites) to survive the journey (e.g. Drake & Lodge, 2004), which may lead to less pronounced bottlenecks and parasite escape. In my analyses, I found little support for this expectation (parasite escape: R = 0.001; p = 0.636; genetic diversity: R = 0.004; p = 0.862; native to non-native ratio: R = 0.084; p = 0.066), except in a couple instances for specific groups: there was a significant positive correlation between the genetic bottleneck index and distance for *Tunicata* ($R^2 = 0.419$; p = 0.043), thus the bottleneck increased with distance, and there was also a significant negative correlation for the non-native to native haplotype diversity ratio and distance for the aquaculture/oyster vector (Figure 7.10); in other words, as distance increased, population-level haplotype diversity was lower in non-native regions compared to native regions. This result is a bit more difficult to explain considering the strong propagule pressure in the oyster vector, thus I would have predicted little effect of distance for this vector type. There were no apparent patterns for parasite escape.

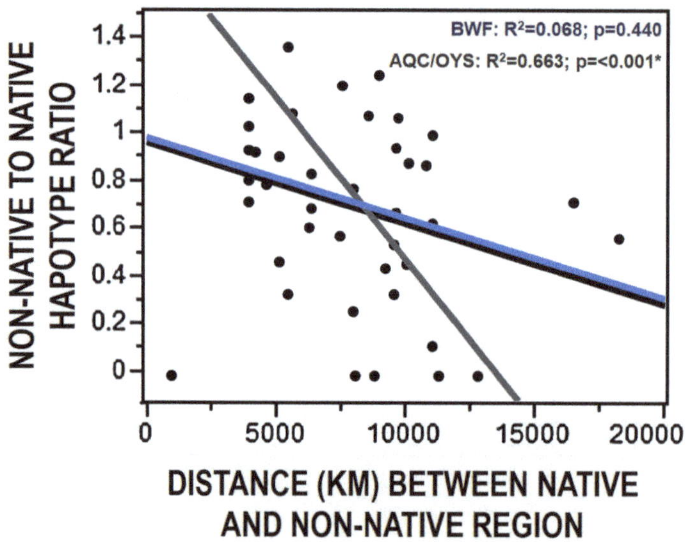

Fig. 7.10: Regression of the ratio of non-native to native haplotype diversity with distance (km) between source and recipient regions, grouped by vector. Altogether, there was a significant negative correlation for the vector, AQC/OYS (gray line), but no correlation for BWF (blue line).

7.3.3.4 Time since Introduction

Finally, I explored the potential effect of time since introduction on genetic or parasite diversity in non-native regions. Time since introduction may be expected to have an effect if older introductions have had more time to accrue more alleles and more parasites than newer introductions (Torchin & Lafferty, 2009). However, I found no effect of introduction timing on either of the two indexes, nor on the ratio of non-native to native haplotype diversity based on the study species in Table 7.1 (data not shown).

7.3.4 Do Parasites Demonstrate Greater Losses of Genetic Diversity in Non-native Regions than their Hosts?

In Figure 7.2B, I proposed that parasites may be more likely to demonstrate genetic bottlenecks compared to their hosts based on their more complex life cycles, which often require multiple suitable hosts. For this analysis, I was only able to compile evidence for four hosts where there was also genetic evidence for their parasites (n = 7). My analysis here based on these four hosts and seven parasites does not appear to support this hypothesis for either response variable ($p = 0.616$ and $p = 0.814$, respectively), albeit the sample size is very small (Figure 7.11). While Blakeslee & Fowler (2012) found some support for greater genetic diversity in aquatic systems (freshwater and marine) for non-native hosts compared to parasites, presently there is too little evidence in marine systems to adequately assess this question. More host-parasite systems need to be analyzed in order to determine whether such a pattern is likely to exist across multiple marine communities.

Fig. 7.11: The genetic bottleneck index (dark shades) and the ratio of non-native to native haplotype diversity (light shades) in hosts versus parasites. This analysis includes four hosts (*B. attramentaria*, *I. obsoleta*, *L. littorea*, and *R. harrisii*) and seven parasites (six trematodes and one rhizocephalan). Altogether, there is no difference between hosts and parasites for either analysis.

7.3.5 Conclusions: Parasite and Genetic Analyses—Implications for their Use in Marine Invasions

Altogether, these results demonstrate the complexity of marine introductions and how they are influenced by several variables associated with their invasion pathway, e.g., vector, source area, NIS taxa, and geography. In particular, I found trends for differences in diversity analyses based on the type of vector and propagule pressure; e.g., intentional vectors demonstrated lessened bottlenecks and parasite escape than did accidental vectors. I also found a significant effect of NIS taxa on diversity indexes, particularly for crustaceans and bivalves, which demonstrated higher versus lower losses of diversity, respectively. I also found geographic distance between source and recipient regions to be a factor for a few vector and taxonomic comparisons.

Even with all this complexity, my analysis continues to support conventional expectations for genetic founder effects and parasite escape in non-native regions, in that both haplotype and parasite richness were significantly lower in non-native versus native regions in this global review. In addition, I found a significant, positive correlation between the index of parasite escape and the genetic bottleneck index, showing that reductions in both parasite and genetic diversity can be closely linked and that they represent strong signatures of invasion—albeit, depending on propagule pressure, the signature may be a lot less apparent for some NIS than for others (Figure 7.2A; Roman & Darling, 2007).

On the whole, these results further emphasize the utility of these two signatures, especially when used together and with other lines of evidence for helping to resolve uncertain invasion histories and those species where invasion status remains uncertain (cryptogenic). These signatures can be especially important when historical information about the species is vague or unknown. For example, originally thought to be native to Europe, the Portuguese oyster (*Crassostrea angulata*) was discovered within the last 15 years to have actually been introduced from Asia as a result of intentional transplantation by Portuguese traders in the 16th century. As a result of its misclassified status, management and conservation plans in Europe had been based upon its incorrectly assigned native status, and there was even concern about the possible impact of the Pacific oyster, *C. gigas*, on the abundance and distribution of *C. angulata* in the region (Huvet *et al.*, 2000). The use of genetics, therefore, was very important in resolving this misconception and the oyster's true origin. Moreover, there are numerous other examples for how genetic data can be a highly important tool in marine investigations (reviewed in Geller *et al.*, 2010), and many other investigations on the use of parasite diversity in better understanding host invasions (reviewed in Blakeslee *et al.*, 2013); however, few investigations have explored these two invasion signatures in concert. Such a combination can be an even more powerful approach for resolving uncertainties and better understanding invasion processes in marine invertebrate systems, particularly gastropods, bivalves, and crustaceans, which are the most commonly introduced marine organisms globally (e.g., Ruiz *et al.*, 2000) and also common hosts for marine parasites (e.g., Torchin *et al.*, 2002). For example, via host and parasite genetic analyses, Miura *et al.* (2006) were able to pinpoint the source area within Asia for *B. attramentaria's* introduction to western North America, and they also discovered that what was originally believed to be a single associated parasite species introduced with the snail was, in actuality, three cryptic species.

In conclusion, while genetic evidence has been recognized as a powerful tool in biological investigations for many years, parasites have been an understudied resource. In this review, I have demonstrated how parasites can be highly valuable to studies of global marine invasions, especially in cases where numerous uncertainties exist—a common reality in many NIS studies that go undetected for years following a successful introduction. In turn, there may be a multitude of questions surrounding an invasion; thus, innovative scientific clues may be required. As argued here, a combination of parasite and genetic evidence, along with other sources of evidence, could in fact provide the needed proof to resolve many of these challenging invasion questions. Moreover, for many invasive species, parasite and genetic diversity losses may in fact correlate, providing even more informative evidence for studies of biological invasions.

7.4 Acknowledgements

I gratefully acknowledge the editor, João Canning Clode, the associate editor, Anssi Vainikka, and two reviewers, Dr. Oliver Verneau and an additional anonymous reviewer, for their guidance and assistance on this chapter. I also thank Dr. Joe Roman and Dr. John Darling for permission to build upon their figure in Roman & Darling (2007).

In a nutshell

- The movement of marine organisms, especially invertebrate species, has rapidly increased with enhanced human globalization. As a result, accidental or intentional introductions of marine NIS have added numerous new species to marine ecosystems, including parasites—less visible associates of invading species that can have major impacts on native communities.
- Two patterns often emerge in species introductions: significant genetic bottlenecks (i.e., founder effects) and significant reductions in parasite diversity (i.e., parasite escape) in founding populations. While both signatures are apparent in some systems, one or both may be less so in others, especially when there have been multiple introductions. Yet few studies have synergistically examined these signatures to determine potential correlations, or if variables associated with invasion pathways influence the patterns.
- Using a meta-analysis of global marine invertebrate introductions with parasite and/or genetic evidence, this study found haplotype and parasite richness to be significantly lower in non-native versus native regions at large scales; additionally, positive correlations were found between the two diversity indexes. Results also demonstrated the complexity of marine introductions and the influence of invasion pathway variables on genetic and parasite diversity patterns, including vector, source area, NIS taxa, and geography.
- While genetic evidence has long been recognized as a powerful tool across biological disciplines, the role that parasites can play in such investigations is much less recognized. This study demonstrates the importance of both signatures in better understanding biological invasions.

7.5 Bibliography

Aguirre-Macedo, M.L., Kennedy C.R. (1999). Diversity of metazoan parasites of the introduced oyster species *Crassostrea gigas* in the Exe estuary. Journal of the Marine Biological Association, 79, 57–63.

Albins, M. A., Hixon, M. A. (2013). Worst case scenario: potential long-term effects of invasive predatory lionfish (*Pterois volitans*) on Atlantic and Caribbean coral-reef communities. Environmental Biology of Fishes, 96, 1151-1157.

Ashton, G.V., Stevens, M.C., Hart, D.H., *et al.* (2008). Mitochondrial DNA reveals multiple Northern Hemisphere introductions of *Caprella mutica* (Crustacea, Amphipoda). Molecular Ecology, 17, 1293–303.

Armonies, W., Reise, K. (1998). On the population development of the introduced razor clam *Ensis americanus* near the island of Sylt (North Sea). Helgoländer Meeresuntersuchungen, 52, 291-300.

Asif, J.H., Krug, P.J. (2011). Lineage distribution and barriers to gene flow among populations of the globally invasive marine mussel *Musculista senhousia*. Biological Invasions, 14, 1431–1444.

Blakeslee, A.M.H. (2007). Native or Invasive? The case history of the marine snail, *Littorina littorea*, in northeast North America. In D. Starkey, P. Holm, M. Barnard (Eds.), Oceans Past: Management Insights from the History of Marine Animal populations (pp.7-24). London, England: Earthscan.

Blakeslee, A.M.H., Byers, J.E. (2008). Using parasites to inform ecological history: Comparisons among three congeneric marine snails. Ecology, 89, 1068–1078.

Blakeslee, A.M.H., Byers, J.E., Lesser, M.P. (2008). Solving cryptogenic histories using host and parasite molecular genetics: the resolution of *Littorina littorea's* North American origin. Molecular Ecology, 17, 3684–3696.

Blakeslee, A.M.H., Keogh, C.L., Byers, J.E., *et al.* (2009). Differential escape from parasites by two competing introduced crabs. Marine Ecology Progress Series, 393, 83–96.

Blakeslee, A.M.H., Fowler, A.E. (2011). Aquatic introductions and genetic founder effects: how do parasites compare to hosts? Genetic Diversity, 1, 315-336.

Blakeslee, A.M.H., Altman, I., Miller, A.W., *et al.* (2012). Parasites and invasions: a biogeographic examination of parasites and hosts in native and introduced ranges. Journal of Biogeography, 39, 609–622.

Blakeslee, A.M.H., Fowler, A.E., Keogh, C.L. (2013). Marine invasions and parasite escape: updates and new perspectives. Advances in Marine Biology, 66, 87-169.

Boudry, P., Heurtebise, S., Collet, B., *et al.* (1998). Differentiation between populations of the Portuguese oyster, *Crassostrea angulata* (Lamark) and the Pacific oyster, *Crassostrea gigas* (Thunberg), revealed by mtDNA RFLP analysis. Journal of Experimental Marine Biology and Ecology, 226, 279–291.

Boyle T., Keith, D., Pfau, R. (2010). Occurence, reproduction, and population genetics of the estuarine mud crab, *Rhithropanopeus harrisii* (Gould) (Decapoda, panopidae) in Texas freshwater reservoirs. Crustaceana, 83, 493–505.

Brant, S.V., Cohen, A.N., James, D., *et al.* (2010). Cercarial dermatitis transmitted by exotic marine snail. Emerging Infectious Diseases, 16, 1357.

Brawley, S.H., Coyer, J.A., Blakeslee, A.M.H., *et al.* (2009). Historical invasions of the intertidal zone of Atlantic North America associated with distinctive patterns of trade and emigration. Proceedings of the National Academy of Sciences, 106, 8239–8244.

Byers, J.E. (2000). Competition between two estuarine snails: Implications for invasions of exotic species. Ecology, 81, 1225–1239.

Byers, J.E. (2009). Competition in marine invasions. In G. Rilov, J. A. Crooks (Eds.), Biological invasions in marine ecosystems: ecological, management, and geographic perspectives (pp.245-260). New York: Springer.

Carlton, J.T. (1979). History, biogeography, and ecology of the introduced marine and estuarine invertebrates of the Pacific coast of North America. PhD Thesis. Davis, CA: University of California.

Carlton, J.T. (1992). Introduced marine and estuarine mollusks of North America: an end-of-the-20th-century perspective. Journal of Shellfish Research, 11, 489–505.

Carlton, J.T., Geller, J.B. (1993). Ecological roulette: the global transport of nonindigenous marine organisms. Science, 261, 78–82.

Carlton, J.T. (1996a). Pattern, process, and prediction in marine invasion ecology. Biological Conservation, 78, 97–106.

Carlton, J.T. (1996b). Biological invasions and cryptogenic species. Ecology, 77, 1653-1655.

Carlton, J.T., Cohen, A.N. (1998). Periwinkle's progress: the Atlantic snail *Littorina saxatilis* (Mollusca: Gastropoda) establishes a colony on a Pacific shore. Veliger, 41, 333–338.

Carlton, J.T. Cohen, A.N. (2003). Episodic global dispersal in shallow water marine organisms: the case history of the European shore crabs *Carcinus maenas* and *C. aestuarii*. Journal of Biogeography, 30, 1809–1820.

Carlton, J. T. (2008). The zebra mussel *Dreissena polymorpha* found in North America in 1986 and 1987. Journal of Great Lakes Research, 34, 770-773.

Chang, A.L., Blakeslee, A.M.H., Miller, A.W., *et al.* (2011). Establishment failure in biological invasions: a case history of *Littorina littorea* in California, USA. PLOS One, 6, e16035.

Chapman, J.W., Carlton, J.T. (1991). A test of criteria for introduced species: the global invasion by the isopod *Synidotea laevidorsalis* (Miers, 1881). Journal of Crustacean Biology, 11, 386–400.

Cohen, A.N., Carlton, J.T. (1995). Nonindigenous aquatic species in a United States estuary: a case study of the biological invasions of the San Francisco Bay and delta. Executive summary report for the United States Fish and Wildlife Service, Washington, DC and the National Sea Grant College Program, Connecticut Sea Grant.

Coles, S.L., DeFelice, R.C., Eldredge, L.G. *et al.* (1999). Historical and recent introductions of non-indigenous marine species into Pearl Harbor, Oahu, Hawaiian Islands. Marine Biology, 135, 147–158.

Couceiro, L., López, L., Ruiz, J.M. *et al.* (2012). Population structure and range expansion: the case of the invasive gastropod *Cyclope neritea* in northwest Iberian Peninsula. Integrative Zoology, 7, 286–298.

Criscione, C.D., Cooper, B., Blouin, M. (2006). Parasite genotypes identify source populations of migratory fish more accurately than fish genotypes. Ecology, 87, 823–826.

Cristescu, M.E.A., Hebert ,P.D.N., Witt, J.D.S. *et al.* (2001). An invasion history for *Cercopagis pengoi* based on mitochondrial gene sequences. Limnology and Oceanography, 46, 224–229.

Cristescu, M.E.A., Witt, J.D.S., Grigorovich, I.A. *et al.* (2004). Dispersal of the Ponto-Caspian amphipod *Echinogammarus ischnus*: invasion waves from the Pleistocene to the present. Heredity, 92, 197–203.

Crowl, T.A., Crist, T.O., Parmenter, R.R. *et al.* (2008). The spread of invasive species and infectious disease as drivers of ecosystem change. Frontiers in Ecology and the Environment, 6, 238–246.

Curtis, L.A. (1997). *Ilyanassa obsoleta* (Gastropoda) as a host for trematodes in Delaware estuaries. The Journal of Parasitology, 83, 793-803.

Dang, C., De Montaudouin, X., Bald, J. *et al.* (2009). Testing the enemy release hypothesis: trematode parasites in the non-indigenous Manila clam *Ruditapes philippinarum*. Hydrobiologia, 630, 139–148.

Darling, J.A., Bagley, M.J., Roman, J. *et al.* (2008). Genetic patterns across multiple introductions of the globally invasive crab genus *Carcinus*. Molecular Ecology, 17, 4992–5007.

Dinamani, P. (1987). Gametogenic patterns in populations of Pacific oyster, *Crassostrea gigas*, in Northland, New Zealand. Aquaculture, 64, 65–76.

Drake, J.M., Lodge, D.M. (2004). Global hot spots of biological invasions: evaluating options for ballast–water management. Proceedings of the Royal Society of London B, Biological Sciences, 271, 575–580.

Elsner, N.O., Jacobsen, S., Thieltges, D.W. *et al.* (2011). Alien parasitic copepods in mussels and oysters of the Wadden Sea. Helgoland Marine Research, 65, 299–307.

English, L.J., Maguire, G.B., Ward, R.D. (2000). Genetic variation of wild and hatchery populations of the Pacific oyster, *Crassostrea gigas*, in Australia. Aquaculture, 187, 283–298.

Epifanio, C.E. (2013). Invasion biology of the Asian shore crab *Hemigrapsus sanguineus*: a review. Journal of Experimental Marine Biology and Ecology, 441, 33-49.

Foighil, D., Gaffney, P., Wilbur, A.E. *et al.* (1998). Mitochondrial cytochrome oxidase I gene sequences support an Asian origin for the Portuguese oyster *Crassostrea angulata*. Marine Biology, 131, 497–503.

Fowler, A.E., Forsström, T., von Numers, M. *et al.* (2013). The North American mud crab *Rhithropanopeus harrisii* (Gould, 1841) in newly colonized Northern Baltic Sea: distribution and ecology. Aquatic Invasions, 8, 89-96.

Geller, J.B., Darling and, J.A., Carlton, J.T. (2010). Genetic perspectives on marine biological invasions. Annual Review of Marine Science, 2, 367–393.

Gérard, K., Bierne, N., Borsa, P. *et al.* (2008). Pleistocene separation of mitochondrial lineages of *Mytilus* spp. mussels from Northern and Southern Hemispheres and strong genetic differentiation among southern populations. Molecular Phylogenetics and Evolution, 49, 84-91.

Ghabooli, S., Shiganova, T.A., Zhan, A. *et al.* (2010). Multiple introductions and invasion pathways for the invasive ctenophore *Mnemiopsis leidyi* in Eurasia. Biological Invasions, 13, 679–690.

Gillis, N.K., Walters, L.J., Fernandes, F.C. *et al.* (2009). Higher genetic diversity in introduced than in native populations of the mussel *Mytella charruana*: evidence of population admixture at introduction sites. Diversity and Distributions, 15, 784–795.

Glasby, T.M., Connell, S.D., Holloway, M.G. *et al.* (2006). Nonindigenous biota on artificial structures: could habitat creation facilitate biological invasions? Marine Biology, 151, 887–895.

Goldstien, S.J., Dupont, L., Viard, F. *et al.* (2011). Global phylogeography of the widely introduced North West Pacific ascidian *Styela clava*. PLOS One, 6, e16755.

Grosberg, R., Cunningham, C.W. (2000). Genetic structure in the sea from populations to communities. In M. Bertness, S. Gaines, M. Hay (Eds.), Marine Community Ecology (pp.61-88). Sunderland, Massachusetts: Sinauer Associates.

Hanson, D., Cooke, S., Hirano, Y. *et al.* (2013). Slipping through the cracks: the taxonomic impediment conceals the origin and dispersal of *Haminoea japonica*, an invasive species with impacts to human health. PLOS One, 8, e77457.

Hawkes, C.R., Meyers, T.R., Shirley, T.C. *et al.* (1986). Prevalence of the parasitic barnacle *Briarosaccus callosus* on king crabs of southeastern Alaska. Transactions of the American Fisheries Society, 115, 252–257.

Hechinger, R.F. (2007). Annotated key to the trematode species infecting *Batillaria attramentaria* (Prosobranchia: Batillariidae) as first intermediate host. Parasitology International, 56, 287–296.

Hemmingsen, W., Jansen, P.A., MacKenzie, K. (2005). Crabs, leeches and trypanosomes: an unholy trinity? Marine Pollution Bulletin, 50, 336–339.

Hoos, P.M., Miller, A.W., Ruiz, G.M. *et al.* (2010). Genetic and historical evidence disagree on likely sources of the Atlantic amethyst gem clam *Gemma gemma* (Totten, 1834) in California. Diversity and Distributions, 16, 582–592.

Hua, C. (1989). Life cycle of *Proctoeces orientalis* sp. nov. in marine bivalves. Acta Zoologica Sinica, 35, 58–65.

Huvet, A., Lapègue, S., Magoulas, A. *et al.* (2000). Mitochondrial and nuclear DNA phylogeography of *Crassostrea angulata* , the Portuguese oyster endangered in Europe. Conservation Genetics, 1, 251–262.

Jansen, P.A., MacKenzie, K., Hemmingsen, W. (1998). Some parasites and commensals of red king crabs, *Paralithodes camtschaticus* (Tilesius), in the Barents Sea. Bulletin of the European Association of Fish Pathology, 18, 46–49.

Jeschke, J.M., Aparicia, L.G., Haider, S. *et al.* (2012). Support for major hypotheses in invasion biology is uneven and declining. NeoBiota, 14, 1–20.

Jokela, J., Lively, C.M. (1995). Parasites, sex, and early reproduction in a mixed population of freshwater snails. Evolution, 49, 1268-1271.

Keane, R.M., Crawley, M.J. (2002). Exotic plant invasions and the enemy release hypothesis. Trends in Ecology and Evolution, 17, 164–170.

Kelly, D.W., Muirhead, J.R., Heath, D.D. *et al.* (2006). Contrasting patterns in genetic diversity following multiple invasions of fresh and brackish waters. Molecular Ecology, 15, 3641–53.

Kolar, C.S., Lodge, D.M. (2001). Progress in invasion biology: predicting invaders. Trends in Ecology and Evolution, 16, 199–204.

Kolar C.S., Chapman, D.C., Courtenay, W.R. Jr. *et al.* (2005). Asian carps of the genus *Hypophthalmichthys* (Pisces, Cyprinidae)—a biological synopsis and environmental risk assessment. Washington D.C.: U.S. Fish and Wildlife Service.

Krakau, M., Thieltges, D.W., Reise, K. (2006). Native parasites adopt introduced bivalves of the North Sea. Biological Invasions, 8, 919–925.

Kruse, I., Hare, M.P. (2007). Genetic diversity and expanding nonindigenous range of the rhizocephalan *Loxothylacus panopaei* parasitizing mud crabs in the western north Atlantic. The Journal of Parasitology, 93, 575–82.

Kruse, I., Hare, M.P., Hines, A.H. (2011). Genetic relationships of the marine invasive crab parasite *Loxothylacus panopaei*: an analysis of DNA sequence variation, host specificity, and distributional range. Biological Invasions, 14, 701–715.

Kuris, A.M., Blau, S.F., Paul, A.J. *et al.* (1991). Infestation by brood symbionts and their impact on egg mortality of the red king crab, *Paralithodes camtschatica*, in Alaska: geographic and temporal variation. Canadian Journal of Fisheries and Aquatic Sciences, 48, 559–568.

Lafferty, K.D., Kuris, A.M. (2009). Parasitic castration: the evolution and ecology of body snatchers. Trends in Parasitology, 25, 564–572.

Lassalle, G., De Montaudouin, X., Soudant, P. *et al.* (2007). Parasite co-infection of two sympatric bivalves, the Manila clam (*Ruditapes philippinarum*) and the cockle (*Cerastoderma edule*) along a latitudinal gradient. Aquatic Living Resources, 20, 33-42.

Lauckner, G. (1987a). Ecological effects of larval trematode infestation on littoral marine invertebrate populations. International Journal of Parasitology, 17, 391–398.

Lauckner, G. (1987b). Effects of parasites on juvenile Wadden Sea invertebrates. In S. Tougaard, S. Asbirk (Eds.), Proceedings of the 5th International Wadden Sea Symposium (pp.103-121). Esbjerg: The National Forest and Nature Agency and the Museum of Fisheries and Shipping.

Lavergne, S., Molofsky, J. (2007). Increased genetic variation and evolutionary potential drive the success of an invasive grass. Proceedings of the National Academy of Sciences, 104, 3883–3888.

Lee, M.K., Cho, B.Y., Lee, S.J. *et al.* (2001). Histopathological lesions of Manila clam, *Tapes philippinarum*, from Hadong and Namhae coastal areas of Korea. Aquaculture, 201, 199–209.

Lejeusne, C., Bock, D.G., Therriault, T.W. *et al.* (2010). Comparative phylogeography of two colonial ascidians reveals contrasting invasion histories in North America. Biological Invasions, 13, 635–650.

Liu, H., Stiling, P. (2006). Testing the enemy release hypothesis: a review and meta-analysis. Biological Invasions, 8, 1535–1545.

Mann, R., Burreson, E.M., Baker, P.K. (1991). The decline of the Virginia oyster fishery in Chesapeake Bay: considerations for introduction of a non-endemic species, *Crassostrea gigas* (Thunberg, 1793). Journal of Shellfish Research, 10, 379–388.

Marcogliese, D.J. (2002). Food webs and the transmission of parasites to marine fish. Parasitology, 124, 83–99.

Martel, C., Viard, F., Bourguet, D. *et al.* (2004). Invasion by the marine gastropod *Ocinebrellus inornatus* in France. Journal of Experimental Marine Biology and Ecology, 305, 155–170.

Mei, C. (1994). The parasitized location of larval *Cercaria elegans* Tang 1992 in *Ruditapes philippinarum* and its histochemistry. Acta Zoologica Sinica, 4, 007.

McDermott, J.J. (2011). Parasites of shore crabs in the genus *Hemigrapsus* (Decapoda: Brachyura: Varunidae) and their status in crabs geographically displaced: a review. Journal of Natural History, 45, 2419-2441.

Miller, A.W. (2000). Assessing the importance of biological attributes for invasion success: Eastern oyster (*Crossostrea virginica*) introductions and associated molluscan invasions of Pacific and Atlantic coastal systems. PhD Thesis. Los Angeles, CA: University of California Los Angeles.

Miller, A.W., Ruiz, G.M. (2009). Differentiating successful and failed invaders: species pools and the importance of defining vector, source and recipient regions. In G. Rilov, J. Crooks (Eds.), Biological Invasions in Marine Ecosystems: Ecological, Management, and Geographic Perspectives (pp.153-170). Berlin: Springer Verlag.

Miller, A., Inglis, G.J., Poulin, R. (2008). Use of the introduced bivalve, *Musculista senhousia*, by generalist parasites of native New Zealand bivalves. NZJ Marine Freshwater Research, 42, 143–151.

Miura, O., Torchin, M.E., Kuris, A.M. *et al.* (2006). Introduced cryptic species of parasites exhibit different invasion pathways. Proceedings of the National Academy of Sciences, 103, 19818–19823.
Moehler, J., Wegner, K.M., Reise, K., *et al.* (2011). Invasion genetics of Pacific oyster *Crassostrea gigas* shaped by aquaculture stocking practices. Journal of Sea Research, 66, 256–262.
Morley, N.J. (2008). The role of the invasive snail *Potamopyrgus antipodarum* in the transmission of trematode parasites in Europe and its implications for ecotoxicological studies. Aquatic Sciences, 70, 107–114.
Panova M., Blakeslee, A.M.H., Miller, A.W. *et al.* (2011). Survival of a North Atlantic marine snail in multiple glacial refugia—implications for phylogeographic patterns. PLOS One, 6, e17511.
Park, K.I., Tsutsumi, H., Hong, J.S. *et al.* (2008). Pathology survey of the short-neck clam *Ruditapes philippinarum* occurring on sandy tidal flats along the coast of Ariake Bay, Kyushu, Japan. Journal of Invertebrate Pathology, 99, 212–219.
Petersen, C. (2006). Range expansion in the Northeast Pacific by an estuary mud crab – a molecular study. Biological Invasions, 8, 565–576.
Petersen K.S., Rasmussen, K.L., Heinemeier, J. *et al.* (1992). Clams before Columbus? Nature, 359, 679.
Pringle, J., Blakeslee, A.M.H., Byers, J.E. *et al.* (2011). Asymmetric dispersal allows an upstream region to control population structure throughout a species' range. Proceedings of the National Academy of Sciences, 108, 15288-15293.
Projecto-Garcia, J., Cabral, H., Schubart , C.D. (2009). High regional differentiation in a North American crab species throughout its native range and invaded European waters: a phylogeographic analysis. Biological Invasions, 12, 253–263.
Pysek, P., Richardson, D.M., Pergl, J.*et al.* (2008). Geographical and taxonomic biases in invasion ecology. Trends in Ecology and Evolution, 23, 237–244.
Ratnasingham S., Hebert, P.D. (2007). The barcode of life data system (http://www.barcodinglife.org). Molecular Ecology Notes, 7, 355–364.
Rilov, G., Galil, B. (2009). Marine bioinvasions in the Mediterranean Sea–history, distribution and ecology. In G. Rilov, J.A. Crooks (Eds.), Biological invasions in marine ecosystems: ecological, management, and geographic perspectives (pp.549-575). New York: Springer.
Roman, J. (2006). Diluting the founder effect: cryptic invasions expand a marine invader's range. Proceedings of the Royal Society B, Biological Sciences, 273, 2453–2459.
Roman, J., Darling, J. A. (2007). Paradox lost: genetic diversity and the success of aquatic invasions. Trends in Ecology and Evolution, 22, 454–464.
Ruesink, J.L., Lenihan, H.S., Trimble, A.C. *et al.* (2005). Introduction of non-native oysters: Ecosystem Effects and Restoration Implications. Annual Review of Ecology, Evolution, and Systematics, 36, 643–689.
Ruiz, G.M., Fofonoff, P.W., Carlton, J.T. *et al.* (2000). Invasion of coastal marine communities in North America: apparent patterns, processes, and biases. Annual Review of Ecology, Evolution, and Systematics, 31, 481–531.
Rybakov, A.V. (1983). Parthenitae and larvae of trematodes in the bivalve mollusc *Ruditapes philippinarum* in the Petr Velikii Bay, Sea of Japan. Biologiya Morya, Vladivostok, USSR, 1, 12–20.
Rybakov, A.V., Mamaev, Y.L. (1987). Helminth fauna of *Batillaria cumingii* (Gastropoda: Potaminidae) in the Peter the Great Bay, Sea of Japan. Gel'minty i vyzyvaemye imi zabolevaniya, 77–87.
Saltonstall, K. (2002). Cryptic invasion by a non-native genotype of the common reed, *Phragmites australis*, into North America. Proceedings of the National Academy of Sciences, 99, 2445–2449.
Selander, E., Møller, L.F., Sundberg, P. *et al.* (2010). Parasitic anemone infects the invasive ctenophore *Mnemiopsis leidyi* in the North East Atlantic. Biological Invasions, 12, 1003–1009.
Simberloff, D. (2009). The role of propagule pressure in biological invasions. Annual Review of Ecology, Evolution, and Systematics, 40, 81–102.

Simberloff, D., Martin, J.L., Genovesi, P. *et al.* (2013). Impacts of biological invasions: what's what and the way forward. Trends in Ecology and Evolution, 28, 58–66.

Simon-Bouhet, B., Garcia-Meunier, P., Viard, F. (2006). Multiple introductions promote range expansion of the mollusk *Cyclope neritea* (Nassariidae) in France: evidence from mitochondrial sequence data. Molecular Ecology, 15, 1699–1711.

Sparks, A.K., Morado, J.F. (1987). A putative carcinoma-like neoplasm in the hindgut of a red king carb, *Paralithodes camtschatica.* Journal of Invertebrate Pathology, 50, 45–52.

Stefaniak, L., Lambert, G., Gittenberger, A. *et al.* (2009.) Genetic conspecificity of the worldwide populations of *Didemnum vexillum* Kott, 2002. Aquatic Invasions, 4, 29-44.

Thieltges, D.W., Krakau, M., Andresen, H. *et al.* (2006). Macroparasite community in molluscs of a tidal basin in the Wadden Sea. Helgoland Marine Research, 60, 307-316.

Thieltges, D.W., Engelsma, M.Y., Wendling, C.C. *et al.* (2013). Parasites in the Wadden Sea food web. Journal of Sea Research, 82, 122-133.

Thomsen, M.S., Wernberg, T., Tuya, F. *et al.* (2010). Ecological performance and possible origin of a ubiquitous but under-studied gastropod. Estuarine and Coastal Shelf Science, 87, 501–509.

Torchin, M.E., Lafferty, K.D., Kuris, A.M. (2001). Release from parasites as natural enemies: increased performance of a globally introduced marine crab. Biological Invasions, 3, 333–345.

Torchin, M.E., Lafferty, K.D., Kuris, A.M. (2002). Parasites and marine invasions. Parasitology, 124, S137–S151.

Torchin, M.E., Lafferty, K.D., Dobson, A.P. *et al.* (2003). Introduced species and their missing parasites. Nature, 421, 628–630.

Torchin, M.E., Mitchell, C.E. (2004). Parasites, pathogens, and invasions by plants and animals. Frontiers in Ecology and the Environment, 2, 183–190.

Torchin, M.E., Byers, J.E., Huspeni, T.C. (2005). Differential parasitism of native and introduced snails: replacement of a parasite fauna. Biological Invasions, 6, 885–894.

Torchin, M.E., Lafferty, K.D. (2009). Escape from parasites. In G. Rilov, J.A. Crooks (Eds.), Biological invasions in marine ecosystems: ecological, management, and geographic perspectives (pp.203-214). New York: Springer.

Troost, K. (2010). Causes and effects of a highly successful marine invasion: Case-study of the introduced Pacific oyster *Crassostrea gigas* in continental NW European estuaries. Journal of Sea Research, 64, 145-165.

Vierna, J., Jensen, K.T., González-Tizón, A.M. *et al.* (2012). Population genetic analysis of *Ensis directus* unveils high genetic variation in the introduced range and reveals a new species from the NW Atlantic. Marine Biology, 159, 2209–2227.

Villalba, A., Mourelle, S.G., Carballal, M.J.*et al.* (1997). Symbionts and diseases of farmed mussels *Mytilus galloprovincialis* throughout the culture process in the Rías of Galicia (NW Spain). Diseases of Aquatic Organisms, 31, 127-139.

Vitousek, P.M., Mooney, H.A., Lubchenco, J. *et al.* (1997). Human domination of Earth's ecosystems. Science, 277, 494–499.

Willis, K.J., Cook, E.J., Lozano-Fernandez, M. *et al.* (2004). First record of the alien caprellid amphipod, *Caprella mutica*, for the UK. Journal of the Marine Biological Association of the UK, 84, 1027-1028.

Yanagida, T., Shirakashi, S., Iwaki, T. *et al.* (2009). Gymnophallid digenean *Parvatrema duboisi* uses Manila clam as the first and second intermediate host. Parasitology International, 58, 308–310.

Zardus, J. D., Hadfield, M. G. (2005). Multiple origins and incursions of the Atlantic barnacle *Chthamalus proteus* in the Pacific. Molecular Ecology, 14, 3719–3733.

Zeitlmeisl, C., Hermann, J., Petney, T. *et al.* (2011). Parasites of the shore crab *Carcinus maenas* (L.): implications for reproductive potential and invasion success. Parasitology, 138, 394–401.

Japo Jussila, Al Vrezec, Jenny Makkonen, Raine Kortet, Harri Kokko

8 Invasive Crayfish and Their Invasive Diseases in Europe with the Focus on the Virulence Evolution of the Crayfish Plague

8.1 Introduction

The high financial and cultural value of the freshwater crayfish in Europe (Lehtonen, 1975; Ackefors, 1998; Edsman, 2004; Jussila & Mannonen, 2004; Jussila *et al.*, 2014a) and the devastation of the native crayfish stocks during 20th century (Alderman, 1996; Souty-Grosset *et al.*, 2006) encouraged fisheries officers and researchers in several European countries to grasp the opportunity to introduce alien freshwater crayfish into Europe (Holdich *et al.*, 2009). The possibility was created by the eradication of the native freshwater crayfish stocks after the introduction of *Aphanomyces astaci*. This parasite was first identified in the River Po valley (Italy) in the 1850s. Afterwards, it gained access to France and Germany, and from there the disease spread rapidly to other European countries (Alderman, 1996). The recommendation that one should always take a cautious approach, stated already at the end of 19th century (e.g. Hubad, 1894), was largely ignored and the bizarre decision was made to introduce alien freshwater crayfish from the geographical region of *A. astaci*'s original distribution (Svärdson, 1965; Westman, 2000; Holdich *et al.*, 2009). Thus, the widespread introduction of different *A. astaci* strains along with their native, comparably *A. astaci*-resistant host species was initiated.

The original introduction of *A. astaci* to Europe was most probably accidental, although its vector is still not known. It took approximately 50 years before astacologists discovered the cause of the mysterious mass mortalities devastating the European crayfish stocks, but by that time, *A. astaci* had spread throughout Europe (Alderman, 1996). The resulting attempts to restore the crayfisheries resulted in an even greater catastrophe: the introduction of several novel alien pathogen strains and new host species that could function as a permanent reservoir for the pathogens. The impact was felt by both the remaining native crayfish stocks and the biodiversity of the European aquatic ecosystems (Souty-Grosset *et al.*, 2006; Nyström, 1999; Nyström *et al.*, 1999; Ruokonen, 2012).

After the turn of the millennium, strategies to deal with alien species were drafted at the EU level, as well as by individual EU member countries (e.g. MMM, 2012; EU, 2013). These strategies clearly tackled the apparent threats posed by the alien invading species. Thus, alien freshwater crayfish (defined as 'detrimental') and their diseases (defined as 'extremely detrimental') were listed as threats to the European ecosystems' biodiversity (MMM, 2012). One should note that in the case of Nordic aquatic ecosystems, a detrimental alien host species (e.g. signal crayfish, *Pacifastacus leniusculus*) is normally a chronic carrier of an extremely detrimental pathogen species (e.g. *A. astaci*). To make matters even more complicated, at least from the academic viewpoint, the alien invaders and their diseases have over time created a novel and complex environmental situation; i.e. not only is there a rapid co-evolution of the native and alien crayfish but there is also their co-adaptations against the disease agent, *A. astaci*. As a result, the resistance of both native European and alien crayfish against the crayfish plague has changed, as has the virulence of the disease agent, *A. astaci* (e.g. Jussila *et al.*, 2011, 2013a, 2014a, b; Makkonen *et al.*, 2012a, b; Makkonen, 2013).

This chapter provides an overview of the introductions of the alien species and speculates on the co-evolution of the parasite and its hosts, in addition to background about the adaptation scenarios. Recent studies on the virulence of *A. astaci* and the relationship between the disease agent and its native European and alien hosts both in the wild and laboratory conditions (Makkonen, 2013; Jussila *et al.*, 2014a,b) make it possible to speculate on the impacts of the introduction of alien species. We feel that the main issue when evaluating the introduction of an alien species should not be their potential financial benefits, but instead the focus should be on avoiding their possible extensive disastrous effects on native ecosystems. By taking a more cautious approach, it should be possible to avoid the devastating consequences presently associated with the introductions. Introducing an alien species into the natural environment is an irreversible act, how hard can it be to understand this simple fact?

8.2 European Crayfish: Indigenous Diversity and Xenodiversity

Before one can understand the spread and evolution of crayfish diseases in Europe, one must first be aware of the past and current patterns in crayfish distributions. The coexistence of crayfish species represents the main route by which a parasite can jump from one host to another. This provides favourable habitat conditions for invasive parasites such as *A. astaci*. In fact, these conditions have been changed extensively during the last 150 years after the first introduction of *A. astaci* to Europe (Alderman, 1996) and they have involved local native crayfish species extinctions, their translocations, and the introduction of new alien species (Souty-Grosset *et al.*, 2006; Holdich *et al.*, 2009; Kouba *et al.*, 2014).

The diversity of the European native crayfish can be stated to consist of five species, at least some of which most probably represent species complexes (Starobogatov, 1995; Largiader *et al.*, 2000; Trontelj *et al.*, 2005; Holdich *et al.*, 2009; Klobučar *et al.*, 2013). As estimated by Souty-Grosset *et al.* (2006), the original distribution ranges of native crayfish species in Europe were to a great extent parapatric, with the overlapping range being mainly confined to South-Eastern Europe (Figure 8.1A). The sympatric occurrence was limited to three species at most, which rarely formed syntopic populations, as indicated by the high displacement rates encountered in the case of introductions even within crayfish plague-free native crayfish species (Stucki & Romer, 2001; Gherardi, 2002). This marked segregation is emphasized by the great genetic phylogeographic divergence of crayfish populations found especially within the *Austropotamobius* genus (Fratini *et al.*, 2005; Trontelj *et al.*, 2005; Klobučar *et al.*, 2013). However, human translocations have significantly changed species ranges and thus have led to compulsory coexistence between species. The first translocations took place in historical times, an example being the translocation of the white-clawed crayfish *(Austropotamobius pallipes)* into Western European countries (Souty-Grosset *et al.*, 2006). After a devastating crayfish plague epidemic that eradicated crayfish populations across Europe, intensive crayfish translocation programs were instigated in the 20th century in many European countries in an attempt to restock the lost crayfish populations (Henttonen & Huner, 1999). These were intracontinental translocations of native species, especially of the more important economic species such as the noble crayfish *(Astacus astacus)* and narrow-clawed crayfish *(A. leptodactylus)* (Souty-Grosset *et al.*, 2006; Schrimpf *et al.*, 2011), but also intercontinental translocations of species that had originated from North America and Australia (Holdich *et al.*, 2009; Kouba *et al.*, 2014).

Unfortunately, these translocations brought together many native and alien crayfish species, which under natural conditions would not coexist. This situation promotes fierce competitive interactions between species, as well as rapid transmission of diseases. Since the crayfish plague epidemics were the main reason behind the new crayfish translocations, the species that have been introduced can generally be divided into *A. astaci*-susceptible species (native *Astacus* and *Austropotamobius* species, the introduced Australian *Cherax* species) and *A. astaci*-resistant North American species such as the signal crayfish *(P. leniusculus)* and several cambarid species, i.e. *Procambarus* and *Orconectes* (Souty-Grosset *et al.*, 2006). The latter group currently encompasses nine species, which are found in natural water bodies across Europe (Kouba *et al.*, 2014). The greatest xenodiversity of this group is confined to Western Europe, where most of the introductions had taken place (Figure 8.1B). In contrast, the diversity of susceptible species, currently including five native species and two *Cherax* species (Kouba *et al.*, 2014), does not show such a specific regional limitation, but a core of diversity is found across continental Europe (Figure 8.1C), which was mainly a consequence of intracontinental translocations of native species. Overall, due to the recent translocations and introductions, the situation in Europe has changed significantly, and today most crayfish species currently inhabiting European waters exhibit sympatric distribution patterns, and in many cases syntopic populations, this being especially true in Western and Central Europe (Figure 8.1D). This novel biogeographic pattern in crayfish fauna in Europe has to be taken into account when dealing with disease transmissions and the evolution of the pathogen with rapid jumping from one host to another, including mixtures of different genotypes.

Here, we will briefly present the status of the alien invasive crayfishes in Europe, with a special emphasis on their potential to colonise and also spread the alien disease agents that they could carry, especially *A. astaci*. The chapter should give enough background to highlight the devastating potential that the alien crayfish pose not only to their native European counterparts but also to aquatic ecosystem biodiversity.

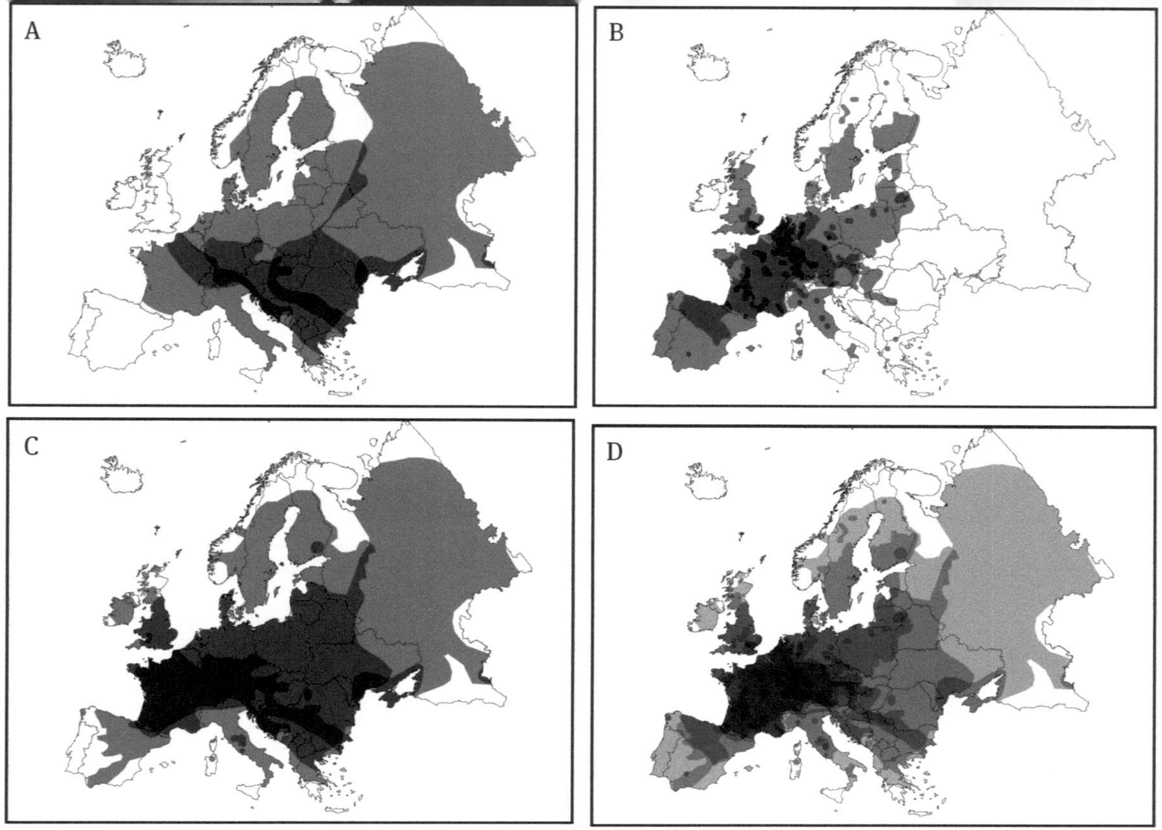

Fig. 8.1: Overlaps in the distribution ranges of crayfish species in Europe as indicated by intensity of grey shadowing (light grey indicates presence of only one crayfish species) with (A) original distribution patterns of native species (*Astacus*, *Austropotamobius*), (B) recent update of distribution patterns of introduced crayfish plague-resistant North American species (*Procambarus*, *Orconectes*, *Pacifastacus*), (C) recent update of distribution patterns of native and introduced crayfish plague-susceptible species (*Astacus*, *Austropotamobius*, *Cherax*), and (D) recent update of overall distribution patterns of all crayfish species currently present in Europe in the wild (data modified from Souty-Grosset *et al.*, 2006; Kouba *et al.*, 2014).

8.2.1 Signal Crayfish (*Pacifastacus leniusculus*)

The rationale for the introductions of the signal crayfish to Europe was the assumption that it could rejuvenate the crayfisheries in those water bodies where crayfish plague prevented the recovery of the wild native crayfish stocks (Fjälling & Fürst, 1985; Westman, 2000; Kirjavainen & Sipponen, 2004; Bohman *et al.*, 2011). The signal crayfish introductions were initiated in the late 1950s with the first experimental shipments arriving in Sweden (Svärdson, 1995; Holdich *et al.*, 2009), followed by an increasing number of introductions in Sweden and Finland. Initially, the signal crayfish were restricted to experimental farms, mostly owned by government research institutes and later to commercial farms where large numbers were transferred into natural water bodies. The introduced signal crayfish had been captured from wild stocks living in the western parts of North America, mainly from Lake Tahoe and Lake Hennessy (Abrahamsson, 1969; Westman, 1973). The introduced stocks had originated from a mixture of sites (Larson *et al.*, 2012) and it has been recently discovered that the species *Pacifastacus leniusculus* is indeed a combination of three subspecies (Agerberg & Jansson, 1995; Larson *et al.*, 2012). In addition to these introductions into Northern Europe, the signal crayfish were also transferred to continental Europe, e.g. into Austria from California and into France from Oregon (Souty-Grosset *et al.*, 2006). Currently the signal crayfish is the most widely distributed alien crayfish in Europe; in the majority of other countries it was introduced from the original Swedish stock dispersed naturally, a process which continues rapidly at the present time (Holdich *et al.*, 2009; Kouba *et al.*, 2014).

It is now generally acknowledged that the signal crayfish acts as a chronic carrier of PsI-genotype *A. astaci* (Alderman *et al.*, 1990; Bohman *et al.*, 2006; Filipová *et al.*, 2013; Viljamaa-Dirks *et al.*, 2013; Jussila *et al.*, 2014a, b) with recent evidence indicating that other *A. astaci* genotypes might also infect these crayfish (Aydin *et al.*, 2014). In addition, PsII-genotype *A. astaci* has been reported to infect the signal crayfish (Huang *et al.*, 1994). Signal crayfish also seem to be affected by opportunistic pathogens when they are living in the Nordic countries (Persson & Söderhäll, 1983; Thörnqvist & Söderhäll, 1993; Edsman *et al.*, 2015).

8.2.2 *Procambarus* Species

The most widespread and invasive *Procambarus* species in Europe is the red swamp crayfish *(Procambarus clarkii)*, which was introduced into Spain from Louisiana in 1973 (Souty-Grosset *et al.*, 2006). Due to its wide ecological adaptation to different types of water bodies, including seasonally flooded wetlands, it has spread rapidly and ultimately has become an important commercial species in Spain. This led to new illegal introductions not only throughout Spain, but also in France and Italy, and later in the Netherlands, Germany, and to many islands including Great Britain,

Azores, Canary Islands, Balearic Islands, Sardinia, Sicily and even Cyprus (Holdich *et al.*, 2009). This species is still spreading rapidly across continental Europe (Kouba *et al.*, 2014), but its northern expansion to colder climates might be restricted since it is a warm water species (Souty-Grosset *et al.*, 2006).

The red swamp crayfish has long been known to be a carrier of *A. astaci* (Diéguez-Uribeondo & Söderhäll, 1993; Aquiloni *et al.*, 2011). The species is thought to be quite tolerant against this disease, but stressful conditions might cause death and extensive sporulation of *A. astaci* (Souty-Grosset *et al.*, 2006). This species has a major influence on aquatic ecosystems not only through disease transmissions to native crayfish populations, but also since it is an efficient polytrophic predator, exerting a detrimental impact on species living at several trophic levels in the aquatic animal community (Renai & Gherardi, 2004) and on macrophytes with indirect effects even at higher trophic levels, i.e. waterbirds (Souty-Grosset *et al.*, 2006).

In Europe at the moment, three more *Procambarus* species from North America have been identified in the wild (Kouba *et al.*, 2014): the marbled crayfish *(P. fallax* f. *virginalis)*, white river crayfish *(P.* cf. *acutus)*, and Florida crayfish *(P. alleni)*. The marbled crayfish (*Procambarus fallax*) was introduced to Europe as an aquarium pet, due to its visual appearance and possibly also for its efficient reproduction, only later discovered to be parthenogenetic (Scholtz *et al.*, 2003; Holdich *et al.*, 2009). The first documented wild population was detected in Germany in 2003 (Marten *et al.*, 2004) and since then there have been reports of small, scattered, introduced and established populations in the Netherlands, Italy and Slovakia (Soes & van Eekelen, 2006; Marzano *et al.*, 2009; Chucholl *et al.*, 2012). The species has also been found in Sweden but it does not seem to be properly established there yet (Bohman *et al.*, 2013). In fact, at least 25 independent introductions have been recorded in Europe, and the probability of expansion of this species has been shown to be very high. For this reason, it is essential that consistent trade regulations be urgently introduced in order to limit its spread and to prohibit further introductions (Chucholl, 2014). The marbled crayfish can be a carrier of *A. astaci* (Scholtz *et al.*, 2003; Holdich *et al.*, 2009) and our experiments have indicated that marbled crayfish may exhibit elevated resistance towards crayfish plague (unpublished data). In summary, it seems that the marbled crayfish is a potential colonizer and dangerous carrier of *A. astaci*. The taxonomic position of the white river crayfish is not yet clear and includes at least the white river crayfish *(P. acutus)* as well as the southern white river crayfish *(P. zonangulus)* (Kouba *et al.*, 2014). The introduction of this taxon in Europe (Spain) in the 1970s failed (Henttonen & Huner, 1999), but an established population was detected in the Netherlands in 2005 (Soes & van Eekelen, 2006) and later in Great Britain (Kouba *et al.*, 2014). At present, little is known about its possible threat to the European native crayfish. The Florida crayfish *(Procambarus alleni)* is freely available through the aquarium trade in Europe, and although individual specimens have been caught at several sites in France and Germany, it is not presently known if there is an established population (Kouba *et al.*, 2014).

8.2.3 *Orconectes* Species

The spiny-cheek crayfish *(Orconectes limosus)* is native to North America, where its original distribution area was in the Northeastern United States (Hamr, 2002). In Europe, *O. limosus* was first introduced to Poland in 1890 (Kossakowski, 1966; Aklehnovich & Razlutskij, 2013) and is thus the oldest alien crayfish species known in Europe. At present, it has been identified in at least 22 countries (Kouba *et al.*, 2014). *O. limosus* is currently spreading towards Northern Europe, having reached Lithuania (Arbačiauskas *et al.*, 2011); towards Eastern Europe in Belarus (Alkenovich & Razlutskij, 2013); and towards South Eastern Europe via the River Danube (Hudina *et al.*, 2009; Pârvulescu *et al.*, 2009). Based on genetic variability studies, there has likely been only a single incident of introduction of *O. limosus* into Europe (Filipová *et al.*, 2009; Filipová *et al.*, 2011) and thereafter the spread has occurred both naturally and by human-mediated translocations.

O. limosus has also been shown to carry *A. astaci* (Kozubíková *et al.*, 2011a; Matasová *et al.*, 2011; Pârvulescu *et al.*, 2012; Schrimpf *et al.*, 2012). However, the prevalence of the infected individuals seems to be very variable in different populations (Kozubíková *et al.*, 2011a; Matasová *et al.*, 2011) and in some cases, it seems that non-infected populations do exist. Some of these populations even coexist with native European crayfish species (Schrimpf *et al.*, 2013a). Similarly to the signal crayfish (Strand *et al.*, 2012), *O. limosus* has been shown to pose a constant threat to native European crayfish species due to the continual release of infectious *A. astaci* spores from infected individuals (Svoboda *et al.*, 2013). Although it has long been known that *O. limosus* can act as a vector for *A. astaci* (Vey *et al.*, 1983), it was only recently discovered that it also carries a novel genotype of *A. astaci* (Kozubíková *et al.*, 2011b). This Or-genotype can be differentiated from other known genotypes by the random amplification of polymorphic DNA (RAPD) technique, or with microsatellite markers (Grandjean *et al.*, 2014). Unfortunately, due to the rather limited availability of these novel strains, there is still very little information about its specific characteristics.

In addition to *O. limosus*, at least three other *Orconectes* species have been introduced into Europe since 1990, and all of them are regarded as non-susceptible *A. astaci* carriers (Schrimpf *et al.*, 2013b; Kouba *et al.*, 2014): calico crayfish *(O. immunis)*, Kentucky River crayfish (*O. juvenilis*), and virile crayfish *(O.* cf. *virilis)*. The calico crayfish was first detected in the wild in 1997 in Germany (Dehus *et al.*, 1999), and since then it has rapidly colonized the Upper Rhine system both up- and downstream, and is now spreading into France (Chucholl, 2012). It is a species that exerts a major impact on the ecosystem. In the River Rhine, *O. immunis* came into contact with the previously established *O. limosus* and has ultimately displaced it (Chucholl, 2012). This represents a new perspective about recent introductions, in the sense that one may have artificially created unsustainable syntopic populations. The species has been shown to be an *A. astaci* carrier with a relatively high infection prevalence (Schrimpf *et al.*, 2013b). The Kentucky River crayfish (*O. juvenilis*) was first discovered in France in 2005 (Chucholl & Daudey, 2008), suspected to have escaped from a pond into the River Dessoubre. Though this alien population is still rather limited, it is considered to represent a threat to native European species due its potential to act as a carrier of *A. astaci*, its rapid life cycle, its high fecundity and its pollution tolerance. One attempt was made to eradicate this restricted established population but was not successful (Kouba *et al.*, 2014).

The identity of the fourth *Orconectes* species established in Europe is less clear, since it belongs to the virile crayfish *(O.* cf. *virilis)* species complex. The species introduction history in Europe is quite long, since the first imports are believed to have happened in France in 1897 and then were repeated in 1960 in Sweden, but apparently both trials failed (Souty-Grosset *et al.*, 2006). However, in 2004 a thriving population was discovered in the Netherlands (Soes & Koese, 2010), which has spread so rapidly that it has even displaced the already established *O. limosus* (Kouba *et al.*, 2014). Furthermore, their appearance in the wild, presumably after escaping from an aquarium, took place in 2004 in London (UK) from where they have started to disperse at a rate 2 km y^{-1} (Holdich *et al.*, 2009). Newly introduced *Orconectes* species have been demonstrated to possess a high invasive potential and dispersal capability, and one can predict that in the future they will contribute to a significant increase in overall crayfish xenodiversity in Europe.

8.2.4 *Cherax* Species

There seems to be two reasons to account for the introduction of the different *Cherax* species throughout Europe: 1) the aquarium trade (Holdich *et al.*, 2009) and 2) the possibility for farming (Souty-Grosset *et al.*, 2006). The putative aquaculture potential of these crayfish was based on experiences from Australia and also from the belief that the *Cherax* species could tolerate elevated temperatures (Morrissy, 1990; King, 1994), which were also benefits in the eyes of aquarium enthusiasts. Currently, at least two species have found their way into the wild in Europe: the yabbie (*Cherax destructor*) and the redclaw (*C. quadricarinatus*). The yabbie has been successfully introduced to Catalonia and Navarra in Spain from stock imported from California in 1983, and to Zaragosa between 1984 and 1985 (Souty-Grosset *et al.*, 2006), and Italy in 2008 (Scalici *et al.*, 2009a, b). The redclaw is extensively sold live across Europe through the aquarium trade as well as for farming (Kouba *et al.*, 2014). In 2009, this species was found to be established in a small thermal oxbow lake in Slovenia, but several single specimens have been caught in different parts of Europe as well (Jaklič & Vrezec, 2011). Both *Cherax* species are so far restricted to isolated water bodies, although their spread potential is a matter of concern.

As an alien species, *Cherax* pose a potential threat to the native European crayfish even though they are regarded as susceptible to *A. astaci* infection. The population of yabbie in Navarra has actually been successfully eradicated by introducing *A. astaci*, and there have been reports of devastating outbreaks of crayfish plague in farmed redclaws at Sicily (Kouba *et al.*, 2014). Furthermore, all *Cherax* species introduced to Europe are fast growing and reproduce efficiently (Lawrence & Jones, 2002), and thus they possess the potential for colonization and to spread further if there is not an immediate instigation of an eradication campaign of existing established populations (Tricarico *et al.*, 2010).

8.3 Signal Crayfish in the Nordic Context

The widespread introduction of the signal crayfish into Europe was initiated in the Nordic Countries in order to revitalise the crayfisheries; a void had been created by the eradication of the majority of the wild noble crayfish stocks by crayfish plague epidemics and changes in aquatic ecosystems. The driving force behind these rather hasty actions was the tradition of crayfish trapping and related cultural festivities (Jussila *et al.*, 2014a). The signal crayfish has now achieved an established position in the Nordic countries and currently accounts for the majority of the annual crayfish catch (Jussila & Mannonen, 2004; Fiskeriverket, 2005). Its commercial value and the general relevance of crayfish as the centrepiece of crayfish parties—an important occasion in the late summer in the Nordic countries—has allowed these alien signal crayfish to effectively take the place of the native noble crayfish both in Nordic aquatic ecosystems and as the crayfish being celebrated and consumed in crayfish parties (Ackefors, 1998). Unfortunately, the negative aspects of the signal crayfish on the aquatic ecosystem (Nyström, 1999; Ruokonen, 2012), and especially on the native crayfish (Bohman *et al.*, 2006), have been largely ignored and the continuous spreading of this invasive alien crayfish has even been encouraged (Jussila *et al.*, 2014a).

In Finland and Sweden, the first signal crayfish introductions were restricted to the southern parts of the countries (Jussila & Mannonen, 2004; Kirjavainen & Sipponen, 2004). There were several reasons for limiting the introductions to the southern parts, e.g. the environmental threats of the signal crayfish and the protection of the existing productive noble crayfish stocks. However, once the introductions of the signal crayfish had started, there were also illegal introductions. Now the signal crayfish has spread very efficiently, paying no heed to various national and regional crayfisheries strategies, legal proceedings against illegal stockings, and intensive and innovative information campaigns (Jussila *et al.*, 2014a). The strong cultural ties between the crayfish and the Nordic peoples have only assisted the spread of the alien signal crayfish, similar to the situation with the native noble crayfish some 150 years earlier (Kilpinen, 2003). This has created a situation where alien crayfish stocks are growing in numbers, crayfish plague is commonplace, and the very existence of the native noble crayfish is threatened.

Contrary to earlier belief, the signal crayfish has been proven to be susceptible to *A. astaci* (Persson & Söderhäll, 1983; Thörnqvist & Söderhäll, 1993; Aydin *et al.*, 2014) to such an extent that several collapses of wild stocks have been reported (Jussila *et al.*, 2014a, b; Sandström *et al.*, 2014). The signal crayfish has been shown to act as a chronic *A. astaci* carrier (Alderman *et al.*, 1990; Bohman *et al.*, 2006; Diéguez-Uribeondo, 2006; Jussila *et al.*, 2014a, b), thus spreading the disease presumed to be the virulent PsI-genotype. This chronic crayfish plague infection has even been noted to decrease their commercial value (e.g. Jussila *et al.*, 2013b). There have been collapses of the signal crayfish populations, but other effects have also been reported (Smith & Söderhäll, 1986; Pakkasmaa, 2006; Jussila *et al.*, 2013b; Edsman *et al.*, 2015; Jussila *et al.*, 2014a, b). The present rate of stock collapses in the natural environment has been estimated to be at least 10% of all known stocks in Finland and Sweden (Sahlin *et al.*, 2010; Jussila *et al.*, 2014a; L. Edsman, personal communication, November 13, 2013). Another stressful factor on the stocks of signal crayfish are the increasingly unpredictable micro-climatic changes, like very rapid water cooling in autumn or the lakes becoming ice-covered but then thawing soon afterwards, with the latter process possibly repeating itself several times during one winter. As the climate becomes warmer, the habitat suitable for the signal crayfish is spreading northwards (Capinha *et al.*, 2013) and the belief that the native noble crayfish stocks would be protected by the long cold winters may be naïve. It is also worth noting that crayfish immune defences seem to have an innate seasonal clock (Gruber *et al.*, 2014b), and the related patterns may be distorted by predicted climatic change. So far, the 65th northern parallel has been taken as the upper limit for the signal crayfish because it cannot reproduce in colder climatic conditions (Heinimaa & Pursiainen, 2010), but milder winters will tend to push that boundary further north.

The signal crayfish is having an adverse effect on the biodiversity of aquatic ecosystems (Nyström, 1999; Ruokonen, 2012; Ercoli, 2014) since it is exerting an impact on macro-invertebrate richness and community composition. The signal crayfish acts as a link between littoral and profundal areas and may even interfere with the abundance of fish parasites. In addition to acting on the ecosystem level, huge signal crayfish populations can interfere with traditional ways of fishing by becoming tangled in fishermen's nets, as some of the best sites for catching whitefish (*Coregonus* sp.) are the stony hard-bottom shallows of the water bodies, which signal crayfish also inhabit.

8.4 The Diseases of the Invasive Crayfish

8.4.1 Crayfish Plague (*Aphanomyces astaci*)

The first mass mortalities of native crayfish were detected in Lombardy (Italy) in 1859 (Cornalia, 1860). During the next three decades, the crayfish plague spread rapidly from France and Germany all over Continental Europe, both east and west, mainly along large river watersheds (Alderman, 1996), reaching Finland via Russia in 1893 (Järvi, 1910). From Finland, the disease then spread to Sweden in 1907 and from Sweden to Norway in 1971, but only in the late 1970s and 1980s did it extend into Spain, Greece and further to Turkey, and to Great Britain and Ireland (Alderman, 1996). During this first wave of epidemics, human activities associated with the crayfish trade were the main reason for the fast spread of the disease (Alderman, 1996). The vector for the first disease wave was most likely an infected crayfish of North American origin (Unestam, 1972; Unestam, 1975a, b). However, the species of this first vector still remains unidentified (Makkonen, 2013). In Europe, there is now a large area where the alien pathogen *A. astaci* has probably been coexisting with crayfish populations for 100 years or more. This area is bounded by France in the west, Italy and the northern Balkans in the south, Russia in the east and Finland and Sweden in the north (Figure 8.2). This represents almost the entire continent of Europe and encompasses the distribution ranges of more or less all native crayfish species, potentially allowing gradual adaptation of the pathogen to native crayfish species populations (e.g. Gruber *et al.*, 2014a).

The first wave of crayfish plague spread was mainly reported as mass mortalities and population eradications in the native crayfish stock. However, there was a second wave of crayfish plague epidemics during the 1990s due to introductions and therapid spread of alien North American crayfish species, which act as *A. astaci* carriers (Persson & Söderhäll, 1983; Huang *et al.*, 1994; Vennerström *et al.*, 1998; Oidtmann *et al.*, 1999; Vogt, 1999; Oidtmann *et al.*, 2006; Kozubíková *et al.*, 2008). Both crayfish plague epidemic waves probably overlapped in time at least in the 1980s and 1990s, which is well documented in many regions, for example in the Czech Republic (Kozubíková *et al.*, 2008).

To illustrate the spread and effects of *A. astaci* in crayfish stocks, we will use two countries as case studies, one from the south and the other from the north of Europe. In Slovenia in southern Europe, the first wave of crayfish plague was detected quite early in 1880, and lasted until 1935 (Franke, 1889; Šulgaj, 1937). The disease spread from the River Danube drainage basin and affected crayfish populations in all the main rivers draining into the River Danube (Figure 8.3). There was no record of any outbreaks of crayfish plague in western Slovenia in the Adriatic drainage or in adjacent parts of northern Italy despite their close vicinity to the first European reported occurrence of crayfish plague in the River Po. As already indicated by Alderman (1996), this may not necessarily reflect some kind of limitation of disease spread to Italy, but it may be a result of poor documentation due to national boundaries in this region, i.e. between Italy and the Austro-Hungarian Empire. However, the Museum of Natural History in Vienna has preserved specimens of the white-clawed crayfish with clear signs of crayfish plague infection; these specimens were collected in 1892 near Gorica (Gorizia) by the River Soča (Isonzo) from the Adriatic drainage region (A. Vrezec & M. Jaklič, unpublished). This indicates that the spread from the River Po was actually more extensive than that described by Alderman (1996). After the outbreaks of crayfish plague, there were collapses in almost all crayfish populations in the large rivers in Slovenia, especially of the noble crayfish, which had previously been an important and heavily trapped species (Šulgaj, 1937) but were never to be revitalized. Numerous attempts at restocking noble crayfish populations were either completely unsuccessful or limited to only smaller streams (Budihna, 1996). In 2003 and 2007, the first alien signal crayfish invasions took place in the River Mura and River Drava from populations introduced into Austria (Bertok *et al.*, 2003; Vrezec *et al.*, 2013), and these animals were known to be *A. astaci* carriers (Kušar *et al.*, 2013). This is the first incident of a crayfish plague disease agent occurring in Slovenia after 1935, although no mass mortality outbreaks have been described so far.

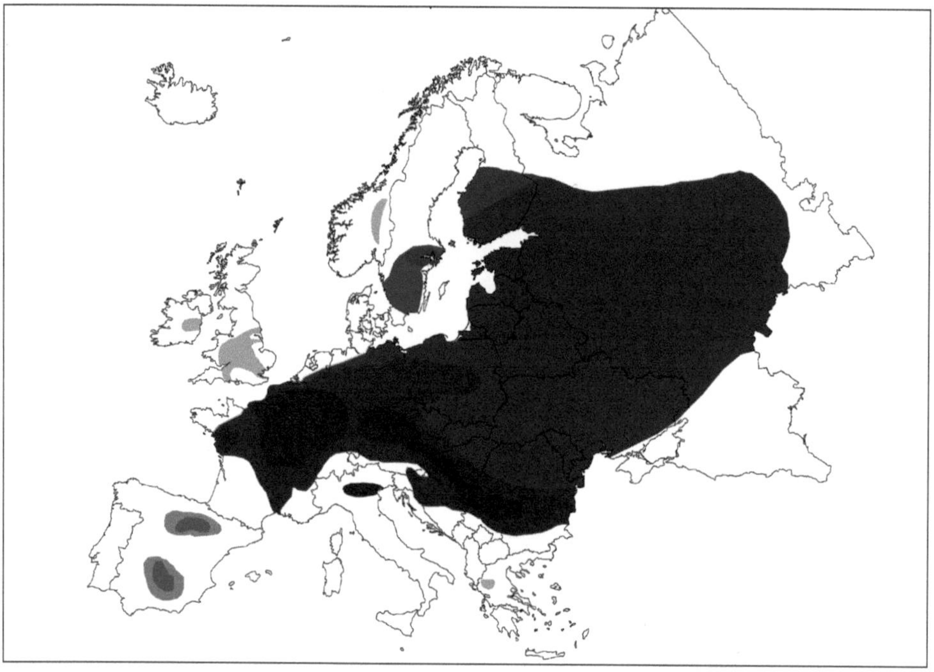

Fig. 8.2: Dynamics of the first wave of crayfish plague expansion across Europe between 1860 and 1995 (modified after Alderman, 1996). The expansion is shown according to the following time periods: 1860–1869 (the darkest), 1870–1879, 1880–1889, 1890–1899, 1900–1909, 1910–1929, 1950–1969, 1970–1979, and 1980–1995 (the brightest).

Our second case is Finland in northern Europe, where the first wave of crayfish plague epidemics represented an enormous setback to crayfish fishermen and the associated trade (Järvi, 1910). Noble crayfish were intensively restocked into the affected lakes, but quite often after the recovery of the population to an exploitable level within 10 to 20 years, the disease struck again, causing a new collapse (Fürst, 1995; Erkamo *et al.*, 2010). As a solution to the chronic crayfish plague infection problem, re-stockings with the supposedly "crayfish plague resistant" signal crayfish were initiated during the 1960s (Westman, 1973). Subsequently it was discovered that the signal crayfish often carried the *A. astaci* infection. Thus, the second and still ongoing wave of outbreaks of crayfish plague in Finland can be traced to the introduction of these alien

North American crayfish species. These species were able to spread the novel strains of *A. astaci* permanently to new areas, when new habitats became colonized. Surprisingly, the distribution and prevalence of *A. astaci* in North America is unclear (Huang *et al.*, 1994; Makkonen *et al.*, 2012a), since the disease agent does not trigger dramatic population collapses there.

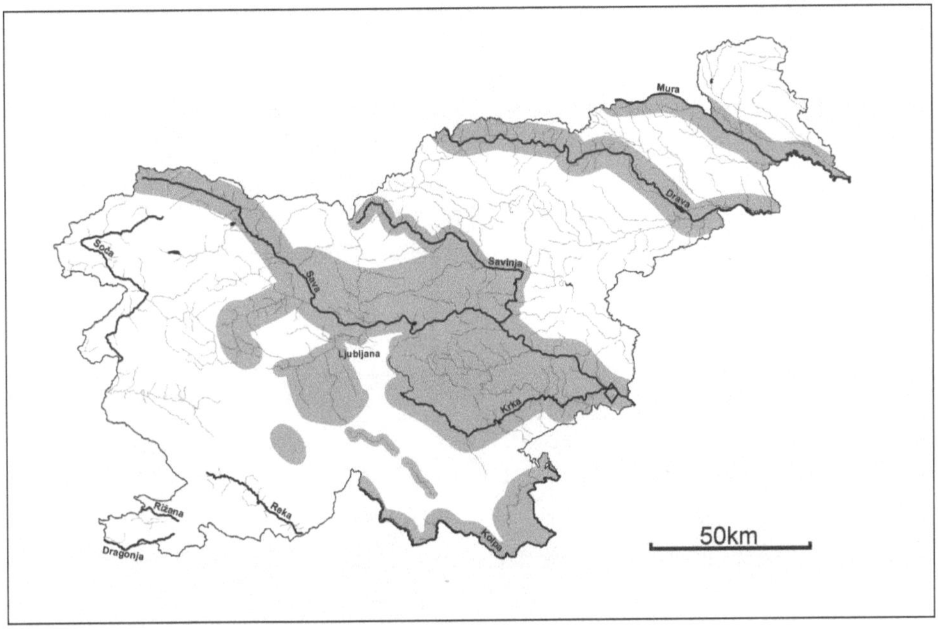

Fig. 8.3: The estimated distribution (light grey) of *Aphanomyces astaci* in the first crayfish plague wave in Slovenia, which took place between 1880 and 1935 (after Kušar *et al.*, 2013).

The infective units of *A. astaci*, zoospores, are viable in temperatures between 2 and 25°C (Unestam, 1969). The Pc-genotype of *A. astaci* is an exception, since it has been shown to have an even higher temperature optimum as an adaptation to warmer environmental conditions (Diéquez-Uribeondo *et al.*, 1995). American crayfish species, which are carriers of *A. astaci*, have been demonstrated to constantly release minor but nonetheless lethal amounts of spores into the ambient water (Diéquez-Uribeondo & Söderhäll, 1993; Strand *et al.*, 2012; Svoboda *et al.*, 2013). There is a continual production of the spores, although the largest numbers of spores are released during molting and at the death of the crayfish (Strand *et al.*, 2012; Svoboda *et al.*, 2013), as had been previously postulated (Oidtmann *et al.*, 2002). Therefore, the presence of introduced species carrying *A. astaci* do pose a serious and ever-present risk to the surrounding native populations (Strand *et al.*, 2012; Svoboda *et al.*, 2013; Strand *et al.*, 2014).

Currently, five genotypes of *A. astaci* are known (Table 8.1) to infect native and alien crayfish in Europe (Huang *et al.*, 1994; Diéguez-Uribeondo *et al.*, 1995; Kozubíková *et al.*, 2011b). The As-genotype has been traced back to the first invasion of *A. astaci*, before the importation of signal crayfish. The PsI-genotype and PsII-genotype have been found not only in the signal crayfish, but also in the native European species after signal crayfish introductions. The PsI-genotype has been attributed to the signal crayfish originating from Lake Tahoe and Lake Hennessey (USA), which were the main sources of the imported animals (Abrahamsson, 1969; Westman, 1973). A single isolate of the PsII-genotype has been linked to the introduction of signal crayfish from Lake Pitt (Canada) to Sweden (Huang *et al.*, 1994) but, based on current knowledge, the PsII-genotype is not widely present in Europe (Söderhäll & Cerenius, 1999). The Pc-genotype was brought here along with the red swamp crayfish (Diéguez-Uribeondo & Söderhäll, 1993; Diéguez-Uribeondo *et al.*, 1995) and the Or-genotype accompanied the spiny-cheek crayfish (Kozubíková *et al.*, 2011b).

Tab. 8.1: Genotypes of *Aphanomyces astaci*.

Genotype	Original host	Reference
As	unknown	Huang *et al.* (1994)
PsI	Signal crayfish (*P. leniusculus*, Lake Tahoe, USA)	Huang *et al.* (1994)
PsII	Signal crayfish (*P. leniusculus*, Lake Pitt, Canada)	Huang *et al.* (1994)
Pc	Red-swamp crayfish (*P. clarkii*, Louisiana, USA)	Diéguez-Uribeondo *et al.* (1995)
Or	Spiny-cheek crayfish (*O. limosus*, USA)	Kozubíková *et al.* (2011a)

Laboratory infection trials have shown extensive variation in the virulence of different *A. astaci* strains (Makkonen *et al.*, 2012b; Makkonen *et al.*, 2014). In general, strains of PsI-genotype seemed to possess considerably higher virulence (Makkonen *et al.*, 2012b; Jussila *et al.*, 2013a; Makkonen *et al.*, 2014; Gruber *et al.*, 2014a) in comparison with the strains of As-genotype which, on the other hand, appear to be more variable in their virulence (Makkonen *et al.*, 2012b; Makkonen *et al.*, 2014). Furthermore, latent crayfish plague infections without mass mortalities have recently been reported in the native European noble crayfish (Jussila *et al.*, 2011; Viljamaa-Dirks *et al.*, 2011), narrow-clawed crayfish (Kokko *et al.*, 2012; Svoboda *et al.*, 2012), and stone crayfish (Kušar *et al.*, 2013) populations, and most recently in the white-clawed crayfish (Manfrin & Pretto, 2014).

8.5 The Adaptation of *Aphanomyces astaci* in Europe

8.5.1 The Virulence Evolution of the *Aphanomyces astaci*

A. astaci has been under high selective pressure to adapt to the European crayfish hosts and its new environmental conditions since its arrival in the 1860s in the River Po valley (Alderman, 1996). The highly susceptible European crayfish species did not allow *A. astaci* to establish a long-term habitat, as these animals were mostly eradicated in crayfish plague epidemics. In this situation, the parasite itself was facing an evolutionary dead-end, as it was destroying the wild native European crayfish stocks, i.e. its habitat. Of course, the European crayfish were also encountering a major challenge to their existence, and they had to find some way to adapt to this novel disease, which was being efficiently spread by human activity (Kilpinen, 2003). Time was quickly running out for both the parasite and its host populations.

Within 100 years, there is indirect evidence to indicate that some of the European crayfish populations may have survived repeated crayfish plague epidemics (Fürst, 1995; Kilpinen, 2003), and there is now new evidence from Turkey that the narrow clawed crayfish stocks under certain circumstances could survive, at least to some extent, a crayfish plague epidemic (Harlioğlu, 2004; Kokko *et al.*, 2012; Svoboda *et al.*, 2012) and even recover afterwards. In this section, we will present some potential scenarios that could explain the newly established balance between the parasite and its hosts, i.e. the *A. astaci* – native European crayfish co-evolutive adaptation process. From an evolutionary point of view, the parasite-host interaction usually results in strong selection pressure for the parasite, and not the host, to favour a low virulent parasite which infects but does not kill the host (Tokeshi, 1999).

8.5.2 The Adaptation of *Aphanomyces astaci* to its Native and Alien Hosts

It has been long presumed that *A. astaci* may have adapted to its European hosts and that its virulence may have declined, but it is only in the last decade that it has been possible to demonstrate this experimentally (Jussila *et al.*, 2011; Makkonen, 2013; Jussila *et al.*, 2014a). Wild native European crayfish stocks, which are viable and producing commercial catch, have been shown to be latent carriers of *A. astaci* (Jussila *et al.*, 2011, Viljamaa-Dirks *et al.*, 2011). Laboratory-scale infection studies have revealed significant virulence differences among *A. astaci* isolates, both among and within the As-genotype and PsI-genotype, and even the existence of very low virulent isolates (Makkonen *et al.*, 2012; Jussila *et al.*, 2013a; Makkonen, 2013).

The alien crayfish, especially the signal crayfish, have been shown to be susceptible to *A. astaci*, which points to the high virulence of the PsI-genotype and possibly a lowered resistance of the signal crayfish towards *A. astaci* (Jussila *et al.*, 2014a; Aydin *et al.*, 2014). Laboratory experiments have demonstrated that the PsI-genotype of

A. astaci, although highly virulent, is also capable of exhibiting significant virulence variation (Jussila *et al.*, 2013a). This indicates that even European based PsI-genotypes could be adapting, while the presence of a permanent host habitat for the PsI-genotype *A. astaci* allows for the maintenance of high virulence without the immediate threat of the parasite's evolutionary suicide due to the outbreak of a devastating crayfish plague epidemic. Thus, the adaptation pressure of *A. astaci* strains capable of infecting North American crayfish species in Europe is significantly different than that of the *A. astaci* strains infecting only native European crayfish species.

8.5.2.1 The Effect of the Host Jumps on the Adaptation Process

The As-genotype *A. astaci*, after its presumed arrival in Europe in the 1850s (Alderman, 1996), had access to a variety of host habitats among the European native crayfish spectrum, a total of five different species (Souty-Grosset *et al.*, 2006). The European crayfish species were all susceptible to *A. astaci*, and the outcome of the crayfish plague epidemic during the first decades was a complete elimination of host populations. If it had not had assistance from humans, the disease might have had a short history in Europe, but the disease agent was unintentionally aided by transferring it to new water bodies and populations. The rapid and efficient spreading allowed for both the constant presence of epidemics and chance to host jump from one European crayfish species to the next and, apparently, to jump back to the species that it had already infected.

The As-genotype *A. astaci* was thus allowed a longer time to adapt to novel conditions and access to crayfish that were highly susceptible to the disease. One could postulate that this scenario was fundamental for the lately discovered putative better balance between *A. astaci* and European crayfish (Jussila *et al.*, 2014a). The theory that there was a chronic crayfish plague infection was proposed in the last century (e.g. Fürst, 1995), but only verified during the past decade (Jussila *et al.*, 2011; Viljamaa-Dirks *et al.*, 2011). Since then, there have been several reports of native European crayfish being apparent latent carriers of *A. astaci* (Kokko *et al.*, 2012; Svoboda *et al.*, 2012; Kušar *et al.*, 2013; Viljamaa-Dirks *et al.*, 2013). In addition to the host jumps as such, the different *A. astaci* strains have been transferred between different aquatic habitats, i.e. different ambient aquatic conditions, which include, for example, differences in water quality and temperature regime. These habitat jumps would have added to the adaptation pressure on *A. astaci*, as perhaps reflected in the differing environmental preferences of the individual *A. astaci* genotypes (e.g. Dieguez-Uribeondo *et al.*, 1995).

The mass introduction of the alien North American crayfish into Europe changed the host species range and habitat for *A. astaci*, i.e. several relatively resistant crayfish species were intentionally introduced all over Europe, creating rather complicated and overlapping alien species distributions (see Fig. 8.1d). Since it seemed that they were often carrying *A. astaci*, some discovered to be novel genotypes (Holdich *et al.*, 2009; Viljamaa-Dirks *et al.*, 2013), the parasites may have felt themselves fortunate now to have been conveniently provided with a permanent host reservoir. There are also theoretical reasons to believe that the As-genotype *A. astaci*, which had probably already adapted to co-existing with the European crayfish, were no longer capable of infecting the newly introduced alien crayfish, despite the fact that they had originated from the same geographical region (i.e. North America) where they could well have had an overlapping distribution. On the other hand, the novel genotypes were able to infect both the native European crayfish and the newly introduced alien counterparts. The former case was responsible for population collapses (e.g. Kirjavainen & Sipponen, 2004; Bohman *et al.*, 2006) but in the latter case, only minor gross symptoms and very rarely increased mortality were originally reported. However, during recent decades there have been several reports of population collapses in the signal crayfish (Sahlin *et al.*, 2010; Jussila *et al.*, 2014a, b; Sandström *et al*, 2014) and it is now obvious that the alien North American crayfish are susceptible to *A. astaci* when they are living under European conditions (Persson & Söderhäll, 1983; Thörnqvist & Söderhäll, 1993; Aydin *et al.*, 2014), i.e. this is evidence of the evolution of a balance, albeit a labile balance, between the disease and its North American hosts.

To complicate matters further, it has been shown that there is considerable genetic variation even within different *A. astaci* genotypes (Makkonen *et al.*, 2012a) and that As-genotype *A. astaci* may be capable of infecting signal crayfish (Aydin *et al.*, 2014). One could hypothesize that the crayfish populations during different crayfish plague epidemics could be expressing different *A. astaci* strains and the *A. astaci* genotype itself would be largely host species-specific. Thus, these host jumps would contribute to the spectrum of differences within and among the *A. astaci* genotypes, thus encouraging the adaptation of the various different *A. astaci* strains to European host habitats.

8.5.3 Selection Pressure by Native Crayfish Hosts in Europe

Currently, the infected host crayfish seem to be expressing certain *A. astaci* genotypes and quite often the analyses indicate that only one genotype is infecting an individual host. As there are several *A. astaci* genotypes and species present in Europe, this could be evidence of some kind of superinfection-type infection (May & Nowak, 1994) occurring during crayfish plague epidemics, with normally only one genotype been detected in the infected crayfish.

There are a few possible scenarios to explain the adaptation pressures on *A. astaci* in its native hosts during an infection (Figure 8.4), and we will present some simplified models here. At present, the most common scenario has been one with a high host mortality, which means that there is no special need for *A. astaci* to adapt, since this situation guarantees optimum conditions for sporulation and thus maximizes the probability of further infections. However, this scenario demands that some host habitat is preserved, since otherwise the maintenance of high virulence could cause complete eradication of the *A. astaci* host habitat. Subpopulation(s) of the host crayfish might survive in specific cases; there may be individuals with elevated resistance or certain environmental conditions may promote survival. This scenario ensures high sporulation, but might accidentally kill all of the hosts and could thus be the type I suicide situation.

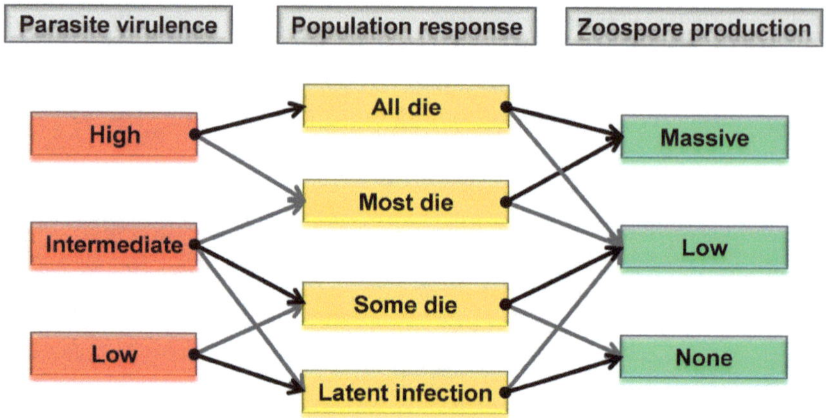

Fig. 8.4: A schematic representation of possible *A. astaci* adaptative scenarios and outcomes when the parasite infects native European crayfish species. Black arrows indicate the most probable outcomes and grey arrows indicate alternative options.

The second scenario would be an *A. astaci* infection with lowered host mortality (Figure 8.4), which would result in the survival of the majority of native host crayfishes and a low overall spore density in the water. This represents the other end of the infection continuum in comparison with the scenario described in the previous paragraph. This would make possible a latent crayfish plague infection in a native host crayfish population. Alternatively, a chronic crayfish plague infection with more severe gross symptoms would also be possible with a slightly higher host crayfish mortality rate. This would require a rather delicately balanced relationship between *A. astaci* and its native host crayfish. There might also be occasional acute *A. astaci* infection phases, as a result of changes in the ambient conditions or different levels of stress in host crayfish leading to a window of opportunity for mass sporulation of *A. astaci*. In this case, *A. astaci* might need to adopt a different tactic for reproduction other than asexual sporulation or, alternatively, it would need to rely on the possibility that a low level sporulation would nonetheless ensure further infections. This scenario raises the risk level of a cessation of *A. astaci* infection and lowered sporulation, and could thus be the type II suicide situation.

The third scenario would be that an *A. astaci* infection would be capable of inducing an intermediate level mortality (Fig. 8.4), where part of the population would survive and mortalities would also be delayed, allowing for a longer time span for sporulation and a lower density of infective spores released at any given time. This would result in lower infective pressure and a situation where perhaps those hosts in better physiological condition or those that were more resistant towards *A. astaci* might survive. This scenario could lead to an acute epidemic followed by a longer latent infection phase. Occasional acute phases might occur and they should ensure sufficient sporulation and thus the avoidance of both type I and type II suicide situations.

8.5.4 Selection Pressure in North American Hosts in Europe

The novel environmental pressures in conjunction with opportunistic European parasites have modified the resistance of the alien signal crayfish to plague disease in Europe and this has greatly altered the relationship between *A. astaci* and its alien crayfish hosts. The outcome has been very different from the unrealistically optimistic predictions. The epidemics of crayfish plague have proved to be more detrimental to alien crayfish than expected and the reasons behind the observed changes have yet to be clarified. The *A. astaci* RAPD-genotypes that have been detected in the alien crayfish species are normally highly virulent even though the PsI-genotype strains do exhibit some variations in their virulence (Jussila *et al.*, 2013a). The adaptation scenarios in this situation are far more limited (Figure 8.5) than those in the *A. astaci* genotypes responsible for epidemics in native European crayfish. There is one key difference compared to *A. astaci* infecting native European crayfish, i.e. there is only a negligible possibility for these *A. astaci* genotypes to adopt the suicide route option.

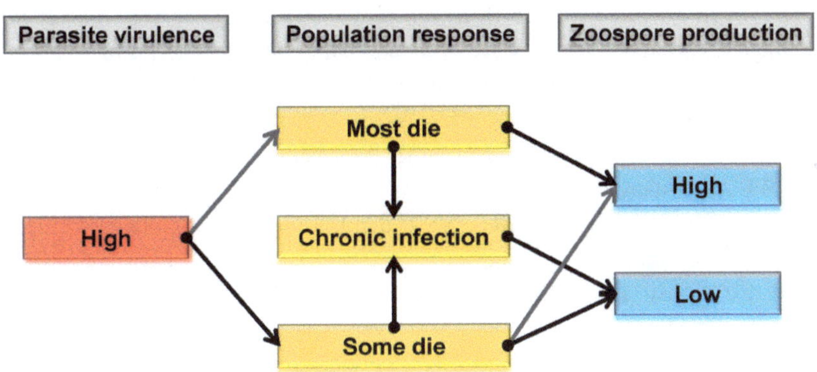

Fig. 8.5: A schematic representation of *A. astaci* adaptative scenarios and outcomes for infecting alien crayfish species in Europe. Black arrows indicate the most probable outcomes and grey arrows indicate alternative options.

The incidents of collapsed signal crayfish populations reported from the Nordic countries (Edgerton & Jussila, 2004; Sahlin *et al.*, 2010; Sandström *et al*, 2014; Jussila *et al.*, 2014b) indicate that the alien crayfish might be losing their advantage in terms of disease resistance as they adapt to European conditions. This could provide *A. astaci* with a novel platform for mass sporulation (Figure 8.4), as it has been shown that the alien crayfish release only small numbers of spores during chronic crayfish plague infections, but there is a peak in sporulation when the condition of the host crayfish starts to deteriorate (Strand *et al.*, 2012). This, together with the well reported role of alien crayfish as a permanent host habitat for *A. astaci* (Holdich *et al.*, 2009), would further increase the risk of the even wider spreading of *A. astaci*. The host jumps, which can occur as a consequence of the close proximity of different populations of crayfish species, are also a factor triggering evolution of *A. astaci*, although the host jumps would only be back and forth between the noble crayfish and the signal crayfish populations, as is happening in the Nordic countries. The network for host jumps, bearing in mind that *A. astaci* has permanent populations of alien crayfish as host habitats, would be even more complex in continental Europe (Figure 8.1).

It is clear that the signal crayfish can act as a platform for *A. astaci* sporulation, quite possibly promoting interactions between different *A. astaci* strains. The sporulation from chronically infected signal crayfish is continual but limited (Figure 8.4; Strand *et al.*, 2012), and there are indications that the signal crayfish may be infected with several different *A. astaci* strains (Aydin *et al.*, 2014). The possibility of multiple infections is extremely interesting, and although this scenario has not been reported, were it to occur it would permit *A. astaci* to exchange genetic information.

The co-infections with different parasites have been reported to cause different syndromes in the signal crayfish (Edsman *et al.*, 2015) and even increased mortality (Thörnqvist & Söderhäll, 1993; Aydin *et al.*, 2014). The crayfish plague epidemics could thus be either superinfections causing elevated mortality or co-infections (May & Nowak, 1995) where the gross symptoms are variable and seem to allow opportunistic parasites to avoid destruction by the weakened immune system in the host (Persson & Söderhäll, 1993). This would raise the possibility of interactions between the parasites during an acute infection and possibility for genetic recombination between closely related species.

There are reports that signal crayfish are only infected with the PsI-genotype *A. astaci* (Jussila *et al.*, 2014b) or, at least, are expressing only that genotype. It has been shown that the PsI-genotype *A. astaci* is highly virulent (Jussila *et al.*, 2013a; Makkonen, 2013; Gruber *et al.*, 2014a) and this species appears to require high virulence in order to effectively complete its life cycle in the signal crayfish via successful mass sporulation. The change in the host-parasite balance between the PsI-genotype *A. astaci* and its signal crayfish hosts favouring *A. astaci* sporulation would create an alternative adaptation platform for the *A. astaci* in Europe. In this case, the less resistant signal crayfish populations would ensure an even more successful spreading of the highly virulent *A. astaci* throughout European aquatic ecosystems and then the native crayfish would face an even greater threat from alien crayfish and their lethal diseases.

8.6 Acknowledgements

We are grateful to have received the strategic funding of the University of Eastern Finland, which allowed us to do much of the work discussed here. LIFE+ CrayMate (LIFE12 INF/FI/233) has provided essential support. We are grateful to Andrej Kapla (Ljubljana, Slovenia) who prepared the maps presented within this chapter. We are thankful to Dr. Ewen MacDonald (University of Eastern Finland) for revising the language. In a bizarre way, we wish to acknowledge the work of those misguided individuals who are still attempting to spread these alien crayfish species – it means that those of us trying to frustrate their efforts will never be out of work or run out of topics about which we can argue. And finally, please consider carefully and always exercise caution before you start to play around with Mother Nature.

In a nutshell

- *Aphanomyces astaci* belongs to the class of Oomycetes, a diverse group of fungus-like organisms, including not only a wide variety of plant and animal pathogens, but also saprophytic species. *A. astaci* itself is a very specific parasite infecting the freshwater crayfish of North America. However, in the European crayfish, it causes a lethal disease known as crayfish plague. The pathogen spreads from host to host by producing swimming zoospores; should a suitable host be found, these then encyst, germinate, and start to grow hyphae into the host tissues. A massive sporulation occurs with the death of the host. In contrast, in the case of the disease-resistant North American crayfish species, there is a continual low level sporulation.
- The introduction in Europe of *A. astaci* and the repeated introductions of its North American host species are a classic example of a man-made ecological disaster, stemming from the naïve belief that the manipulation of an ecosystem would be straightforward. Currently, the native European crayfish is on the brink of extinction and the alien crayfish species which were supposed to replace the eradicated native stocks display many signs and symptoms of maladaptation. *A. astaci* itself has apparently adapted rather well to European conditions, and seems to be currently co-evolving at multiple levels, while maintaining contact with its relatively resistant hosts as new crayfish stocks of North American origin are imported into Europe. One can presume that this multiple-host-species pattern is likely to guarantee *A. astaci*'s survival even in the most severe crayfish plague epidemics. One could also predict that this will inevitably lead to the total eradication of the remaining native European crayfish stocks.
- The introduction of the *A. astaci* to Europe, though it was purely accidental, has not only seriously devastated native crayfish populations throughout Europe, but also resulted in further damage due to misguided management attempts to rectify the situation.

- According to data about wild crayfish populations, it seems that the adaptation of *A. astaci* allowing it to sometimes coexist with the native European crayfish has taken less than 150 years. The indirect evidence for this proposal, originally obtained from the records of chronic crayfish plague infections of native European crayfish populations, especially the noble crayfish *(Astacus astacus)*, has recently been verified using qPCR techniques.
- In the Nordic countries, there is much anecdotal evidence describing possible population collapses of the signal crayfish *(Pacifastacus leniusculus)* due to crayfish plague epidemics. Previously, the signal crayfish had been thought to be highly resistant to crayfish plague infection. Recent observations from the wild populations indicate that crayfish plague and subsequent infections by opportunistic parasites, such as the *Fusarium* sp. complex, may lower the reproduction potential of the wild signal crayfish stocks. Alarmingly, it has also been shown that the signal crayfish could be susceptible to *A. astaci* of both the PsI-genotype and As-genotype, which means that associated population collapses could well be possible in these crayfish. These recent findings seem to confirm the long-established theories about the reduced *A. astaci* resistance in signal crayfish.
- These conclusions, which are no longer merely theoretical but are now based on research findings, clearly indicate that further introduction of alien crayfish species, and especially their diseases, are predicted to pose severe problems to the receiving ecosystems, and the consequences would not be limited to certain target species, but would often spread through the whole ecosystem. The fundamental concept of always erring on the side of caution is unfortunately all too often ignored. Although this creates circumstances that are very interesting from an academic view point, they are very detrimental to the organisms and biodiversity in these ecosystems.

8.7 Bibliography

Abrahamsson, S. (1969). Signalkräftan — erfarenheter från USA och aspekter på dess inplantering i Sverige. Fauna och Flora, 64, 109-116. (In Swedish)

Ackefors, H. (1998). The culture and capture crayfish fisheries in Europe. World Aquaculture Magazine, 29 (2), 18-67.

Agerberg, A., Jansson, H. (1995). Allozymic comparisons between three subspecies of the freshwater crayfish *Pacifastacus leniusculus* (Dana), and between populations introduced to Sweden. Heredity, 122, 33-39.

Aklehnovich, A., Razlutskij, V. (2013). Distribution and spread of spiny-cheek crayfish *Orconectes limosus* (Rafinesque, 1817) in Belarus. BioInvasions Records, 2 (3), 221-225.

Alderman, D.J. (1996). Geographical spread of bacterial and fungal diseases of crustaceans. Revue Scientifique et Technique, 15, 603–632.

Alderman, D.J., Holdich, D., Reeve, I. (1990). Signal crayfish as crayfish plague vectors in Britain. Aquaculture, 86 (1), 3-6.

Aquiloni L., Martin M.P., Gherardi F. *et al.* (2011). The North American crayfish *Procambarus clarkii* is the carrier of the oomycete *Aphanomyces astaci* in Italy. Biological Invasions, 13, 359-367.

Arbaćiauskas, K., Visinskien, G., Smilgeviien, S. (2011). Non-indigenous macroinvertebrate species in Lithuanian freshwaters. Knowledge and Management of Aquatic Ecosystems, 402, 13.

Aydin, H., Kokko, H., Makkonen *et al.* (2014). The signal crayfish is vulnerable to both the As and the PsI-isolates of the crayfish plague. Knowledge and Management of Aquatic Ecosystems, 413, 03.

Bertok, M., Budihna, N., Povž, M. (2003). Strokovne osnove za vzpostavljanje omrežja Natura 2000, Ribe (Pisces), Piškurji (Cyclostomata), Raki deseteronožci (Decapoda). Zavod za ribištvo Slovenije, Ljubljana. (in Slovene)

Bohman, P., Nordwall, F., Edsman, L. (2006). The effect of the large-scale introduction of signal crayfish on the spread of crayfish plague in Sweden. *Bulletin Français* de la Pêche et de la Pisciculture, 380-381, 1291-1302.

Bohman, P., Degerman, E., Edsman, L. *et al.* (2011). Exponential increase of signal crayfish in running waters in Sweden – due to illegal introductions? Knowledge and Management of Aquatic Ecosystems, 402, 23.

Bohman, P., Edsman, L., Martin, P. *et al.* (2013). The first Marmorkrebs (Decapoda: Astacida: Cambaridae) in Scandinavia. BioInvasions Research, 2, 227–232.

Budihna, N. (1996). Potočni raki (Astacidae). In: J. Gregori, A. Martinčič, K. Tarman *et al.* (Eds.), Narava Slovenije, stanje in perspektive. Društvo ekologov Slovenije, Ljubljana, 228-233. (in Slovene)

Capinha, C., Larson, E.C., Tricarino, E. *et al.* (2013). Effects of climate change, invasive species, and disease on the distribution of native European crayfishes. Conservation Biology, 27 (4), 731-740.

Chucholl, C. (2012). Understanding invasion success: life-history traits and feeding habits of the alien crayfish *Orconectes immunis* (Decapoda, Astacida, Cambaridae). Knowledge and Management of Aquatic Ecosystems, 404, 4.

Chucholl, C. (2014). Predicting the risk of introduction and establishment of an exotic aquarium animal in Europe: insights from one decade of Marmorkrebs (Crustacea, Astacida, Cambaridae) releases. Management of Biological Invasions, 5, In Print.

Chucholl, C., Daudey, T. (2008). First record of *Orconectes juvenilis* (Hagen, 1870) in eastern France: update to the species identity of a recently introduced orconectid crayfish (Crustacea: Astacida). Aquatic Invasions, 3 (1), 105-107

Chucholl, C., Morawetz, K., Groß, H. (2012). The clones are coming – strong increase in Marmorkrebs [*Procambarus fallax* (Hagen, 1870) f. *virginalis*] records from Europe. Aquatic Invasions, 7, 4, 511-519.

Cornalia, E. (1860). Sulla malattia dei gamberi. Atti della Societa Italiana di Scienze Naturali e del Museo Civili di storia naturale, 2, 334-336. (in Italian)

Dehus, P., Dussling, U., Hoffmann, C. (1999). Notes on the occurrence of the calico crayfish *(Orconectes immunis)* in Germany. Freshwater Crayfish, 12, 786–790.

Diéguez-Uribeondo, J. (2006). The dispersion of the *Aphanomyces astaci*-carrier *Pacifastacus leniusculus* by humans represent the main cause of disappearance of the indigenous crayfish *Austropotamobius pallipes* in Navarra. *Bulletin Français* de la Pêche et de la Pisciculture, 380-381, 1303-1312.

Diéguez-Uribeondo, J., Söderhäll, K. (1993). *Procambarus clarkii* Girard as a vector for the crayfish plague fungus, *Aphanomyces astaci* Schikora. Aquaculture and Fisheries Management, 24 (6), 761-765.

Diéguez-Uribeondo, J., Huang, T-S., Cerenius, L. *et al.* (1995). Physiological adaptation of an *Aphanomyces astaci* strain isolated from the freshwater crayfish *Procambarus clarkii*. Mycological Research, 99 (5), 574-578.

Edgerton, B.F., Jussila, J. (2004). Keynote presentation and rountable session 4. Crayfish pathology in Europe: past, present and a programme for the future. *Bulletin Français de la Pêche et de la Pisciculture*, 372-373, 473-482.

Edsman, L. (2004). The Swedish story about import of live crayfish. *Bulletin Français* de la Pêche et de la Pisciculture, 372-373, 281-288.

Edsman, E., Nyström, P., Sandström, A. *et al.* (2015). Eroded swimmeret syndrome (ESS) in female signal crayfish (*Pacifastacus leniusculus*) is caused by multiple infection of *Aphanomyces astaci* and *Fusarium* sp. Diseases of Aquatic Organisms, 112, 219-228.

Ercoli, F. (2014) A comparison of the impacts of introduced signal crayfish and native noble crayfish in boreal lake ecosystems. Jyväskylä, Finland: Doctoral dissertation, the University of Jyväskylä.

Erkamo, E., Ruokonen, T., Alapassi, T. *et al.* (2010). Evaluation of crayfish stocking success in Finland. Freshwater Crayfish, 17, 77-83.

EU (2013). Invasive alien species. Retreived from http://ec.europa.eu/environment/nature/invasivealien/. Accessed 10.3.2014.

Filipová, L., Kozubíková, E., Petrusek, A. (2009). Allozyme variation in Czech populations of the invasive spiny-cheek crayfish Orconectes limosus (Cambaridae). Knowledge and Management of Aquatic Ecosystems, (394-395), 10.

Filipová, L., Lieb, D. A., Grandjean, F. *et al.* (2011). Haplotype variation in the spiny-cheek crayfish Orconectes limosus: Colonization of Europe and genetic diversity of native stocks. Journal of the North American Benthological Society, 30 (4), 871-881.

Filipová, L., Petrusek, A., Matasová, K. *et al.* (2013). Prevalence of the crayfish plague pathogen Aphanomyces astaci in populations of the signal crayfish Pacifastacus leniusculus in France: evaluating the threat to native crayfish. PLOS One, 8 (7), e70157.

Fiskeriverket (2005). Fiske 2005. En undersökning om svenskarnas fritidsfiske. Stockholm, Sweden: Fiskeriverket och Statistiska centralbyrån. (in Swedish)

Fjälling, A., Fürst, M. (1985). The introduction of the signal crayfish Pacifastacus leniusculus into Swedish waters: 1969–1984. Information from the Institute of Freshwater Research, Drottningholm, 8, 1–29.

Franke, J. (1889). Zur Krebsfrage in Krain. Mitteilungen des österreichischen Fischvereins, Wien, 2-7. (in German)

Fratini, S., Zaccara, S., Barbaresi, S. *et al.* (2005). Phylogeography of the threatened crayfish (genus Austropotamobius) in Italy: implications for its taxonomy and conservation. Heredity, 94, 108–118.

Fürst, M. (1995). On the recovery of *Astacus astacus* L. populations after an epizootic of the crayfish plague (*Aphanomyces astaci* Shikora). Freshwater Crayfish, 8, 565-576.

Gherardi, F. (2002). Behaviour. In D.M. Holdich (Ed.), Biology of Freshwater Crayfish (pp.258-290). Oxford, UK: Blackwell Science.

Grandjean, F., Vrålstad, T., Diéguez-Uribeondo, J. *et al.* (2014). Microsatellite markers for direct genotyping of the crayfish plague pathogen *Aphanomyces astaci* (Oomycetes) from infected host tissues. Veterinary Microbiology, 170(3-4), 317-324.

Gruber, C., Kortet, R., Vainikka, A. *et al.* (2014a). Variation in resistance to the invasive crayfish plague and immune defence in the native noble crayfish. Annales Zoologici Fennici, 51 (4), 371-389.

Gruber, C., Vainikka, A., Hirvonen H. *et al.* (2014b). Endogenous seasonal variation in the encapsulation response of the noble crayfish (*Astacus astacus*). Annales Zoologici Fennici, 51 (5), 433-444.

Hamr, P. (2002). *Orconectes*. In: Holdich D.M. (Ed.), Biology of freshwater crayfish (pp.585-608). Oxford: Blackwell Science.

Harlioğlu, M.M. (2004). The present situation of freshwater crayfish, *Astacus leptodactylus* (Eschscholtz, 1823) in Turkey. Aquaculture, 230, 181-187.

Heinimaa, S. & Pursiainen, M. (2010). Joki- ja täpläravun elinkierto ja levinneisyys. Helsinki, Suomi-Finland: Riista- ja kalatalouden tutkimuslaitos, selvityksiä 6/2010.

Henttonen, P., Huner, J.V. (1999). The introduction of alien species of crayfish in Europe: A historical introduction. In: F. Gherardi, D.M. Holdich (Eds.), Crayfish in Europe as alien species – how to make the best of a bad situation? (pp.13-22). Rotterdam, Netherlans: A.A. Balkema.

Holdich, D.M., Reynolds, J.D., Souty-Grosset, C. *et al.* (2009). A review of the ever increasing threat to European crayfish from non-indigenous crayfish species. Knowledge and Management of Aquatic Ecosystems, 394-395, 11.

Huang, T-S., Cerenius, L., Söderhäll, K. (1994). Analysis of genetic diversity in the crayfish plague fungus, *Aphanomyces astaci*, by random amplification of polymorphic DNA. Aquaculture, 126 (1-2), 1-9.

Hubad, J. (1894) O račji kugi. Izvestje c. kr. državne nižje gimnazije v Ljubljani, Ljubljana, p. 15-23. (in Slovene)

Hudina, S., Faller, M., Lucić, A. *et al.* (2009). Distribution and dispersal of two invasive crayfish species in the Drava River basin, Croatia. Knowledge and Management of Aquatic Ecosystems, 394-395, 09.

Jaklič, T., Vrezec, A. (2011). The first tropical alien crayfish species in European waters: the redclaw *Cherax quadricarinatus* (Von Martens, 1868) (Decapoda, Parastacidae). Crustaceana, 84 (5), 651-665.

Jussila, J., Mannonen, A. (2004). Crayfisheries in Finland, a short overview. *Bulletin Français* de la Pêche et de la Pisciculture, 372-373, 263-273.

Jussila, J., Makkonen, J., Vainikka, A. *et al.* (2011). Latent crayfish plague (*Aphanomyces astaci*) infection in a robust wild noble crayfish (*Astacus astacus*) population. Aquaculture, 321 (1-2), 17-20.

Jussila, J., Kokko, H., Kortet, R. *et al.* (2013a). *Aphanomyces astaci* PsI-genotype isolates from different Finnish signal crayfish stocks show variation in their virulence but still kill fast. Knowledge and Management of Aquatic Ecosystems, 411, 10.

Jussila, J., Tiitinen, V., Fotedar, R. *et al.* (2013b). A simple and efficient cooling method for post-harvest transport of the commercial crayfish catch. Freshwater Crayfish, 19 (1), 15-19.

Jussila, J., Maguire, I., Kokko, H. *et al.* (2014a). Chaos and adaptation in the host-pathogen relationship in relation to the conservation: the case of the crayfish plague and the noble crayfish. In T. Kawai, Z. Faulker, G. Scholtz (Eds.), Freshwater Crayfish: Global Overview (pp.1-14). USA: Science Publishers.

Jussila, J., Makkonen, J., Vainikka, A. *et al.* (2014b). Crayfish plague dilemma: how to be a courteous killer? Boreal Environmental Research, 19, 235-244.

Järvi, T.H. (1910). Uber den Krebs (*Astacus fluviatilis* Rond.) und die Krebsepidemieen in Finland. *Acta* Societatis pro *Fauna* et *Flora* Fennica, 33 (3), 1–41. (in German)

Kilpinen, K. (2003). Suomen rapu: ravun nousu, tuho ja tulevaisuus. Helsinki, Suomi-Finland: Edita Publishing Oy. (In Finnish)

King, C.R. (1994). Growth and survival of redclaw crayfish hatchlings (*Cherax quadricarinatus* von Martens) in relation to temperature, with comments on the relative suitabitity of *Cherax quadricarinatus* and *Cherax destructor* for culture in Queensland. Aquaculture, 122 (1), 75-80.

Kirjavainen, J., Sipponen, M. (2004). Environmental benefit of different crayfish management strategies in Finland. Fisheries Management and Ecology, 11, 213-218.

Klobučar, G.I.V., Podnar, M., Jelić, M. *et al.* (2013). Role of the Dinaric Karst (western Balkans) in shaping the phylogeographic structure of the threatened crayfish *Austropotamobius torrentium*. Freshwater Biology, 58, 1089-1105.

Kokko, H., Koistinen, L., Harlioğlu, M.M. *et al.* (2012). Recovering Turkish narrow clawed crayfish (*Astacus leptodactylus*) populations carry *Aphanomyces astaci*. Knowledge and Management of Aquatic Ecosystems, 404, 12.

Kossakowski J. (1966). Raki [Crayfish]. Warszawa: Państwowe Wydawnictwo Rolnicze i Leśne. (In Polish)

Kouba, A., Petrusek, A., Kozak, P. (2014). Continental-wide distribution of crayfish species in Europe: update and maps. Knowledge and Management of Aquatic Ecosystems, 413, 5.

Kozubíková, E., Petrusek, A., Ďuriš, Z. *et al.* (2008). The old menace is back: Recent crayfish plague outbreaks in the Czech Republic. Aquaculture, 274, 208-217.

Kozubiková, E., Vrålstad, T., Filipová, L. *et al.* (2011a). Re-examination of the prevalence of *Aphanomyces astaci* in North American crayfish populations in Central Europe by TaqMan MGB real-time PCR. Diseases of Aquatic Organisms, 97, 113-125.

Kozubíková, E., Viljamaa-Dirks, S., Heinikainen, S. *et al.* (2011b). Spiny-cheek crayfish *Orconectes limosus* carry a novel genotype of the crayfish plague pathogen *Aphanomyces astaci*. Journal of Invertebrate Pathology, 108 (3), 214-216.

Kušar, D., Vrezec, A., Ocepek, M. *et al.* (2013). *Aphanomyces astaci* in wild crayfish populations in Slovenia: first report of persistent infection in a stone crayfish *Austropotamobius torrentium* population. *Diseases of Aquatic Organisms*, 103, 157-169.

Largiader, C.R., Herger, F., Lörtscher, M. *et al.* (2000). Assessment of natural and artificial propagation of the white-clawed crayfish *(Austropotamobius pallipes* species complex) in the Alpine region with nuclear and mitochondrial markers. Molecular Ecology, 9, 25-37.

Larson, E.R., Abbott, C.L., Usio, N. *et al.* (2012). The signal crayfish is not a single species: cryptic diversity and invasions in the Pacific Northwest range of *Pacifastacus leniusculus*. Freshwater Biology, 57, 1823-1838.

Lawrence, C., Jones, C. (2002). Cherax. In: D.M. Holdich (Ed.), Biology of Freshwater Crayfish (pp.635-669). Oxford, UK: Blackwell Science Ltd.

Lehtonen, J.U.E. (1975). Kansanomainen ravustus ja rapujen hyväksikäyttö Suomessa. Tapiola, Suomi-Finland: Oy Weilin+Göös Ab. (In Finnish)

Makkonen, J. (2013). The crayfish plague pathogen *Aphanomyces astaci*. Genetic diversity and adaptation to the host species. Kuopio, Finland: Doctoral dissertation, University of Eastern Finland.

Makkonen, J., Jussila J., Kokko H. (2012a). The diversity of the pathogenic oomycete *(Aphanomyces astaci)* chitinase genes within the genotypes indicate adaptation to its hosts. Fungal Genetics and Biology, 49 (8), 635-642.

Makkonen, J., Jussila, J., Kortet, R. *et al.* (2012b). Differing virulence of *Aphanomyces astaci* isolates and elevated resistance of noble crayfish *Astacus astacus* against crayfish plague. Diseases of Aquatic Organisms, 102, 129-136.

Makkonen, J., Kokko, H., Kortet, R. *et al.* (2014). Dose-dependent mortality of the noble crayfish (*Astacus astacus*) to different strains of the crayfish plague (*Aphanomyces astaci*). Journal of Invertebrate Pathology, 115, 86-91.

Manfrin, A., Pretto, T. (2014). Aspects of health and disease prevention. In RARITY. Eradicate invasive Louisiana red swamp and preserve native white clawed crayfish in Friuli Venezia Giulia (pp.123-125). RARITY project LIFE10 NAT/IT/000239.

Marten, M., Werth, C., Marten D. (2004). Der Marmorkrebs *Procambarus* sp. (Cambaridae, Decapoda) in Deutschland – ein weitreres Neozoon in Rheineinzugsgebeit. *Lauterbonia*, 50, 17-23.

Marzano, F.N., Scalici, M., Chiesa, S. *et al.* (2009). The first record of the marbled crayfish adds further threats to freshwaters in Italy. Aquatic Invasions, 4 (2), 401–404.

Matasová, K., Kozubíková, E., Svoboda, J. *et al.* (2011). Temporal variation in the prevalence of the crayfish plague pathogen, *Aphanomyces astaci*, in three Czech spiny-cheek crayfish populations. Knowledge and Management of Aquatic Ecosystems, 401, 14.

May, R.M., Nowak, M.A. (1994). Superinfection, metapopulation dynamics, and the evolution of diversity. Journal of Theoretical Biology, 170, 95-114.

May, R.M., Nowak, M.A. (1995). Coinfection and the evolution of parasite virulence. Proceedings of the Royal Society B, Biological Sciences, 261, 209-215.

MMM (2012). Finland's national strategy on invasive alien species. Retreived from http://www.mmm.fi/en/index/frontpage/natural_resources/invasive_alien_species.html. Accessed 16.3.2014.

Morrissy, N.M. (1990). Optimum and favourable temperatures for growth of *Cherax tenuimanus* (Smith 1912) (Decapoda: Parastoacidae). *Australian Journal of Marine and Freshwater Research,* 41 (6), *735 – 746.*

Nyström, P. (1999). Ecological impact of introduced and native crayfish on freshwater communities: European perspectives. In F. Gherardi F., D.M. Holdich (Eds.), Crayfish in Europe as alien species - How to make the best of bad situation? (pp.63-85). Rotterdam, Netherlans: A.A. Balkema.

Nyström, P., Brönmark, C., Granéli, W. (1999). Influence of an exotic and a native crayfish species on a littoral benthic community. Oikos, 85 (3), 545-553.

Oidtmann, B., Cerenius, L., Schmid, I. *et al.* (1999). Crayfish plaque epizootics in Germany - classification of two German isolates of the crayfish plaque fungus *Aphanomyces astaci* by random amplification of polymorphic DNA. Diseases of Aquatic Organisms, 35, 235-238.

Oidtmann, B., Heitz, E., Rogers, D. *et al.* (2002). Transmission of crayfish plague. Diseases of Aquatic Organisms, 52, 159-167.

Oidtmann, B., Geiger, S., Steinbauer, P. *et al.* (2006). Detection ofAphanomyces astaci in North American crayfish by polymerase chain reaction. Diseases of Aquatic Organisms, 72, 53-64.

Pakkasmaa, S. (2006). Ruotsin rapukansa koolla Jönköpingissä. Suomen kalastuslehti, 5, 22-23. (In Finnish)

Pârvulescu, L., Paloş, C., Molnar, P. (2009) First record of the spiny-cheek crayfish *Orconectes limosus* (Rafinesque, 1817) (Crustacea: Decapoda: Cambaridae) in Romania. North-West Journal of Zoology, 5, 424-428.

Pârvulescu, L., Schrimpf, A., Kozubíková, E. *et al.* (2012). Invasive crayfish and crayfish plague on the move: first detection of the plague agent *Aphanomyces astaci* in the Romanian Danube. Diseases of Aquatic Organisms, 98, 85–94.

Persson, M., Söderhäll, K. (1983). *Pacifastacus leniusculus* Dana and its resistance to the parasitic fungus *Aphanomyces astaci* Schikora. Freshwater Crayfish, 5, 292-298.

Renai, B., Gherardi, F. (2004). Predatory efficiency of crayfish: comparison between indigenous and non-indigenous species. Biological Invasions, 6, 89-99.

Ruokonen, T. (2012). Ecological impacts of invasive signal crayfish in large boreal lakes. Jyväskylä, Finland: Doctoral dissertation, the University of Jyväskylä.

Sahlin, U., Smith, H.G., Edsman, L. *et al.* (2010). Time to establishment success for introduced signal crayfish in Sweden – a statistical evaluation when success is partially known. Journal of Applied Ecology, 47, 1044-1052.

Sandström, A., Andersson, M., Asp, A. *et al.* (2014). Population collapses in introduced non-indigenous crayfish. Biological Invasions, 16, 1961-1977.

Scalici, M, Pitzalis, M., Gibertini, G. (2009a). Crayfish distribution updating in central Italy. Knowledge and Management of Aquatic Ecosystems, 394-395, 06.

Scalici, M., Chiesa, S., Gherardi, F. *et al.* (2009b). The new threat to Italian inland waters from the alien crayfish "gang": the Australian *Cherax destructor* Clark, 1936. Hydrobiologia, 632, 341–345.

Scholtz, G., Braband, A., Tolley, L. *et al.* (2003). Parthenogenesis in an outsider crayfish. Nature, 421, 806.

Schrimpf, A., Schulz, H.K., Theissinger, K. *et al.* (2011). The first large-scale genetic analysis of the vulnerable noble crayfish *Astacus astacus* reveals low haplotype diversity in central European populations. Knowledge and Management of Aquatic Ecosystems, 401, 35.

Schrimpf, A., Pârvulescu, L., Copilaş-Ciocianu, D. *et al.* (2012). Crayfish plague pathogen detected in the Danube delta - A potential threat to freshwater biodiversity in Southeastern Europe. Aquatic Invasions, 7 (4), 503-510.

Schrimpf, A., Maiwald, T., Vrålstad, T. *et al.* (2013a). Absence of the crayfish plague pathogen *(Aphanomyces astaci)* facilitates coexistence of European and American crayfish in central Europe. Freshwater Biology, 58 (6), 1116-1125.

Schrimpf, A., Chucholl, C., Schmidt, T. *et al.* (2013b). Crayfish plague agent detected in populations of the invasive North American crayfish *Orconectes immunis* (Hagen, 1870) in the Rhine River, Germany. Aquatic Invasions, 8 (1), 103-109.

Smith, V., Söderhäll, K. (1986). Crayfish pathology: an overview. *Freshwater Crayfish*, 6, 199-211.

Soes, D.M., van Eekelen, R. (2006). Rivierkreeften een opruuend problem? De Levende Natuur, 107 (2), 56–59.

Soes, M., Koese, B. (2010). Invasive freshwater crayfish in the Netherlands: a preliminary risk analyses. Culemborg, the Netherlands: Bureau Waardenburg bv.

Souty-Grosset, C., Holdich, D.M., Noël, P.Y. *et al.* (Eds.) (2006). Atlas of crayfish in Europe. Paris, France: Muséum national d'Historie naturelle.

Strand, D.A., Jussila, J., Viljamaa-Dirks, S. *et al.* (2012). Monitoring the spore dynamics of Aphanomyces astaci in the ambient water of latent carrier crayfish. Veterinary Microbiology, 160, 99-107.

Strand, D.A., Jussila, J., Johnsen, S.I. *et al.* (2014). Detection of crayfish plague spores in large freshwater systems. Journal of Applied Ecology, 51 (2), 544-553.

Stucki, T.P. & Romer, J. (2001). Will Astacus leptodactylus displace Astacus astacus and Austropotamobius torrentium in Lake Äegeri, Switzerland? Aquatic Sciences, 63, 477-489.

Šulgaj, A. (1937). Naš potočni rak. Zveza ribarskih društev Dravske banovine, Ljubljana (in Slovene)

Svoboda, J., Kozubíková, E., Kozák, P. *et al.* (2012). PCR detection of the crayfish plague pathogen in narrow-clawed crayfish inhabiting Lake Eğirdir in Turkey. Diseases of Aquatic Organisms, 98, 255-259.

Svoboda, J., Kozubíková-Balcarová, E., Kouba, A. *et al.* (2013). Temporal dynamics of spore release of the crayfish plague pathogen from its natural host, American spiny-cheek crayfish (Orconectes limosus), evaluated by transmission experiments. Parasitology, 140, 792-801.

Svärdson, G. (1965). The American crayfish Pacifastacus leniusculus Dana introduced into Sweden. Reports of the Institute of Freshwater Research, Drottningholm, 46, 90-94.

Svärdson, G. (1995). The early history of signal crayfish inroduction to Europe. Freshwater Crayfish, 8, 68-77.

Söderhäll, K., Cerenius, L. (1999). The crayfish plague fungus: history and recent advances. Freshwater Crayfish, 12, 11-35.

Starobogatov, Y.I. (1995). Taxonomy and geographical distribution of crayfishes of Asia and East Europe (Crustacea Decapoda Astacoidei). Arthropoda Selecta, 4–3, 3–25.

Thörnqvist, P.-O. & Söderhäll K. (1993). Psorospermium haeckeli and its interaction with the crayfish defence system. Aquaculture, 117, 205-213.

Tokeshi, M. (1999). Species coexistence. Ecological and evolutionary perspectives. London: Blackwell Science.

Tricarico, E., Vilizzi, L., Gherardi, F. *et al.* (2010). Calibration of FI-ISK, an invasiveness screening tool for nonnative freshwater invertebrates. Risk Analysis, 30, 285–292.

Trontelj, P., Machino, Y., Sket, B. (2005). Phylogenetic and phylogeographic relationships in the crayfish genus *Austropotamobius* inferred from mitochondrial COI gene sequences. Molecular Phylogenetics and Evolution, 34, 212–226.

Unestam, T. (1969). On the physiology of zoospore production in *Aphanomyces astaci*. Physiologia Plantarum, 22, 236-246.

Unestam, T. (1972). On the host range and origin of the crayfish plague fungus. Reports for the Institute of Freshwater Research, Drottningholm, 52, 192–198.

Unestam, T. (1975a). Defence reactions in and susceptibility of Australian and New Guinean freshwater crayfish to European-crayfish-plague fungus. Australian Journal of Experimental Biology and Medical Science, 53 (5), 349-359.

Unestam, T. (1975b). The dangers of introducing new crayfish species. Freshwater Crayfish, 2, 557-561.

Vennerström, P., Söderhäll, K., Cerenius, L. (1998). The origin of two crayfish plague *(Aphanomyces astaci)* epizootics in Finland on noble crayfish, *Astacus astacus*. Annales Zoologici Fennici, 35, 43-46.

Vey, A., Söderhäll, K., Ajaxon, R. (1983). Susceptibility of *Orconectes limosus* Raff. to the crayfish plague, *Aphanomyces astaci* Schikora. Freshwater Crayfish, 5, 284-291.

Viljamaa-Dirks, S., Heinikainen, S., Nieminen, M. *et al.* (2011). Persistent infection by crayfish plague *Aphanomyces astaci* in a noble crayfish population - A case report. Bulletin of European Association of Fish Pathologists, 31, 182-188.

Viljamaa-Dirks, S., Heinikainen, S., Torssonen, H. *et al.* (2013). Distribution and epidemiology of genotypes of the crayfish plague agent *Aphanomyces astaci* from noble crayfish *Astacus astacus* in Finland. Diseases of Aquatic Organisms, 103, 199-208.

Vogt, G. (1999). Diseases of European freshwater crayfish, with particular emphasis on interspecific transmission of pathogens. In F. Gherardi, D. Holdich (Eds.), Crayfish in Europe as alien species. How to make the best of a bad situation? (pp.87-102). Rotterdam, Netherlands: A.A. Balkema.

Vrezec, A., Jaklič, M. & Govedič, M. (2013). Distribution patterns of indigenous and non-indigenous crayfish in Slovenia. In Book of abstracts, Regional crayfish meeting CrayCro: 26th–28th September, 2013, Rovinj, Croatia, 31.

Westman, K. (1973). The population of the crayfish, *Astacus astacus* L. in Finland and the introduction of the American crayfish *Pacifastacus leniusculus* Dana. Freshwater Crayfish, 1, 41-55.

Westman, K. (2000). Comparison of the crayfish *Pacifastacus leniusculus* Dana, a species introduced into Finland, with the native species, *Astacus astacus L.*, in allopatry and sympatry. Helsinki, Finland: Doctoral Dissertation, University of Helsinki.

Laura Härkönen, Arja Kaitala

9 Host Dynamics and Ectoparasite Life Histories of Invasive And Non-Invasive Deer Ked Populations

9.1 Introduction to Ectoparasite Invasions

A considerable fraction of biological invasions, including terrestrial ectoparasites, is a consequence of human activity. The most important environment for arthropod ectoparasites is a vertebrate host – the host as an environment can be highly heterogeneous both quantitatively (abundance of host individuals) and qualitatively (different genotypes, host races, populations and species). Ectoparasites are vectors for harmful pathogens, whereas ectoparasite infestations as such rarely cause substantial fitness costs or health issues for their host (Lehane, 2005; Muirhead-Thomson, 1982). Accordingly, ectoparasites are often considered in the context of transmitting invasive pathogens (Lehane, 2005), whereas ectoparasite invasions among host populations, or invasive expansion of geographical range have received less attention (Pyšek *et al.*, 2008). However, understanding the natural history and evolutionary ecology of parasites should become a key part of public health planning (Renaud *et al.*, 2005). Here, we review the main factors behind the invasion of a blood-feeding fly, the deer ked (*Lipoptena cervi*; Hippoboscidae), in a host population coupled with substantial expansion of geographical distribution.

9.1.1 Host-parasite Interactions

Warm-blooded vertebrates provide a necessary habitat and resources for a variety of terrestrial ectoparasites. True ectoparasites, such as ticks (Parasitiformes: Acari), fleas (Siphonaptera), lice (Phthiraptera) and many louse flies (superfamily Hippoboscoidea: Diptera), are specialized in inhabiting the host surface for at least part of their life. During the obligate host-dependent stage(s) they consume host-derived resources (e.g. blood, secretions, keratin, skin cells; Lehane, 2005). Due to the tight host-dependency, factors related to host dynamics, i.e. spatio-temporal variation in host abundance, and co-evolutionary interactions, are the main factors driving ectoparasite invasions (Holt *et al.*, 2003; Meier *et al.*, 2014).

A typical relationship between an ectoparasite and its host is characterized by co-evolution. Ectoparasites adopt a variety of morphological, physiological, and behavioural adaptations to exploit hosts for living and reproduction. Hosts, in turn, evolve adaptations to regulate the number of parasites and their negative effects. They may either behaviourally avoid contact with ectoparasites, or use defensive mechanisms to prevent or minimize infections and/or the infection can be counteracted by the immune system (Agnew *et al.*, 2000; Poulin, 2007). The more competent the host defence systems, the fewer the chances for the ectoparasite to exploit the host and spread effectively in the host population. Accordingly, well-documented cases of invasions by terrestrial ectoparasites have been associated with hosts that have not yet evolved efficient defences against the invaders [e.g. the winter tick *Dermacentor albipictus* invasion of moose *Alces americanus* in Northern America (Samuel *et al.*, 2000) and the deer ked invasion of European moose *Alces alces* in Finland (Kaitala *et al.*, 2009)].

The geographical range size of ectoparasites is often smaller than the range(s) of its host(s) owing to narrower environmental tolerance of the parasite, and smaller range sizes are more frequently found among host-opportunistic than in host-specific ectoparasites (e.g. in fleas; Shenbrot *et al.*, 2007). Accordingly, the most efficient way for an ectoparasite to adapt to a host environment is to actively select only the most suitable hosts. However, several potential host species may be available in a given environment, and depending on the degree of host specificity, ectoparasites exhibit varying performances across the different hosts (Valimaki *et al.*, 2011). Host-specialist ectoparasites may use their host more efficiently, but a high degree of specialization often leads to low performance across the other potential host species (Poulin, 2007). On the contrary, a generalist parasite should invest in counter-adaptations against several host species. A lower degree of specificity often results in lower performance across all potential species, even on an optimal host species. However, ectoparasites may shift their host preferences according to local host availability (Shenbrot *et al.*, 2007), and an adaptation to geographically different host populations or species may be an important determinant of an ectoparasite's ability to invade new host species and areas (Härkönen *et al.*, 2015).

Ectoparasites may encounter novel, but potentially suitable host species when colonizing new geographical areas. A recently infected host is often poorly adapted to the novel parasite, either because it has not yet developed an immunological defence or because of fortuitous vulnerability, and hence there is potential for the parasite to spread rapidly in a new host population (Poulin, 2007). The overall performance of specialist ectoparasites on a novel host may remain low due to specialization in native hosts or because the resources derived from novel hosts are of lower quality than from native hosts (Gandon & van Zandt, 1998). The attempts to reproduce on low-quality hosts may be so frequent that the invasion process ceases. In other words, high abundance of low-quality novel hosts qualifies as an ecological trap for ectoparasites (Battin, 2004; Robertson & Hutto, 2006; Robertson *et al.*, 2013).

9.1.2 Invasion Process in Temperate Ectoparasites

Ectoparasites' ability to fly and migrate long distances is often negligible for active dispersal. Instead, the vertebrate host usually migrates long distances, and thus the hosts are efficient carriers of ectoparasite propagules into new areas. Colonization of new geographical areas is affected by the extent of propagule pressure, referring to the number of parasites carried by the host to new environments (Lockwood *et al.*, 2005). Thus, high host density and high ectoparasite prevalence on the hosts increase both propagule pressure and genetic diversity in the introduced founder population, and may hence facilitate the establishment of the ectoparasite population in a new area (Davis, 2009).

During transportation phase, temperate ectoparasites are protected from unfavourable external temperatures by the thermoregulatory abilities of their warm-blooded hosts (Wharton, 1999), but establishment in novel geographical areas depends on the survival of ectoparasites through the free-living stage(s) (Härkönen *et al.*, 2010; Khatchikian *et al.*, 2012). Since ectoparasites are rarely able to actively choose their environments, adaptations to the local abiotic (off-host) or biotic (host) environment are predicted to only occur when the dispersal rate of the parasite is very low (Kirkpatrick & Barton, 1997). Instead, many ectoparasites have adopted a plastic behaviour and life history that allows them to tolerate conditions in different off-host habitats (Gandon & van Zandt, 1998). When survival requires phenotypic plasticity, wide ecological requirements and high ecological tolerance, the same characteristics may facilitate the invasion process across different off-host environments (Agrawal *et al.*, 2001; Yeh & Price, 2004).

An essential requisite for persistence of an ectoparasite population in a new area is an individuals' ability to complete its life-cycle, i.e. ectoparasites must find a host from the new environment during the free-living infective stage. Most true ectoparasites exploit a *sit-and-wait* host-ambushing strategy instead of an active 'cruising' strategy, and thus the infection strongly depends on the probability of close host contact (Fenton & Rands, 2004). High local host density is particularly important for host-ambushing ectoparasites to increase the likelihood of host encounter. Host densities are likely to decline towards the host's distribution range edge, which could limit ectoparasites' range expansion by decreasing the probability of host encounter. In addition, since ectoparasites are often restricted to using the same host through the reproductive lifespan, a low number of infective-stage ectoparasites on an individual host animal reduces the probability of male-female pair formation and subsequent reproductive success (Gascoigne *et al.*, 2009; Kramer *et al.*, 2009). Thus, low ectoparasite density in recently colonized areas can prevent further spread of invasive species because of the Allee effect (Lewis & Kareiva, 1993; Tobin *et al.*, 2011).

Ectoparasite persistence in a given environment also depends on the interactions between individual life-history and the experienced abiotic environment (Härkönen *et al.*, 2010; Stearns & Hoekstra, 2005). Towards higher latitudes the severity of winter increases, and the length and temperature of the growth season decline (Bale & Haywards, 2010; Crozier, 2004a,b). Survival and development during off-host stages are dependent on seasonal temperature variation, and hence colder climate commonly increases winter mortality, prolongs the developmental period, and may thus limit range expansion of ectoparasites (Härkönen *et al.*, 2010). Timing of the infective stage is particularly important in the north, where the potential transmission period is often short due to earlier arrival of winter (Härkönen *et al.*, 2010; Samuel *et al.*, 2000). Thus, the prolonged development due to cold climate, and thereby delayed emergence of the infective stage, may lead to a mismatch with host availability and decrease the probability of host encounter.

In theory, when suitable hosts are numerously available beyond the distribution range limits and abiotic environmental factors do not restrict parasite numbers, ectoparasites may spread rapidly in a host population and across geographical areas in an invasive manner. However, due to geographical differences in absolute and relative densities of differently suitable host species, the ability of ectoparasites to survive and complete their life-cycle may vary markedly, resulting in population differences in invasion potential (Härkönen *et al.*, 2015). Particularly among host-specialist ectoparasites, relative rather than absolute density of primary host species is likely to determine the invasion potential of an ectoparasite.

9.2 Effects of Host Dynamics and Deer Ked Life Histories on Population Invasiveness

The blood-sucking deer ked (*Lipoptena cervi*, Hippoboscidae) is an Old World dipteran ectoparasite of European moose (*Alces alces*) and other boreal cervids (Cervidae). Its distribution extends from the British Isles through Europe and Russia to Northern China and the Korean peninsula (e.g. Haarløv, 1964; Sleeman, 1983; Kim *et al.*, 2010; Välimäki *et al.*, 2010). In addition, the deer ked was apparently introduced in North America in the late 1800s by an unknown European deer species, and currently infests native cervids of the North-eastern United States (Samuel *et al.*, 2012).

In Fennoscandia (Northern Europe), the deer ked has undergone significant geographical expansion in distribution during recent decades (Välimäki *et al.*, 2010; Figure 9.1). In addition, a moose has been found to host over 17,000 deer keds (Paakkonen *et al.*, 2010), demonstrating the intensity of fly infestation in recently colonised hosts. The contributing factors for the invasion potential of deer ked among the Fennoscandian cervid populations and their range expansion towards high latitudes have been studied intensively (Härkönen, 2012; Härkönen *et al.*, 2010, 2012; 2013; Kaitala *et al.*, 2009; Kaunisto, 2012; Kaunisto *et al.*, 2011, 2012; Madslien, 2013; Madslien *et al.*, 2012; Meier *et al.*, 2014; Nieminen *et al.* 2012, 2013; Paakkonen, 2012; Paakkonen *et al.*, 2010, 2012; Välimäki *et al.*, 2010, 2011). The importance of host dynamics and life-history attributes in deer ked invasion are reviewed here.

9.2.1 Invasion of Fennoscandian Cervids

The distribution of the deer ked in Fennoscandia is divided by the Baltic Sea into two distinct, western (Sweden and Norway) and eastern (Finland) ranges (Fig. 9.1.).

The deer ked is a host-ambushing ectoparasite and requires close contact with a potential host for infection. Thus, high abundance of suitable hosts is of particular importance for deer ked invasion. A relatively wide range of cervid species is available for the deer ked in Fennoscandia (Välimäki *et al.*, 2011). The European moose (*Alces alces*) is currently the only host species that promotes high reproductive success of the deer ked in both western and eastern parts of Fennoscandia (Välimäki *et al.*, 2011). Reproduction on roe deer (*Capreolus capreolus*) has proved to be successful only in the western range; in the eastern range, successful offspring production has been observed only once (Härkönen *et al.*, 2015). There is as yet no data on reproduction on less common cervids, such as the fallow deer (*Dama dama*) or the red deer (*Cervus elaphus*), in the western range. In the eastern range, deer ked reproduction – with low success – has been reported on reindeer subspecies [semi-domesticated reindeer (*Rangifer tarandus tarandus*; Kaunisto *et al.*, 2009) and wild forest reindeer (*R. t. fennicus*; Välimäki *et al.*, 2011). In the southern parts of eastern range, the white-tailed deer (*Odocoileus virginianus*) is relatively common in dense deer ked areas, but no evidence of infestation has been found (Välimäki *et al.*, 2011).

Abundances of many large vertebrates, including cervids, have been regulated by man, which has exerted immediate consequences on their parasite community and distribution (reviewed by Thomas *et al.*, 2009). Correspondingly, many Fennoscandian cervid species have undergone remarkable population declines during the last centuries due to human activities, such as hunting and forest management (Liberg *et al.*, 2010; Tiilikainen *et al.*, 2012). Today harvest policies and other management actions are similar in both the western and eastern parts of Fennoscandia (Luoma, 2002). In particular, moose and roe deer population sizes are maintained in artificially large numbers for recreational hunting purposes, facilitating outbreaks of deer ked (Meier *et al.*, 2014; Figure 9.2).

The deer ked was first reported by Carl Linneaus in Sweden, Western Fennoscandia (i.e. Scandinavian Peninsula; Linneaus, 1758). This western deer ked population originates from Central Europe, but its arrival time in Sweden is not exactly known. The colonization of the Scandinavian Peninsula has been poorly documented, but the deer ked has been resident in southern and central parts of Sweden since at least the 18th century (below the latitude of 62°N; Fig. 9.1). The abundances of the two main hosts in the western deer ked range, the moose and the roe deer, have varied independently from near extinction to very large numbers (Lavsund *et al.*, 2003; Liberg *et al.*, 2010). The Scandinavian roe deer was hunted nearly to extinction before the remnant population was declared protected in Southern Sweden around 1840. The roe deer quickly recovered and recolonized most of the Scandinavian Peninsula (Liberg *et al.*, 2010; Thulin, 2006). The Scandinavian moose population began to recover markedly later, around the 1930s. It is likely that the roe deer maintained the western deer ked population during the threatened moose extinction. Since then, the western range of the deer ked has remained relatively stable and the population is considered non-invasive (Välimäki *et al.*, 2010). However, a minor range expansion started in the 1980s when the deer ked spread westwards across the Norwegian border and invaded South-eastern Norway (Figure 9.1).

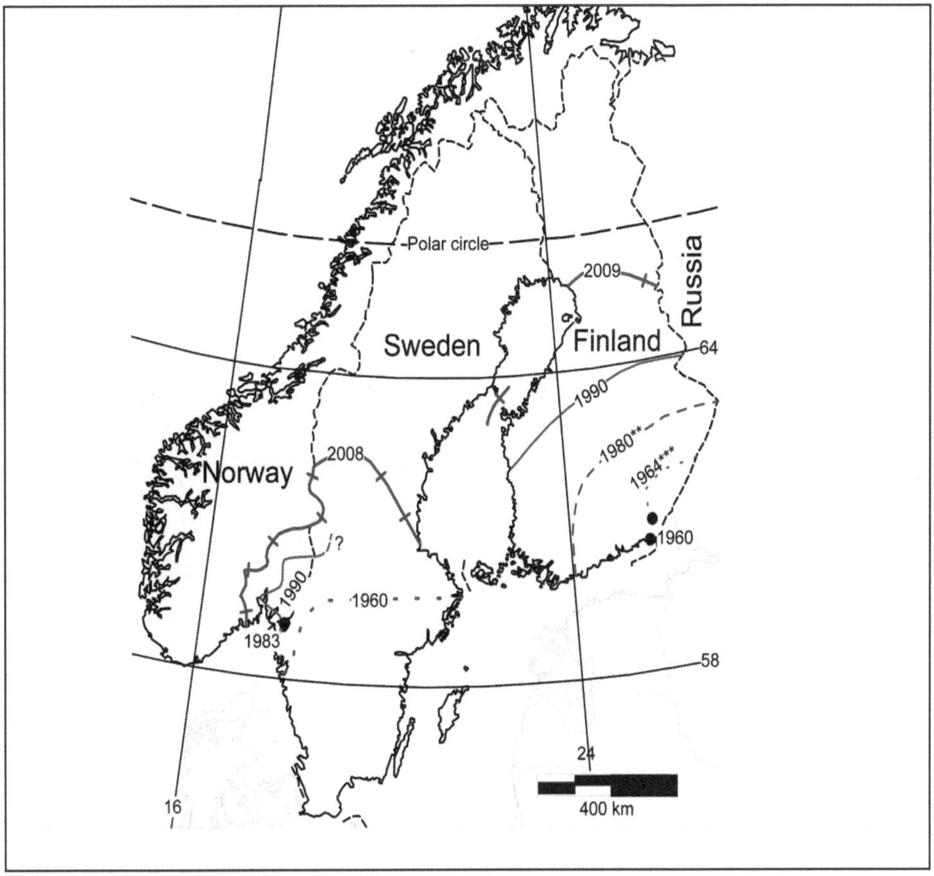

Fig. 9.1: Illustration of the bimodal distribution of the deer ked in Northern Europe from the early 1960s to the present. The first verified observations (in Norway and Finland) are indicated by black dots. The northern range limit is based on scarce observations of the northernmost individuals, which do not correspond the established range limit. The western and eastern deer ked populations likely separated thousands of years ago, but there are no genetic differences between the populations (Jaakola *et al.*, 2015). Reprinted from Välimäki *et al.* (2010), with kind permission from Springer Science and Business Media.

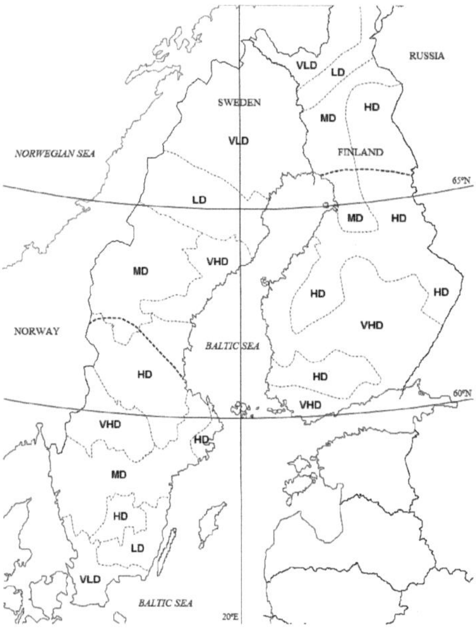

Fig. 9.2 (A): Current densities of the moose (VHD = very high density, HD = high density, MD = medium density, LD = low density, VLD = very low density) in Sweden (Kindberg *et al.*, 2008) and Finland (FGFRI, 2011). The northernmost observations of the deer ked are presented as a black dashed line (according to Välimäki *et al.*, 2010).

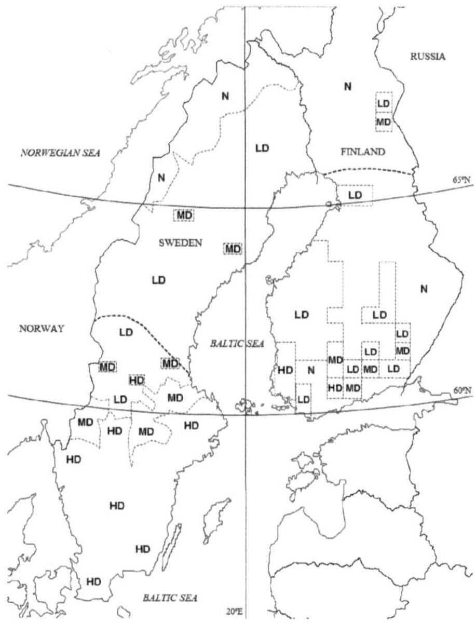

Fig. 9.2 (B): Current densities of the roe deer (HD = high density, MD = medium density, LD = low density, N = no observations) in Sweden (Liberg *et al.*, 2010; harvesting numbers in 2004) and Finland (FGFRI, 2013; Snow-track counting). The northernmost observations of the deer ked are presented as a black dashed line (according to Välimäki *et al.*, 2010)

The first records of the deer ked in Eastern Fennoscandia are from 1960 in Southeastern Finland, where it arrived from Russia (former U.S.S.R.; Hackman *et al.*, 1983). Systematic moose management in Finland began in 1969–1971 when the nearly extinct moose was declared protected, after which moose numbers showed an exponential population increase (Luoma, 2002). Accordingly, the spread of the deer ked began from the latitude of 60°N in the 1970s, and during the following twenty years its range extended to western and central parts of Finland (Hackman, 1977). In other words, the spatial and temporal dynamics of the deer ked outbreak are closely associated with the extremely rapidly expanded distribution and increased densities of moose (Meier *et al.*, 2014). The range expansion towards higher latitudes was rapid until the northern range limit was recently reached, along the latitude of 65°N where the southern border of reindeer herding territory is located (Välimäki *et al.*, 2010). Altogether, the eastern range extended over five latitudes and two climatic zones in less than 50 generations (i.e. 50 years). Hence, the eastern deer ked population has been considered an invasive population.

Taken together, the western distribution has remained at markedly more southern latitudes for centuries compared to the distribution of eastern population. There are no definite figures of the present total number of moose in Sweden and Finland, but the harvest level and local surveys indicate that winter populations (post-harvest) are around 200,000 individuals in Sweden (2005) and 80–90,000 in Finland (2006) (Liberg *et al.*, 2010). In Sweden, the roe deer population has been estimated to consist of 400–500,000 individuals (post-harvest in 2005–2007; Liberg *et al.*, 2010). The present roe deer number in Finland is significantly lower, being approximately 15,000 individuals (Burbaité & Csânyi, 2009). Moreover, these cervids are available for the deer ked beyond the current range limits (Figure 9.2) indicating that the latitudinal differences are not simply explained by abundance of suitable hosts. Instead, the differences in invasiveness of deer ked populations likely derive from co-evolutionary or environmental factors driving the population dynamics through off-host survival (Härkönen *et al.*, 2015).

9.2.2 Life-history Variation in the Deer Ked

9.2.2.1 Reproduction and off-host Survival

A cervid host offers a constantly favourable habitat for the deer ked – feeding, mating and reproduction occur on the host all year round (see Figure 9.3). The deer ked is a live-bearing, viviparous insect: the development of an embryo and egg-hatching occurs inside the female, where the larva is nourished through "milk glands" until it reaches the final instar. The mother gives birth to one pre-pupa at a time, which completes its pupation immediately after birth. The total offspring number of a female has been estimated as few dozen at maximum (Ivanov, 1981). The new-born pupae drop off from the host to undergo an obligate free-living period, including phases of diapause (i.e. winter dormancy that varies in its duration according to the birth date), pupal development, and host search as an adult (Härkönen & Kaitala, 2013).

Viviparous females provide a *safe harbor* for juvenile development, and by producing well-developed offspring, the survival prospects of each offspring are relatively high (Stearns, 1992). In the deer ked, offspring survival after birth correlates strongly with its size as large size increases physiological tolerance to environmental adversities and starvation resistance during the off-host period (Härkönen *et al.*, 2012, 2013). Offspring survival thus depends on the resources their mother transfers from the host's blood during larval incubation, but offspring size also varies with respect to host condition (Härkönen *et al.*, 2013). When a host is in good condition, it may be able to increase its immunological resistance against parasites (Tschirren *et al.*, 2007). As the winter progresses, the condition of a moose declines, which increases the deer ked's ability to exploit the host for offspring production — the offspring size, and thereby also off-host survival, increase steadily from autumn towards the spring and the end of the reproductive period (Härkönen *et al.*, 2012, 2013).

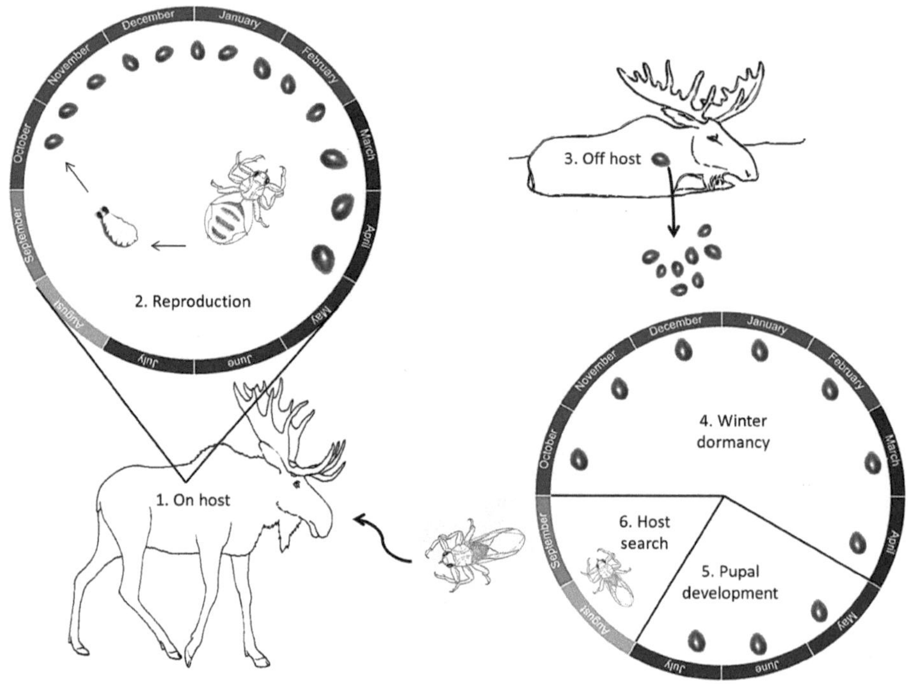

Fig. 9.3: Illustration of the deer ked life-cycle. (1) Once an adult finds and attaches on a host, it loses its wings, exploiting the same host until senescence. (2) Viviparous blood-sucking females give birth to one full-grown larva at a time. Offspring production starts in autumn and lasts at least until the following spring (even until July; Härkönen, 2012). (3) The larva pupates immediately after birth and the pupa falls off from the host, often on a host bedding site. (4) On the ground, the pupae first over-winter at diapause. (5) Once the temperature rises in the spring, diapause ends and pupal development begins. After approx. three months of development, adults emerge in high local synchrony. (6) The adults may ambush hosts until late autumn.

In addition to host-dependent offspring quality, off-host survival is directly affected by environmental harshness and the duration of the period outside the host. Due to viviparity, the period of offspring production is exceptionally long (Figure 9.3). In seasonally challenging environments, early-born small autumn pupae have a low survival probability due to higher risk of starvation during long winter dormancy (Härkönen *et al.* 2013). Since offspring are produced throughout winter months some of them will risk freezing, and thus the roles of cold tolerance and insulating snow cover for winter survival are of great importance in areas with frequent harsh frosts (Härkönen *et al.*, 2012). The newborn pupae are highly tolerant against short-term frosts when they drop off from the warm host: even without cold acclimation they survive briefly in temperature as low as -26°C by supercooling. An increase in offspring size enhances tolerance of pupae to long-term harsh frosts — especially large pupae can tolerate three days of direct exposure at -20°C but are likely to die if frost period is longer. Late-born large pupae in spring overwinter only briefly and do not experience harsh winter conditions (frosts, lack of snow, predation) (Kaunisto, 2012), and are likely to survive until the infective stage (Härkönen *et al.*, 2012, 2013). Taken together, long reproductive period and production of large well-developed offspring increases the probability that some of the introduced propagules will experience favourable conditions in a novel off-host environment and survive long enough to find a host.

The deer ked adults emerge in late summer and search for a host through the autumn. A host-ambushing adult relies on very simple host location cues; it does not discriminate between potential hosts, instead attacking all large, dark, moving objects, including humans, dogs, cows, and horses (Kortet *et al.*, 2010). The adults are able to survive without feeding for a few months while sitting-and-waiting for a potential host to pass by. In addition to risk of starvation before host contact, arrival of winter may cease the host search period. However, the adult stage may briefly tolerate frost below -15°C and host search may continue as long as the daily temperatures remain high enough for flying (Härkönen *et al.*, 2012; Nieminen *et al.*, 2012). The ability to tolerate autumnal frosts may extend the host search time and thereby facilitate the range expansion northward.

9.2.2.2 Host-specific Life History Variation

Ectoparasite life histories at the margin distribution areas determine the invasion potential of a population, i.e. the ability to spread beyond established range limits (Bridle & Vines, 2006; Hill *et al.*, 2011; Sexton *et al.*, 2009). Large size and high physiological tolerance are generally among the best predictors of invasion success (Davis, 2009). Both these characteristics in the deer ked are closely associated with host-derived resources (Härkönen *et al.*, 2012). However, distinct co-evolutionary and invasion histories of the Fennoscandian populations have modified deer ked life history characteristics, and nowadays the local host species and populations differ in their quality for deer ked reproduction (Härkönen *et al.*, 2015).

The most important difference in deer ked characteristics between the noninvasive western and invasive eastern populations is body size. The pupae and adults in the invasive population may be as much as twice the size of the non-invasive population (Välimäki *et al.*, 2011). This difference has likely emerged as a result of long co-evolution with different hosts. For example, a positive relationship has been observed between ectoparasite size and host mass (Poulin, 1998). Large size and inability to infect multiple host species indicate that the invasive eastern population is well-adapted to exploit the largest cervid, i.e. the moose. Although the joint history of moose and the invasive deer ked in eastern Fennoscandia is short, the moose has already become the native host species in its origin areas in Eastern Europe and Russia (Ivanov, 1981). Co-occurrence of the western deer ked with roe deer goes back to Central Europe, where the deer ked has exploited medium- or small-sized cervids as its native hosts (Haarløv, 1964). Joint history with the moose is shorter as it presumably started when deer ked colonized Western Scandinavia. Simultaneous co-evolution with several cervids, and particularly with small-sized cervids, may have selected for small body size in the western deer ked population (Härkönen *et al.*, 2015).

Reciprocal influence of the host resistance and the deer ked female's ability to invest resources in developing larva varies with respect to invasion histories and with different host species. In other words, variation in offspring size and cold tolerance indicate immediate host effects on off-host survival owing to local variation in host condition or quality (Härkönen *et al.*, 2012, 2013). Within the well-established western range, offspring size, survival, and cold-hardening capacity are greater on pupae produced on roe deer than on moose in the same area, indicating a higher ability to exploit roe deer for reproduction. On the other hand, the moose differ in quality for reproduction between the eastern and western populations. We have observed production of offspring with markedly larger size, higher supercooling capacity and survival on moose in the invasive eastern than in the stable western population (Harkonen *et al.*, 2015). Such differences in performance on the same host species in different geographical areas indicate that the invasive, moose-specialist deer ked exploits the moose more efficiently for offspring production. When parasites become locally adapted on one host species they often lose the ability to infect other hosts (Gandon & Van Zandt, 1998). Accordingly, the reproductive performance of the eastern population on any other local cervid species (roe deer, wild forest reindeer or semi-domesticated reindeer) is very low (Kaunisto *et al.*, 2009; Kynkäänniemi *et al.*, 2010; Välimäki *et al.*, 2011). Interestingly, our preliminary findings suggest that the deer keds in the recently expanded western subpopulation in Norway exhibit a substantial increase in size and viability compared to the old, well-established core population in Sweden. The recently invaded moose population in Norway may still have low ability to resist the deer keds, and thus the deer keds may hold the potential to spread in that area owing to higher ability to survive in adverse off-host environments.

9.2.3 Current State of Deer Ked Invasion

The potential for further invasion of the deer ked would primarily depend on host availability and their movements and secondarily on off-host survival beyond the current range limits. In theory, both Fennoscandian deer ked populations have the potential to spread further north because a high number of cervids are also available far beyond the current northern range limits. Long distance migrations of cervids facilitate spreading of the deer ked to new areas. For example, moose are highly mobile and may move for hundreds of kilometres between seasonal ranges during the deer ked's reproductive life span (Heikkinen, 2000). However, current distribution limits correlate with the density of that host species on which the off-host survival probability is the highest. In the western deer ked range, the roe deer is currently the most abundant cervid species in the area (Liberg *et al.*, 2010) – it is common in the southern and central parts of Scandinavia but rarer further north (Figure 9.2B).

Accordingly, the well-established deer ked range limit follows the high density roe deer distribution, rather than moose distribution which is high also beyond the current northern range limit. Correspondingly the invasive eastern population has followed spatio-temporal variation in moose densities. Northward range expansion has also exposed the semi-domesticated reindeer populations to a novel ectoparasite (Kynkäänniemi *et al.* 2010, 2014). Once the deer ked range extended to southern reindeer herding areas, where the densities of reindeer are tenfold compared to moose densities in the area, its invasion ceased. Along the current northern range limit, the deer ked has been reported to infect reindeer but the reproductive success is low (Kaunisto *et al.*, 2009; Kynkäänniemi *et al.*, 2010), suggesting that low quality hosts may function as an ecological trap for the deer ked, preventing further spread (Aikio & Kaitala, unpublished data). Taken together, instead of absolute density of the host species, the relative densities of different host species play an important role in deer ked invasion potential in the future.

The deer ked pupae dropping off from the host during seasonal migrations are effectively distributed over a wide range of habitats and climatic conditions. Due to viviparous production of large, high-quality offspring, the deer ked off-host stages are highly tolerant to northern adversities prevailing outside the host. However, in order to establish even further north, free-living stages should not only tolerate low temperatures but also adjust the life-cycle to shorten the growth season. Although the eastern invasive deer keds have better ability to survive in the north due to larger size and higher cold tolerance, off-host survival of the western deer keds does not differ so dramatically that it would consecutively explain their southern range limits. Under identical conditions, however, adults from the stable western population emerge markedly later than in the invasive eastern population (Välimäki *et al.*, 2011). Longer development time indicates that the emergence of the infective stage in the western population is adjusted to the relatively longer and warmer summer prevailing in Southern Fennoscandia and thus, there may be a mismatch between emergence time and ability to ambush hosts under colder northern climate conditions, and thereby a limit to northwards invasion. On the contrary, faster pupal development rate in the eastern range has facilitated rapid northwards invasion (Kaunisto *et al.*, 2011). Furthermore, a transplant experiment conducted using the invasive eastern population showed that the deer ked is able to complete its pupal development at the latitude of 70°N, i.e. 500 kilometres to north from the current range limit (65°N). Further northward invasion may still be unlikely because lower spring and summer temperatures restrict the parasite numbers substantially by decreasing pupal survival, delaying adult emergence, and shortening the time window for host search (Härkönen *et al.*, 2010).

9.2.3.1 Socio-economical Consequences of Deer Ked Invasion

The detrimental effects of deer ked on host health have been contradictory – severe alopecia has been associated with massive deer ked infestation in Norway (Madslien *et al.*, 2011) whereas no physiological effects, or only minor effects, have been reported in moose in Finland (Paakkonen *et al.*, 2012). However, the blood-feeding deer ked is a potential vector for transmitting various diseases (Rantanen *et al.*, 1982; Dehio *et al.*, 2004). The deer ked is known to harbour haemotrophic *Bartonella schoenbuchensis*. The same *Bartonella* species have been identified, with high prevalence, in moose in areas with deer ked infestation (Duodu *et al.*, 2013; Korhonen *et al.*, 2015). It is yet unknown whether chronic bacteremia with *Bartonella* has any impact on the health of the moose, or on other species that the deer ked may attack.

The outbreak of the deer ked has been accompanied by public health issues for humans, particularly in high density areas in southern and central parts of Finland. Due to simple host location cues, the deer ked often mistakenly attacks humans (Kortet *et al.*, 2010). The deer keds readily accept humans as a host, but they have never been reported to reproduce on humans. Instead, an increasing number of people are suffering from chronic, long lasting dermatitis associated with deer ked bites (Laukkanen *et al.*, 2005). The incidental infestation is a nuisance in that it affects participation in autumnal outdoor activities, such as hunting and berry picking (Härkönen *et al.*, 2009). Moreover, given that deer keds may take blood meals from humans, there is also substantial risk for occasional transmission of *B. schoenbuchensis* to humans, but its direct health risks for human have not yet been demonstrated (Dehio *et al.*, 2004; Duodu *et al.*, 2013; Korhonen *et al.*, 2015).

9.3 Acknowledgements

We thank Oliver Verneau and an anonymous reviewer for their comments on the manuscript, and Sirpa Kaunisto and Sirpa Niinimäki for their suggestions on an earlier draft of this chapter. Sirpa Kaunisto is also acknowledged for preparing the host density maps, and Anna Honkanen for sketching a picture of the deer ked.

In a nutshell

– A considerable fraction of biological invasions is a consequence of human activity. For terrestrial ectoparasites, those activities include artificial regulation of the numbers of host animals. Ectoparasites may spread in an invasive manner when suitable hosts are numerously available beyond the distribution range limits, and the abiotic environment does not restrict parasite numbers. However, management of invasive ectoparasites may be challenging because of host-parasite co-evolution and their interactions with the abiotic environment.

- Harvesting and regulation of cervid species have been a major determinant of deer ked invasion across Fennoscandia. In Finland, Eastern Fennoscandia, the invasion of blood-feeding deer ked of European moose has been intense once the host population recovered from over-harvesting. During the recent decades the geographical range expansion of the ectoparasite towards higher latitudes has also been rapid. In Sweden, Western Fennoscandia, the deer ked have alternately exploited two host species – the moose and roe deer – for centuries, and the ectoparasite population has been stable and remained at markedly lower latitudes. However, a small invasive frontier has recently emerged from the stable population after which the distribution expanded westwards to Norway.
- The deer ked life histories across Fennoscandia have been modified by co-evolutionary specialization on different local hosts. The invasive eastern deer ked only exploits moose efficiently, and females produce large and cold-tolerant offspring with high survival probability outside the host. High success on moose has been explained by short co-existence and specialization in only one host species. The stable western deer ked population has undergone long co-evolution with both moose and roe deer. At present, the western deer ked exploits the roe deer more efficiently than moose, but as a cost of having two main host species, particularly the offspring size, and consequently the off-host survival is lower than in the eastern population.
- In theory, both Fennoscandian deer ked populations have high invasion potential since current host abundances are high and also available beyond current range limits. However, the host effects on the deer ked life histories and off-host performance varies between the populations due to co-evolutionary interactions. Invasion of deer keds seems to be limited by high density of low-quality hosts. Thus it is the relative, not absolute, densities of the local host species which determines the population differences in invasion potential.

9.4 Bibliography

Agnew, P., Koella, J. C., Michalakis, Y. (2000). Host lifehistory responses to parasitism. Microbes and Infection, 2, 891-896.

Agrawal, A. A. (2001). Phenotypic plasticity in the interactions and evolution of species. Science, 294, 321-326.

Bale, J. S., Hayward, S. A. L. (2010). Insect overwintering in a changing climate. Journal of Experimental Biology, 213, 980-994.

Battin, J. (2004). When good animals love bad habitats: ecological traps and the conservation of animal populations. Conservation Biology, 18, 1482-1491.

Bridle, J. R., Vines, T. H. (2006). Limits to evolution at range margins: when and why does adaptation fail? Trends in Ecology and Evolution, 22, 140-147.

Burbaitė, L., Csányi, S. (2009). Roe deer population and harvest changes in Europe. Estonian Journal of Ecology, 58, 169-180.

Crozier, L. (2004a). Field transplants reveal summer constraints on a butterfly range expansion. Oecologia, 141, 148-157.

Crozier, L. (2004b). Warmer winters drive butterfly range expansion by increasing survivorship. Ecology, 85, 231-241.
Davis, M. A. (2009). Invasion Biology. New York, NY: Oxford University Press.
Dehio, C., Sauder, U., Hiestand, R. (2004). Isolation of *Bartonella schoenbuchensis* from *Lipoptena cervi*, a blood-sucking arthropod causing deer ked dermatitis. Journal of Clinical Microbiology, 42, 5320-5323.
Duodu, S., Harris, P. D., Hjelm, E., *et al.* (2013). Bartonella infections in deer keds (*Lipoptena cervi*) and moose (*Alces alces*) in Norway. Applied Environmental Microbiology, 79, 322-327.
Fenton, A., Rands, S. A. (2004). Optimal parasite infection strategies: A state-dependent approach._ International Journal for Parasitology, 34, 813-821.
FGFRI (2011). Hirvi. Kannan koko ja vasatuotto vuonna 2011. [Finnish Game and Fisheries Research Institute. Moose: population size and calf production in 2011]. Retrieved from: http://www.rktl.fi/riista/hirvielaimet/hirvi/
FGFRI (2013). Lumijälkilaskenta: metsäkauris. [Finnish Game and Fisheries Research Institute. Snow track counts: European roe deer]. Retrieved from: http://www.rktl.fi/riista/pienriista/lumijalkilaskenta/metsakauris.html
Gandon, S., Van Zandt, P. A. (1998). Local adaptation and host-parasite interactions. Trends in Ecology and Evolution, 13, 1-3.
Gascoigne, J., Berec, L., Gregory, S., *et al.* (2009). Dangerously few liaisons: a review of mate-finding Allee effects. Population Ecology, 51, 355-372.
Haarløv, N. (1964). Life cycle and distribution pattern of *Lipoptena cervi* (L.) (Dipt., Hippobosc.) on Danish deer. Oikos, 15, 93-129.
Hackman, W. (1977). Hirven täikärpänen ja sen levittäytyminen Suomeen [Moose louse-fly and its immigration in Finland]. Luonnon Tutkija, 81, 75-77.
Hackman, W., Rantanen, T., & Vuojolahti, P. (1983). Immigration of *Lipoptena cervi* (Diptera, Hippoboscidae) in Finland, with notes on its biology and medical significance. Notulae Enlomologicae, 63, 53-59.
Härkönen, L. (2012). Seasonal variation in the life histories of a viviparous ectoparasite, the deer ked. Doctoral dissertation, University of Oulu, Oulu, Finland.
Härkönen, L., Kaitala, A. (2013). Months of asynchrony in offspring production but synchronous adult emergence: the role of diapause in an ectoparasite's life cycle. Environmental Entomology, 42, 1408-1414.
Härkönen, L., Härkönen, S., Kaitala, A., *et al.* (2010). Predicting range expansion of an ectoparasite – the effect of spring and summer temperatures on deer ked (*Lipoptena cervi*) performance along a latitudinal gradient. Ecography, 33, 906-912.
Härkönen, L., Kaitala, A., Kaunisto, S., *et al.* (2012). High cold tolerance through four seasons and all free-living stages in an ectoparasite. Parasitology, 139, 926-933.
Härkönen, L., Hurme, E., Kaitala, A. (2013). Unexpected seasonal variation in offspring size and performance of a viviparous ectoparasite. Parasitology, 140, 229-236.
Härkönen, L., Kaunisto, S., Månsson, J., Hurme, E. & Kaitala, A. 2015. Host-specific variation in off-host performance of a temperate ectoparasite. Biological Journal of Linnean Society, DOI: 10.1111/bij.12628 [Early view: Article first published online: 3 AUG 2015]
Härkönen, S., Laine, M., Vornanen, M., *et al.* (2009). Deer ked (*Lipoptena cervi*) dermatitis in humans – an increasing nuisance in Finland. Alces, 45, 73-79.
Härkönen, L., Kaunisto, S., Månsson, J., Hurme, E. & Kaitala, A. 2015. Host-specific variation in off-host performance of a temperate ectoparasite. Biological Journal of Linnean Society, DOI: 10.1111/bij.12628 [Early view: Article first published online: 3 AUG 2015]
Heikkinen, S. (2000). Hirven vuosi. Suomen Riista, 46, 82-91.
Hill, J. K., Griffiths, H. M., Thomas, C. D. (2011). Climate chance and evolutionary adaptation at species' range margins. Annual Review of Entomology, 56, 143-159.

Holt, R. D., Dobson, A. P., Begon, M., *et al.* (2003). Parasite establishment in host communities. Ecology Letters, 6, 837-842.

Ivanov, V. I. (1981). Spread of the deer ked *Lipoptena cervi* L. (Diptera, Hippoboscidae) in Belarus, its biology and deleterious effects [In Russian]. Doctoral dissertation, Veterinary Academy of Moscow, Moscow, Soviet Union.

Jaakola, M., Ojanen, M., Hurme, E., Mutanen, M., Wahlberg, N., Välimäki, & Kaitala, A. 2015. Morphological variation between populations of the expanding ectoparasitic deer ked *Lipoptena cervi* (Diptera: Hippoboscidae) in Fennoscandia. Biological Journal of the Linnean Society, 116, 432-448. DOI: 10.1111/bij.12587

Kaitala, A., Kortet, R., Härkönen, S., *et al.* (2009). Deer ked, an ectoparasite of moose in Finland: A brief review of its biology and invasion. Alces, 45, 85-88.

Kaunisto, S. (2012). An invasive ectoparasite of cervids, the deer ked: dispersion, cold tolerance and predation. Doctoral dissertation, University of Eastern Finland, Joensuu, Finland.

Kaunisto, S., Kortet, R., Härkönen, L., *et al.* (2009). New bedding site examination-based method to analyse deer ked (*Lipoptena cervi*) infection in cervids. Parasitology Research, 104, 919-925.

Kaunisto, S., Härkönen, L., Niemelä, P., *et al.* (2011). Northward invasion of the parasitic deer ked (*Lipoptena cervi*), is there geographical variation in pupal size and development duration? Parasitology, 138, 354-363.

Kaunisto, S., Välimäki, P., Kortet, R., *et al.* (2012). Avian predation on a parasitic fly of cervids during winter: Can host-related cues increase the predation risk? Biological Journal of the Linnean Society, 106, 275-286.

Khatchikian, C. E., Prusinski, M., Backenson, P. B., *et al.* (2012). Geographical and environmental factors driving the increase in the Lyme disease vector *Ixodes scapularis*. Ecosphere, 3, 1-18.

Kim, H. C., Chong, S. T., Chae, J.-S., *et al.* (2010). New record of *Lipoptena cervi* and updated checklist of the louse flies (Diptera: Hippoboscidae) of the Republic of Korea. *Journal* of Medical Entomology, 47, 1227-1230.

Kindberg, J., Holmqvist, N., Bergqvist, G. (2008). Årsrapport Viltövervakningen 2006/2007. Svenska Jägareförbundet. Viltforum, 2, 1-28.

Kirkpatrick, M., Barton, N. H. (1997). Evolution of a species' range. American Naturalist, 150, 1-23.

Korhonen, E. M., Pérez Vera, C., Pulliainen, a T., *et al.* (2015). Molecular detection of Bartonella spp. in deer ked pupae, adult keds and moose blood in Finland. Epidemiology and Infection, 143, 578-585.

Kortet, R., Härkönen, L., Hokkanen, P., *et al.* (2010). Experiments on the ectoparasitic deer ked that often attacks humans; preferences for body parts, colour and temperature. Bulletin of Entomological Research, 100, 279-285.

Kramer A. M., Dennis, B., Liebhold, A. M., *et al.* (2009). The evidence for Allee effects. Population Ecology, 51, 341-354.

Kynkäänniemi, S.-M., Kortet, R., Härkönen, L., *et al.* (2010). Threat of an invasive parasitic fly, the deer ked (*Lipoptena cervi*), to the reindeer (*Rangifer tarandus tarandus*): experimental infection and treatment. Annales Zoologici Fennici, 47, 28-36.

Kynkäänniemi, S.-M., Kettu, M., Kortet, R., *et al.* (2014). Acute impacts of the deer ked (*Lipoptena cervi*) infestation on reindeer (*Rangifer tarandus tarandus*) behaviour. Parasitology Research, 113, 1489-1497.

Laukkanen, A., Ruoppi, P., Mäki-Kiljunen, S. (2005). Deer ked-induced occupational allergic rhinoconjunctivitis. Annals of Allergy, Asthma and Immunology, 94, 604-608.

Lavsund, S., Nygrén, T., Solberg, E. (2003). Status of moose populations and challenges to moose management in Fennoscandia. Alces, 39, 109-130.

Lehane, M. J. (2005). The biology of blood-sucking insects. Cambridge, UK: Cambridge University Press.

Lewis, M. A., Kareiva, P. (1993). Allee dynamics and the spread of invading organisms. Theoretical Population Biology, 43, 141-158.

Liberg, O., Bergstrom, R., Kindberg, J., *et al.* (2010). Ungulates and their management in Sweden. In M. Apollonio, R. Andersen, & R. Putman (Eds.), European ungulates and their management in the 21st century (pp.37-70). Cambridge, UK: Cambridge University Press.

Linnaeus, C. (1758). Systema naturae per regna tria naturae: secundum classes, ordines, genera, species, cum characteribus, differentiis, synonymis, locis. Stockholm (Holmiae), Sweden: Laurentii Salvii.

Lockwood, J. L., Cassey, P., Blackburn, T. (2005). The role of propagule pressure in explaining species invasions. Trends in Ecology and Evolution, 20, 223-228.

Luoma, A. (2002). Moose hunting in Finland – management of a heavily harvested population. Doctoral dissertation, University of Helsinki, Helsinki, Finland.

Madslien, K. (2013). Deer ked (*Lipoptena cervi*) and moose (*Alces alces*) in Norway – interactions between an invading ectoparasite, its host and the environment. Doctoral dissertation, Norwegian School of Veterinary Science, Oslo, Norway.

Madslien, K., Ytrehus, B., Vikøren, T., *et al.* (2011). Hair-loss epizootic in moose (*Alces alces*) associated with massive deer ked (*Lipoptena cervi*) infestation. Journal of Wildlife Diseases, 47, 893-906.

Madslien, K., Ytrehus, B., Viljugrein, H., *et al.* (2012). Factors affecting deer ked (Lipoptena cervi) prevalence and infestation intensity in moose (*Alces alces*) in Norway. Parasites and Vectors, 5, 251.

Meier, C., Bonte, D., Kaitala, A., *et al.* (2014). Invasion rate of deer ked depends on spatio-temporal variation in host density. Bulletin of Entomological Research, 104, 314-322.

Muirhead-Thomson, E. C. (1982). Behaviour patterns of blood-sucking flies. New York, NY: Pergamon Press.

Nieminen, P., Paakkonen, T., Eerilä, H., *et al.* (2012). Freezing tolerance and low molecular weight cryoprotectants in an invasive parasitic fly, the deer ked (*Lipoptena cervi*). Journal of Experimental Zoology Part A: Ecological Genetics and Physiology, 317, 1-8.

Nieminen, P., Paakkonen, T., Halonen, T., *et al.* (2013). Fatty acid modifications during autumnal cold-hardening in an obligatory ectoparasite, the deer ked (*Lipoptena cervi*). Journal of Insect Physiology, 59, 631-637.

Paakkonen, T. (2012). Ecophysiology of the deer ked (*Lipoptena cervi*) and its hosts. Doctoral dissertation, University of Eastern Finland, Joensuu, Finland.

Paakkonen, T., Mustonen, A.-M. Roininen, H., *et al.* (2010). Parasitism of the deer ked, *Lipoptena cervi*, on the moose, *Alces alces*, in eastern Finland. Medical and Veterinary Entomology, 24, 411-417.

Paakkonen, T., Mustonen, A.-M., Käkelä, R., *et al.* (2011). Experimental infection of the deer ked (*Lipoptena cervi*) has no negative effects on the physiology of the captive reindeer (*Rangifer tarandus tarandus*). Veterinary parasitology, 179, 180-188.

Paakkonen, T., Mustonen, A.-M., Käkelä, R., *et al.* (2012).The effects of an abundant ectoparasite, the deer ked (*Lipoptena cervi*), on the health of moose (*Alces alces*) in Finland. Parasitology Research, 111, 1223-1232.

Poulin, R. (1998). Host and environmental correlates of body size in ticks (Acari: Argasidae and Ixodidae). Canadian Journal of Zoology, 76, 925-930.

Poulin, R. (2007). Evolutionary ecology of parasites. Princeton NJ: Princeton University Press.

Pyšek, P., Richardson, D. M., Pergl, J., *et al.* (2008). Geographical and taxonomical biases in invasion ecology. Trends in Ecology and Evolution, 23, 237-244.

Rantanen, T., Reunala, T., Vuojolahti, P., *et al.* (1982). Persistent pruritic papules from deer ked bites. Acta Dermato-Venereologica, 62, 307-311.

Renaud, F., De Meeus, T., Read, A. F. (2005). Parasitism in man-made ecosystems. In F. Thomas, F. Renaud, & J.-F. Guegan (Eds), Parasitism and ecosystems (pp.155-170). Oxford UK: Oxford University Press.
Robertson, B. A., Hutto, R. L. (2006). A framework for under- standing ecological traps and an evaluation of existing evidence. Ecology, 87, 1075-1085.
Robertson, B. A., Rehage, J., Sih, A. (2013). Ecological novelty and the emergence of evolutionary traps. Trends in Ecology and Evolution, 28, 552-560.
Samuel, W. M., Mooring, M. S., Aalangdong, O. I. (2000). Adaptations of winter ticks to invade moose and moose to evade ticks. Alces, 36, 183-195.
Samuel, B., Madslien, K., Gonynor-McGuire, J. (2012). Review of deer ked (*Lipoptena cervi*) on moose in Scandinavia with implications for North America. Alces, 48, 27-33.
Sexton, J. P., McIntyre, P. J., Angert, A. L., *et al.* (2009). Evolution and ecology of species range limits. Annual Review of Ecology and Systematics, 40, 415-436.
Shenbrot, G., Krasnov, B., Lu, L. (2007). Geographical range size and host specificity in ectoparasites: a case study with *Amphipsylla* fleas and rodent hosts. Journal of Biogeography, 34, 1679-1690.
Sleeman, D. P. (1983). Parasites of deer in Ireland. Journal of Life Sciences Royal Dublin Society, 42, 203-210.
Stearns, S. C. (1992). The evolution of life histories. Oxford, UK: Oxford University Press.
Stearns, S. C., Hoekstra, R. F. (2005). Evolution: An Introduction. New York, NY: Oxford University Press.
Thomas, F., Guégan, J.-F., Renaud, F. (2009). Ecology and evolution of parasitism. Oxford UK: Oxford University Press.
Thulin, C-F. (2006). Microsatellite investigation of roe deer (*Capreolus capreolus*) in Scandinavia reveals genetic differentiation of a Baltic Sea Island population. European Journal of Wildlife Research, 52, 228-235.
Tiilikainen, R., Solberg, E. J., Nygrén, T., *et al.* (2012). Spatio-temporal relationship between calf body mass and population productivity in Fennoscandian moose *Alces alces*. Wildlife Biology, 18, 304-317.
Tobin, P. C., Berec, L., Liebhold, A. M. (2011). Exploiting Allee effects for managing biological invasions. Ecology Letters, 14, 615-624.
Tschirren, B., Bischoff, L. L., Saladin, V., *et al.* (2007). Host condition and host immunity affect parasite fitness in a bird–ectoparasite system. Functional Ecology, 21, 372-378.
Välimäki, P., Kaitala, A., Madslien, K., *et al.* (2011). Geographical variation in host use of a blood-feeding ectoparasitic fly: implications for population invasiveness. Oecologia, 166, 985-995.
Välimäki, P., Madslien, K., Malmsten, J., *et al.* (2010). Fennoscandian distribution of an important parasite of cervids, the deer ked (*Lipoptena cervi*), revisited. Parasitology Research, 107, 117-125.
Wharton, D. A. (1999). Parasites and low temperatures. Parasitology, 119, S7–S17.
Yeh, P. J., Price, T. D. (2004). Adaptive phenotypic plasticity and the successful colonization of a novel environment. American Naturalist, 164, 531-542.

Ane T. Laugen, Johan Hollander, Matthias Obst, Åsa Strand

10 The Pacific Oyster (*Crassostrea gigas*) Invasion in Scandinavian Coastal Waters: Impact on Local Ecosystem Services

10.1 Introduction

Shellfish have always played an important role in subsistence and income for coastal communities. Shallow-water bivalve beds are easily accessible and have been exploited and overexploited for centuries. Depletion of stocks has lead to farming and aquaculture of many species. Few species, if any, have been as popular and successful as the the Pacific oyster (*C. gigas*). Its highly efficient filter feeding, high growth rates, massive repoductive output, and tolerance of a large range of abiotic conditions, has not only made it one of the world's 20 most cultured species, but also one of the most invasive. This chapter tracks the Pacific oyster on its way towards the North-eastern corner of its European distribution, predicts its future distribution, and discusses the implications for local ecosystems.

10.2 Invasion History of Pacific Oysters: from Deliberate Introduction to Establishment of Feral Populations

Crassostrea gigas (hereafter *C. gigas* or oysters) originates from the coastlines of the Japanese Sea. Due to its suitability for aquaculture, it has been introduced to several geographical areas, and now has a world-wide distribution (Ruesink *et al.*, 2005). In the mid-1960s, European shellfish farmers started oyster production using *C. gigas*, and by the mid-90s, feral populations emerged from the French Atlantic coast to the German and Danish Wadden Sea (Héral, 1989; Drinkwaard, 1999; Reise *et al.*, 2005; Couzens, 2006; Kochmann *et al.*, 2008). In this section, we will briefly review the current status of the species in Scandinavia.

Denmark. From the early 1970s spat was introduced to several Danish fjords for aquaculture, and production continued until the late 1990s. Oysters were left in the sea when farming discontinued under the assumption that natural reproduction would not occur in the present environmental conditions, but feral populations appeared both around the trial areas and at new sites (for current known distribution, see Figure 10.1). Common habitats are blue mussel beds and mudflats in sheltered intertidal areas. Densities in three well-studied areas range from very high (500-2000 individuals × m^{-2} in the Wadden Sea), to medium (50 individuals × m^{-2} in the Limfjord), to low (max 0.11 individuals × m^{-2} in the Isefjord) (Dolmer *et al.*, 2014).

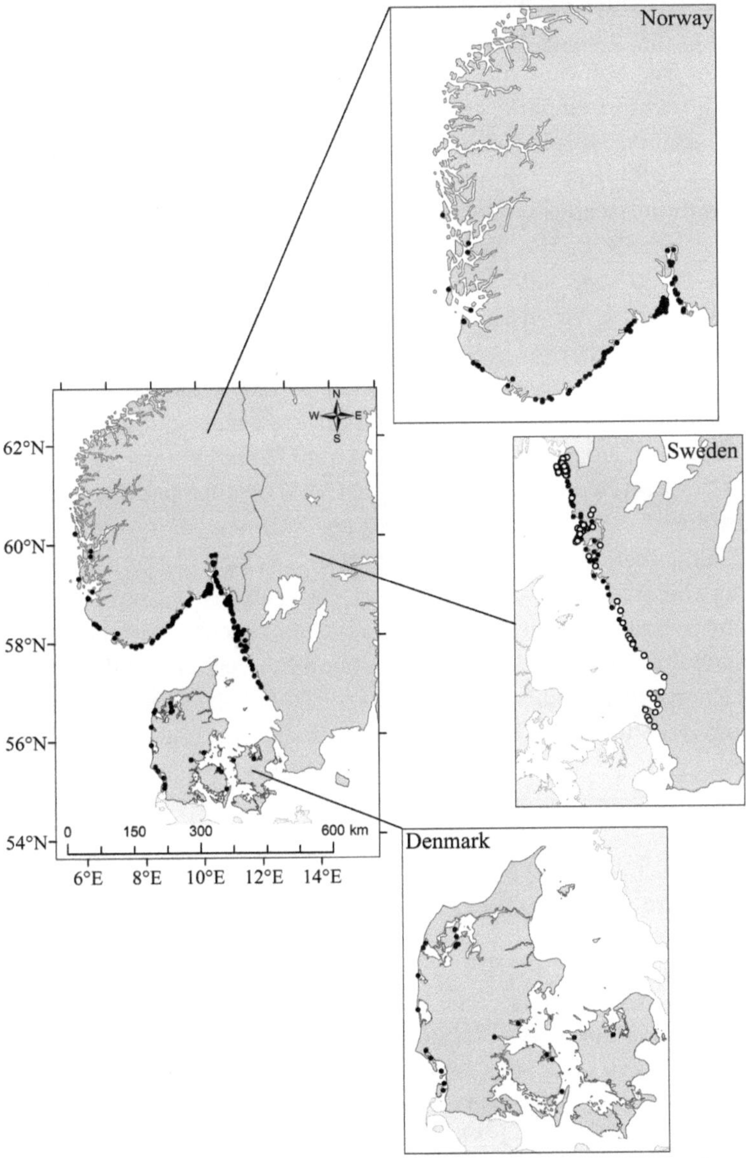

Fig. 10.1: *C. gigas* distribution in Scandinavia (filled circles) and monitored stations (open circles). Figure credit: Mark Wejlemann Holm (Roskilde University).

Sweden. Aquacultural trials in the vicinity of Tjärnö between 1973 and 1976 introduced oysters to the Swedish west coast. When the trials ended, oysters were left in the sea. For the next three decades the oysters were absent, until the public reported several independent sightings of the species in 2007. Presently, we know of about 250 sites with a total live biomass of 100 000 – 500 000 tonnes (Strand & Lindegart, 2014), along the Swedish west coast, mostly concentrated north of Gothenburg (Figure 10.1). In contrast to Denmark, vast shallow areas are scarce and tidal variations are low. The oysters instead settle in small, shallow bays, narrow sounds, and short beaches — sites with high current velocities or high water exchange rates. Densities have increased dramatically since the initial colonization event and now range from 0.06 to 1170 individuals × m^{-2} at colonized sites. Currently, at least eight different cohorts exist in Sweden (oysters settled 2006–2013). Recent reports about increasing densities of oysters in deeper sub-tidal areas indicate that the species may now be extending its depth range.

Norway. Oyster observations started in 2002, and by 2012 more than 100 oyster locations had been registered along the coast. The oysters grow mostly in habitats very similar to those in Sweden, along the southern coast from Østfold in the east to Rogaland in south-west Norway (Figure 10.1). Dolmer *et al.* (2014) reported population densities up to 91 individuals × m^{-2}, and increasing densities in recent years. Additionally, Bodvin *et al.* (2013) described several new populations in Rogaland in 2012 and 2013.

The sources of the Danish oyster populations are probably a combination of larval drift from the German Wadden sea and abandoned aquaculture trials (Dolmer *et al.*, 2014). Meanwhile, the origins of the Swedish and Norwegian populations are still unknown. The prevailing hypothesis is that larvae of the species hitch-hiked with ocean currents from Denmark to the Swedish west coast and then onward to Norway. In Box 10.1 we present results from two research approaches that, together, aim to reveal the source and the pathways of the intra-Scandinavian dispersal.

Box 10.1. Origin and Dispersal of *Crassostrea gigas* Larvae in Scandinavia

Oceanographic modelling. Based on the prevailing currents in the North Sea, Skagerrak and Kattegat, the distance between the Danish and Swedish west coasts, and required size of the source population, the most likely origin of the oysters in Sweden is the Limfjord population at the Danish north-western coast. To evaluate the likelihood of dispersal via the oceanographic coastal circulation, we used a biophysical model combining ocean current velocities with a particle-tracking routine to simulate dispersal of oyster larvae at three different depth intervals between four sampling sites (Figure 10.2); Wadden Sea (Wa; south-west Denmark), Limfjord (Li), Kristineberg (Kr), and Tjärnö (Tj; north-western Swedish coast).

We based the biophysical model on the 3D oceanographic circulation model BaltiX (Hordoir *et al.*, 2013), a NEMO ocean engine-based (Madec, 2010) regional model covering the Baltic Sea and the North Sea (see Hordoir *et al.*, 2013 for a detailed model description with preliminary validations). To simulate larval dispersal we used a particle-tracking model that calculates transport of particles using temporal and spatial interpolation of flow-field data from the BaltiX circulation model using a time step of 15 min. We released particles (70500 in total) from 4 sites (Fig. B1A) during August and September for 8 years (1995–2002). To accout for development time from spawning to settling, we set the pelagic larval phase to 20–30 days, which is common in this area. Finally, we explored three depth intervals; 0–2, 10–12, and 24–26 m. To estimate dispersal probability between the sampling sites we calculated the proportion of particles released from the focal site that ended up at one of the other sites. In the figure, the colours indicate areas to which there is a certain probability that larvae will disperse if drifting at a depth of 10–12 m. Increasing probability is indicated by a transition from blue to red color. The calculation of connectivities was performed by Per Jonsson (BioEnv, GU).

Our model results indicate that the Danish Wadden Sea is relatively isolated from the northern populations because northward larval dispersal was only possible in surface water (Figure 10.2A). The Limfjord appears to be better connected to the surrounding water masses; Limfjord larvae dispersed to Kristineberg and Tjärnö with surface water, and to Kristineberg at medium water depth (Figure 10.2B). It is therefore highly unlikely that the Pacific oysters in Sweden arrived from the Danish Wadden Sea, while dispersal from the Limfjord to Sweden is possible. Larvae may disperse both north and south along the Swedish west coast, although the northward dispersal is more likely (Figure 10.2B). Furthermore, dispersal from the Swedish west coast to Norway is also probable, indicating that the populations in Norway may be enriched by drifting larvae from Swedish populations (Figure 10.2C, D).

Adult oysters already existed in Norway in 2007 when the main invasion event in Sweden occurred (Dolmer *et al.*, 2014), thus the origin of the Norwegian oysters is still unclear and the relative importance of dispersal from Sweden versus other origins for the dynamics of Norwegian populations between the Pacific oysters in Sweden and Norway need to be studied further.

Molecular genetics. Source populations can also be found or validated by comparing molecular signatures of the target population and a range of possible source populations. A study using microsatellite genetic markers to compare the genetic make-up of six Scandinavian populations (Wadden Sea, Limfjord, Kristineberg, Tjärnö, Hui, and Tromlingene, Figure 10.2A) with French oysters found that the Scandinavian populations are different enough to exclude France as a direct source (Meurling, 2013; Strand & Lindegarth, 2014; Strand, unpublished data). The data also indicated that the majority of the investigated Scandinavian populations either have significant gene flow between them or that they were very recently separated.

Earlier studies of the genetic structure of the Pacific oysters in Europe (not including Scandinavia) have shown very little genetic differentiation between populations, with the exception of a population outside the island of Sylt in the northern Wadden Sea (North-Western Germany; Moehler *et al.*, 2011). The introduction history may explain the differentiation between the south European populations and the Sylt population. Most of the European populations descend from individuals directly imported from the aquaculture gene pool in British Columbia. The direct origin of the Sylt population, however, are farmed oysters from the British Isles.

The Scandinavian populations may have the same origin as the Sylt population (Meurling, 2013). However, one of the Norwegian populations, Hui, did not resemble any of the other populations in the study, indicating a different source (Strand & Lindegarth, 2014).

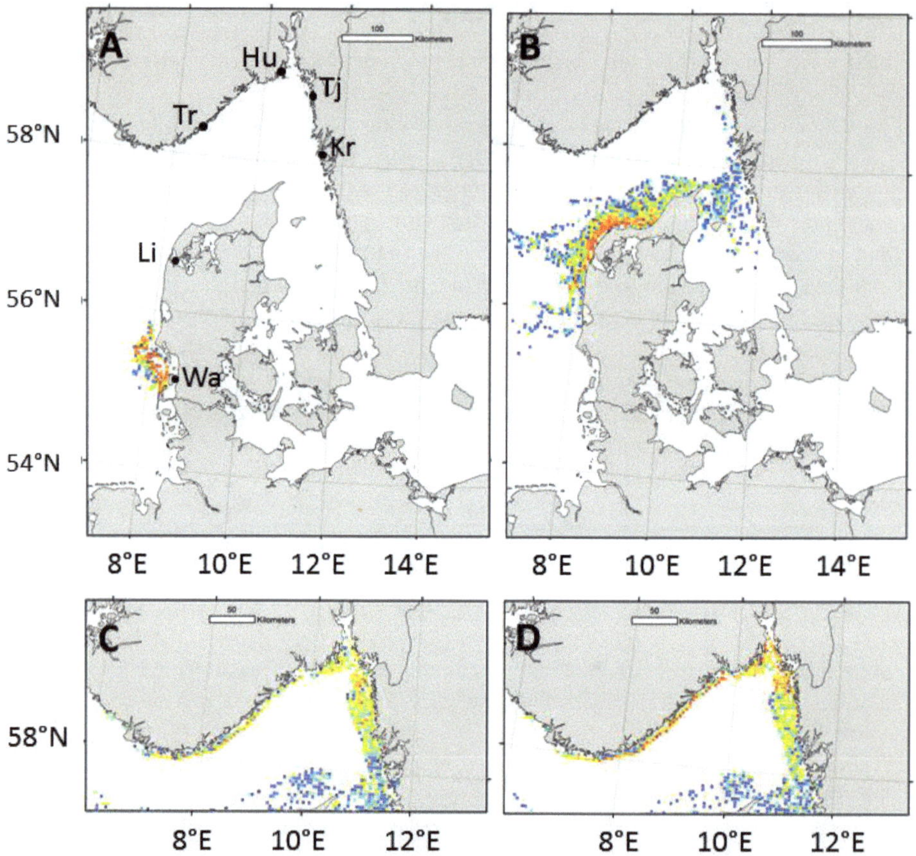

Fig. 10.2: Source populations for oceanographic modelling; Wadden Sea (Wa), Limfjord (Li), Kristineberg (Kr) and Tjärnö (Tj), and genetic analyses; Wa, Li, Kr, Tj, Hui (Hu) and Tromlingene (Tr), as well as dispersal probabilities from the Danish Wadden Sea (A), larval dispersal probabilities from the Limfjord population (B), larval dispersal probabilities from the Kristineberg population (C), and larval dispersal probabilities from the Tjärnö population (D). The colours indicate areas to which there is a certain probability that larvae will disperse if drifting at a depth of 10–12 m. Increasing probability is indicated by a transition from blue to red color.

10.3 Invasion Future of Pacific Oysters: from Present to Forthcoming Distribution

Spatiotemporal variation in distribution limits of a species depends on both biotic and abiotic environmental factors. Each species has its own unique optimal environmental range in which its fitness is optimized, and can survive only limited periods of time when the environmental conditions approach the species' critical tolerance limits. *C. gigas* is – like many invasive species – very tolerant to varying conditions. As an intertidal species, *C. gigas* has a very broad temperature range for survival, from sub-zero degrees to 30°C (Quayle, 1969; Walne, 1974; Le Gall & Raillard, 1988; Bougrier *et al.*, 1995; Diederich *et al.*, 2005; Diederich, 2006; Carrasco & Baron, 2010; Strand *et al.*, 2011). Normal temperature ranges in Scandinavia are -1 ± 24°C and -12 ± 30°C for surface seawater and air temperature, respectively (Strand *et al.*, 2012). This indicates that neither surface seawater nor air temperature in Scandinavia should cause any problems for survival and continued dispersal of the species under normal conditions. However, long periods of extremely cold conditions may push the oysters beyond their lower thermal limits. For instance, the extreme conditions during the winter of 2009–2010 killed large proportions of the oysters at many Scandinavian loactions (Strand *et al.*, 2012). In Box 10.2, we present results from the first effort to use an ecological niche modelling approach to find suitable habitats for *C. gigas* in Scandinavia.

The Intergovernmental Panel on Climate Change (IPCC) predicts that global mean temperature will increase over the next 100 years, with extreme meteorological conditions becoming more common (IPCC, 2007). As the geographical distribution of species is highly dependent upon abiotic factors, species distributions will shift in response to climatic change. A meta-analysis of the effects of climate change on marine ecosystems (Poloczanska *et al.*, 2013) found that widespread systemic shifts in processes such as distribution of species and phenology are comparable to or greater than those observed on land. Moreover, the front-line of marine species distributions is moving toward the poles 12 times faster than that of terrestrial species. In Box 10.2, we present forecasts for changes in suitable oyster habitats in Scandinavia from 2013 to 2050 and 2100.

The dynamics of the geographical range of a species not only depend on a suitable abiotic environment, but also on the mode of dispersal and capability of overcoming dispersal barriers. Due to the current trends in globalization, the range expansion of many species has increased dramatically through human-mediated vector transfer (for more details, please refer to Chapter 1). Indeed, global human-assisted distribution of *C. gigas* started almost a century ago and will likely continue in the future as global warming facilitates aquaculture in new areas (see also section 10.4). The topic of range limits have been reviewed repeatedly (see Brown *et al.*, 1996; Gaston, 1996; Gaston, 2003; Gaston, 2009; Parmesan *et al.*, 2005) and will not be discussed further in this chapter.

Box 10.2. Current and Predicted Habitat Suitability for *Crassostrea gigas* in Scandinavia

Ecological niche modelling methods are widely used to decribe the potential range shifts of species from current and future projections of suitable habitat. We developed species distribution models for the invasive oyster using openModeller webservices (Muñoz *et al.*, 2011) and following the methods described by Leidenberger *et al.* (2015). We obtained *C. gigas* occurrence data (Figure 10.3A) from collaborators and the Global Biodiversity Information Facility (http://gbif.org), and filtered these for environmentally unique points with the BioClim algorithm for the environmental layers specified in Table 10.1. The BioClim algorithm (Busby, 1986; Nix, 1986) also calculated the range for each environmental factor shown in Table 10.1. Environmental layers were obtained from Bio-Oracle at 5 arc-minutes (Tyberghein *et al.*, 2012), and AquaMaps (Kaschner *et al.*, 2010) at 30 arc-minutes. For the projections of 2050 and 2100 climate conditions, we used layers from the ECHAM5 A1B climate change scenario (Jungclaus *et al.*, 2006; IPCC, 2007).

To calculate habitat suitability, we used all environmentally unique points available from the global *C. gigas* distribution and the environmental layers (Table 10.1) in the Ecological Niche Modelling workflow (ENM), applying the Mahalanobis distance method (Mahalanobis, 1936; Farber and Kadmon, 2003). Model performance was assessed using 10-fold cross-validation, measuring AUC (Area Under the Curve), which was calculated using the proportional area approach (Phillips *et al.*, 2006), and omission error, which was calculated using the lowest presence threshold (LPT). Model testing resulted in Mean AUC = 0.92 ± 0.02 and omission error = 1.77%.

We projected three climatic scenarios (2013, 2050, and 2100) into 50 km coastal bands of Norway, Sweden, and Denmark. Extent of potential species distribution was calculated with the ENM Statistical Workflow (Leidenberger *et al.*, 2015). The workflow computes the coverage and overall intensity of suitable habitat, as well as the difference between these values for two climate scenarios in a given area, using the R statistical environment (R Core Team, 2013). Overall coverage was computed as the percentage of raster cells with values > 0, while overall intensity was computed as the sum of all cell values divided by the number of raster cells. Habitat suitability values are presented as a heat map (Figure 10.3B) ranging from white (not suitable) to red (suitable). The calculation of difference values represents the maximum possible change between two scenarios and are presented as eight positive and negative classes on a heat map (Figure 10.3C, D). Cells with colours from white–red indicate increasing habitat suitability, while cells with colours from white–blue indicate decreasing habitat suitability.

The suitable habitat as inferred by the model matches the known distribution of *C. gigas* in Scandinavia with one exception (Figure 10.3A, B). Along the western coast of Norway, suitable habitats exist as far as to the Lofoten Islands, while sustained populations are only reported to latitudes of around Bergen, approximately 800 km south of the Lofoten Islands. This indicates that the current range distribution of *C. gigas* in Scandinavia may increase northward along the Norwegian coastline, and that the invasion may still be in an initial stage in Norway.

The most affected Scandinavian coastlines are those of Norway and Denmark, where suitable habitat currently covers 90.41% and 34.08% of the coastal band, respectively (Table 10.1). We found no indications of suitable habitats in the Baltic (Figure 10.3A, B), while the transitional area between the North Sea and the Baltic showed decreasing habitat suitability values from the Skagerrak towards the Kattegat and into the Belt region. Consequently, only minor parts of the Swedish coastline currently provide suitable habitat (6.72%) for *C. gigas* (Table 10.1).

Our model scenarios for 2050 and 2100 (Figure 10.3C, D) predict that *C. gigas* will shift its Scandinavian range towards the north-east along the coastline of the Scandinavian shield. We found no indications of suitable habitat in the Baltic in any of the climatic scenarios. The projected range shift has highest impact in Norway, where both coverage and intensity of suitable habitat increase in the coastal band facing the Atlantic until 2100 at 4.75% and 27.87%, respectively (Table 10.1). In Sweden and Denmark, the extension of the species declines slightly under the projected range shift because of less suitable habitat in the Kattegat region. It is worth noting, however, that the modelling approach relies solely on current distribution of the focal species and does not account for possible evolutionary adaptation or increased phenotypic plasticity in response to new environmental conditions.

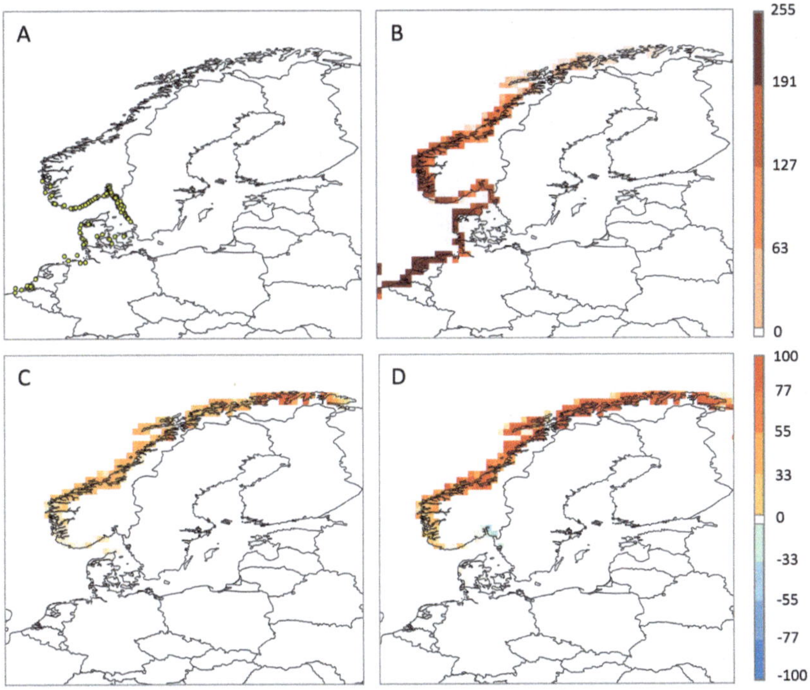

Fig. 10.3: Distribution of confirmed occurrence points (A), present distribution of suitable habitat (B), changes in habitat suitability between 2013 and 2050 (C), and changes in habitat suitability between 2013 and 2100 (D). See text for description of colour coding.

Tab. 10.1: Summary of occurrence data and environmental parameters (including their range) used for modelling, as well as statistical output of 12 spatio-temporal projections for *Crassostrea gigas*; n = number of environmental unique points, MD = mean depth, DL = distance to land, SIC = sea ice cover, SSS = sea surface salinity, SST = sea surface temperature. Asterisks indicate environmental factors used for modelling.

	Global	Denmark	50 km coastline Sweden	Norway
Data				
Occurrence points	1249			
n	338			
DL [km]*	0 – 92			
MD [m]*	1 – 1738			
SIC*	0 – 0.2			
SST [°C]*	-1.34 – 30.17			
SSS [PSU]*	11.86 – 35.64			
pH	7.88 – 8.31			
Projection				
DIFF statistic [%]				
Coverage 2013	-	34.08	6.72	90.41
Coverage 2050	-	34.08	6.72	95.12
Coverage 2100	-	31.83	6.14	95.16
2100-2013	-	-2.24	-0.57	4.75
Intensity 2013	-	10.12	1.37	32.59
Intensity 2050	-	10.71	1.42	45.42
Intensity 2100	-	10.27	1.26	60.47
2100-2013	-	0.15	-0.10	27.87

10.4 Invasion Impacts of Pacific Oysters: from Ecosystem Engineering to Community Restructuring

A recent risk assessment report (Dolmer *et al.*, 2014) evaluated the potential impacts of the *C. gigas* invasion on four different habitat types common to Scandinavian waters: low energy rock, littoral sand and mud, littoral biogenic reefs, and sub-littoral sediments. The report concluded that the biogenic reefs and sub-littoral sediments are under moderate to high risk of an invasion in the future – especially on high-energy sites. This section outlines the possible impacts such an invasion would have on species and communities in these ecosystems.

10.4.1 Ecosystem Engineers

C. gigas is what Jones *et al.* (1994) describe as an ecosystem engineer. Ecosystem engineers are species with the ability to directly or indirectly modify, create or destruct habitats. There are both negative and positive effects ecosystem engineers can have on species richness and species abundance at a small scale, but on a larger scale, Jones *et al.* (1997) argue that the net effect is probably mainly positive.

C. gigas reefs, often with shells tightly packed on top of each other, may occupy hundreds of meters of shallow bottoms where they trap sediment, restrict water movement, and contribute to sediment stability. Moreover, oyster reefs may increase the organic richness of sediments through biodeposition and influence nutrition levels through benthic-pelagic coupling (Ruesink *et al.*, 2005; Grabowski & Peterson, 2007). In Sweden, Norling *et al.* (2015) have observed sediment enrichment in live oyster beds. Moreover, in areas with dense populations of oysters, water movement has been restricted with increasing sedimentation as a result. The physical structures of the bivalves may also be of great importance as post-mortem ecosystem engineers. This has been demonstrated both by Guo and Pennings (2012) for salt march plant communities, and by Norling *et al.* (2015) for epifauna in Swedish shallow sublittoral habitats. In the following section, we will describe the effects of *C. gigas* on inter-species interactions and communities.

10.4.2 Impacts on Species-level Interactions

Mytilus edulis. There is always a risk that new invading species will compete with indigenous fauna. Both in the Wadden Sea (Diederich, 2005) and in Scandinavia (Dolmer *et al.*, 2014), *C. gigas* prefers similar habitats to the native blue mussel *Mytilus edulis.* Negative effects include competition for space and food (Nehring, 2003; Nehls *et al.*, 2006). Competition for space is probably not a limiting factor for the blue mussels in the Wadden Sea for two reasons: large areas are still unoccupied, and blue mussels utilize oyster reefs for settlement and protection from predation (Troost, 2009). Similar results have been obtained in Sweden (Hollander *et al.*, 2015; Norling *et al.*, 2015). Competition or interference for food between blue mussels and oysters cannot, however, be excluded (Troost *et al.*, 2009). Furthermore, because the oyster filtering capacity is high (Ruesink *et al.*, 2005; Troost, 2009), there is a risk that species with planktonic larval stages will be reduced if oysters become abundant. Data from the Wadden Sea indicate that even though the recruitment of blue mussels may decline due to climate change (Diederich *et al.*, 2005; Nehls *et al.*, 2006), larviphagy, and food limitations for the adults and larvae (Troost *et al.*, 2008a, b; Troost *et al.*, 2009), local extinction of blue mussels is unlikely. To summarize, there are strong indications that the two species will be able to coexist (Diederich, 2005; Troost, 2009; 2010).

Ostrea edulis. The ecological distribution of the native flat oyster (*Ostrea edulis*), still occurring in Scandinavia, is somewhat different from *C. gigas*. Flat oysters prefer deeper habitats, mainly the subtidal zone, while *C. gigas* most often reside in the intertidal. Many therefore assume that the interference seen between *C. gigas* and the blue mussel would not occur between *C. gigas* and the flat oyster (Reise, 1998). However, the ecological distribution pattern among *C. gigas* populations varies geographically. For instance, as noted above, *C. gigas* occupy much deeper waters in Norway and Sweden than along the Danish Wadden Sea coast. In addition, new data indicate that *O. edulis* may have a more shallow distribution than previously thought (L. Thorngren, pers. comm.). Moreover, in addition to being attracted to adult conspecifics (Kochmann *et al.*, 2008; Wilkie *et al.*, 2013), *C. gigas* larvae also appear to approach flat oyster beds (L. Thorngren, pers. comm.). If this is a consistent pattern, there is an imminent risk that *C. gigas* may, when populations in Scandinavia grow larger and more sustainable, interfere with the native flat oyster.

10.4.3 Impacts on Community-level Interactions

The oyster reef accommodates various types of organisms, such as infauna, sessile and mobile epifauna, and necton. The formation of oyster reefs on soft and sandy subtrates leads to much stronger changes in substrate modification and habitat complexity than reefs forming on hard bottoms and will therefore have more pronounced effects on local biodiversity. In addition, while epifauna species thrive on oyster beds (Gutierrez *et al.*, 2003), infauna species are usually not successful in colonizing sediments covered by oysters. Thus, emerging oyster reefs will eventually replace soft-bottom communities with hard-substrate communities (Troost, 2009).

Numerous studies have examined how bivalves affect the community structure. Bivalve beds tend to have higher species richness, and higher total abundance or biomass of both infauna, epifauna (mobile and sessile) and necton than bare sediment habitats (Van Broekhoven, 2005; Hosack *et al.*, 2006; Royer *et al.*, 2006; Kochmann *et al.*, 2008; Troost, 2010; Lejart & Hily, 2011). Oyster reefs provide a hard biogenic substrate that increases habitat complexity, provides living space for other species, and offers shelter, foraging grounds, and nesting sites (Crooks, 2002; Escapa *et al.*, 2004; Ruesink *et al.*, 2005; Green *et al.*, 2012; Kingsley-Smith *et al.*, 2012). In addition, the extended shell surface area contributes to increased attatchment substrate for sessile species and larger access to microbial films for grazing species (Kochmann *et al.*, 2008; Markert *et al.*, 2010). The infauna community will be dominated by predators and deposit feeders, whereas in bare sediment suspension feeders dominate (Norling & Kautsky, 2007; Mendonça *et al.*, 2009; Markert *et al.*, 2010; Lejart & Hily, 2011). Similar effects are also found in post-mortem structures of oysters (Guo & Pennings, 2012; Norling *et al.*, 2015). Increased epifauna and infauna abundance in bivalve beds and in post mortem structures of the species has also been observed in Sweden (Hollander *et al.*, 2015; Norling *et al.*, 2015).

For instance, resident fish species prefer both oyster shells and live mussels, while small crustaceans prefer blue mussel shells (Norling *et al.*, 2015).

Bivalves also influence trophic structure (reviewed by Grabowski & Peterson, 2007). By removing plankton from the water column, they promote benthic flora and fauna, and prevent nutrients from entering and staying in lower trophic levels. Moreover, by enhancing the nutrient transfer to bentic invertebrates, the nutrient flux also increases to higher trophic levels such as crabs, demersal fish, and ultimately to predatory fish, many of which are fished commercially.

Bivalve beds also provide prey for other species. In European estuaries and coastlines there are two main predators of invertebrate shellfish: the shore crab (*Carcinus maenas*) and the common starfish (*Asterias rubens*, Dare *et al.*, 1983; O'Neill, 1983; Cohen *et al.*, 1995). Both species are abundant in shallow waters and estuaries along Scandinavian coastlines, and may cause severe damage to wild and commercial bivalve stocks (Hancock, 1955; Walne & Davies, 1977; Allen, 1983). Experimental evidence (Y. Fredriksson and Å. Strand, unpubl. data) suggests that both shore crabs (*Carcinus maenas*) and starfish (*Asterias rubens*) in Sweden identify and utilize *C. gigas* as prey, and have the ability to open live oysters. In addition, land-based predators such as herring gulls (*Larus argentatus*) or oystercatchers (*Haematopus ostralegus*) may also take advantage of oysters as a new food source (Cadée, 2008a, b; Scheiffarth *et al.*, 2007).

Increased presence of *C. gigas* will thus affect local species interactions, communities, and ecosystems. In the next section we will review how these potential changes will affect ecosystem services and future management decisions.

10.5 Invasion Consequences of Pacific Oysters: from Perceived Nuisance to Valuable Resource?

Since the arrival of *C. gigas* in Scandinavia, the only two factors believed to be able to stop the oysters from establishing permanently were harsh winter conditions and disease (Nyberg, 2010; Wrange *et al.*, 2007). Both these factors have now been discarded as potential threats to the oysters in the region. First, the unusually severe winter conditions of 2009–2010 caused high mortalities (Strand *et al.*, 2012), but not enough to prevent subsequent reproduction. Second, in 2014 high summer mortalities affected populations in Sweden and Norway (see Box 10.3 for details), but many large populations remain. Thus, natural factors seem inadequate to eradicate the species from the region. On the contrary, they will likely disperse further (Box 10.2). Moreover, because larvae are attracted to already existing populations (Kochmann *et al.*, 2008; Wilkie *et al.*, 2013), the number of dense aggregations of oysters and reef formations may increase. Finally, the Norwegian populations may receive larvae from Swedish populations, which in turn may be enhanced by Danish larvae (Box 10.1, 10.2) This makes management of the invasion and its impact on local ecosystem services an urgent and international issue.

Box 10.3. Summer Mortality 2014

Abnormal mortalities in *C. gigas* during the summer months have occurred for more than five decades in major oyster production areas across the world (reviewed in Samain and McCombie, 2008; EFSA 2010). The etiology underlying these recurrent episodes has not been clearly defined, but the events have been denoted "summer mortalities" due to the typical seasonal occurrence (Samain & McCombie, 2008; EFSA 2010). Suggested explanations include physiological stress during spawning (Cho & Jeong, 2005) in combination with environmental stressors such as reduced salinity (Luna-Gonzalez et al., 2008), eutrophic conditions (Malham et al., 2009), decrease in dissolved oxygen (Cheney *et al.*, 2000), toxicity of sediments (Soletchnik *et al.*, 2005) and contamination with anthropogenic pollution (Luna-Acosta *et al.*, 2010). Biotic stressors often associated with summer mortalities in *C. gigas* are the bacteria *Vibrio splendidus* and *V. aestuarianus* (Lacoste et al., 2001; Garnier et al., 2007; Samain & McCombie, 2008), and the virus OsHV-1 μvar (Renault *et al.*, 1995; Segarra *et al.*, 2010; Renault *et al.*, 2014).

Scandinavian populations of *C. gigas* have hithereto appeared healthy, but in mid-September 2014 mass-mortalities in oyster populations were observed along the coasts of Vestfold and Østfold in Norway and in Bohuslän in Sweden. The initial observations were followed by several independent sightings of mass mortalities in wild and cultivated oysters exceeding 90%. By the end of 2014, no mortalities had been reported from Denmark. Initial demographic examinations of four *C. gigas* populations in the Koster-Strömstad-Tjärnö archipelago indicated that 60–90% of the individuals were affected. The pattern was complex with only spat and juveniles being affected at some sites, while at other sites all individuals died. Provisional inspection of *O. edulis* and *M. edulis* in affected habitats suggests that these native bivalves were not affected.

Preliminary reports show that oyster herpes virus OsHV-1 μvar was present in affected populations in Scandinavia (Mortensen *et al.*, 2014; Anders Alfjorden pers. com.). Although the virus has not yet (as of December 2014) been causally linked to the mortality, its presence in Scandinavian waters causes concern as it may also infect other bivalves (Farley et al., 1972; Hine & Thorne, 1997; Arzul *et al.*, 2001), such as the native flat oyster, *O. edulis* (Comps & Cochennec, 1993).

The mortality pattern reported in Sweden (high mortalities of both juveniles and adults) is not consistent with herpes-related mortality in Europe (high mortalities among juveniles and low lethality among adults) (Segarra *et al.*, 2010; Dégremont, 2011). Furthermore, preliminary observations indicate an extremely patchy occurrence of mortalities, approaching 100% in populations at some sites while neighboring stocks (approximately 1 km away) are unaffected. This may indicate that factors other than pathogens are involved in the observed mortalities, and ongoing research should aim at elucidating the importance other factors, including pathogens and environmental conditions.

Despite high mortalities in affected populations, large unaffected oyster populations still remain in Scandinavia. Thus we expect only a similar temporary setback as observed after the winter mortality of 2009–2010 (Strand *et al.*, 2012), followed by recovery.

Marine ecosystem services refer to benefits that people obtain from marine ecosystems, including the open ocean, coastal areas and estuaries. Supporting services (e.g. primary production and nutrient cycling) are the basis for the three following service categories: regulating services (e.g. climate regulation, natural hazards control, and water quality), provisioning services (e.g. food, job opportunities, building materials and pharmaceutical compounds), and cultural services (e.g. recreational, traditions, esthetic and spiritual benefits). Oysters in general (not just *C. gigas*) contribute to all four service categories (Ruesink *et al.*, 2005; Grabowski & Peterson, 2007; Grabowski *et al.*, 2012). At the same time, invasives such as *C. gigas* may also have some negative effects, which may cause substantial economic loss to the region where it establishes. In the following section we will discuss the impacts of *C. gigas* may have on local ecosystem services in Scandinavia, and the implications for coastal management.

Nehring (2011) outlines the main drawbacks of a *C. gigas* invasion. These include displacement of native species by competition for food and space, altered benthic-pelagic and food-web interactions, habitat modifications, hybridization with local oyster species, and transfer of parasites, diseases and pests. Currently, the main concern for managers and commercial stakeholders is competition with native blue mussels and flat oysters. The concern for the blue mussel may be less well founded than previously thought, as successful coexistence with oysters is now considered a reality. For flat oysters, the situation is different. International treaties (e.g. OSPAR) now regard the flat oyster as a species of particular importance, thus forcing governmental stakeholders to label the potential competition from *C. gigas* as a severe concern. If recent results from the Swedish west coast – that flat oysters grow in shallower waters than previously thought – also apply to similar habitats along the Norwegian coast, this might indeed be a real concern for flat oysters.

Large oyster reefs already exist in many places. Especially in shallow, narrow sounds and bays, the reefs alter currents and increase sedimentation. After the winter of 2009–2010 that killed a substantial part of the Swedish oysters, reefs became a sanitary problem when several thousand newly dead oysters decomposed. At public beaches, reefs hamper recreational activities. Moreover, fouling of boats, docks, drainage pipes, and other underwater constructs is an increasing problem.

Notwithstanding the previously discussed negative impacts, *C. gigas* provides a range of different ecosystem goods and services (Graborowski & Peterson, 2007). Some of the most obvious services are the contributions to rural development and economy throught provisioning and cultural services. The species is one of the top 20 most cultured species in the world (FAO), and as such is an important food resource and source of income. In Denmark and Sweden, the interest in exploiting the species is growing. Hand picking oysters is popular in both countries, and there are ongoing discussions about possibilities for aquaculture. However, in Sweden exploitation is hampered by the old fishing law, attributing the rights to the oysters to the coastal land owners. Furthermore, aquaculture may not be allowed due to national and international regulations of invasive species. Nevertheless, the exploitation of wild oyster populations could become a valuable source of income for many coastal communities. *C. gigas* is being exploited not only for commercial food production, but also by tourist enterprises organising day trips to easily accessible sites for hand picking of oysters. Regular commercial and touristic harvesting of existing oyster populations could be a realistic way to reduce the risk of reef formation, but it is highly unlikely that such management actions will be socioeconomically possible unless commercial interests are involved.

In addition to contributing to provisioning and cultural services directly, increased presence of bivalves may indirectly boost local fisheries (Peterson *et al.*, 2003). If the community re-structuring discussed in section 10.4 increases the nutrient flux to higher trophic levels, which includes commercially important fish, then commercial, recreational, and tourist fishing may benefit from more *C. gigas*. This may be further enhanced by the suitability of bivalve beds as nursery areas for fish and decapods, which feed on the higher abundance of macrofauna and find refuge from predators in the complex structures (Jansson *et al.*, 1985; Tolley & Volety, 2005; Hosack *et al.*, 2006; Mendonça *et al.*, 2009). Juvenile transient fish species prey upon resident fish (Coen & Luckenbach 2000; Coen *et al.*, 2007), making bivalve beds important nursery areas for young stages of fish that support coastal communities and commercial and recreational fisheries (Ruesink *et al.*, 2005).

Eutrophication of shallow costal areas is, like in many other regions, a major concern in Scandinavia (Diaz & Rosenberg, 2008). Bivalve filter feeders have the potential to alleviate this situation (Officer *et al.*, 1982; Coen & Luckenbach, 2000; Rice, 2001; Newell, 2004; Grabowski & Peterson, 2007). Accordingly, bivalve cultures are currently being presented and evaluated in Sweden as a possible management option for restoring eutrophicated coastal areas to healthy environments (Lindahl *et al.*, 2005; Bergström *et al.*, 2013; Bergström, 2014). In addition, being a very efficient filter feeder with high growth rates and dense populations in shallow areas (Ren *et al.*, 2000; Ruesink *et al.*, 2005), *C. gigas* has the potential to, without much human intervention, improve the status of eutrophicated ecosystems (Ruesink *et al.*, 2005).

Oysters provide additional important services. Due to their reef-building abilities, oysters have been proposed as a tool for reducing erosion of intertidal flats in the Wadden Sea (Troost, 2010), and reducing erosion from boat wakes (Beck *et al.*, 2009). Because reefs reintroduce structural complexity to areas where native communities such as seagrass meadows, blue mussel beds and *O. edulis* beds may have been lost, *C. gigas* beds may replace the function of the lost communities. Indeed, oysters may even increase the occurrence of such habitats. Seagrass meadows may benefit from reduced water turbidity and increased nutrient content of sediments (Beck *et al.*, 2009; Grabowski & Peterson, 2007), and blue mussels benefit from the shelter and a settlement substrate (Troost, 2009). Finally, blue mussel meal is considered a realistic, environmentally friendly and healthy alternative to the use of fish meal in the fish feed used in aquaculture and in chicken feed for egg production (Duinker *et al.*, 2005; Kollberg &

Lindahl, 2006). If technological solutions for the utilization of oyster meat as a similar product can be developed, the proportion of environmentally friendly components in fish feed could increase even further.

Despite the potential positive impacts on many direct and indirect ecosystem services (e.g. filtering capacity, benthic–pelagic coupling, nutrient dynamics, sediment stabilization, provision of habitat, etc.), the values of these services have been largely ignored or underestimated in management decisions (Coen & Luckenbach, 2000). This is also the case in Scandinavia. The general management discussion on *C. gigas* has focused on the oysters status as invasive, as well as the governmental responsibilities related to national and international regulations of invasive species.

Climate change presents an additional challenge for management. Because species distributions are continuously altered by the changing environmental framework, agencies are increasingly looking to modeled projections of species' distributions under future climate scenarios to be able to make informed management decisions. Species distribution models such as the one presented in Box 10.2 are numerical tools that combine observations of species occurrence or abundance with estimates of environmental parameters. In the case of *C. gigas* in Scandinavia, predictions from the ecological niche models highlight the urgent need for informed management actions along the Norwegian coast.

Established populations of *C. gigas* will undoubtedly affect the ecosystem in many different ways. How its presence will be valued will depend on the point of view and agenda of the observer. Regardless of stakeholder interests, *C. gigas* will likely be a substantial part of the Scandinavian ecosystem in the future. Thus, there is an urgent need to consolidate stakeholders, develop research and monitoring programmes, and create evidence-based management plans for how to handle the species.

10.6 Acknowledgements

The authors thank assistants, students, and collaborators who contributed to the work presented in this chapter. Mark Wejlemann Holm produced Figure 10.1 and Per Jonsson (Department of Marine Science, University of Gothenburg) perfomed the oceanographic modelling in Box 10.1. Comments and suggestions from Oliver Verneau and an anonymous reviewer greatly improved the manuscript. ATL acknowledges funding from the Royal Swedish Academy of Sciences and Wåhlströms Foundation. MO acknowledges funding from the Swedish Research Council through the Swedish LifeWatch program (www.slu.se/lifewatch, grant no. 829-2009-6278) and by the Framework Program project BioVeL (www.biovel.eu, grant no. 283359). ÅS received funding from Formas (Dnr 2011-1132), the Swedish Agency for Marine and Water Management (former Swedish Environmental Protection Agency) through Havsmiljöanslaget Dnr. 309-1719-09, Wåhlströms Foundation and the foundation Olle Engkvist Byggmästare.

In a nutshell

- The Pacific oyster originates from the Pacific Ocean, but has been the subject of widespread introductions all over the world. In addition to being deliberately transferred in large numbers for farming, Pacific oysters have also dispersed by hitch-hiking with long-distance vessels and local and regional ocean currents. Natural spatfall has established feral populations on almost all continents with sometimes-large ecological and socioeconomic consequences.
- Since the mid 2000s, *Crassostrea gigas* has been observed in Scandinavia and is now forming dense populations in both Denmark, Sweden and Norway. In this chapter we outline existing knowledge of the species origin and dispersal routes in Scandinavia, predicted future distribution in relation to climate change, and observed alterations to habitat structure and ecosystem function. We related this information to ecosystem services provided by the oysters, and to expected changes in community ecosystem services, with special emphasis on Scandinavian ecosystems.
- We anticipate that *C. gigas* will affect the Scandinavian ecosystem in both negative and positive ways, and the valuation of *C. gigas* will depend on the point of view and agenda of the observer. Despite the potential positive impact on many direct and indirect ecosystem services (e.g. source of food and income, filtering capacity, benthic-pelagic coupling, nutrient dynamics, sediment stabilization, provision of habitat), the values of these services have so far been largely ignored or underestimated by management.
- Regardless of stakeholder interests, *C. gigas* will likely be a substantial part of the Scandinavian ecosystem in the future. Thus, there is an urgent need to consolidate stakeholders, develop research and monitoring programmes, and create evidence-based management plans for how to handle the species.

10.7 Bibliography

Allen, P. L. (1983). Feeding behaviour of *Asterias rubens* (l.) on soft bottom bivalves: a study in selective predation. Journal of Experimental Marine Biology and Ecology, 70 (1), 79-90.

Arzul, I., Renault, T., Lipart, C., Davison, A. J. (2001). Evidence for interspecies transmission of oyster herpes virus in marine bivalves. Journal of General Virology, 82 (4), 865–870.

Beck, M. W., Brumbaugh, R. D., Airoldi, L. *et al.* (2009). Shellfish reefs at risk. A global analysis of problems and solutions. Arlington, VA: The Nature Conservancy.

Bergström, P., Lindegarth, M. och Lindegarth, S. (2013). Restaurering av övergödda havsvikar med hjälp av miljömusselodling. Hav möter land, Länsstyrelsen i Västra Götaland.

Bergström, P. (2014). Blue oceans with blue mussels. Managing and planning of mussel farming in coastal ecosystems. Doctoral thesis. University of Gothenburg.

Bodvin, T., Jelmert, A., Moy, F. *et al.* (2013). First reef of *Crassostrea gigas* in the Oslo fjord – a warning for future developments? Poster, Nordic Marine Science Conference 2013.

Bougrier, S., Geairon, P., Deslous-Paoli, J. *et al.* (1995). Allometric relationships and effects of temperature on clearance and oxygen consumption rates of *Crassostrea gigas* (Thunberg). Aquaculture, 134 (1-2), 143-154.

Brown, J. H., Stevens, G. C. & Kaufman D. M. (1996). The geographical range: size, shape, boundaries, and internal structure. Annual Review of Ecology and Systematics, 27, 597-623.

Busby, J. R. (1986). A biogeoclimatic analysis of *Nothofagus cunninghamii* (Hook.) Oerst. in southeastern Australia. Australian Journal of Ecology, 11 (1), 1–7.

Cadée, G. C. (2008a). Oystercathers *Haematopus ostralegus* cathing Pacific oysters *Crassostrea gigas*. Basteria, 72 (1-3), 25-31.

Cadée, G. C. (2008b). Herring gulls feeding again on Pacific oysters *Crassostrea gigas* in the Dutch Wadden Sea near Texel. Basteria, 72 (1-3), 33-36.

Carrasco, M. F., Barón, P.J. (2010). Analysis of the potential geographic range of the Pacific oyster *Crassostrea gigas* (Thunberg, 1793) based on surface seawater temperature satellite data and climate charts: the coast of South America as a study case. Biological Invasions, 12 (8), 2597–2607.

Cheney, D. P., Macdonald, B. F., Elston R. A. (2000). Summer mortality of Pacific oysters, *Crassostrea gigas* (Thunberg): initial findings on multiple environmental stressors in Puget Sound, Washington, 1998. Journal of Shellfish Research, 19 (1), 353-359.

Cho, S.-M. & Jeong W.-G. (2005). Spawning impact on lysosomal stability of the Pacific oyster, *Crassostrea gigas*. Aquaculture, 244 (1-4), 383-387.

Coen, L. D. & Luckenbach, M. W (2000). Developing success criteria and goals for evaluating oyster reef restoration: ecological function or resource exploitation? Ecological Engineering, 15 (3-4), 323–343.

Coen, L. D., Brumbaugh, R. D., Bushek, D. *et al.* (2007). Ecosystem services related to oyster restoration. Marine Ecology Progress Series, 341, 303–307.

Cohen, A. N., Carlton, J. T., Fountain, M. C. (1995). Introduction, dispersal and potential impacts of the green crab *Carcinus maenas* in San Francisco Bay, California. Marine Biology, 122 (2), 225-237.

Comps, M., Cochennec, N. (1993). A herpes-like virus from the European oyster *Ostrea edulis*. Journal of Invertebrate Pathology, 62 (2), 201–203.

Couzens, G. (2006). The distribution and abundance of the non-native Pacific oyster, *Crassostrea gigas*, in West Devon - A result of climate change? Shellfish News, 22, 5–7.

Crooks, J. A. (2002). Characterizing ecosystem-level consequences of biological invasions: the role of ecosystem engineers. Oikos, 97 (2), 153-166

Dare, P. J., Davies, G. & Edwards, D. B. (1983). Predation on juvenile Pacific oysters (*Crassostrea gigas* Thunberg) and mussels (*Mytilus edulis* L.) by shore crabs (*Carcinus maenas* L.). Fisheries research technical repport. Lowestoft, UK: Directorate of fisheries research, Ministry of agriculture, fisheries and food.

Dégremont, L. (2011). Evidence of herpesvirus (OsHV-1) resistance in juvenile *Crassostrea gigas* selected for high resistance to the summer mortality phenomenon. Aquaculture, 317 (1-4), 94-98.

Diaz & Rosenberg 2008. Spreading dead zones and consequences for marine ecosystems. Science, 321 (5891), 926-929

Diederich, S. (2005). Differential recruitment of introduced Pacific oysters and native mussels at the North Sea coast: coexistence possible? Journal of Sea Research, 53 (4), 269-281.

Diederich, S. (2006). High survival and growth rates of introduced Pacific oysters may cause restrictions on habitat use by native mussels in the Wadden Sea. Journal of Experimental Marine Biology and Ecology, 328 (2), 211-227.

Diederich, S., Nehls, G., van Beusekom, J. *et al.* (2005). Introduced Pacific oysters (*Crassostrea gigas*) in the norhtern Wadden Sea: invasion accelerated by warm summers? Helgoland Marine Research, 59 (2), 97-106.

Dolmer, P., Holm, M. W., Strand, Å. *et al.* (2014). The invasive Pacific oyster, *Crassostrea gigas*, in Scandinavian coastal waters: a risk assessment on the impact in different habitats and climate conditions. Project report. Bergen, Norway: Institute for Marine Research.

Drinkwaard, A. C. (1999). Introductions and developments of oysters in the North Sea area: a review. Helgoländer Meeresunters, 52 (3-4), 301-308.

Duinker, A., Moen, A.-G., Nortvedt, R. *et al.* (2005). Utvidet kunnskap om blåskjell som fiskefôr-ressurs. Sluttrapport till Norges forskningsråd, Prosjekt nr. 150109.

EFSA Panel on Animal Health and Welfare (AHAW) (2010). Scientific Opinion on the increased mortality events in Pacific oyster (*Crassostrea gigas*). EFSA Journal, 8 (11), 1894.

Escapa, M., Isacch, J. P., Daleo, P. *et al.* (2004). The distribution and ecological effects of the introduce Pacific oyster, *Crassostrea gigas* (Thunberg, 1793) in northern Patagonia. Journal of Shellfish Research, 23 (3), 765–772.

FAO (2014). Aquatic Species Distribution Map Viewer. http://www.fao.org/figis/geoserver/factsheets/species.html, accessed on 13 December 2014).

Farber, O., Kadmon, R. (2003). Assessment of alternative approaches for bioclimatic modeling with special emphasis on the Mahalanobis distance. Ecological Modelling, 160 (1-2), 115-130.

Farley, C. A., Banfield, W. G., Kasnic, G. Jr. *et al.* (1972). Oyster herpes-type virus. Science, 178 (4062), 759–760

Garnier, M., Labreuche, Y., Garcia, C. *et al.* (2007). Evidence for the involvementof pathogenic bacteria in summer mortalities of the Pacific oyster *Crassostrea gigas*. Microbial Ecology, 53 (2), 187-96.

Gaston, K. J. (1996). Species-range-distributions: patterns, mechanisms and implications. Trends in Ecology and Evolution, 11 (5), 197-201.

Gaston, K. J. (2003). The structure and dynamics of geographic ranges. Oxford, UK: Oxford University Press.

Gaston, K. J. (2009). Geographic range limits of species. Proceedings of the Royal Society B: Biological Sciences, 276 (1661), 1391–1393

Grabowski, J.H., Peterson, C. H. (2007). Restoring oyster reefs to recover ecosystem services. In K. Cuddington, J. E. Byers, W. G. Wilson *et al.* (Eds.), Ecosystem Engineers: Plants to Protists (pp.281–298). Burlington, MA: Elsevier.

Grabowski, J.,H., Brumbaugh, R.,D., Conrad, R.,F. *et al.* (2012). Economic valuation of ecosystem services provided by oyster reefs. Bioscience, 62 (10), 900-909.

Green, D. S., Boots, B., Crowe, T. P. (2012). Effects of non-indigenous oysters on microbial diversity and ecosystem functioning. PLoS ONE, 7 (10), e48410.

Guo, H, Pennings, S. C. (2012). Post-mortem ecosystem engineering by oysters creates habitat for a rare marsh plant. Oecologia, 170 (3), 789–798.

Gutierrez, J. L., Jones, C. G., Strayer, D. L. *et al.* (2003). Mollusks as ecosystem engineers: the role of shell production in aquatic habitats. Oikos, 101 (1), 79-90.

Hancock, D. A. (1955). The feeding behaviour of starfish on Essex oyster beds. Journal of the Marine Biological Association of the United Kingdom, 34 (2), 313-331.

Héral, M. (1989). Traditional oyster culture in France. In G. Barnabe, J. F. de l. B. Solbe (Eds.), Aquaculture: biology and ecology of cultured species (pp.342-387). Ellis Horwood series in aquaculture and fisheries support. Chichester, UK: Wiley and Sons.

Hine PM, Thorne T (1997). Replication of herpes-like viruses in haemocytes of adult flat oysters *Ostrea angasi*: an ultrastructural study. Diseases of Aquatic Organisms, 29, 189–196.

Hollander, J., Blomfeldt, J., Carlsson, P. *et al.* (2015). Effects of the alien Pacific oyster (*Crassostrea gigas*) on subtidal macrozoobenthos communities. Marine Biology, 162, 547–555.

Hordoir, R., Dieterich, C., Basu, C. *et al.* (2013). Freshwater outflow of the Baltic Sea and transport in the Norwegian current: a statistical correlation analysis based on a numerical experiment. Continental Shelf Research, 64, 1–9.

Hosac, G. R., Dumbaul, B. R., Ruesink, J. L. *et al.* (2006). Habitat associations of estuarine species: comparisons of intertidal mudflat, seagrass (*Zostera marina*), and oyster (*Crassostrea gigas*) habitats. Estuaries and Coasts, 29 (6B), 1150–1160.

IPCC (2007). Climate Change 2007: Synthesis Report. In Core Writing Team, R.K. Pachauri, A. Reisinger (Eds.), Contribution of Working Groups I, II and III to the Fourth Assessment Report of the Intergovernmental Panel on Climate Change (pp.104). Geneva, Switzerland: IPCC.

Jansson, B.-O., Aneer, G., Nellbring, S. (1985). Spatial and temporal distribution of the demersal fish fauna in a Baltic archipelago as estimated by SCUBA census. Marine Ecology Progress Series, 23, 31–43.

Jones, C. G., Lawton, J. H., Shachak, M. (1994). Organisms as ecosystem engineers. Oikos, 69 (3), 373-386.

Jones, C. G., Lawton, J. H., Shachak, M. (1997). Positive and negative effects of organisms as physical ecosystem engineers. Ecology, 78 (7), 1946–1957.

Jungclaus, J. H., Keenlyside, N., Botzet, M. *et al.* (2006). Ocean circulation and tropical variability in the coupled ECHAM5/MPI-OM. Journal of Climatology, 19 (16), 3952- 3972.

Kaschner, K., Ready, J. S., Agbayani, E. *et al.* (2010). Editors 2008 AquaMaps Environmental Dataset: Half-Degree Cells Authority File (HCAF). World Wide Web electronic publication, www.aquamaps.org/data, Version 07/2010.

Kingsley-Smith, P. R., Joyce, R. E., Arnott, S. A. *et al.* (2012). Habitat use of intertidal eastern oyster (*Crassostrea virginica*) reefs by nekton in South Carolina estuaries. Journal of Shellfish Research, 31 (4), 1009–1021.

Kochmann, J., Buschbaum, C., Volkenborn, N. *et al.* (2008). Shift from native mussels to alien oysters: differential effects of ecosystem engineers. Journal of Experimental Marine Biology and Ecology, 364 (1), 1-10.

Kollberg, S., Lindahl, O. (2006). Musselmjöl istället för fiskmjöl i ekologiskt foder. Ekhaga projekt: D. nr: 2004-55.

Lacoste, A., Jalabert, F., Malham, S. *et al.* (2001). A *Vibrio splendidus* strain associated with Summer Mortality of juvenile oysters *Crassostrea gigas* in the bay of Morlaix (North Brittany, France). Diseases of Aquatic Organisms, 46 (2), 139-145.

Le Gall, J., Raillard, O. (1988). Influence de la temperature sur la physiologie de l'huitre. Océanis, 14 (5), 603-608.

Lejart, M., Hily, C. (2011). Differential response of benthic macrofauna to the formation of novel oyster reefs (*Crassostrea gigas*, Thunberg) on soft and rocky substrate in the intertidal of the Bay of Brest, France. Journal of Sea Research, 65 (1), 84-93.

Leidenberger, S., De Giovanni, R., Kulawik, R. *et al.* (2015). Mapping present and future predicted distribution patterns for a meso-grazer guild in the Baltic Sea. Journal of Biogeography, 42 (1), 241-254, doi: 10.1111/jbi.12395.

Lindahl, O., Hart, R., Hernroth, B. *et al.* (2005). Improving marine water quality by mussel farming: a profitable solution for Swedish society. AMBIO, 34 (2), 131-138.

Luna-Acosta, A., Bustamante, P., Godefroy, J. *et al.* (2010). Seasonal variation of pollution biomarkers to assess the impact on health status of juvenile Pacific oysters *Crassostrea gigas* exposed in situ. Environmental Science and Pollution Research, 17 (4), 999-1008.

Luna-Gonzalez, A., Romero-Geraldo, M de J, Campa-Córdova A. *et al.* (2008). Seasonal variations in the immunological and physiological parameters of the Pacific oyster *Crassostrea gigas* cultured in the bahia de Macapule (Sinaloa, Mexico). Aquaculture Research, 39 (14), 1488-1497.

Madec, G. (2010). Nemo ocean engine, v. 3.3. Technical report. Paris, France: IPSL.

Malham, S. K., Cotter, E., O'Keeffe, S., Lynch, S. *et al.* (2009). Summer mortality of the Pacific oyster, *Crassostrea gigas*, in the Irish Sea: the influence of temperature and nutrients on health and survival. Aquaculture, 287 (1-2), 128-138.

Markert, A., Wehrmann, A., Kroencke, I. (2010). Recently established *Crassostrea*-reefs versus native *Mytilus*-beds: differences in ecosystem engineering affects the macrofaunal communities (Wadden Sea of Lower Saxony, southern German Bight). Biological Invasions, 12 (1), 15-32.

Mendonça, V. M., Raffaelli, D. G., Boyle, P. R. *et al.* (2009). Trophodynamics in a shallow lagoon off northwestern Europe (Culbin Sands, Moray Firth): spatial and temporal variability of epibenthic communities, their diets, and consumption efficiency. Zoological Studies, 48 (2), 196-214.

Meurling, S. (2013). The origin of the invasive species *Crassostrea gigas* in Sweden. Master thesis work. Department of Biological and Environmental Studies, University of Gothenburg.

Mortensen, S., Bodvin, T., Skår, C. S. *et al.* (2014). Massedød av stillehavsøsters, *Crassostrea gigas*, i Sverige og Norge, September 2014. Undersøkelser av østers fra Vestfold og funn av østers herpesvirus (OsHV-1 µvar). Bergen, Norway: Institute of Marine Research.

Moehler, J., Wegner, K. M., Reise, K. *et al.* (2011). Invasion genetics of Pacific oyster *Crassostrea gigas* shaped by aquaculture stocking practices. Journal of Sea Research, 66 (3), 256-262.

Muñoz, M. E. S., Giovanni, R., Siqueira, M. F. *et al.* (2011). openModeller: a generic approach to species' potential distribution modelling. Geoinformatica, 15 (1), 111-135.

Mahalanobis, P. C. (1936). On the generalized distance in statistics. Proceeding of the National Institute of Science of India, 12 (1), 49-55.

Nehls, G., Diederich, S., Thieltges, D.W. *et al.* (2006). Wadden Sea mussel beds invaded by oysters and slipper limpets: competition or climate control? Helgoland Marine Research, 60 (2), 135-143.

Nehring, S. (2003). Pacific oysters in the European Wadden Sea: an irreversible impact in a highly protected ecosystem. Aliens, 17, 20-21.

Nehring, S. (2011): NOBANIS – Invasive Alien Species Fact Sheet. *Crassostrea gigas*. European Network on Invasive Alien Species – NOBANIS, http://www.nobanis.org/Factsheets.asp, accessed on 13 December 2014.

Newell, R. I. E. (2004). Ecosystem influences of natural and cultivated populations of suspension-feeding bivalve molluscs: a review. Journal of Shellfish Research, 23 (1), 51-61.

Nix, H. A. (1986). A biogeographic analysis of Australian elapid snakes. In R. Longmore (Ed.), Atlas of Australian elapid snakes (p. 415). Australian Flora and Fauna Series 7. Canberra, Australia: Australian Government Publishing Service.

Norling, P., Kautsky, N. (2007). Structural and functional effects of *Mytilus edulis* on diversity of associated species and ecosystem functioning. Marine Ecology Progress Series, 351, 163-175.

Norling, P., Lindegarth, M., Lindegarth, S. *et al.* (2015). Effects of live and post-mortem shell structures of invasive Pacific oysters and native blue mussels on macrofauna and fish. Marine Ecology Progress Series, 518, 123-138.

Nyberg, C. (2010). Japanska jätteostronet (*Crassostrea gigas*) på västkusten 2009. Stoxkholm, Sweden: Swedish Environmental Protection Agency.

O'Neill, S. M., Sutterlin, A. M., Aggett, D. (1983). The effects of size-selective feeding by starfish, *Asterias vulgaris*, on the production of mussels, *Mytilus edulis*, cultured on nets. Aquaculture, 38, 211-220.

Officer, C. B., Smayda, T. J., Mann, R. (1982). Benthic filter feeding: a natural eutrophication control. Marine Ecology Progress Series, 9, 203-210.

OSPAR. http://www.ospar.org.

Parmesan, C., Gaines, S. D., Gonzalez, L. *et al.* (2005). Empirical perspectives on species borders: from traditional biogeography to global change. Oikos, 108 (1), 58–75.

Peterson, C. H., Grabowski, J. H., Powers, S. P. (2003). Estimated enhancement of fish production resulting from restoring oyster reef habitat: quantitative valuation. Marine Ecology Progress Series, 264, 249-264.

Phillips, S. J., Anderson, R. P., Schapire, R. E. (2006). Maximum entropy modelling of species geographic distributions. Ecological Modelling, 190 (3-4), 231-259.

Poloczanska, E. S., Brown, C. J., Sydeman, W. J. *et al.* (2013). Global imprint of climate change on marine life. Nature Climate Change, 3, 919-925.

Quayle, D. B. (1969). Pacific oyster culture in British Columbia. Bull. 169, Fisheries Research Board of Canada.

R Core Team (2013). R: A language and environment for statistical computing. R Foundation for Statistical Computing, Vienna, Austria.

Reise, K. (1998). Pacific oysters invade mussel beds in European Wadden Sea. Senckenbergiana Maritima, 28 (4/6), 167-175.

Reise, K., Dankers, N., Essink, K. (2005). Introduced species. In: K. Essink, C. Dettmann, H. Farke *et al.* (Eds.), Wadden Sea quality status report (pp.155–161). Wilhelmshaven, Germany: Common Wadden Sea Secretariat.

Ren, J. S., Ross, A. H., Schiel, D. R. (2000). Functional desriptions of feeding and energetics of the Pacific oyster *Crassostrea gigas* in New Zealand. Marine Ecology Progress Series, 208, 119-130.

Renault, T., Le Deuff, R. M., Cochennec, N. *et al.* (1995). Herpes-like viruses associated with high mortality levels in larvae and spat of Pacific oysters, *Crassostrea gigas*: a comparative study, the thermal effects on virus detection in hatchery-reared larvae, reproduction of the disease in axenic larvae. Veterinary Research, 26 (5-6), 539-543.

Renault, T., Bouquet, A. L., Maurice, J.-T. *et al.* (2014). Ostreid herpesvirus 1 infection among Pacific oyster (*Crassostrea gigas*) spat: relevance of water temperature to virus replication and circulation prior to the onset of mortality. Applied and Environmental Microbiology, 80 (17), 5419-5426.

Rice, M. A. (2001). Environmental impacts of shellfish aquaculture: filter feeding to control. In Tlusty, M. F., D. A. Bengston, H. O. Halvorson *et al.* (Eds.), Marine Aquaculture and the Environment: A Meeting for Stakeholders in the Northeast. Falmouth, MA: Cape Cod Press.

Royer, J., Ropert, M., Mathieu, M. *et al.* (2006). Presence of spionid worms and other epibionts in Pacific oysters (*Crassostrea gigas*) cultured in Normandy, France. Aquaculture, 253 (1-4), 461–474.

Ruesink, J. L., Lenihan, H. S., Trimble, A. C. *et al.* (2005). Introduction of non-native oyster: ecosystem effects and restoration implications. Annual Review of Ecology and Systematics, 36, 643-89.

Scheiffarth, G., Ens, B., Schmidt, A. (2007). What will happen to birds when Pacific oysters take over the mussel beds in the Wadden Sea? Wadden Sea Newsletter, 1, 10-14.

Samain, J. F., McCombie, H. (2008). Summer Mortality of the Pacific oyster *Crassostrea gigas*. Versailles, France: The Morest Project, Editions Quae.

Segarra, A., Pepin, J. F., Arzul, I. *et al.* (2010). Detection and description of a particular Ostreid herpesvirus 1 genotype associated with massive mortality outbreaks of Pacific oysters, *Crassostrea gigas*. Virus Research, 153 (1), 92–95.

Soletchnik, P., Lambert, C., Costil, K. (2005). Summer mortality of *Crassostrea gigas* (Thunberg) in relation to environmental rearing conditions. Journal of Shellfish Research, 24 (1), 197-207.

Strand, Å., Blanda, E., Bodvin, T. *et al.* (2012). Impact of an icy winter on the *C. gigas* (*Crassostrea gigas* Thunberg, 1793) populations in Scandinavia. Aquatic Invasions, 7 (3), 433-440.

Strand, Å., Waenerlund, A., Lindegarth S. (2011). High tolerance of the Pacific oyster (*Crassostrea gigas*, Thunberg) to low temperatures. Journal of Shellfish Research, 30 (3), 733–735.

Strand, Å., Lindegarth, S. (2014). Japanska ostron i svenska vatten-en främmande art som är här för att stanna. Rapport No. 2, Vattenbrukscentrum Väst, Göteborgs universitet. Bosås, Sweden: Responstryck.

Tolley, S. G., Volety, A. K. (2005). The role of oysters in habitat use of oyster reefs by resident fishes and decapod crustaceans. Journal of Shellfish Research, 24 (4), 1007-1012.

Troost, K., Kamermans, P., Wolff, W. J. (2008a). Larviphagy in native bivalves and an introduced oyster. Journal of Sea Research, 60 (3), 157-163.

Troost, K., Veldhuizen, R., Stamhuis, E. J. *et al.* (2008b). Can bivalve veligers escape feeding currents of adult bivalves? Journal of Experimental Marine Biology and Ecology, 358 (2), 185-196.

Troost, K., Stamhuis, E. J., van Duren, L. A. *Et al.* (2009). Feeding current characteristics of three morphologically different bivalve suspension feeders, *Crassostrea gigas*, *Mytilus edulis* and *Cerastoderma edule*, in relation to food competition. Marine Biology, 156 (3), 355-372.

Troost, K. (2009). Pacific oysters in Dutch estuaries. Causes of success and consequences for native bivalves. Doctoral thesis. University of Groningen.

Troost, K. (2010). Causes and effects of a highly successful marine invasion: case-study of the introduced *Crassostrea gigas* in continental NW European estuaries. Journal of Sea Research, 64 (3), 145-165.

Tyberghein, L., Verbruggen, H., Pauly, K. *et al.* (2012). Bio-Oracle: a global environmental dataset for marine species distribution modelling. Global Ecology and Biogeography, 21 (2), 272-281.

Van Broekhoven, W. (2005). Macrofaunal diversity on beds of the Pacific oyster (*Crassostrea gigas*) in the Oosterschelde estuary. MSc thesis, 36 pp.

Walne, P. R. (1974). Culture of bivalve mollusca: 50 years experience at Conwy. Fishing news. London, UK: Whitefriars Press.

Walne, P. R., Davies, G. (1977). The effect of mesh covers on the survival and growth of *Crassostrea gigas* Thunberg grown on the sea bed. Aquaculture, 11 (4), 313-321.

Wilkie, E. M., Bishop, M. J., O'Connor, W. A. (2013). The density and spatial arrangement of the invasive oyster *Crassostrea gigas* determines its impact on settlement of native oyster larvae. Ecology and Evolution, 3 (15), 4851–4860.

Wrange, A.-L., Valero, J., Harkestad, L. S. *et al.* (2009). Massive settlements of the Pacific oyster, Crassostrea gigas, in Scandinavia. Biological Invasions, 12 (5), 1145–1152.

Antonella Petrocelli, Ester Cecere

11 Invasive Seaweeds: Impacts and Management Actions

11.1 Introduction

In the world, many alien species are seaweeds. Alien species have been reported for all the three taxonomical divisions, with Rhodophyta more than twice as much as both Ochrophyta and Chlorophyta (Williams & Smith, 2007), even though they are the least studied. The highest number of alien seaweeds has been reported in the Mediterranean, mainly coming from the Northwest Pacific and Indo-Pacific regions (Klein *et al.*, 2005; Williams & Smith, 2007).

Most alien seaweeds were accidentally introduced (Hewitt *et al.*, 2007), with only a small percentage introduced intentionally, mainly for aquaculture purposes in past times when knowledge of risks deriving from the introduction of alien species was low (Pickering *et al.*, 2007). Some species seem more likely to become invasive due to distinctive features (e.g. capacity for successful spread), but it is not always a sure thing that, once introduced, they will successfully establish in the new area or become harmful (Boudouresque & Verlaque, 2002b). For this reason, one seaweed species cannot be defined as invasive in an absolute sense (Inderjit *et al.*, 2006) and when invasive it can show different behaviours; that is, it may have diverse impacts in different areas and on different scales (Schaffelke *et al.*, 2006, Williams & Smith, 2007, Thomsen *et al.*, 2009a).

Since biological invasions by seaweeds can cause irreversible damage to the biodiversity, structure, and functioning of receiving ecosystems, once an introduced species is detected, the assessment of its real distribution and of its impact at each trophic level should be of primary importance in ecological studies (Bulleri *et al.*, 2012). The planning of either its possible eradication or its management should follow (Aguilar-Rosas *et al.*, 2013). However, the finding of an invasive species is often tardy compared with its arrival in a given environment, such that it can be difficult to disentangle its impact from other impacts due to pollution, climate change, or habitat destruction (Junqueira, 2013).

According to available literature, about 280 species of introduced seaweeds are currently present in the world's seas (Williams & Smith, 2007). The majority did not show any visibly high invasiveness until now (Johnson, 2007); after all, only few were deeply studied concerning their invasion patterns and impacts (Lyons & Scheibling, 2009), even though their capacity for invasion, even a long period after their introduction, was already known (Smith *et al.*, 2004).

The aim of this paper is to take stock of the situation regarding the distribution and impact of three of the most spread invasive seaweeds around the world, one for each taxonomic division: the chlorophycean *Codium fragile* (Suringar) Hariot ssp. *fragile*, the rhodophycean *Gracilaria vermiculophylla* (Ohmi) Papenfuss, and the phaeophycean *Undaria pinnatifida* (Harvey) Suringar.

Information about the most common vectors of introduction of alien seaweed species, management actions, as well as the present laws regulating the transfer of imported organisms and possible precautionary measures were also analysed.

11.2 Most Widespread Invasive Seaweeds

11.2.1 *Codium fragile* ssp. *fragile* (Chlorophyta, Bryopsidales) (Figure 11.1)

Fig. 11.1: Thallus of *Codium fragile* ssp. *fragile* in the Mar Piccolo of Taranto. 1 cm = 6 mm.

Codium fragile ssp. *fragile* (hereafter *C. fragile*) is a worldwide introduced species (Provan *et al.*, 2005) (Figure 11.2). It ranks first among the top five hazardous invasive seaweeds, due to dispersal and establishment ability as well as ecological and economic impact (Nyberg & Wallentinus, 2005; Provan *et al.*, 2005). Its possible impacts vary from the reduction of biodiversity in the invaded communities to fouling of fishing gear and damage to shellfish aquaculture activities (Bridgwood, 2010) (Table 11.1). One of the nicknames of *C. fragile* is “oyster thief”, because it commonly fouls shellfish and can sweep them away, causing considerable economic losses (Trowbridge, 1999). In Chile, the invasion of *C. fragile* caused substantial economic damage to seaweed farms, since alien thalli remained entangled with cultivated plants of *Gracilaria chilensis* Bird, McLachlan et Oliveira causing them to sink before harvesting. The burden of work and time imposed by having to remove the invader even bankrupted a farm (Neill *et al.*, 2006). In Nova Scotia, a marked competition with local seaweed species, mainly kelp, was observed: the presence of well-structured kelp communities did not allow *C. fragile* settlement, while dense populations of *C. fragile* prevented kelp settlement (Scheibling & Gagnon, 2006). In a lagoon in Eastern Canada, a negative impact of *C. fragile* on the eelgrass *Zostera marina* Linnaeus was observed in manipulative experiments: higher *C. fragile* biomass values matched lower density of eelgrass shoots and lower values of leaf length. However, the observations performed in the field did not support the entirety of the experimental results (Drouin *et al.*, 2012).

Tab. 11.1: Impact (positive or negative) of the alien seaweeds *Codium fragile*, *Gracilaria vermiculophylla* and *Undaria pinnatifida* on biodiversity, structure and function of ecosystems or economic. O = observed; E = experimental; S = supposed.

Species	Impact	Locality	Reference
Codium fragile			
O	negative, economic: "oyster thief" fouling and sweeping of reared shellfish	Australia	Trowbridge, 1999
O	negative, biodiversity: replacement of native canopy species	Atlantic Ocean, Gulf of Maine, USA	Harris & Tyrrel, 2001
E	negative, biodiversity and functioning: death of fed sea-urchins	Atlantic Ocean, Nova Scotia, Canada	Scheibling & Anthony, 2001
O	negative, structure: reduction of kelp abundances	Atlantic Ocean, Canada	Chapman *et al.*, 2002
O, E	negative, biodiversity: replacement of native kelps	Atlantic Ocean, Gulf of Maine, USA	Levin *et al.*, 2002
O, E	negative, biodiversity and structure: “eelgrass thief” removing shoots and rhizomes of Z. marina	Atlantic Ocean, Prince Edward Island and Nova Scotia, Canada	Garbary *et al.*, 2004

continued **Tab. 11.1:** Impact (positive or negative) of the alien seaweeds *Codium fragile*, *Gracilaria vermiculophylla* and *Undaria pinnatifida* on biodiversity, structure and function of ecosystems or economic. O = observed; E = experimental; S = supposed.

Species	Impact	Locality	Reference
O, E	positive, biodiversity and functioning: favouring of mussel recruitment	Mediterranean Sea, Adriatic Sea, Italy	Bulleri *et al.*, 2006
O	negative, economic: decrease of cultivated Gracilaria chilensis yeld	Pacific Ocean, Chile	Neill *et al.*, 2006
E	negative, biodiversity and structure: prevention of kelp colonization	Atlantic Ocean, Nova Scotia, Canada	Scheibling & Gagnon, 2006
E	negative, biodiversity and structure: decrease of eelgrass shoot density	Atlantic Ocean, Canada	Drouin *et al.*, 2012
Gracilaria vermiculophylla			
O	positive, biodiversity: increase of filamentous seaweeds	Atlantic Ocean, Virginia, USA	Thomsen *et al.*, 2006
O	positive, biodiversity: increase of animal abundances	Kattegat, Sweden	Nyberg *et al.*, 2009
O	positive, biodiversity: increase of animal abundances	Atlantic Ocean, Virginia, USA	Nyberg *et al.*, 2009
E	negative, biodiversity and function: survival of Z. marina	Baltic Sea, Isle of Fyn, Denmark	Martínez-Lüscher & Holmer, 2010
E	positive, biodiversity: increase of associated fauna	Baltic Sea, Denmark	Thomsen, 2010
S	positive, economic: production of good quality food grade agar	Atlantic Ocean, Portugal	Villanueva *et al.*, 2010
E	positive, function and economic: bioremediation	Atlantic Ocean, Portugal	Abreu *et al.*, 2011
E	positive, biodiversity and structure: enhancement of epifaunal densities	Atlantic Ocean, Georgia and South Carolina, USA	Byers *et al.*, 2012
E	positive, structure and function: fostering survival of native blue crab	Atlantic Ocean, Chesapeake Bay, USA	Johnston & Lipcius, 2012
E	negative, biodiversity and structure: grazer avoidance against native species	Baltic Sea, Denmark	Nejrup *et al.*, 2012
E	negative, biodiversity and structure: reduction of native Fucus growth	Baltic Sea, Kiel Fjord, Germany	Hamman *et al.*, 2013a
E	positive, biodiversity: increase of invertebrates abundance	Odense Fjord, Denmark	Thomsen *et al.*, 2013

continued **Tab. 11.1:** Impact (positive or negative) of the alien seaweeds *Codium fragile*, *Gracilaria vermiculophylla* and *Undaria pinnatifida* on biodiversity, structure and function of ecosystems or economic. O = observed; E = experimental; S = supposed.

Species	Impact	Locality	Reference
O	positive, structure: increase of egg capsule deposition of invertebrates	Atlantic Ocean, Rhode Island, USA	Guidone *et al.*, 2014
E	positive, structure and function: reduction of predation on invertebrates	Atlantic Ocean, Georgia, USA	Wright *et al.*, 2014
Undaria pinnatifida			
O	negative, biodiversity and structure: decrease of native species total cover	Mediterranean Sea, Venice, Italy	Curiel *et al.*, 2001
O	positive, biodiversity: increase of refuges for cryptic fauna	Mediterranean Sea, Mar Piccolo of Taranto, Italy	Cecere *et al.*, 2003
E	negative, biodiversity: decrease of native seaweed species richness	Atlantic Ocean, Nuevo Gulf, Argentina	Casas *et al.*, 2004
E, S	positive, function and economic: bioremediation	Atlantic Ocean, Patagonia, Argentina	Torres *et al.*, 2004
E	neutral, biodiversity and structure: no variation in associated flora and fauna	Atlantic Ocean, Cracker Bay, Argentina	Raffo *et al.*, 2009
O	negative, biodiversity and structure: reduction of rocky-reef fishes	Atlantic Ocean, Nuevo Gulf, Argentina	Irigoyen *et al.*, 2011a
O	negative, economic: obstruction of the entrance of fish holes	Atlantic Ocean, Nuevo Gulf, Argentina	Irigoyen *et al.*, 2011a
E	positive, biodiversity and structure: increase of invertebrate species richness	Atlantic Ocean, Nuevo Gulf, Argentina	Irigoyen *et al.*, 2011b
O	negative, economic: obstacle for local navigation	Mediterranean Sea, Venice, Italy	Sfriso & Facca, 2013

By contrast, the interaction between *C. fragile* and *Mytilus galloprovincialis* Lamarck on artificial structures dipped in the Adriatic Sea showed a benign effect. The presence of both germlings and canopy of the macroalga favoured the settlement of the mussel recruits, while on the bare surfaces the number of these recruits was much lower. Contrarily, the presence of a well-developed mussel bed reduced the abundance of *C. fragile* (Bulleri *et al.*, 2006).

Fig. 11.2: Worldwide distribution of *Codium fragile* ssp. *fragile*. Green star indicates the type locality; green circles indicate native distribution; red circles indicate alien distribution.

In the Mar Piccolo of Taranto (southern Italy, Mediterranean Sea), a small number of thalli of *C. fragile* were found for the first time in July 2002 and successively in 2003, in a zone characterised by the presence of several seafood shops. No other thalli were found until 2009, when a new finding was registered in the same zone, possibly due to a new introduction event. Since then, only a few thalli have appeared on pebbles in the same zone each summer with no negative impact (Petrocelli *et al.*, 2013).

Several features could justify the high invasiveness of *C. fragile* around the world:

1. High tolerance to chemical-physical variability (Thomsen & McGlathery, 2007);
2. Sexual, vegetative, and parthenogenetical reproduction (Bridgwood, 2010);
3. Opportunistic behaviour. In its native region, where the dominant species were removed, *C. fragile* predominated as a canopy-forming species; where the canopy species were well developed, it was an understory species (Chavanich *et al.*, 2006);
4. High dispersal potential. Notwithstanding the absence of specialised structures for floating, *C. fragile* thalli have a notable capacity for buoyancy due to the accumulation of gas bubbles deriving from the photosynthetic process within the thallus, particularly at the tip level (Gagnon *et al.*, 2011). Laboratory experiments showed that *C. fragile* (as *C. fragile* ssp. *tomentosoides*) can live up to 90 days of emersion in a dry environment, entangled on anchors or fishing nets during vessel travel, recovering its photosynthetic capacity after re-submersion (Schaffelke & Deane, 2005). Moreover, besides easily spreading through man-mediated activities, *C. fragile* can also spread naturally through drifting vegetative thallus fragments, buds, and

detached fertile thalli. Due to the capacity for reattachment of these structures, the species can colonize new areas at great distances from the initial introduction site. The presence of turf algae enhances their settlement (Watanabe *et al.*, 2009);

5. Unpalatability for most grazers. The production of dimethylsulfoniopropionate and its derivatives, experimentally determined in *C. fragile* (as var. *tomentosoides*) from Nova Scotia, favours the alien's success by reducing its palatability for sea urchins (Lyons *et al.*, 2007). A partial natural control by the snail *Littorina littorea* (Linnaeus, 1758) on *C. fragile* populations was observed. The snail actively grazed on the alien seaweed, but only on new plantlets and residual basal parts, damaging thalli growth; adult healthy thalli did not suffer this grazing (Scheibling *et al.*, 2008).

A genetic molecular analysis was performed on the plastid genome of *C. fragile* (as ssp. *tomentosoides*) collected in the native range in Japan as well as the Mediterranean, Northern Europe, North Atlantic, and South Pacific. It showed that the spread of this invasive species was due to two different introduction events, one into the Mediterranean and the other to the rest of the world. Therefore, only two alien haplotypes are present worldwide (Provan *et al.*, 2005).

Presumably, *C. fragile* was mainly introduced around the world through the importation of shellfish, but also through fouling of ships and boat hulls as a possible vector (GISD, 2014).

Eradication of *C. fragile* was not effective in Australia, either by chemical methods or by manual removal (Trowbridge, 1999). No other attempt has been performed anywhere in the world, since the morphological and physiological features of the species would surely have made them unsuccessful (GISD, 2014).

11.2.2 *Gracilaria vermiculophylla* (Rhodophyta, Gracilariales) (Figure 11.3)

Fig. 11.3: Thallus of *Gracilaria vermiculophylla* in the Venice Lagoon (courtesy of A. Sfriso). 1 mm = 3 mm.

Gracilaria vermiculophylla is native to East Asia, and in less than two lustra invaded the coasts of other continents such as Europe, North America and, recently, North Africa (Figure 11.4). It became one of the main invasive seaweeds, especially in estuarine and lagoon environments, where it commonly lives unattached, partially embedded in the mud, and less frequently as attached (Kim *et al.*, 2010; Abreu *et al.*, 2011; Sfriso *et al.*, 2012; Hammann *et al.*, 2013b). In two years, *G. vermiculophylla* spread for about 150 km along the Swedish coasts, with a larger expansion range than other invasive seaweeds, such as *U. pinnatifida* and *Sargassum muticum* (Yendo) Fensholt, neither of which reached 50 km per year (Nyberg *et al.*, 2009).

Fig. 11.4: Worldwide distribution of *Gracilaria vermiculophylla*. Green star indicates the type locality; green circles indicate native distribution; red circles indicate alien distribution.

A recent review summarized the main impacts recorded after *G. vermiculophylla* invasions around the world (Hu & Juan, 2014) (Table 11.1). In the Baltic Sea, considerable unattached biomasses of *G. vermiculophylla* drifted on soft bottoms, so high interference with both the settlement of plantlets and the growth of adults of native *Fucus vesiculosus* Linnaeus occurred. Moreover, *Gracilaria vermiculophylla* threatened *F. vesiculosus*'s survival, giving hospitality to grazers greedy for this species (Hammann *et al.*, 2013a). In Danish coastal communities, both field observations and lab experiments showed that the prevalence of *G. vermiculophylla* was promoted by the lack of grazing by local herbivores, which preferred the short-lived Ulvales. *G. vermiculophylla* may produce secondary metabolites that deter grazer activity (Nejrup *et al.*, 2012). Mesocosm experiments showed that the presence of considerable biomasses of *G. vermiculophylla* reduced net photosynthesis of *Z. marina* leaves (Martínez-Lüscher & Holmer, 2010). Considering that successive lab experiments showed a high sensitivity of *Z. marina* growth to high temperature (Hoffle *et al.*, 2011), it was hypothesised that, in a future warmer world, the combined effect of higher temperatures and *G. vermiculophylla* presence could cause eelgrass disappearance (Hoffle *et al.*, 2011).

However, some cases of positive impacts of this alien on biodiversity were also recorded. Field experiments demonstrated a positive influence of *G. vermiculophylla* on the faunal assemblages in a *Z. marina* meadow in Denmark, probably through the increase of refuges from predators, of food for herbivores, and of attachment space for epibionts (Thomsen, 2010). In Swedish waters, a high diversity of associated fauna and flora was observed on both attached and unattached biomass of *G. vermiculophylla* (Nyberg *et al.*, 2009). In the Adriatic Sea, association with molluscs, tunicates, and worms was reported (Sfriso *et al.*, 2012). The presence of *G. vermiculophylla* in a lagoon in Virginia (USA) proved to be beneficial for overall local biodiversity. In particular, the biomass of epiphytic filamentous algal species positively correlated with that of this alien seaweed, which served as a hard substratum for the attachment in a place characterised by soft bottoms (Thomsen *et al.*, 2006). Moreover, the invasive *G. vermiculophylla* in Ria de Aveiro (Portugal) was found to produce a good quality of food grade agar (Villanueva *et al.*, 2010). Therefore, in the case of the eradication of threating biomasses, a useful by-product could be obtained.

Gracilaria vermiculophylla adapts well to estuarine and lagoon conditions due to (Nyberg & Wallentinus, 2009; Abreu *et al.*, 2011):

1. Tolerance to high variation in salinity and temperature. Experiments carried out in Denmark, with variously combined values of light and temperature, showed that *G. vermiculophylla* responds with great plasticity to these variations, reaching high growth rates. This could explain its recent spread in the Scandinavian waters (Nejrup *et al.*, 2013);
2. Capacity to grow well on muddy and sandy bottoms;
3. Capability of surviving long periods of darkness;
4. Ability to vegetatively propagate through thallus fragmentation;
5. Resistance to grazing and desiccation.

The low palatability of the alien plants of *G. vermiculophylla* for *Littorina littorea* could explain the success of this species in Germany (Hammann *et al.*, 2013b).

Japanese oysters have been considered the main vector for the introduction of *G. vermiculophylla* into Western Atlantic waters; but, the vicinity of harbours to several zones of first observation suggests that shipping from Japan, Korea and Russia may also have acted as a source (Kim *et al.*, 2010). Indeed, from the results of molecular analysis, it is clear that multiple introductions from different geographical areas have occurred (Gulbransen *et al.*, 2012). For Swedish waters, a likely vector of introduction could have been the dredges used for the excavation of the Gothenburg harbour chartered from the Netherlands (Rueness, 2005). In the lagoons of the North Adriatic Sea (Italy), *G. vermiculophylla* was most probably introduced through the importation of the Manila clam *Venerupis philippinarum* (Adams & Reeve, 1850). Afterwards, high nutrient concentrations and moderate salinity were the environmental factors that most likely favoured its establishment and spread (Sfriso *et al.*, 2012). *Gracilaria vermiculophylla* was also observed in the unattached form in some salt marshes in Virginia (USA), where seaweeds are typically absent. Its most likely origin was from nearby lagoons, where it lives in tight association with the tubeworm *Diopatra cuprea* (Bosc, 1802) (Thomsen *et al.*, 2009b). No information about any attempt of *G. vermiculophylla* eradication is available to date.

11.2.3 *Undaria pinnatifida* (Ochrophyta, Laminariales) (Figure 11.5)

Fig. 11.5: Thallus of *Undaria pinnatifida* from the Mar Piccolo of Taranto. 1 cm = 1.7 cm.

Undaria pinnatifida is native to Japan. It has been introduced along the coasts of all the continents except for Africa and Antarctica (Figure 11.6), generally found in sheltered zones (Aguilar-Rosas *et al.*, 2004). In Europe, it is considered the third most invasive seaweed (ICES, 2007; Báez *et al.*, 2010).

When introduced, the behaviour of *U. pinnatifida* can differ, generally depending on the environmental conditions of the recipient system (Table 11.1). Where the species retains its typical seasonal cycle, it can be controlled by native species regrowth during summer, when alien sporophytes die (Zabin *et al.*, 2009).

In contrast, where *U. pinnatifida* endures year round, it can most likely outcompete native species, so invasion can have negative consequences at a biodiversity level, causing a reduction of local species, and also at an economic level if it invades communities of commercial species (Casas *et al.*, 2004). Therefore, when possible, eradication is advisable. Indeed, in Nuevo Gulf (Argentina), the rocky coast has been almost completely and continuously colonised by this alien since 1992. Its experimental removal triggered a large increase (+175%) in the number of native seaweeds (Casas *et al.*, 2004).

Fig. 11.6: Worldwide distribution of *Undaria pinnatifida*. Green star indicates the type locality; green circles indicate native distribution; red circles indicate alien distribution.

In New Zealand, the results of a risk assessment model showed that *U. pinnatifida* has the potential for high negative impact in High Value Areas (Campbell & Hewitt, 2013). In Tasmania, manipulation experiments in the field demonstrated that any already-present disturbance of the natural ecosystems favours the establishment of *U. pinnatifida* populations, and continuous disturbance seems necessary for its persistence (Valentine & Johnson, 2003; 2005). In the Venice Lagoon, *U. pinnatifida* is one of the two major invasive seaweeds, together with *S. muticum*. It is present from autumn to spring, in different sites, with very high biomass and cover values on different hard substrata (Sfriso & Facca, 2013), and competes with the native species for the substratum, causing the reduction of their cover index rather than of their number (Curiel *et al.*, 1998). Conversely, it does not compete with other alien seaweeds, which are preferentially floating and distributed on mobile bottoms. Due to its large dimensions, *U. pinnatifida* can represent an obstacle for local navigation along the canals, but its biomass is negligible in comparison with that of all the seaweeds present in the Lagoon. It does not cause any anoxic crises since, after detachment, it either is carried away to the sea or is run aground (Sfriso & Facca, 2013). In the Mar Piccolo of Taranto, *U. pinnatifida* was observed for the first time in April 1998. After an initial increase in population density (Cecere *et al.*, 2003), it completely disappeared within ten years, most likely due to the inability of microscopic gametophytes to overcome the high summer temperatures reached by the basin seawater (Cecere & Petrocelli, 2009). However, the small size of the founder population should not be undervalued (Báez *et al.*, 2010). No negative impact was registered in that period.

Besides its ecological negative effects on coastal systems, *U. pinnatifida* can also interfere with some recreational human activities, such as diving and angling. In Argentina, detached and drifting old thalli were observed clinging to the rocky reefs, obstructing the entrance of fish holes (Irigoyen *et al.*, 2011a).

However, this species could have also a positive impact. For example, it houses many epibionts, since its morphology seems to enhance the availability of refuges for cryptic benthic fauna, as occurred in the Mar Piccolo of Taranto (Cecere *et al.*, 2003). Moreover, *U. pinnatifida* is a food resource for some animals and can enhance consumer populations (Irigoyen *et al.*, 2011b).

Undaria pinnatifida can be considered an opportunistic species, which succeeds in invading spaces due to the following characteristics (Valentine & Johnson, 2003; ICES, 2007):

1. Easy settling on artificial substrates, including in disturbed zones;
2. Tolerance to wide variations in both temperature and salinity;
3. Fast growing, including in extreme conditions of turbidity and pollution;
4. Survival of gametophytes out of seawater for up to one month;
5. Year-round reproduction in some localities, and production of a huge quantity of zoospores transported by the currents.

Except for Brittany, where it was intentionally introduced for cultivation purposes, *U. pinnatifida* was accidentally introduced around the world either by fouling boats and ship hulls or by oyster transportation (ICES, 2007). In both Atlantic and Mediterranean France, the introduction of this species was probably due to the massive importation of the oyster *Crassostrea gigas* (Thunberg, 1793) from Japan (Boudouresque *et al.*, 1985). In the Venice Lagoon (Italy), the first report of *U. pinnatifida* was from Chioggia, where the importation of edible molluscs from northern Europe and the Mediterranean Sea is common (Curiel & Marzocchi, 2010). In the Mar Piccolo of Taranto (Ionian Sea, southern Italy), the introduction was most likely due to the importation of Japanese oysters (*C. gigas*) from France. To keep imported molluscs hydrated, they were transported covered with seaweed blades, which were presumably later thrown into the seawater and attached to the surrounding docks (Cecere *et al.*, 2000). Boats are the most probable vector for the introduction of *U. pinnatifida* into British waters, since some plants were observed attached to the hulls of recreational vessels moored at marinas in several ports (Fletcher & Farrell, 1999; Farrell & Fletcher, 2006). In Todos Santos Island (Mexico), this alien seaweed was probably introduced via commercial and touristic sailing, but also by recreational boats (Aguilar-Rosas *et al.*, 2004). Several possible vectors could have favoured *U. pinnatifida* introduction in central Patagonia, e.g. ballast waters, fouling of cargo ships or fishing boats from Japan or Korea (Casas *et al.*, 2004).

For prevention and control of *U. pinnatifida* introduction, boats and ship hulls should be continuously checked and cleaned out of water, taking care that when present, fertile specimens have to be disposed of and not re-immersed. Cargo ship ballast water must be treated with high temperatures before being discharged to avoid the release of any *U. pinnatifida* gametophytes, since they can survive at temperatures near to 30°C for long periods. All the structures in marinas and ports where *U. pinnatifida* thalli are found have to be carefully scraped. Moreover, a continuous monitoring of not-yet-colonised zones, especially in close proximity to already colonised areas, is necessary to avoid new settlements. The cultivation of *U. pinnatifida* in areas where it is not present must also be avoided, as well as its maintenance in aquaria where flow-through systems are used (ICES, 2007).

In New Zealand, mussel farming was considered the first vector for the spread of *U. pinnatifida*, by way of seeded ropes and mussel seeds. Therefore, careful cleaning was suggested, through a first washing followed by a second treatment by means of an environmentally friendly system such as high pressure, air-drying, freshwater, hot water (Forrest & Blakemore, 2006).

Eradication of this kelp is only possible at an early stage and in narrow colonised areas (Aguilar-Rosas *et al.*, 2004). Up to now, few attempts have been carried out. The only documented effective eradication was in the Chatham Islands (New Zealand), where *U. pinnatifida* was completely removed from a sunken ship at a depth of 20 m, through a heat treatment method (Wotton *et al.*, 2004). In the Venice Lagoon, eradication was unsuccessful when performed both during and after the reproductive period (Curiel *et al.*, 2001). In British waters, a manual eradication was initially attempted, but was unsuccessful since many of the removed thalli were already fertile (Fletcher & Farrell, 1999). In a Marine Reserve in Tasmania, a monthly manual eradication of *U. pinnatifida* sporophytes was carried out. As a result, the next generation, developed by zoospores or microscopic stages, consisted of a considerably reduced number of smaller thalli, few of which succeeded in maturing (Hewitt *et al.*, 2005).

11.3 Vectors

The transport mechanisms of alien seaweeds throughout the world are numerous. Hull fouling is considered the most ancient vector for the introduction of alien species in the marine system (Boudouresque & Verlaque, 2002a). Seaweeds can attach to vessels as juveniles, as encrusting and filamentous thalli, or as large developed thalli, and are able to survive the highly variable journey conditions. Mineur *et al.*, (2007) studied the hulls of 22 ships arriving in the commercial harbour of Sète (France, Mediterranean Sea) and found 31 seaweeds, mainly cosmopolitan species. The importance of recreational vessels in bays and coastal environments was investigated in California, since this region has been a hot spot for the introduction of alien species since the 1960s, suffering economic damage in excess of 2 trillion US dollars by 2010 (Ashton *et al.*, 2012). However, the most recent investigations showed that, contrary to what has been observed for alien animals (Canning-Clode *et al.*, 2013), hull fouling seems less important than aquaculture for the introduction of alien macroalgae. Nonetheless, the use of modern non-toxic paints and the high number of vessels mooring in marinas all over the world could enhance the risk of dispersal of these species after their introduction (Mineur *et al.*, 2008). Indeed, some of the more dessication-resistant species can survive transport attached to anchors, ropes, and chains (Hewitt *et al.*, 2007).

Today, the most likely vector for the introduction of alien seaweeds seems to be the importation of aquaculture organisms for different purposes (Hewitt *et al.*, 2007).

Ballast water is indicated as the main vector for the introduction of plankton species, but microscopic stages, propagules, and vegetative fragments of seaweeds are also able to survive the stress linked to ballast transport such as uptake, the ballast pump, and prolonged darkness (Flagella *et al.*, 2007). Ballast sediment is a less probable vector (Hewitt *et al.*, 2007).

Aquarium species, even when carefully controlled with quarantine periods, can accidentally escape from tanks and settle in the surrounding environment, as occurred for *Caulerpa taxifolia* (M. Vahl) C. Agardh in the Mediterranean Sea, California, and Australia (Hewitt *et al.*, 2007).

Finally, another vector is packing material, namely thalli used to maintain mollusc and live bait moisture during long routes. Once these thalli are thrown into seawater, they can settle and form new populations (Hewitt *et al.*, 2007).

11.4 Control and Management

Prediction of future invasions is not possible, but if suitable prevention and management are not implemented, the number of alien seaweeds will increase in coming years (Ashton *et al.*, 2012). Indeed, prevention of introduction is the most effective method in limiting biological invasions (Doelle *et al.*, 2007), but the correct management of human activities directly implicated in the spread of invasive species will be a strong constraint on further propagation (Lyons & Scheibling, 2009).

Different management actions are possible for intentional and unintentional introductions. In the first case, a precautionary risk assessment is necessary to fulfil the requirements of the ICES Code of Practice for Introductions and Transfers of Marine Organisms, in order to evaluate the possible damage that the introduced species and any associated alien species can cause. The knowledge of their biological and ecological features could allow us to avoid possible new damaging introductions and to make provisions for the possible spread of these species (Meinesz, 2007). Before introduction into the field, the first step must be a quarantine period – specimens must be held in segregation from which they cannot escape (Pickering *et al.*, 2007).

Concerning accidental introductions, the detection of possible candidate sites (e.g. harbours, marinas, aquaculture plants, public aquaria) and their successive monitoring should be regularly performed, since the early finding of alien seaweed species is important for effective management of the problem (Meinesz, 2007). Indeed, when these organisms have not yet formed consistent reproductive and spreading populations, it is almost certainly easier to eradicate them (Ashton *et al.*, 2012). For example, management of recreational boats, which are also a source of economic entries, should go beyond the common activities performed to avoid fouling settlement on hulls, and also inspect all the gear associated with the boat (Ashton *et al.*, 2012). Generally, boat owners are neither acquainted nor interested in the problem of alien introduction, so they do not the necessary precautions in their boat management (Ashton *et al.*, 2012). In this respect, the need for an adequate information campaign aimed at sea users and the general populace is clear, to make them aware of the problem and of its risks at all levels, facilitating the early detection of new introduced species (Aguilar-Rosas *et al.*, 2013). As an example, the prompt reply of a fisherman, informed through a brochure circulated to the population, led to the first detection of *Caulerpa taxifolia* in Tunisia (Johnson & Chapman, 2007; Meinesz, 2007). However, few examples of this kind of informed activity have been found. In California, some sporadic awareness campaigns were carried out after the finding of *Undaria pinnatifida* in some marinas (Ashton *et al.*, 2012). The Hawai'i Department of Land and Natural Resources, together with the U.S. Environmental Protection Agency, made a set of waterproof cards to hand out, not only to sea stakeholders but also to scholars, to help them in the identification of alien seaweeds (http://www.epa.gov/region09/water/oce/seaweed/alien.html). In Italy, the research project "Individuation and Monitoring of Alien Species in the Taranto seas (IMSAT)" produced a pamphlet about several categories of marine alien species, including seaweeds, which was circulated to all the Italian captaincies and to local stakeholders, to raise their awareness of this problem (Cecere *et al.*, 2005). RAC/SPA worked out an Action Plan concerning invasive species, creating informative booklets with guidelines for the control of introductions, including some of the more threatening seaweed species (http://www.rac-spa.org/publiclations#en11).

11.4.1 Policy and Laws

For effective prevention of bioinvasions, all current laws and practices (e.g. quarantine for imported live products, control of ballast water discharge, ban of potential invasive species) have to be fully implemented and enforced (Boudouresque & Verlaque, 2002b). For seaweeds specifically, there are currently no laws; only some general guidelines are present at a global scale, aimed at regulating the intentional introduction of some economically important species (Pickering *et al.*, 2007). However, several measures concerning alien species in general are present, and they can be effective for alien seaweed control. The Ballast Water Management Convention, issued in 2004 by the International Maritime Organization (IMO), addresses the hazardous introduction of "harmful aquatic organisms" by ship ballast water (Doelle *et al.*, 2007). For the control of hull fouling, only the adoption of anti-fouling paints and the cleaning of hulls out of water are recommended (Hewitt *et al.*, 2007). Australia and New Zealand were the first nations that realised the importance of a healthy sea, and formulated the governance of their maritime districts based on Ecologically Sustainable Development. Here, the quarantine of imported species and risk assessment became primary principles of sea management (Ashton *et al.*, 2012). At this moment, in the USA, the National Aquatic Invasive Species Act 2005 (NAISA) is effective for the management of aliens through partnership between the government and private stakeholders (Godwin *et al.*, 2006). In Europe, within the Marine Strategy Framework Directive (2008/CE/56, MSFD), alien species are considered one of the descriptors to be used in monitoring programs aimed at achieving the Good Environmental State (GES) designation by 2020. The rules about the introduction of alien species for aquaculture purposes (CE 708/2007) have been present for some time. Moreover, on 29th September 2014, the European Council adopted an ordinary legislative procedure, "Regulation of the European Parliament and of the Council on the prevention and management of the introduction and spread of invasive alien species" (COD 2013/0307). It was published in the Official Journal on 5th November (N. 1143/2014) and entered into force on 1st January 2015. Article 4 of the Regulation provides for drawing up *a list of invasive alien species of Union concern* to be reviewed every six years, in which all the species meeting fixed criteria have to be included (http://eur-lex.europa.eu/legal-content/EN/TXT/?qid=1415116378291&uri=OJ:JOL_2014_317_R_0003).

11.5 Conclusion

Over the last 30 years, the increase in commercial and touristic trade and the change in economic activities led to the rise of introductions of alien seaweeds, which have had, on balance, a negative impact on receiving systems (Schaffelke & Hewitt, 2007).

Nonetheless, from the analysis of the current available literature on three of the most spread alien seaweeds, the scarcity of pluriannual studies in the field to assess their actual negative or positive impacts on native communities is evident. Indeed, most of the studies were carried out in the laboratory or in mesocosms, and the reported impact was only a speculative extrapolation of results.

In addition, despite the heavy impact substantiated for a few alien invasive seaweeds, no real effective solution has been found for the prevention and the management of their introduction, either from science or policy. However, the noticeable proliferation of practices (e.g. increase of commercial trade, use of non-native aquaculture organisms) that have favoured the introduction of invasive seaweeds in most of the world seas underlines the urgent necessity of regulating such activities, not only at a national level but also and above all at an international level (Hewitt & Campbell, 2007). This is more valid in Europe where, among the state members, the free circulation of goods is warranted. In this way, goods (and thus alien species) coming from extra-European states, once entered into an EU state, can reach all others. According to descriptor 2 of the EU MSFD, *aliens must maintain a level at which they do not adversely alter the ecosystem*. However, the final goal should be to avoid their introduction in the first place, since the introduction of aliens is considered an irreversible phenomenon that, in the case of invasive species, can have effects on a geological scale (Boudouresque *et al.*, 2005).

11.6 Acknowledgements

Research carried out within the framework of the Flagship Project "RITMARE-Italian Research for the sea" (financed by the Italian Ministry of Instruction, University and Research) and within the Italian network LTER.

We thank the reviewers and the Editors for their valuable comments.

In a nutshell

- The number of introductions of alien seaweeds is continuously rising due to the expansion of commercial transoceanic trade.
- This phenomenon can be intentional, mainly concerning economically important species introduced for cultivation purposes, or accidental, concerning species either associated with other imported organisms or attached to vessel hulls.
- The main vector for the accidental introduction of seaweeds are molluscs transferred throughout the world for both aquaculture and food purposes.
- Introduced seaweeds, which have a negative impact, are called invasive. Their biological invasion can cause damage to native biodiversity, ecosystem function, and human health.
- No seaweed species can be defined as invasive in an absolute sense, because their behaviour changes in time and in space.
- Introduced seaweeds can also have positive effects, such as increasing epibiont diversity.
- Biological invasions by seaweeds can be effectively limited through the prevention of introduction and effective management of human activities that contribute to the spread of invasive species.
- No effective solution has been found for the prevention and management of alien seaweed introductions, either from science or policy.
- There is an urgent need for regulation at both national and, far more importantly, international levels.
- The education of both sea stakeholders and the general populace is strongly advisable to raise ecological awareness and vigilance.

11.7 Bibliography

Abreu, M.H., Pereira, R., Sousa-Pinto, I., *et al.* (2011). Ecophysiological studies of the non-indigenous species *Gracilaria vermiculophylla* (Rhodophyta) and its abundance patterns in Ria de Aveiro lagoon, Portugal. European Journal of Phycology, 46, 453-464.

Aguilar-Rosas, R., Aguilar-Rosas, L.E., Ávila-Serrano, G., *et al.* (2004). First record of *Undaria pinnatifida* (Harvey) Suringar (Laminariales, Phaeophyta) on the Pacific coast of Mexico. Botanica Marina, 47, 255-258.

Aguilar-Rosas, L.E., Núñez-Cebrero, F., & Aguilar-Rosas, C.V. (2013). Introduced marine macroalgae in the Port of Ensenada, Baja California, Mexico: Biological contamination. Procedia Environmental Sciences, 18, 836-843.

Ashton, G., Zabin, C., Davidson, I., *et al.* (2012). Aquatic Invasive Species Vector Risk Assessments: Recreational vessels as vectors for non-native marine species in California. Final report submitted to the California Ocean Science Trust.

Báez, J.C., Olivero, J., Peteiro, C., *et al.* (2010). Macro-environmental modelling of the current distribution of *Undaria pinnatifida* (Laminariales, Ochrophyta) in northern Iberia. Biological Invasions, 12, 2131-2139.

Boudouresque, C. F., & Verlaque, M. (2002a). Assessing scale and impact of ship-transported alien macrophytes in the Mediterranean Sea. CIESM Workshop Monographs, 2, 53-61.
Boudouresque, Ch.F., & Verlaque, M. (2002b). Biological pollution in the Mediterranean Sea: invasive versus introduced macrophytes. Marine Pollution Bulletin, 44, 32-38.
Boudouresque, Ch.F., Gerbal, M., Knoepffler-Peguy, M. (1985). L'algue japonnaise *Undaria pinnatifida* (Phaeophyceae, Laminariales) en Méditerranée. Phycologia, 24, 364-366.
Boudouresque, Ch.F., Ruitton, S., & Verlaque, M. (2005). Large-scale disturbances, regime shift and recovery in littoral systems subject to biological invasions. In V. Velikova, N. Chipev (Eds.), UNESCO-Roste/BAS Workshop on regime shifts, 14-16 June 2005, Varna, Bulgaria (pp. 85-101).
Bridgwood, S. (2010). Codium fragile *ssp.* fragile *(Suringar) Hariot summary document.* 2010. Fisheries Research Report No. 202. Department of Fisheries, Western Australia.
Bulleri, F., Airoldi, L., Branca, G.M., *et al.* (2006). Positive effects of the introduced green alga, *Codium fragile* ssp. *tomentosoides*, on recruitment and survival of mussels. Marine Biology, 148, 1213-1220.
Bulleri, F., Mant, R., Benedetti-Cecchi, L., *et al.* (2012). The effects of exotic seaweeds on native benthic assemblages: variability between trophic levels and influence of background environmental and biological conditions. Environmental Evidence, 1, 8.
Byers, J.E., Gribben, P.E., Yeager, C., *et al.* (2012). Impacts of an abundant introduced ecosystem engineer within mudflats of the southeastern US coast. Biological Invasions, 14, 2587-2600.
Campbell, M.L., & Hewitt, C.L. (2013). Protecting high-value areas from introduced marine species. Management of Biological Invasions, 4, 171-189.
Canning-Clode, J., Fofonoff, P., McCann, L., *et al.* (2013). Marine invasions on a subtropical island: fouling studies and new records in a recent marina on Madeira Island (Eastern Atlantic Ocean. Aquatic Invasions, 8, 261-270.
Casas, G., Scrosati, R., & Piriz, M.L. (2004). The invasive kelp *Undaria pinnatifida* (Phaeophyceae, Laminariales) reduces native seaweed diversity in Nuevo Gulf (Patagonia, Argentina). Biological Invasions, 6, 411-416.
Cecere, E., & Petrocelli, A. (2009). The Mar Piccolo of Taranto. In E. Cecere *et al.* (Eds), Flora and Vegetation of the Italian Transitional Water Systems (pp.195-227). Spinea, Venezia: CoRiLa, Stampa Multigraf.
Cecere, E., Petrocelli, A., & Saracino, O.D. (2000). *Undaria pinnatifida* (Fucophyceae, Laminariales) spread in the central Mediterranean: its occurrence in the Mar Piccolo of Taranto (Ionian Sea, southern Italy). Crytpogamie, Algologie, 21, 305-309.
Cecere, E., Fanelli, G., Petrocelli, A. (2003). *Undaria pinnatifida* (Laminariales, Phaeophyta): an exotic "threatener" or "enhancer" of the biodiversity in the Mar Piccolo of Taranto (southern Italy, Mediterranean Sea)? 38th European Marine Biology Symposium (pp. 179-180), Abstract book, September 8-12, Aveiro, Portugal.
Cecere, E., Fanelli, G., Petrocelli, A., *et al.* (2005). Individuazione e Monitoraggio di Specie Alloctone nei Mari di Taranto. Mottola (Taranto): StampaSud, Italy. http://www.societabotanicaitaliana.it/uploaded/1461.pdf, accessed 10 February 2014.
Chapman, A.S., Scheibling, R.E., Chapman, A.R.O. (2002). Species introductions and changes in the marine vegetation of Atlantic Canada. In R. Claudi, P. Nantel & E. Muckle-Jeffs (Eds.), Alien invaders in Canada's waters, wetlands and forests (pp. 133-148). Ottawa: Natural Resources Canada, Canadian Forest Service, Science Branch.
Chavanich, S., Harris, L.G., Je, J.-G., *et al.* (2006). Distribution pattern of the green alga *Codium fragile* (Suringar) Hariot, 1889 in its native range, Korea. Aquatic Invasions, 1, 99-108.
Curiel, D., & Marzocchi, M. (2010). Stato delle conoscenze nella laguna di Venezia di due alien species: *Undaria pinnatifida* e *Sargassum muticum*. Lavori della Società Veneta di Scienze Naturali, 35, 93-106.

Curiel, D., Bellemo, G., Marzocchi, M., *et al.* (1998). Distribution of introduced Japanese macroalgae *Undaria pinnatifida*, *Sargassum muticum* and *Antithamnion pectinatum* (Rhodophyta) in the Lagoon of Venice. Hydrobiologia, 385, 17-22.

Curiel, D., Guidetti, P., Bellemo, G., *et al.* (2001). The introduced alga *Undaria pinnatifida* (Laminariales, Alariaceae) in the lagoon of Venice. Hydrobiologia, 477, 209-219.

Doelle, M., McConnell, M.L., & VanderZwaag, D.L. (2007). Invasive seaweeds: global and regional law and policy responses. Botanica Marina, 50, 438-450.

Drouin, A., McKindsey, C.W., & Johnson, L.E. (2012). Detecting the impacts of notorious invaders: experiments versus observations in the invasion of eelgrass meadows by the green seaweed *Codium fragile*. Oecologia, 168, 491-502.

Farrell, P., & Fletcher, R.L. (2006). An investigation of dispersal of the introduced brown alga *Undaria pinnatifida* (Harvey) Suringar and its competition with some species on the man-made structures of Torquay Marina (Devon, UK). Journal of Experimental Marine Biology and Ecology, 334, 236-243.

Flagella, M.M., Verlaque, M., Soria, A., *et al.* (2007). Macroalgal survival in ballast water tanks. Marine Pollution Bulletin, 54, 1395-1401.

Fletcher, R.L., & Farrell, P. (1999). Introduced brown algae in the North East Atlantic, with particular respect to *Undaria pinnatifida* (Harvey) Suringar. Helgoländer Meeresuchungen, 52, 259-275.

Forrest, B.M., & Blakemore, K.A. (2006). Evaluation of treatments to reduce the spread of a marine plant pest with aquaculture transfers. Aquaculture, 257, 333-345.

Gagnon,K., McKindsey, C.W., &Johnson, L.E. (2011). Dispersal potential of invasive algae: the determinants of buoyancy in *Codium fragile* ssp. *fragile*. Marine Biology, 158, 2449-2458.

Garbary, D.J., Fraser, S.J., Hubbard, C., *et al.* (2004). *Codium fragile*: rhizomatous growth in the *Zostera* thief of eastern Canada. Helgoland Marine Research, 58, 141-146.

GISD (2014). *Codium fragile* ssp. *tomentosoides*. Global Invasive Species Database. http://www.issg.org/database/species/ ecology. asp?si=796&fr=1&sts=sss&lang=EN. Accessed 17 January 2014.

Godwin, S., Rodgers, K.S., Jokiel, P.L. (2006). *Reducing potential impact of invasive marine species in the northwestern Hawaiian Islands Marine National Monument*. Report submitted to Northwest Hawaiian Islands Marine National Monument Administration.

Guidone, M., Newton, C., Thornber, C.S. (2014). Utilization of the invasive alga *Gracilaria vermiculophylla* (Ohmi) Papenfuss by the native mud snail *Ilyanassa obsolete* (Say). Journal of Experimental Marine Biology and Ecology, 452, 119-124.

Gulbransen, D.J., McGlathery, K.J., Marklund, M. *et al.* (2012). *Gracilaria vermiculophylla* (Rhodophyta, Gracilariales) in the Virginia coastal bays, USA: cox1 analysis reveals high genetic richness of an introduced macroalga. Journal of Phycology, 48, 1278-1283.

Hammann, M., Buchholz, B., Karez, R. *et al.* (2013a). Direct and indirect effects of *Gracilaria vermiculophylla* on native *Fucus vesiculosus*. Aquatic Invasions, 8, 121-132.

Hammann, M., Wang, G., Rickert, E., *et al.* (2013b). Invasion success of the seaweed *Gracilaria vermiculophylla* correlates with low palatability. Marine Ecology Progress Series, 486, 93-103.

Harris, L.G., Tyrrell, M.C. (2001). Changing community states in the Gulf of Maine: synergism between invaders, overfishing and climate change. Biological Invasions, 3, 9-21.

Hewitt C.L., Campbell M.L. (2007). Mechanisms for the prevention of marine bioinvasions for better biosecurity. Marine Pollution Bulletin, 55, 395-401.

Hewitt, C.L., Campbell, M.L., McEnnulty, F., *et al.* (2005). Efficacy of physical removal of a marine pest: the introduced kelp *Undaria pinnatifida* in a Tasmanian Marine Reserve. Biological Invasions, 7, 251-263.

Hewitt, C.L., Campbell, M.L., Schaffelke, B. (2007). Introductions of seaweeds: accidental transfer pathways and mechanisms. Botanica Marina, 50, 326-337.

Hoffle, H., Thomsen, M.S. Holmer, M. (2011). High mortality of *Zostera marina* under high temperature regimes but minor effects of the invasive macroalgae *Gracilaria vermiculophylla*. Estuarine, Coastal and Shelf Science, 92, 35-46.

Hu, Z.-M., Juan, L.-B. (2014). Adaptation mechanisms and ecological consequences of seaweed invasions: a review case of agarophyte *Gracilaria vermiculophylla*. Biological Invasions, 16, 967-976.

ICES (2007). Alien species alert: *Undaria pinnatifida* (wakame or Japanes kelp). ICES Cooperative Research Report n° 283.

Inderjit, D.C., Raneletti, M., Kaushik, S. 2006. Invasive marine algae: an ecological perspective. The Botanical Review, 72, 153-178.

Irigoyen, A.J., Eyras, C., Parma, A.M. (2011a). Alien algae *Undaria pinnatifida* causes habitat loss for rocky reef fishes in north Patagonia. Biological Invasions, 13, 17-24.

Irigoyen, A.J., Trobbiani, G., Sgarlatta, M.P., *et al.* (2011b). Effects of the alien algae *Undaria pinnatifida* (Phaeophyceae, Laminariales) on the diversity and abundance of benthic macrofauna in Golfo Nuevo (Patagonia, Argentina): potential implications for local food webs. Biological Invasions, 13, 1521-1532.

Johnson, C.R. (2007). Seaweed invasions: conclusions and future directions. Botanica Marina, 50, 451-457.

Johnson, C.R., Chapman, A.R.O. (2007). Seaweed invasions: introduction and scope. Botanica Marina, 50, 321-325.

Johnston, C.A., Lipcius, R.N. (2012). Exotic macroalga *Gracilaria vermiculophylla* provides superior nursery habitat for native blue crab in Chesapeake Bay. Marine Ecology Progress Series, 467, 137-146.

Junqueira, A.O. (2013). Guilty or innocent? The need to improve the assessment of impacts caused by non-native species. Aquatic Conservation: Marine and Freshwater Ecosytems, 23, 641-645.

Kim, S.Y., Weinberger, F. Boo, S.M. (2010). Genetic data hint a common donor region for invasive Atlantic and Pacific populations of *Gracilaria vermiculophylla* (Gracilariales, Rhodophyta). Journal of Phycology, 46, 1346-134.

Klein, J., Ruitton, S., Verlaque, M., *et al.* (2005). Species introductions, diversity and disturbances in marine macrophyte assemblages of the nortwestern Mediterranean Sea. Marine Ecology Progress Series, 290, 79-88.

Levin, P.S., Coyer, J.A., Petrik, R., *et al.* (2002). Community-wide effects of nonindigenous species on temperate rocky reefs. Ecology, 83, 3182-3193.

Lyons, D.A., Scheibling, R.E. (2009). Range expansion by invasive marine algae: rates and patterns of spread at a regional scale. Diversity and Distributions, 15, 762-775.

Lyons, D.A., Van Alstyne, K.L., Scheibling, R.E. (2007). Anti-grazing activity and seasonal variation of dimethylsulfoniopropionate-associated compounds in the invasive alga *Codium fragile* ssp. *tomentosoides*. Marine Biology, 153, 179-188.

Martínez-Lüscher, J., Holmer, M. (2010). Potential effects of the invasive species *Gracilaria vermiculophylla* on *Zostera marina* metabolism and survival. Marine Environmental Research, 69, 345-349.

Meinesz, A. (2007). Methods for identifying and tracking seaweed invasions. Botanica Marina, 50, 373-384.

Mineur, F., Johnson, M. P., Maggs, C. A., *et al.* (2007). Hull fouling on commercial ships as a vector of macroalgal introduction. Marine Biology, 151, 1299-1307.

Mineur, F., Johnson M.P., & Maggs C.A. (2008). Macroalgal introductions by hull fouling on recreational vessels: seaweeds and sailors. Environmental Management, 42, 667-676.

Neill, P.E., Alcade, O., Faugeron, S., *et al.* (2006). Invasion of *Codium fragile* ssp. *tomentosoides* in northern Chile: A new threat for *Gracilaria* farming. Aquaculture, 259, 202-210.

Nejrup, L.B., Pedersen, M.F., Vinzent, J. (2012). Grazer avoidance may explain the invasiveness of the red alga *Gracilaria vermiculophylla* in Scandinavian waters. Marine Biology, 159, 1703-1712.

Nejrup, L.B., Staehr, P.A., Thomsen, M.S. (2013). Temperature- and light-dependent growth and metabolism of the invasive red algae *Gracilaria vermiculophylla* – a comparison with two native macroalgae. European Journal of Phycology, 48, 295-308.

Nyberg, C.D., Wallentinus, I. (2005). Can species traits be used to predict marine macroalgal introductions? Biological Invasions, 7, 265-279.

Nyberg, C.D. Wallentinus, I. (2009). Long-term survival of an introduced red alga in adverse conditions. Marine Biology Research, 5, 304-308.

Nyberg, C., Thomsen, M.S., Wallentinus, I. (2009). Flora and fauna associated with the introduced red alga *Gracilaria vermiculophylla*. European Journal of Phycology, 44, 395-403.

Petrocelli, A., Cecere, E., Verlaque, M. (2013). Alien marine macrophytes in transitional water systems: new entries and reappearances in a Mediterranean coastal basin. BioInvasions Records, 2, 177-184.

Pickering, T.D., Skelton, P. Sulu, R.J. (2007). Intentional introductions of commercially harvested alien seaweeds. Botanica Marina, 50, 338-350.

Provan, J., Murphy, S., Maggs, C.A. (2005). Tracking the invasive history of the green alga *Codium fragile* ssp. *tomentosoides*. Marine Ecology, 14, 189-194.

Raffo, M.P., Eyras, M.C., Iribarne, O.O. (2009). The invasion of *Undaria pinnatifida* to a *Macrocystis pyrifera* kelp in Patagonia (Argentina, south-west Atlantic). Journal of the Marine Biological Association of the United Kingdom, 89, 1571-1580.

Rueness, J. (2005). Life history and molecular sequences of *Gracilaria vermiculophylla* (Gracilariales, Rhodophyta), a new introduction to European waters. Phycologia, 44, 120-128.

Schaffelke, B., Deane, D. (2005). Desiccation tolerance of the introduced marine green alga *Codium fragile* ssp. *tomentosoides* – clues for likely transport vectors? Biological Invasions, 7, 557-565.

Schaffelke, B., Hewitt, C.L. (2007). Impacts of introduced seaweeds. Botanica Marina, 50, 397-417.

Schaffelke, B., Smith J.E., Hewitt C.L. 2006. Introduced macroalgae – a growing concern. Journal of Applied Phycology, 18, 529-541.

Scheibling, R.E., Anthony, S.X. (2001). Feeding, growth and reproduction of sea urchins (*Strongylocentrotus droebachiensis*) on single and mixed diets of kelp (*Laminaria* spp.) and the invasive alga *Codium fragile* ssp. *tomentosoides*. Marine Biology, 139, 139-146.

Scheibling, R.E., Gagnon, P. (2006). Competitive interactions between the invasive green alga *Codium fragile* ssp. *tomentosoides* and native canopy-forming seaweeds in Nova Scotia (Canada). Marine Ecology Progress Series, 325, 1-14.

Scheibling, R.E., Lyons, D.A., Sumi, B.T. (2008). Grazing of the invasive alga *Codium fragile* ssp. *tomentosoides* by the common periwinkle *Littorina littorea*: Effects of thallus size, age and condition. Journal of Experimental Marine Biology and Ecology, 355, 103-113.

Sfriso, A., Facca, C. (2013). Annual growth and environmental relationships of the invasive species *Sargassum muticum* and *Undaria pinnatifida* in the lagoon of Venice. Estuarine, Coastal and Shelf Science, 129, 162-172.

Sfriso, A., Wolf, M.A., Maistro, S. *et al.* (2012). Spreading and autoecology of the invasive species *Gracilaria vermiculophylla* (Gracilariales, Rhodophyta) in the lagoons of the nort-western Adriatic Sea (Mediterranean Sea, Italy). Estuarine, Coastal and Shelf Science, 114, 192-198.

Smith, J.E., Hunter, C.L., Conklin, E.J. *et al.* (2004). Ecology of the invasive red alga *Gracilaria salicornia* (Rhodophyta) on O'ahu, Hawai'i. Pacific Science, 58, 325-343.

Thomsen, M.S. (2010). Experimental evidence for positive effects of invasive seaweed on native invertebrates via habitat-formation in a seagrass bed. Aquatic Invasions, 5, 341-346.

Thomsen, M.S., McGlathery, K.J. (2007). Stress tolerance of the invasive macroalgae *Codium fragile* and *Gracilaria vermiculophylla* in a soft-bottom turbid lagoon. Biological Invasions, 9, 499-513.

Thomsen, M.S., McGlathery, K.J., Tyler, A.C. (2006). Macroalgal distribution patterns in a shallow, soft-bottom lagoon, with emphasis on the non-native *Gracilaria vermiculophylla* and *Codium fragile*. Estuaries and Coasts, 29, 465-473.
Thomsen, M.S., Wernberg, T., Tuya, F. *et al.* (2009a). Evidence for impacts of nonindigenous macroalgae: a meta-analysis of experimental field studies. Journal of Phycology, 45, 812-819.
Thomsen, M.S., McGlathery, K.J., Schwarzschild, A. *et al.* (2009b). Distribution and ecological role of the non-native macroalga *Gracilaria vermiculophylla* in Virginia salt marshes. Biological Invasions, 11, 2303-2316.
Thomsen, M.S., Stær, P.A., Nejrup, L., *et al.* (2013). Effects of the invasive macroalgae *Gracilaria vermiculophylla* on two co-occurring foundation species and associated invertebrates. Aquatic Invasions, 8, 133-145.
Torres, A.I., Gil, M.N., Esteves, J.L. (2004). Nutrient uptake rates by the alien alga *Undaria pinnatifida* (Phaeophyta) (Nuevo Gulf, Patagonia, Argentina) when exposed to diluted sewage effluent. Hydrobiologia, 520, 1-6.
Trowbridge, C.D. (1999). An assessment of the potential spread and options for control of the introduced green macroalga *Codium fragile* ssp. *tomentosoides* on Australian shores. Consultancy Report. Australia: Centre for Research on Introduced Marine Pests.
Valentine, J.P., Johnson, C.R. (2003). Establishment of the introduced kelp *Undaria pinnatifida* in Tasmania depends on disturbance to native algal assemblages. Journal of Experimental Marine Biology and Ecology, 29, 63-90.
Valentine, J.P., Johnson, C.R. (2005). Persistence of the exotic kelp *Undaria pinnatifida* does not depend on sea urchin grazing. Marine Ecology Progress Series, 285, 43-55.
Villanueva, R.D., Sousa, A.M.M., Gonçalves, M.P. *et al.* (2010). Production and properties of agar from the invasive marine alga, *Gracilaria vermiculopylla* (Gracilariales, Rhodophyta). Journal of Applied Phycology, 22, 211-220.
Watanabe, S., Metaxas, A., Scheibling, R.E. (2009). Dispersal potential of the invasive green alga *Codium fragile* ssp. *fragile*. Journal of Experimental Marine Biology and Ecology, 381, 114-125.
Williams, S.L. Smith, J.E. (2007). A global review of the distribution, taxonomy, and impacts of introduced seaweeds. Annual Review of Ecology, Evolution and Systematics, 38, 327-359.
Wright, J.T., Byers, J.E., DeVore, J.L., *et al.* (2014). Engineering or food? mechanisms of facilitation by a habitat-forming invasive seaweed. Ecology, 95, 2699-2706.
Wotton, D.M., O'Brien, C., Stuart, M.D. *et al.* (2004). Eradication success down under: heat treatment of a sunken trawler to kill the invasive seaweed *Undaria pinnatifida*. Marine Pollution Bulletin, 49, 844-849.
Zabin, C.J., Ashton, G.V., Brown, C.W., *et al.* (2009). Northern range expansion of the Asian kelp *Undaria pinnatifida* (Harvey) Suringar (Laminariales, Phaeophyceae) in western North America. Aquatic Invasions, 4, 429-434.

Part III. Management and Control of Biological Invasions

João Canning-Clode, Filipa Paiva

Summary of Part III

We have seen in previous sections that both accidental and intentional human-mediated introductions of species around the world have resulted in severe impacts for different ecosystems and significant global change. These changes tend to occur fast once non-native species have arrived and established in new habitats and/or ecosystems. In many cases, economic costs and other harm induced by a non-native species encourage spending vast amounts of time, money and effort to manage and control the invader. In some cases those efforts succeed, but in most cases, invaders tend to spread despite all control efforts. In this section, four chapters will cover different responses to well known invasive species in aquatic and terrestrial systems, and further discuss possible management tools and future directions to prevent and control these global environmental threats.

This section opens with an account by Frederico Cardigos and co-authors (**Chapter 12**) describing a variety of tools employed in attempts to eradicate the green algae *Caulerpa webbiana* in a small harbor located in an Azorean island. Its early detection in 2002 was disregarded, and 3 years later it was already too late to control the spread of this algae. The response to this non-native species was only initiated in 2008 when its spread was already out of control. Here, the authors describe the invasion history of *C. webbiana* and further discuss the importance of early detections of non-native species in eradication and management strategies.

Native to South and Central America, the toxic cane toad *Bufo* [*Rhinella*] *marina* is now an iconic invasive species after being transported to Northeastern Australia during the 1930s as a biocontrol for insect pests. After introduction, cane toads started to spread and become a threat to natural biodiversity due to their toxicity. **Chapter 13,** authored by Richard Shine, discusses recent research on the invasion of cane toads in Australia and reviews several attempts to control and mitigate the ecological impact of this well known invader.

The invasion of pines is considered a relevant ecological and economic issue in several regions of our planet. In **Chapter 14**, Aníbal Pauchard and co-authors evaluate the impacts, management, and policy context of pine invasions in South America. The authors encourage a more complete methodology to control pine invasions with the use of tools like prevention, early detection, containment, and population management, restoration, and the inclusion of society in all steps of this process.

Finally, the greatest proportion of marine invasions has been facilitated by the international commercial shipping industry, through the transportation of large numbers of animal and plant species in ballast water and through fouling. The breaking down of natural barriers is possible through this accidental pathway and gives species the chance to cross entire oceans easily. Ship traffic is immense, and the number of species and individual organisms transported in ballast water is enormous. The most effective form of ballast water management is through ballast water exchange. In **Chapter 15**, Mark Minton and co-authors review the current understanding of ballast water delivery and management in the United States, with particular emphasis on overseas arrivals and discharge. With these data, authors are able to describe the magnitude of ballast water discharge, its variation by geographic source and recipient regions, and differences among ship classes.

Frederico Cardigos, João Monteiro, Jorge Fontes, Paola Parretti, Ricardo Serrão Santos

12 Fighting Invasions in the Marine Realm, a Case Study with *Caulerpa webbiana* in the Azores

12.1 Invasive Marine Species of the Azores

The Azores archipelago, Portugal, is composed of nine small Atlantic islands located in the Northeast Atlantic Ocean over the Mid Atlantic Ridge. The climate is temperate and main currents include a branch of the Gulf Stream.

The first marine non-indigenous species present in the Azores probably arrived with early settlers in the 14th century. The slow sailing ships, with their wooden hulls, were highly prone to fouling by numerous species, including small invertebrates and algae. It is quite likely that the species that have long settled in these islands, and the cryptogenic species, such as the polychaete *Sabella spalanzani*, may have been passively introduced this way. Nevertheless, it is hard to determine how and when they established in the Azores due to lack of records and knowledge of existing fauna and flora in the region prior to the first settlers.

In the late 19th and early 20th centuries, the growing interest in ocean exploration produced baseline information on species diversity and distribution. In the Azores, historical and modern assessments of marine faunal and floral diversity (e.g. Santos *et al.*, 1995; Cardigos *et al.*, 2006) make it possible to trace back arrivals over the last seven decades. This is the case for the algae from the *Asparagopsis* spp. group, which arrived in the Azores around the same time period as when it was first recorded in Europe (late 1920s). Like many other marine species, it is likely that the global increase of maritime traffic in the period between the World Wars contributed to the geographic spreading of this alga.

When considering the likely causes, the first surge of marine non-indigenous organisms arrived with the first settlers, the second movement happened in the beginning of the 20th century associated with the increase of maritime traffic, and the most recent "third wave" of marine non-indigenous species arriving in the Azorean islands is probably linked to the development and growth of world cruise yachting over the last four decades. Unlike the species conveyed in the first two, which were more Europe-centric, species arriving in this third wave have a Caribbean faunal affinity.

Species reaching the Azores in the 21st century seem to be correlated with climatic changes and tropicalization processes (Afonso *et al.*, 2013).

Non-indigenous marine species in the Azores have been listed by Cardigos *et al* (2006). Over the last years, several authors (*e.g.* Afonso *et al.*, 2013; Amat & Tempera,

2009; Cordeiro *et al.*, 2013; Malaquias *et al.*, 2009; Pola *et al.*, 2006; Torres *et al.*, 2010; Torres *et al.*, 2012) identified new occurrences that will certainly increase the list of "alien" species in the Azores, and have studied their routes (Micael *et al.*, 2014).

12.2 History of an Invasion - *Caulerpa webbiana*

The green algae *C. webbiana* (Figure 12.1) was first collected in Faial in 2002. As it was rare and unfamiliar, no special attention was given. Only in 2005, after a significant increase in the total area it covered, a research plan to investigate its biology and ecology was implemented (Amat *et al.*, 2008).

Fig. 12.1: *Caulerpa webbiana* in Faial island (J. Fontes).

Even though this species is not considered invasive elsewhere, the invasive reputation of the genus (Lowe, 2000) and the exponential growth of *Caulerpa* patches made it a matter of concern to local scientists and decision makers. Surveys conducted in the archipelago (2005–2007) suggested that *Caulerpa* was restricted to Horta Bay, in Faial Island. The presence of *C. webbiana* in this limited area surrounding the main harbor and marina of Faial Island, and the higher abundance of the algae in the immediate proximity of the harbor, suggested that this non-indigenous alga arrived using a sailboat or ship as vector (Amat *et al.*, 2008). The preliminary use of molecular tools and genetic analysis did not clarify the origin of the *Caulerpa webbiana* found in Faial (Carreira, unpublished data). The local success of *Caulerpa webbiana* in Faial Island is probably related to tolerance of local temperature ranges, fast growth rate, high re-colonization rate, and its anti-grazer toxicity (caulerpenine) (Amat *et al.*, 2008).

Between 2002, when *Caulerpa webbiana* was first seen, and 2008, the distribution of the algae significantly increased, indicating fast spread and efficient proliferation. In just a few years, *C. webbiana* expanded its boundaries from a few small patches in the outer breakwater of the harbor to over 9,900 m^2. Abundance and cover ranged from thick dominating carpets (where it first settled and in adjacent areas) to sparse variable size patches (in the limits of the local distribution). Observations and data from comprehensive surveys suggested that the alga was not only out-competing and over growing other sessile organisms, but also expanding its earlier distribution limits, in depths ranging from 5 m to 50 m. Facing the possibility of having a scenario similar to that of *Caulerpa* proliferation in the Mediterranean, the need to take action against *Caulerpa webbiana* was clear. The risk and stakes were too high to be ignored. In early 2009, a program for the active control or eradication of the *C. webbiana* population in the Azores was set in motion.

12.3 Planning the Mitigation

Developing a response to a scenario where *C. webbiana* became a threat to local organisms started in 2005, when the newly arrived alga was first clearly identified. The difficulties of eradicating exotic species, especially if one considers the additional challenge of working in the underwater environment, made it clear that an effective response with the hope of eliminating *Caulerpa webbiana* would be labor intensive, time consuming, and with a steep price. Despite the limitations, efforts and resources were allocated to the planning of a strategy and a response program.

This response program was developed to actively intervene and remove *Caulerpa webbiana* from infested areas. Even though the program was primarily focused on tackling the growth and expansion of *Caulerpa webbiana*, it was clear that researchers and scientists had to be involved in the battle. While commercial and scientific divers tested and developed different methods to remove and, if possible, terminate *Caulerpa*, biologists verified the efficiency of the methods and, with the help of volunteer divers, checked the progress of the operations and monitored the infestation and distribution of *C. webbiana* in Horta's Bay and surrounding areas. The framework of the response program reflected its focus: to **mitigate the impact of the proliferation of a newly arrived species,** with the most efficient approaches and with the lowest possible impact on local fauna and flora.

The program was divided into two major work packages, each of them with two main tasks:

Work package I – Intervention

Task 1: Technological and method development

The development and adaptation of available approaches provided technical tools and guidelines to safely remove or terminate *Caulerpa webbiana* from infested areas.

Task 2: Mitigation

Intervention and actions to remove *Caulerpa* from infested areas, making use of and applying tools and methods developed in WPI-T1. During the program, priority areas for intervention were determined based on multiple factors, including available resources, weather and maritime conditions, and the periodic assessment of the *C. webbiana* population (WPII-T2)

Work package II – Monitoring

Task 1: Efficiency and impact of techniques

Custom-designed experimentation tested the efficiency of different approaches and techniques in removing *C. webbiana* and provided a basic assessment of impacts on other organisms.

Task 2: Proliferation status of *Caulerpa webbiana*

Periodic monitoring and mapping of the presence of *Caulerpa webbiana* in infested and surrounding areas to assess the progress of mitigation actions and the local distribution and abundance of *Caulerpa* (Figure 12.2). Results were considered relevant in deciding priority areas of intervention in WPI-T2.

Fig. 12.2: Monitoring *Caulerpa webbiana* using 25 × 25 cm fixed photoquadrats.

Since the very early stages it became clear that it would not be possible to assure the eradication of *C. webbiana* from Faial. The assessment of *C. webbiana* distribution and abundance was crucial in determining where and when to allocate effort in removing *Caulerpa* from a given area. The speed at which *Caulerpa* was growing and spreading to surrounding areas was overwhelming, and the considerable amount of manpower and resources used in removing and terminating the alga was simply not enough.

The periodic mapping and monitoring of *Caulerpa webbiana* in areas surrounding Horta relied to a great extent on the work of volunteer divers and information provided by local stakeholders. By establishing the boundaries of *Caulerpa* and areas of exponential growth, it was possible to continuously manage the available resources and assign them to different "battle fronts": (1) inside the harbor, to avoid the spread of the algae to other islands through maritime traffic; (2) the marine protected area of Monte da Guia; and (3) areas outside previously known limits of distribution, where new reported colonies were quickly terminated in a struggle to prevent the spreading of the *Caulerpa* infestation.

12.4 Mitigating and Tech Development

The first approach to *Caulerpa webbiana* removal included the analysis of methods and principles described in relevant literature and several other empirical methods. Of these, some were later rejected and others improved. The main constraints on the application of some of the techniques, as explained in detail afterwards, were the resistance of the algae and the feasibility of the inherent operation.

Manual removal coupled with suction was the first method tested. It was rapidly rejected due to low removal rates and the extensive fragmentation and fragment dispersal that resulted.

Following the current approach to *Caulerpa* eradication used in NSW Australia (NSW, 2004), **hyaline shock** experiments were performed under laboratory conditions to test the survival rates of *Caulerpa webbiana* in concentrations ranging from 0 to 100‰. None of the tested treatments resulted in mortality of the algae after a week of exposure.

Like any other alga, *Caulerpa webbiana* depends on sunlight as its main energy source; thus, permanently preventing sunlight from reaching the alga should result in its elimination. Although preliminary tests using **smothering** showed that *C. webbiana* would not survive two weeks of light blocking, this approach proved to be unpractical in the field due to the difficulty of covering large areas of roughed high relief underwater substrates with tarp. Even worse was the difficulty of holding the tarps in place for two or more weeks due to the strong ocean agitation present throughout most of the year in the Azores. Though this method was abandoned early on, this principle was later used with a different approach (see "Sand cover").

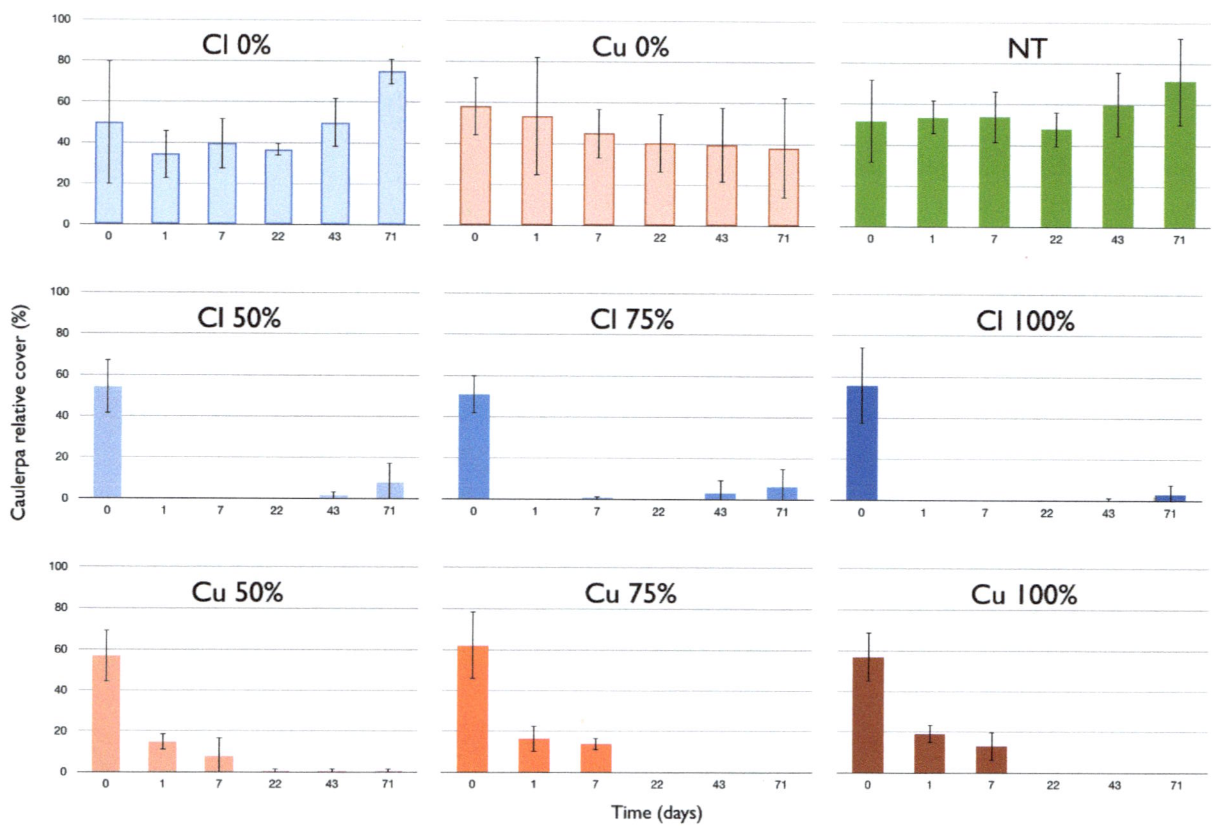

Fig. 12.6: Efficiency of CS and Chlorine treatments for long term removal (t_0 = 0 days; T_1=1 day; t_2=7 days; t_3=22 days; t_4=43 days and t_5=71 days) of *C. webbiana* at variable concentrations. Cl 0% and Cu 0% refer to control, 25%, 50% and 100% of Cl represent treatment at constant concentration (ten 50g tablets) at 2.5, 5, and 10 minutes exposure respectively. 50%, 75% and 100% Cu represent 90 g, 135 g and 180 g Cu per capsule respectively.

The rationale behind the **thermal stress** method is to expose the algae to temperature shock. Several rocks with patches of *C. webbiana* were removed from their natural locations and exposed to sea water at 68°C for approximately 10 seconds. They were immediately returned to their original location along with the control patches. After a week, and in contrast to the control patches, the treated rocks were completely free of the algae as well as any other macroalgae or invertebrates.

This experience has shown that thermal stress has the potential to be used to remove *C. webbiana* with minimal handling and reduced risk of fragmentation. Large scale application of this method would, however, require substantial technological development. Our search on commercially available technology was unable to find an applicable or adaptable off-the-shelf solution. The most important technical challenge lies in the difficulty of generating a large enough volume of warm seawater and to move it into the correct depth with low heat loss and low pressure to avoid fragmentation. To overcome these challenges, designing and producing a prototype with appropriate characteristics would be required. For this, a partnership with a commercial or technological partner would be needed and the necessary funds raised.

Like most invasive organisms, native *C. webbiana* density is controlled by herbivores that coevolved with the algae and can tolerate the anti-herbivory toxin it produces. The key concept of **biological control** includes the use of organisms that feed on the invading organism and keep its density low by controlling its biomass.

The information available indicates that there are no Azorean native organisms that could ingest significant amounts of *C. webbiana*. However, previous *in vitro* short duration experiments revealed that the sea urchin *Paracentrotus lividus* ingests *Caulerpa taxifolia* (Ganteaume *et al.*, 1998). Boudouresque *et al.* (1994) found that the alga *Caulerpa prolifera* intake is not avoided during winter-spring. With this information in mind, an experiment was designed to test if *Paracentrotus lividus* and *Sphaerechinus granularis,* naturally present in the Azores, would feed on *C. webbiana* when no other algae were available. The inclusion of *S. granularis* was due to the fact that this species has been observed in *Caulerpa*-dominated areas, and they were suspected of feeding on them. In order to test this hypothesis, sea urchins were placed in three cages (225 cm^2) and three blank cages with no sea urchins were also placed on the bottom (negative control). *Paracentrotus lividus* showed some consumption of *Caulerpa*, but preferred to scrape the substrate under the alga cover by lifting patches of *Caulerpa*. After 2-3 weeks all urchins began to die (with mortality rates up to 50%). The option of using biological control with these species of sea urchins was dismissed after this experience.

The first method selected to remove *Caulerpa webbiana* in the Azores was the CNRS (*Centre National de la Recherche Scientifique*) patented method, developed to control the expansion of *Caulerpa taxifolia* in Mediterranean marine protected areas (Uchimura *et al.*, 2000). This method is based on the lethal effect of **copper sulfide** (CS) on the Caulerpales (Uchimura *et al.*, 2000; Guillén *et al.*, 2003). This decision was supported by the promising results of pilot experiments conducted in the Azores between 2007 and 2008.

The pilot study and evaluation stage was followed by the scaling up stage. At this point it became clear that scaling up the *Caulerpa webbiana* removal using this methodology would be very difficult to achieve due to the high labor intensity involved in the cover recovery, preparation, and the divers' limited carrying capacity and high exposure to copper sulfide plumes during deployment (Figure 12.3). Other drawbacks were the difficulty of packing covers in water-tight bags containing minimal air (to be handled underwater) and the massive production of plastic waste.

Fig. 12.3: Plume released while positioning the blanket impregnated with copper sulfate.

In order to overcome these limitations and drawbacks, multiple innovations were tested and introduced. To reduce cover preparation labor, increase diver carrying capacity, and reduce exposure to copper sulfide plumes, the copper sulfide solution was replaced with dry micronized copper sulfide contained in watertight capsules that were activated under the cover by the divers only after the cover was secured in place (with velcro). Each capsule was filled with approximately 180 g of micronized CS, and one capsule was used per square meter of cover.

Once activated, the capsules became permeable to seawater and the copper sulfide started to slowly dissolve, resulting in a CS-rich atmosphere under the tarp. After four hours, the covers could be redeployed over the next patch of *Caulerpa* or at a different site with no need for additional preparation other than repositioning a refilled set of capsules. Used capsules were collected and brought to the lab to be refilled with CS, closed, and packed in mesh bags for the next deployment. The "refills" were made from 7 cm diameter PVC pipe with 10 holes (3 mm) in both ends across its diameter, and closed on one end by a nylon lid and on the other end by a screw-on PVC lid (Figure 12.4). Two wide rubber rings (sections of reused bicycle air tube) covered the holes, making the capsules watertight. These rings were displaced by sliding them to the center section when placed under the tarp, allowing seawater to flood the capsule and slowly release CS. The covers were composed of two main materials: the upper face and the structural section is made of traction-resistant plastic tarp sewed to a geotextile (used in construction) section in the lower face. The edges of the cover were fitted with several stainless steel washers.

Fig. 12.4: Bags with copper sulfide "refills".

These innovations resulted in a significant increase in productivity and greatly reduced diver exposure to CS. Diver carrying capacity increased 5 times for experienced divers, from 4 m^2/dive to 20 m^2/dive.

In addition, this method allows the manipulation of larger covers, which can be custom made depending on the characteristics of the area and size of *Caulerpa webbiana* patches. Various sizes were used, ranging from 0.5 × 0.5 m to 4 × 18 m. Simultaneously, an anchoring system was developed to effectively and quickly secure the covers to irregular, high relief rocky substrates. These elastics lines have metal hooks at the ends for easy anchoring to irregularities on the substrate, and a stretcher to adjust the elastics' length according to the distance between anchoring points.

In soft rock areas, steel nails with rubber washers applied on reinforced eyelets along the edges were used to secure the covers. Rocks or other heavy objects could also be opportunistically used to help secure the covers.

In summary, this method significantly reduced complexity and increased productivity and safety when compared to the CNRS method. The basic steps can be summarized as follows: 1) securing tarps over the treatment area using steel nails or elastics; 2) inserting and securing activated (holes exposed) CS capsules under the covers using the velcro attachments; 3) after 4 or more hours, recovering empty capsules and moveing covers to the next area; 4) securing the covers over untreated patches and placing refilled CS capsules. This procedure can be repeated indefinitely in large, heavily infested areas, reusing covers with no need to surface for cover preparation between deployments. Preparation between deployments is essentially limited to refilling capsules with CS.

Although the CS cover method was very effective for large infested areas and large isolated patches, this was not the case for dispersed small patches in unmapped areas. To deal with small and dispersed patches of CW a precision removal method was developed based on the California *Caulerpa* eradication experience. In the California case study, **chlorine** was used with success as a lethal agent and primary eradication method by covering large areas colonized by the algae and pumping in chlorine from land reservoirs (Williams & Schroeder, 2004).

After testing the lethal effect of chlorine on *Caulerpa webbiana*, it was concluded that it could be efficiently used as lethal agent. Based on this, an apparatus capable of delivering a controlled and localized stream of chlorine-saturated seawater to small patches was developed and used extensively (Figure 12.5). The apparatus, a chlorine pump, consisted of a pressurized container where solid chlorine tablets (ten 50 g tablets) were added to seawater. The chlorine tablets dissolve gradually in sea water resulting in hypochlorite. The flow of chlorine was controlled by the diver and the controlled flow was applied one centimeter above the patch, spaying the desired area until the alga started to bleach. The pressurized chlorine reservoir allowed a constant flow of chlorine at any depth without effort. The pressurization was obtained by connecting a 4 liter SCUBA tank with compressed air (up to 200 bar). The air pressure was reduced in two stages. First, a SCUBA first stage diving regulator reduced pressure to 9 bar, then a second pressure gauge (used in industrial butane gas facilities) further reduced the pressure to the final 3 bar. Additionally, there was a stopcock to prevent the entry of chlorine in the compressed air system when depressurized.

The chlorine reservoir could be recharged during the dive as many times as necessary by closing the air inlet, opening the reservoir and allowing seawater in and air out. Additional chlorine tablets could be added when previous tablets had lost 80% of their initial volume. Additional tablets were carried in 50 cm long PVC capsules with PVC screw caps on both ends. The continued use of these devices, in combination with the corrosive characteristics and oxidative chlorine from sea water, requires regular inspection and maintenance.

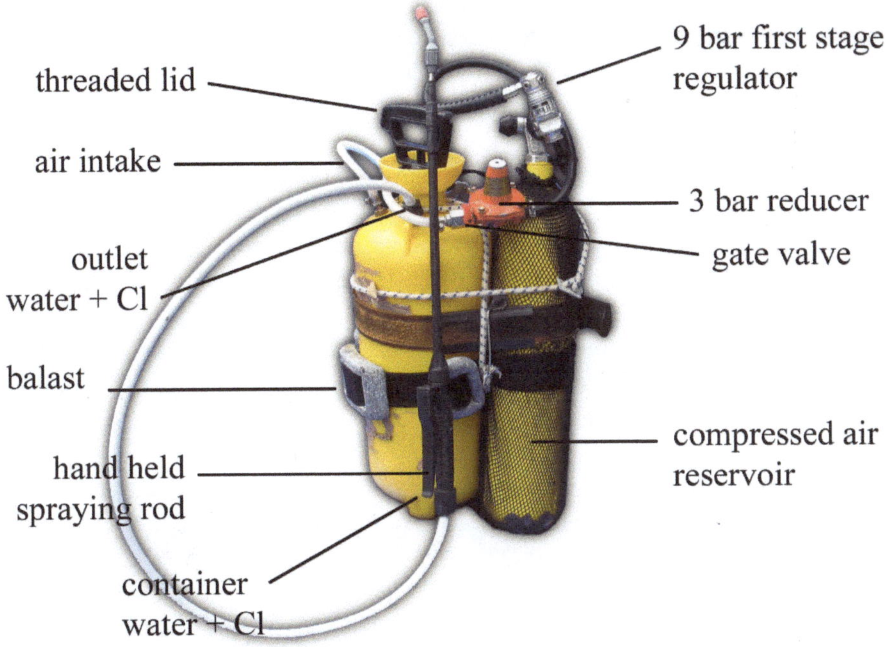

Fig. 12.5: Chlorine pump schematic.

Due to the toxic and corrosive characteristics of chlorine and the increased risk of exposure to this chemical, it is necessary to provide divers with adequate means of protection. The use of full body wet or dry suits, gloves, and full face masks reduces exposed area and protects divers from rashes. Likewise, it is also necessary to use adequate protection when moving, preparing, and servicing chlorine pumps and related equipment in air. For **safety** reasons, the handling of chlorine and chlorine equipment requires the use of gloves, a full body impermeable suit, and a full face mask with specific filtration for chlorine compounds.

With the continued use, the seals in the system, particularly on the reservoir lids, may allow some chlorine gas to escape during transport. This issue may be tackled with the use of a more robust and pressure-resistant container (aluminum or Plexiglas, for example).

Efficiency of different concentrations of both CS and Chlorine were tested in order to determine the lowest possible concentration that still produced lethal effects (Figure 12.6). These experiments led to a 50% reduction of the initial amount of CS used per cover area while still achieving lethal effects. The initial concentration was equivalent to the amount of CS used in the solution necessary to spray a given area of cover with the CNRS method.

The use of **sediment** (mostly marine sand from dredging) to smother *Caulerpa webbiana* over large infested areas was also tested as a potential method. The idea was based on the simple principle that, like any alga, *Caulerpa* requires light to survive. The concept is to simply cover large areas infested with *C. webbiana* with sand and other aggregates, blocking the sunlight until it dies. This method was effective in removing *Caulerpa* over large areas using cheap dredged sand.

Approximately 3,920 m^3 of sand were deposited in an area of 3,000 m^2. A visual assessment of the site before, immediately after, and four weeks after the deposition of sand showed that the method effectively eliminated *C. webbiana* from low relief areas, but was not equally efficient in high relief areas, because the sand did not cover boulder tops. High relief areas require greater amounts of sand delivered per unit area. The abrasion caused by the settling of dropped sand does not appear to cause significant fragmentation of the seaweed (although this risk must be considered).

In summary, this method is potentially useful if the following conditions are met: i) sheltered area; ii) flat areas; iii) densely colonized by the alga.

12.5 Public Involvement

There was an effort to inform the general public about alien species and, in particular, about the *Caulerpa webbiana* issue. The main objectives were:

- Emphasize early detection of newly invaded areas or the arrival of new alien species;
- Inform about the different tasks that were being carried out, thus also justifying the financial effort;
- Raise awareness about marine environmental problems.

Posters, placards, public sessions, and a webpage were produced over the years. In order to disseminate the objectives and results, there was also an effort to communicate at an international level.

Multiple versions of posters advertising the problem of invasive species were developed, always trying to be as precise, objective and engaging as possible. Distribution started with the beginning of the bathing season, taking advantage of a larger target population and greater public attention to this topic in order to maximize the outreach and its usefulness. Posters were also affixed in various public places such as cafes, clubs, etc. Enterprises and associations related to underwater activities and local communities were also used as communication and distribution vectors, using posters and graphic information.

Placards were placed near SCUBA diving and whale watching enterprises. Dive center receptions held informal public awareness sessions addressed mainly to employees and included the distribution of advertising material. Sessions were held on several islands of the Azores.

Drawing on the potential for dissemination through the Internet, a webpage was created and some videos of the species and the work carried out were shared. All content was made open access. The website was created at www.caulerpawebbiana.com to facilitate the communication of observations of *C. webbiana* to the community. The information was updated with news and images that gave an account of the developments, major events, and progress of eradication efforts. The messages sent to the site were automatically directed to the coordinator of the project in order to optimize the speed of response. Despite the fact that the priority was to monitor, explore, and test new eradication methods and eradicate the algae, effort was still made to ensure the website was regularly updated.

The team involved in the eradication activities participated in the World Conference of Marine Biovdiversity 2011 that took place in Aberdeen. The 5-minute video shown during this event summarized the objectives, methods, and results achieved so far. Material was also provided for a documentary produced for “Sentinelles de la Nature”, featuring Cecille Favier and Aymeric Alardet and produced by the Films Concept Associés series.

12.6 Fighting *Caulerpa webbiana*

After the first few months of intervention to eradicate the green algae *Caulerpa webbiana*, the boundaries of distribution, both North and South, had been substantially retracted. In the South, the decrease was due almost exclusively to the intensive action of the intervention team, which focused its activity on this front. In the North, the retreat of the invasive algae was due to natural causes, probably a combination of relatively low temperatures (14–15°C) and the severe and persistent wave propagation caused by the particularly harsh winter. Probably for the same reason, there was a very significant reduction in the biomass of *C. webbiana* on the North coast.

During winter, when the weather and sea conditions make it impossible to work in more exposed areas, the team focused its efforts on the dock. The effort put into eradicating the invasive *Caulerpa webbiana* inside the port of Horta resulted in a drastic reduction of the percentage coverage in higher density areas, which included the docking area of deeper draft vessels (potential vectors for dispersion). Contrary to initial expectations, even in relatively protected locations, the strong sea waves that were felt during the entire winter and early spring prevented any mission to eradicate off port. It was not possible to resume eradication efforts in the outer areas until the beginning of April.

After the improvement of weather conditions in the spring, the work was concentrated in the outer areas. During the four summer months, the work was, in general, fruitful.

An important portion of the infected area was located on a slope ranging from 30 to 43 meters deep. There, the progression and treatment of work has been very slow.

Areas previously treated were recolonized and the biomass of *C. webbiana* has increased again. There were portions treated three consecutive times since, and after each passage of time, new small colonies, more difficult to locate, have emerged. In just two weeks after the application of treatments, it was possible to find new colonies with coverage of about 20–30 cm^2. The alga has demonstrated its great vitality by resettling in only a few months in areas where it had been naturally eliminated during the winter period.

Due to this process of reappearance and recolonization (see Figure 12.7), and in accordance with the strategy and priorities established at the beginning of the control and eradication program, efforts and available resources were focused on the survey and eradication of new colonies that appeared in the extreme of the distribution range. By the end of 2010 it was possible to eliminate all new colonies identified, pushing the distribution to the previous year's limits.

Some volunteer work was used, especially in the actions concerning the detection of new colonies. The objective of the exploratory missions was to inspect vast areas within the limit of distribution and beyond, and in areas of intervention from the previous year. Colonies of *Caulerpa* were marked and mapped for future treatment.

Missions had to be meticulously planned because of the size of the zone, the irregular bottom (homogenous, with only a few elements useful for orientation), and dispersed colonies of *Caulerpa*. In these areas, mobile guide lines were placed to create "corridors" that divers inspected. The 100 m lines were placed perpendicular to the depth gradient and, as the "corridors" were inspected and treated, the mark lines were transferred from deeper to shallower zones. This method, although time consuming, proved to be effective.

Sand was tested as a smother control method for *Caulerpa* during the summer of 2011. The method has limitations, especially in sloppy areas. Although it is effective when it fully and permanently covers *Caulerpa* colonies in flat bottoms, colonies covering large blocks survived in the test zone.

After this experiment, by the end of 2012, 137,000 m^3 of sand was used to cover an area of 30,000 m^2 that was highly contaminated with *Caulerpa webbiana*. As expected, the alga disappeared in the covered area. As it represented an important part of the total area, the contamination was greatly reduced. The impact on the community was very high; nevertheless, the overall outcome was positive.

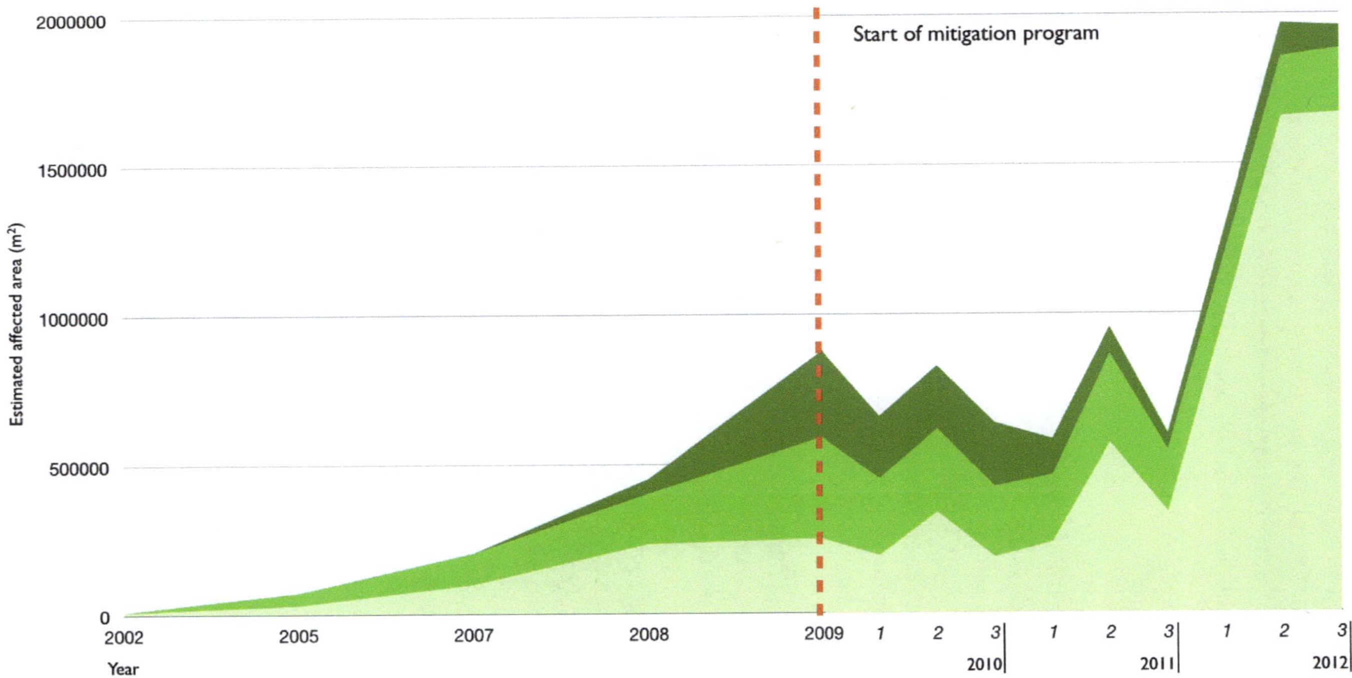

Fig. 12.7: Variation of *Caulerpa webbiana* abundance in Faial over time. In the first months of 2011 there was intervention against the alga. In 2012, the decision to cover the alga with sand was made (light green represents low density areas, medium green stands for medium density areas and dark green represents high concentration areas.

12.7 Lessons Learned

According to Amat *et al.* (2008), the green algae *Caulerpa webbiana* presents no invasive characteristics in other places where it occurs (eg. Madeira and the Canary Islands). There is probably a natural control factor in these other locales, such as predation, that is absent in the Azores. Without human intervention, the expansion and increase in coverage of infested areas of the Azores could achieve disastrous proportions (Figure 12.8, 12.9).

This program was started in order to control and, if possible, eradicate *C. webbiana* in the Bay of Horta and surrounding areas. As shown, the growth rate, resistance, toxin production, and vegetative reproduction dramatically enhance the expansion of the distribution of this alga, making complete eradication effectively impossible.

As observed in the Canary Islands (Haroun *et al.*, 1984), this seaweed has a seasonal growth cycle and is more vulnerable in winter and more resistant during the summer time – this is also true in the Azores. This seasonal cycle has a clear impact on its capacity for expansion and growth, with obvious increases at the height of spring and summer.

The aim of eradicating the algae became increasingly difficult, and to control its growth demands continuous action and high human and material resources. Throughout the initial phase of the program, it was found that, with the available means and techniques, control of *C. webbiana* on Faial Island was beyond reach. After introducing several technical improvements, the program was successful in mitigating and containing this invasive alga in the vicinity of the Bay of Horta.

In 2012, several tons of sand were placed over the densest area of *Caulerpa webbiana*. This action, combined with the previous actions that had constrained the green algae to the limits of the surroundings of Horta bay and, probably, the low water temperatures registered in the winter of 2012/2013, was crucial to the substantial reduction of this local population.

This program was crucial to limit the expansion of *Caulerpa webbiana* on Faial Island. From this perspective, despite the failure to eradicate the algae within the Bay of Horta, the program played a key role in mitigating the problem and in helping the preservation of local biodiversity and biotopes, as well as in controlling the expansion of the area affected by this proliferating algae.

If the program is to continue, especially if the pattern of recolonization continued, it would be mandatory to verify if there was any contamination or accumulation of chlorine or copper along the trophic chain or substrate, and if the permanent use of these chemicals would influence the condition and ecological succession of biotopes. The methods used were those showing the best results and proved most suitable for the eradication and control of *Caulerpa*, but the toxicity and pollutant nature of chlorine and copper sulfide should not be neglected.

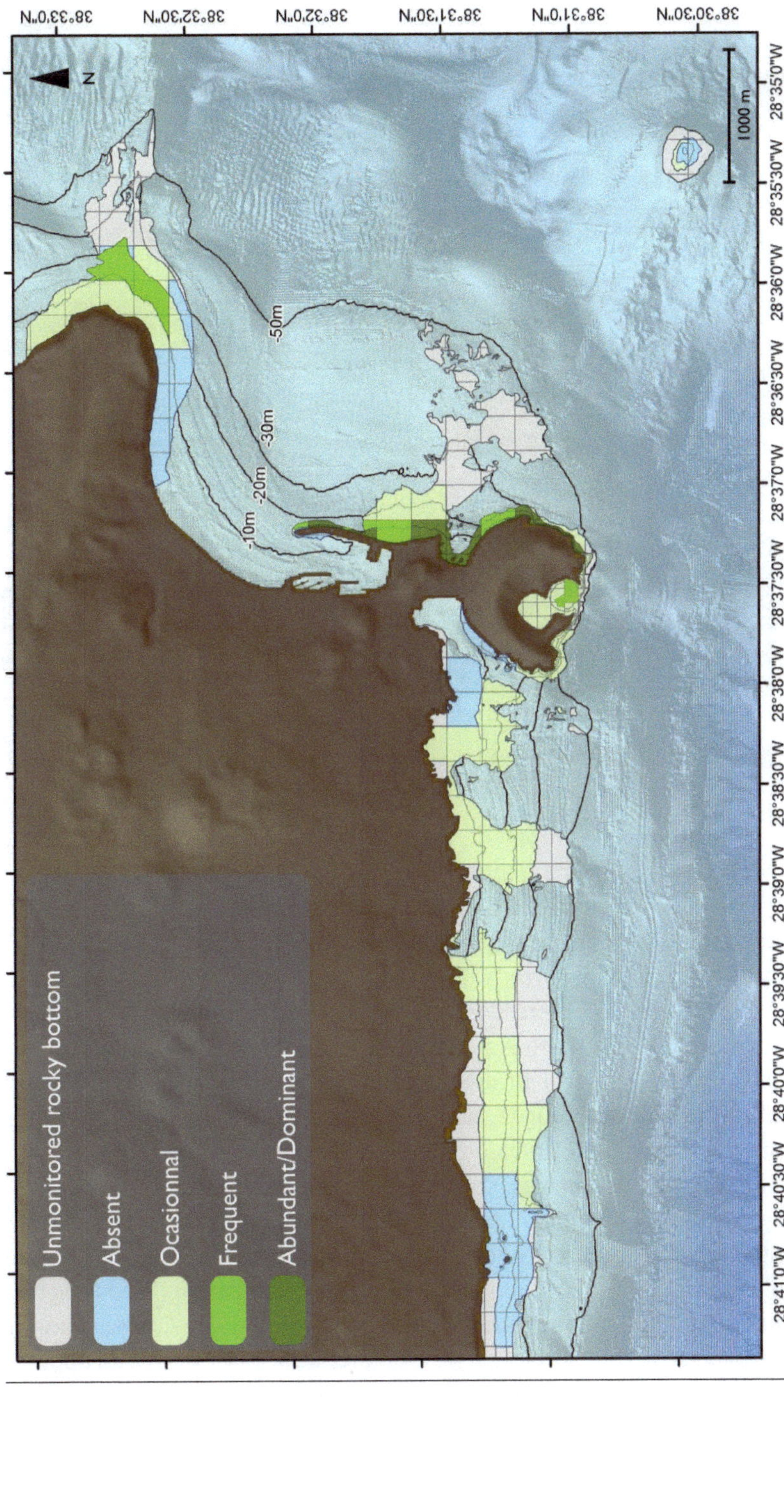

Fig. 12.8: Southeast Faial Island and *Caulerpa webbiana* invasion maximum.

Fig. 12.9: *Caulerpa webbiana* expanding.

Working in the marine environment, the rate of growth and recovery of the alga, the limited work possible during winter time, and the lack of appropriate technologies were obvious constraints to the control and eradication of *Caulperpa webbiana* in Faial. Nevertheless, the greatest handicap this program faced was lack of resources. The availability of more resources would have eliminated these limitations and would have enabled the appropriate response to the scale of the problem.

It is important to note that in similar cases of proliferation and invasion by algae of the genus *Caulerpa*, almost all similar initiatives were less successful in controlling the growth of proliferating algae than this program. There are two exceptions: the case of *Caulerpa taxifolia* introduction in Agua Hedionda Lagoon (California), and in Huntington Harbor, Los Angeles (California), in the late 1980s.

The program has implemented operational improvements in order to adjust and adapt to the existing resources and has developed new methods and techniques that tackled the needs, conditions, and fulfilled the main mitigation objectives. It must be understood that a mitigation plan is not intended to eliminate *Caulerpa* and that the problem will persist. Therefore, such a program should be viewed as an approach that should be maintained continuously.

Their toxicity as pollutants and the other unhealthy characteristics of the chemicals used as algaecides point towards the need for research and technological development of effective alternative methods. This should be considered a priority. Thus, it is recommend that support be given to a project, parallel to and independent of other initiatives, solely focused on the research and development of new approaches and tools to combat biological invasions.

Based on the experience gained during the program, it is possible to compile general high level recommendations to define the strategy and future plans of action against species that constitute a threat to biodiversity and local biotopes with associated environmental, ecological, or socio-economic consequences:

a) Invest in early warning;
b) Discard non-intervention;
c) Do not ignore the risks associated with the proliferation of exotic species;
d) Consider the response and treat bioinvasions as a priority;
e) Adapt the existing means and available resources while creating and implementing an operational plan for eradication.

Morphology, weather conditions (including ocean dynamics), remoteness, and isolation are extreme in the Azores. This means that the strategies employed and lessons learned in this archipelago are, most probably, effective elsewhere. In particular, the high level considerations stated earlier should be used regardless of geography or ecosystem.

12.8 Next Steps

Fortune has played a major role in this matter. The availability of sand and the extreme winter weather between 2012 and 2013 were fundamental to the observed decrease in *Caulerpa webbiana*.

The future must necessarily include a continued monitoring plan and action must be taken to actively control invasive species. These are things repeatedly mentioned by scientists (Micael *et al.*, 2014).

As in many places around the world, and partially as a result of the delay in taking action against *Caulerpa webbiana*, decision makers in the Azores are presently taking invasive species seriously. Recommendations, such as the ones offered by Ojaveer *et al.* (2014), are adapted here and should be followed in the Azores, including:

- The Regional Government should have staff with understanding of marine invasion processes. This is valuable and should be maintained;
- Outsourcing and hiring experts and consultants whenever needed;
- Guidelines to deal with NIS should be developed involving decision makers, local authorities, scientists, and stakeholders. Early detection and monitoring should be one of the main targets of these guidelines, which will promote standardized

approaches for data collection, compilation, and information systems, to allow fast response and action when dealing with new NIS;
- Among the usual vectors that promote the entrance of NIS, only vessels and ships seem to be particularly important in the Azores. Nevertheless, the movement of live bait for tuna fisheries might be responsible for the entrance of *Diplodus vulgaris* in the Azores (Afonso *et al.*, 2013). Those should be particularly carefully inspected;
- The work carried out in the Azores, despite being chosen by the scientific community and available funding for research, covers the use of the three indicators expressed in the Commission Decision 2010/477/EU. Namely, (1) there is an initiative for the surveillance of marine alien species; (2) accurate species lists for fish, macroalgae and mollusks are being created; and (3) there is an effort dedicated to detaining invasive species, as was the case with the *Caulerpa webbiana*. Continuous support should be given to this work;
- In the Azores, all the typical and foreseen vectors for marine species introductions are occasionally studied by marine ecology, marine biology, and fisheries-related research (Cardigos *et al.*, 2006; Afonso *et al.*, 2013; Cardigos *et al.*, 2013; Micael *et al.*, 2014, among others). This includes ballast water analysis, boat hull incrustation identification, biogeographic studies, and others. A large-scale monitoring program was proposed, but is not yet funded;
- There is a movement in the Regional Government of the Azores to grant all the authority over maritime matters to only one regional directorate. These efforts are being made to centralize information and decision making processes concerning marine and maritime matters, which will provide better response times when dealing with NIS and marine invasion threats. This effort should be kept and reinforced.

12.9 Acknowledgements

Jaen Nieto Amat was a consultant in the development of the *Caulerpa webbiana* eradication program of the Azores and we are grateful for her assistance and for the work she produced. We respectfully acknowledge the support and commitment provided by all the professional and voluntary divers that have helped implement the "brigade anti-*Caulerpa*" monitoring program and helped fight this invasive alga on Faial island. Emanuel Arand and *Imag*DOP/UAz developed Figure 12.5. All photos are from the *Imag*DOP/UAz photo image bank. We thank the Harbour of Horta administration and collaborators, including the President, Eng. Fernando Nascimento, and Eng. António João.

In a nutshell

- The fight against *Caulerpa webbiana* started in 2008 in the Azores.
- If awareness and proper reaction methods were in place, the fight could have started several years earlier.
- Even with a huge effort, it now seems too late to eradicate the new population. This alga will likely remain on Faial Island, and the resultant ecological damage is significant.
- Only control actions preventing spreading to surrounding areas, seems feasible.
- This calls for an increase in awareness, in the implementation of early warning and rapid response, and in financial commitment to fight marine invasions.
- Fighting against *Caulerpa webbiana* can be done efficiently using different methods according to the specific context (localization and density, mainly) as discussed in this chapter.

12.10 Bibliography

Afonso, P., F.M. Porteiro, J. Fontes, *et al.* (2013). New and rare coastal fishes in the Azores islands: occasional events or tropicalization process? Journal of Fish Biology, 83 (2), 272-294.

Amat, J.N., Tempera, F. (2009). *Zoobotryon verticillatum* Della Chiaje, 1822 (Bryozoa), a new occurrence in the archipelago of the Azores (North-eastern Atlantic). Marine Pollution Bulletin, 58, 761-764.

Amat, J.N., Cardigos, F., Santos, R.S. (2008). The recent northern introduction of the seaweed Caulerpa webbiana (Caulerpales, Chlorophyta) in Faial, Azores Islands (North-Eastern Atlantic). Aquatic Invasions, 3 (4), 417-422.

Boudouresque, C.F., Rodriguez-Prieto, C., Arrighi, F. (1994). Place de l'espèce de Caulerpa indigène de Méditerranée, *Caulerpa prolifera*, dans les préférences alimentaires de l'oursin *Paracentrotus lividus*. In C.F.Boudouresque, A. Meinesz, V. Gravez (Eds.), First international Workshop on *Caulerpa taxifolia* (pp. 233-237). France: GIS Posidonie publ.

Cardigos, F., Tempera, F., Ávila, S., *et al.* (2006). Non-indigenous marine species of the Azores. Helgoland Marine Research, 60, 160-169.

Cardigos, F., Tempera, F., Fontes, J., *et al.* (2013). Relatório sobre a presença de uma nova espécie no Norte da Ilha do Faial. Ponta Delgada: IMAR/DOP Universidade dos Açores.

Cordeiro, R., Malaquias, M.A., Masa, G., *et al.* (2013). New records for the opisthobranch fauna of the Archipelago of the Azores (north-east Atlantic Ocean). Marine Biodiversity Records, 6, e28.

Ganteaume, A., Gobert, J., Malestroit, P., *et al.* (1998). In vitro consumption of *Caulerpa taxifolia* by accustomed and non accustomed *Paracentrotus lividus* (Echinoid): seasonal variations. Journal of the Marine Biological Association of the United Kingdom, 78, 239-248.

Guillén, J.E., Jiménez, S., Martínez, J. *et al.* (2003). Expansion of the invasive algae *Caulerpa racemosa* var. *cylindracea* (Sonder) Verlaque, Huisman & Boudouresque, 2003 on the region of Valencia seabed. Thalassas, 26 (2), 135-149.

Haroun Tabraue, R.J., Gil-Rodríguez, M.C., Afonso-Carrillo, J., *et al.* (1984) Estudio ecológico y fenológico de algunas especies del género Caulerpa Lamouroux presentes en el archipiélago

Canario. In V. Monteiro Marques (Ed.), Actas IV Simposio Ibérico de Estudios Benthos Marinho (pp.75-82). Lisbon: Faculdade de Ciencias de Lisboa.

Lowe, S., Browne, M., Boudjelas, S., *et al.* (2000). 100 of the World's Worst Invasive Alien Species. A selection from the Global Invasive Species Database. Auckland: Invasive Species Specialist Group of the Species Survival Commission (SSC) of the World Conservation Union (IUCN).

Malaquias, M.A.E., Calado, G.P., Padula, V., *et al.* (2009). Molluscan diversity in the North Atlantic Ocean: new records of opisthobranch gastropods from the Archipelago of the Azores. Marine Biodiversity Records, 2, e38.

Micael, J., Parente, M.I., & Costa, A.C. (2014). Tracking macroalgae introductions in North Atlantic oceanic islands. Helgoland Marine Research, 68 (2), 209-219.

NSW 2004. NSW control plan for the noxious marine weed *Caulerpa taxifolia* in NSW waters. Orange: The New South Wales State.

Ojaveer, H., Galil, B.S., Minchin, D., *et al.* (2014). Ten recommendations for advancing the assessment and management of non-indigenous species in marine ecosystems. Marine Policy, 44, 160–165.

Pola, M., Lucas Cervera, J., Gosliner, T.M. (2006). Taxonomic revision and phylogenetic analysis of the genus *Tambja* Burn, 1962 (Mollusca, Nudibranchia, Polyceridae). Zoologica Scripta, 35 (5), 491-530.

Santos, R.S., Hawkins, S., Monteiro, L.R., *et al.* (1995). Marine research, resources and conservation in the Azores. Aquatic Conservation: Marine and Freshwater Ecosystems, 5 (4), 311-354.

Torres, P., Lopes, C., Dionísio, M.A., *et al.* (2010). Espécies exóticas invasoras marinhas da ilha de Santa Maria, Açores. XIV Expedição Científica do Departamento de Biologia - Santa Maria 2009. Relatórios e Comunicações de Depatramento de Biolgia, 36, 103-111.

Torres, P., Costa, A.C., Dionísio, M.A. (2012). New alien barnacles in the Azores and some remarks on the invasive potential of Balanidae. Helgoland Marine Research, 66, 513-522.

Uchimura, M., Rival, A., Nato, A., *et al.* (2000). Potential Use of Cu^{2+}, K^{+} and Na^{+} for the Destruction of *Caulerpa taxifolia*: Differential Effects on Photosynthetic Parameters. Journal of Applied Phycology, 12 (1), 15-23.

Williams, S.L., Schroeder, S.L. (2004). Eradication of the Invasive Seaweed *Caulerpa taxifolia* by chlorine bleach. Marine Ecology Progress Series, 272, 69-76.

Richard Shine

13 Reducing the Ecological Impact of Invasive Cane Toads

13.1 Introduction

Many invaders cause ecological devastation and huge economic costs, stimulating the expenditure of vast amounts of time and effort in attempts to control the invader. Those efforts sometimes succeed – for example, eradication of feral rodents from offshore islands has enabled native taxa to recover (Howald *et al.*, 2007). In most cases, however, invaders have continued to spread despite intensive efforts to control them. Part of the problem is the inherent difficulty of the challenge; regrettably, many efforts to control invasive species have been undertaken without any clear understanding of the invader's biology (Saunders *et al.*, 2010). Although it is politically attractive to combat an alien's arrival by killing as many of the foreigners as possible, many invasive species reproduce at rates so high that simply culling adults has no significant long-term impact on invader abundance.

The key to effective invader control is to understand the enemy. By analogy, military strategists confronted with an invading army would try to understand why that army was invading, what resources it relied upon, what impacts it was having, and what lines of communication were important in facilitating its onward march. Killing enemy soldiers might be part of the military response, but it would not be the only component. To conquer an invading army, we need to understand that army's ways of functioning. Military history abounds with examples whereby deciphering the communication code used by the enemy was key to eventual victory.

In this chapter I explore recent research on the invasion of cane toads (*Bufo* [*Rhinella*] *marina*) through Australia, and review our attempts to control the numbers and spread of the invader, and mitigate its ecological impact. Much of the work is ongoing, and most of it has yet to proceed to the stage of landscape-scale deployment. Nonetheless, we have developed several novel approaches that work well in laboratory trials and in small-scale field studies, and can claim the first successes in enabling native biota to persist despite the invader's presence. An ecological approach, based on comprehensive basic research into toad biology and impact, has generated new methods that have already achieved clear-cut positive results in a system where the prior expenditure of vast resources has achieved very little. Scientific efforts to mitigate the ecological impact of invasive cane toads thus provide a valuable case history of the application of ecological, evolutionary, and behavioral research to achieve conservation outcomes.

13.2 Background of the Study System

Cane toads are large (up to > 2 kg, but typically around 200 to 400 g as adults; see Figure 13.1A) bufonid anurans, native to an extensive area of South America, Central America, and Mexico (Lever, 2001). Reflecting their large body size, and a phylogenetically conservative tendency for bufonids to produce larger clutches of small eggs, the fecundity of cane toads is extraordinary. Clutch sizes in excess of 30,000 eggs have been reported (Lever, 2001). The eggs are laid in long strings in shallow ponds, fertilized externally by an amplexing male, and hatch within one to two days (depending on water temperature). The small black tadpoles often form schools in shallow water, and grow rapidly. They metamorphose into tiny (approx. 0.1 g, 10 mm long) toadlets after a brief larval phase (thermally-dependent, but typically 1 to 2 months). Toadlets that emerge during the long dry season of the wet-dry tropics are restricted to the margins of their natal pond until seasonal rains allow them to disperse; during this phase, a few of the earlier-transforming metamorphs may grow large enough to ingest their smaller relatives, and the larger toads become specialist cannibals (Pizzatto & Shine, 2008). Toadlets are diurnally active, thereby reducing vulnerability to their cannibalistic nocturnal larger brethren (Pizzatto *et al.*, 2008).

Fig. 13.1: Cane toads (*Rhinella marina*) are large, heavy-bodied bufonids (A) that can reach startlingly high abundances at the invasion front (B). Photographs by Matt Greenlees (A) and Ruchira Somaweera (B), with permission.

Cane toads feed on a diverse array of prey types, but especially on small insects such as ants and beetles. They exploit local opportunities, such as the congregation of insects attracted to artificial lights at night (Zug & Zug, 1979; González-Bernal *et al.*, 2011). Growth rates can be high, with sexual maturation at a few months of age in tropical Australia, and probably around a year of age in cooler climates (Brown *et al.*, 2013a). Reproductive frequency of females in the wild is not well documented.

In the era before pesticides were developed, toads of various species were widely translocated in attempts to control agricultural insect pests (Turvey, 2013). As commercial sugar plantations were established across the tropics, cane toads were brought in to consume insects such as scarab beetles (Anonymous, 1934; Turvey, 2013). In this way, cane toads were brought from their native range in French Guinea to plantations in Puerto Rico, and thence to Hawaii, and thence to Australia (Turvey, 2013). In 1935, 101 toads collected in the Honolulu region were shipped to an agricultural experimental station in a sugar-growing region of northeastern Australia (Mareeba, Queensland), where they were maintained and bred in captivity. Thousands of their offspring were released in the cane fields.

The toads gradually expanded their range westwards across Australia into drier country, at around 10 to 15 km per year (Urban *et al.*, 2007). The rate of invasion accelerated as the toads spread through tropical Queensland, into and through the Northern Territory, and into Western Australia (Urban *et al.*, 2007). By the time the toad invasion front reached the city of Darwin, the invasion was moving at around 50 km per annum. Although human-assisted translocations continue to play a role in extending the toad's distribution in Australia (van Beurden, 1981; White & Shine, 2009), the acceleration of the tropical front is due to evolved shifts in toad behavior and morphology (Phillips *et al.*, 2006; 2010a). Radio-tracking shows that toads at the invasion front disperse almost ten times faster than do their conspecifics in long-colonized areas (Alford *et al.*, 2009; Lindström *et al.*, 2013). Even when raised under standardized conditions, the progeny of these toads inherit their parents' dispersal rates (Phillips *et al.*, 2010a).

The cane toad's march across tropical and subtropical Australia has had devastating impacts on native predators (such as lizards, snakes, crocodiles and carnivorous marsupials) that attempt to consume the newcomers (Shine, 2010). Like many bufonids, cane toads possess potent defensive chemicals (bufadienalides, bufotoxins) that can rapidly be fatal if ingested by a predator (Hayes *et al.*, 2009; Shine, 2010; Ujvari *et al.*, 2013). Minor genetic changes can increase the predator's physiological resistance to the toxins, and such modifications are common in the predator fauna of areas that contain native toads (Ujvari *et al.*, 2013). Because Australia contains no endemic bufonids, however, many Australian taxa are evolutionarily naïve to bufotoxins; and as a result, the invasion of cane toads has been accompanied by a wave of predator mortality (Shine, 2010). In several areas, researchers have documented >90% mortality of large varanid lizards within the first few months of toad invasion (Doody *et al.*, 2009; Brown *et al.*, 2013b). The same is true of freshwater crocodiles in some but not all populations (Letnic *et al.*, 2008; Somaweera *et al.*, 2013), and for large dasyurid marsupials (quolls) (O'Donnell *et al.*, 2010).

Although these cases of "death by toad" have aroused public outrage, and stimulated governments to fund attempts to control toads, recent research has provided a more nuanced view of toad impact. First, many native predators (such as most birds) are capable of surviving the toad invasion, either because they are physiologically resistant to the toad's toxins (reflecting an evolutionary origin in parts of the world containing native bufonids) or because they rapidly learn not to eat the toxic toads. For example, a diverse array of fishes, frogs, lizards, crocodiles, and small marsupials all exhibit conditioned taste aversion. Initial consumption of a small toad results in non-lethal illness; thereafter, the predator deletes toads from its diet (O'Donnell *et al.*, 2010). Some individual predators die, but most survive and can coexist with toads thereafter (Shine, 2010). Taste aversion does not protect larger predator species because the toad invasion front consists entirely of large (and thus, highly mobile) toads; consuming one of these is fatal, giving the predator no opportunity to learn (because toxin content increases rapidly with increasing toad body size) (Phillips & Shine, 2006). High toad abundances at the invasion front (Fig. 13.1b) mean that most or all local predators soon encounter a large toad.

Cane toads may also affect Australian ecosystems by preying on native taxa (especially insects); by competing for food with other insectivores; and by modifying the behavior of other species. For example, some frogs reduce their activity levels when toads are present (Greenlees *et al.*, 2007), move away from toads (Mayer *et al.*, 2015), and change the structure of the male advertisement call in response to toad calls (Bleach *et al.*, 2014). Predators also modify their attack strategies as a result of conditioned taste aversion (Webb *et al.*, 2008) and may ignore palatable native prey that resembles the toxic invader (Nelson *et al.*, 2010). More important, however, are indirect ecological impacts of cane toads, mediated via changes in the abundance of apex predators. The virtual elimination of large varanid lizards due to lethal toxic ingestion of toads, for example, has powerful flow-on effects to the myriad taxa that were previously consumed by these giant reptiles (Brown *et al.*, 2013b). As a result, the invasion of cane toads is often followed – counter-intuitively – by increasing rather than decreasing abundance of many native species (Brown *et al.*, 2011). Such effects may be short-term, due to the eventual recovery of varanid populations. A goanna species whose populations crash dramatically when toads first arrive (*Varanus panoptes*, the floodplain monitor) is common in coastal Queensland, where toads have been present for almost 80 years. Lizards in this long-colonized area refuse to consume dead toads, but readily consume dead frogs (Llewelyn *et al.*, 2014). Thus, the primary ecological impacts of cane toads may be ephemeral, on a timescale of decades, and decrease as the native fauna evolves (or learns) toad-avoidance. Another significant impact is a positive one: native taxa capable of consuming toads without ill effect (notably, invertebrates such as ants and water-beetles) increase in number because of the enhanced food supply. Perhaps for that reason, high toad densities are achieved mostly in the years immediately following invasion; after that, numbers (and perhaps body sizes) fall appreciably (Freeland *et al.*, 1986; Phillips & Shine, 2005).

13.3 Classical Approaches to Toad Control

Widespread revulsion for cane toads among the general public, and an often-exaggerated notion of the breadth and duration of the toads' ecological impacts, have stimulated vigorous but largely ineffective attempts to curtail toad abundance. Collecting toads at night, with the aid of a flashlight, is a common activity for rural residents across the Australian tropics. In several regions, people have banded together to form "toad-busting" associations that undertake organized large-scale culls. The leaders of such groups often make media statements about the threats posed by toads, and the need for their groups to be suitably resourced to deal with that threat. As a result, millions of dollars of government as well as private funding, as well as many thousands of hours of volunteer effort, have been devoted to collecting adult toads and humanely euthanizing them.

The methods used by toad-busting groups (both volunteer and government-funded) have been straightforward. Most groups primarily focus on hand-collecting, but a few also use lighted traps that attract insects, and thus toads. Unfortunately, commercially available toad traps are very inefficient: they catch only a small proportion of the local toads, and inflict significant mortality on native species (R. Shine, unpubl. data). Current research at James Cook University is exploring ways to enhance trapping success by using more appropriate lighting (UV rather than visible spectrum) and including toad advertisement calls (L. Schwarzkopf, pers. comm.).

Despite collecting massive numbers of toads, the toad-busting groups have not conducted any research to evaluate the effectiveness of their activities. Available data are not encouraging. In the most detailed analysis, a major community-group toad-cull on the shores of Lake Argyle depressed toad abundances only briefly; with the next wet season, toad numbers soon surpassed those present prior to the culling operation (Somaweera & Shine, 2012). Mark-recapture studies, and observations during community toad-busts in north-eastern New South Wales, also show that volunteer collectors typically catch only a small proportion of the total toad population (M. Greenlees, pers. comm.).

The fundamental problem with direct culling of adult toads is the immense fecundity of this species. If a single pair of adults can produce 30,000 eggs in a clutch, even a few survivors of the control program can rapidly repopulate an area. Mathematical models show that it would be impossible to exterminate toads from an area, or even to seriously reduce their densities, without achieving unrealistically high monthly rates of capture (McCallum, 2006).

Sadly, the predictions from those models are borne out by the available data. The most clear-cut evidence for the ineffectiveness of current toad-culling operations is that the rate of toad invasion has continued unabated across the regions (near Darwin, NT and Kununurra, WA) that are home to the largest, best-funded, and most highly organized toad-busting groups (Peacock, 2007). In the course of routine collections around the city of Darwin over several years when toad-busts were frequently carried out, my research team always encountered abundant toads even in sites that were "busted" on a regular basis (G. P. Brown, pers. comm.).

The largest expenditure of funds on toad control (more than 11 million Australian dollars) was given by the federal government to the CSIRO (Commonwealth Scientific and Industrial Research Organization) to search for biocontrol approaches. Those funds supported research both in Australia and in the toad's native range in Venezuela. Potentially lethal viruses were found within South America, but also proved lethal to Australian frogs. The work then shifted towards an attempt to create a genetically-engineered virus that would kill toads at metamorphosis (Shanmuganathan *et al.*, 2010). That program was eventually abandoned due to technical difficulties as well as the growing realization that such a virus, even if toad-specific, could wreak havoc on populations of bufonids in other parts of the world, where they are an important component of the native fauna.

In summary, despite vast effort and financial expenditure, attempts to control cane toads have managed only to reduce the densities of adult toads in selected areas for short periods of time. Control efforts have failed because they centered on a single aim (to reduce the number of toads in a local area) and a simple approach (killing as many toads as possible). Unfortunately, the methods developed in this respect (such as traps and hand-collecting) fail to account for the toads' massive fecundity, which make simple removal of animals ineffective. More sophisticated biological control (including genetic-modification) methods can overcome these problems, but face immense technical hurdles as well as near-insuperable political and ethical obstacles to implementation. Is there an alternative approach that avoids these flaws? Recent research has identified several promising new directions, which I will explore below. Most of these new ideas arose from a broad-ranging research program (funded as basic, not applied, research by the Australian Research Council) that set out to understand the biology and impact of the invader before devising novel methods for toad control.

13.4 Manipulation of Habitat Suitability

In severely arid margins of the toad's Australian range, these animals need to find a moist substrate (or standing water) to replenish their water balance every two or three nights during dry conditions. Thus, fencing off a farm dam in arid country can spell rapid death for any toads outside the fence (Florance *et al.*, 2011). However, the method is ineffective in wetter times of year, or in landscapes where alternative water sources are available (Florance *et al.*, 2011).

Eliminating the water source would have similar effects. A recent paper has suggested closing off water bodies as a way to curtail the toads' further expansion into the arid Pilbara region of Western Australia (Tingley *et al.*, 2013). In this harsh landscape, stock watering-points (artificial dams and troughs) offer the only surface water for most of the year; thus, toads may be unable to move through such country without access to those dams (Florance *et al.*, 2011). Tingley *et al.* (2013) identified a hyper-arid region where toad colonization would occur only through a narrow coastal corridor; shutting down the dams in that corridor might prevent cane toads from invading a large area of Western Australia. The feasibility of implementing that suggestion is still under discussion.

Other types of manipulations can create habitats that are avoided by toads. For example, removing domestic stock can enable dense regrowth of vegetation, unattractive to cane toads as foraging sites (Zug & Zug, 1979) or as calling sites for reproductive males (Hagman and Shine, 2006; Semeniuk *et al.*, 2007). Steep banks and deep water also discourage toad-breeding (Hagman & Shine, 2006; Semeniuk *et al.*, 2007). In some regions, local councils require residents who construct farm dams to surround them with dense vegetation to discourage toad breeding.

13.4.1 Manipulation of Biotic Interactions

The increasingly fast pace of the toad invasion has left behind some of their pathogens, including a lungworm (*Rhabdias pseudosphaerocephala*) that was brought to Australia from South America with the original toads (Dubey & Shine, 2008). The parasite is thus absent from the toad invasion front (Phillips *et al.*, 2010b). Infection with the lungworm can kill metamorph toads, and reduces growth rate in adult as well as juvenile toads (Kelehear *et al.*, 2009; 2011). Could we then translocate the parasite to the invasion front, shortening the pathogen-free interval? Unfortunately, further research revealed two flaws in this scheme (which has already been implemented unintentionally by community groups, who have spread the parasite to the invasion front through poor quarantine protocols: Kelehear *et al.*, 2012a). First, the parasite is readily transmitted to a native frog species (*Litoria splendida*), where it causes high mortality (Pizzatto & Shine, 2012). Second, low host densities at the toad invasion front reduce parasite transmission rates, so that the lungworms would be unlikely to flourish (Kelehear *et al.*, 2012b). To overcome the latter problem, we could provide additional hosts by experimentally infecting a "Typhoid Mary" species of native frog, *Litoria caerulea*, that can carry the parasite but is not harmed by it (Pizzatto & Shine, 2011, 2012).

Could we instead increase the abundance of native species that act as predators or competitors of cane toads? Contrary to oft-expressed opinions, many native Australian animals can consume cane toads without ill effect. For example, many birds and rodents are physiologically tolerant of the toads' toxins, and often eat the invasive anurans (Beckmann & Shine, 2009; Shine, 2010). More importantly, many types of aquatic insects not only prey on cane toad larvae, but attack these small slow-swimming tadpoles more readily and effectively than they attack native anuran larvae (Cabrera-Guzmán *et al.*, 2012). By restoring pond habitats in the peri-urban areas where cane toads breed, we may be able to build up healthy populations of these miniature toad-enemies.

Native taxa can reduce toad recruitment by competition as well as by predation. The tadpoles of native frogs can outcompete cane toad tadpoles, in natural water bodies as well as in laboratory trials (Cabrera-Guzmán *et al.*, 2011), so that encouraging recolonization of native frogs in anthropogenically disturbed areas might substantially reduce toad tadpole survival (Cabrera-Guzmán *et al.*, 2011).

13.4.2 Manipulation of Toad Abundance

Previous attempts to control cane toads have focused almost entirely on culling adult animals – with no long-term effect, because of the high fecundity of the invader (above). We first need to eradicate recruitment by killing eggs or tadpoles; only then will a cull of adult toads have a long-term impact on toad abundance. This is a simple and obvious point, so why has almost all of the previous control effort been focused on the terrestrial stages of the toad's life-history? In terms of community activities, people simply are not aware of tadpoles in turbid water bodies; it is the large adult toads in their backyards that upset them, and that they want to remove. In contrast, the lack of scientific research on tadpole control is based on a key assumption: that strong density-dependence within the larval phase makes tadpole culling unproductive. Killing a few thousand tadpoles may just improve conditions for the survivors, and has no net effect on recruitment (Crossland *et al.*, 2009). If we accept the assumption of strong density-dependence in tadpole survival rates, mathematical models identify the adult phase as the best one to target for toad control (Lampo & DeLeo, 1998). The difficulty of precisely measuring the form of density-dependence is a problem, because this parameter strongly affects the output of models that predict population-level responses to given levels of mortality (Thresher & Bax, 2006).

Issues of density-dependence are irrelevant, however, if we are able to totally eradicate tadpoles from water bodies. Toads often select small shallow open ponds, close to human habitation, and breed in only a small proportion of ponds in an area (Williamson, 1999; Hagman & Shine, 2006); thus, the eggs and tadpoles are far more concentrated in space and time than are adult toads. As a result, an effective weapon against tadpoles would be relatively easy to implement. Studies by Michael Crossland and colleagues have revealed exciting opportunities in this respect, based upon exploiting intraspecific competitive mechanisms that facilitate targeted control. Like many invasive species, cane toads thrive in disturbed habitats where native fauna are already extirpated; thus, the toads compete most strongly against conspecifics. The mechanisms they have evolved to fulfill this function can be co-opted to provide novel and species-specific weapons for toad control. For example, cane toad tadpoles seek out and kill (consume) newly-laid eggs of their own species, thereby eliminating future competitors (Crossland & Shine, 2011).

They locate those clutches by detecting chemicals emitted by the eggs, specifically the bufagenin component of the toad's toxins (Crossland *et al.*, 2012). Funnel traps baited with those toxins (easily collected from the parotoid glands of adult cane toads) capture thousands of toad tadpoles with virtually no bycatch, and enable eradication of toad larvae from natural breeding ponds (Crossland *et al.*, 2012; Figure 13.2).

Fig. 13.2: Funnel traps baited with toad parotoid gland secretions attract thousands of toad tadpoles but repel the tadpoles of native frogs.

In current work, we are exploring an even more powerful weapon for toad control. Again reflecting the advantages of suppressing the viability of younger conspecifics, toad tadpoles produce chemicals that disrupt early development of eggs laid in their pond. Even a few hours' exposure to a low concentration of those chemicals generally induces 100% mortality (Crossland & Shine, 2012; Clarke *et al.*, 2015a,b). If we can identify the nature of that chemical, it has great promise as a simple, easily-deployable and highly species-specific method to prevent toad recruitment, and hence to achieve the key first step necessary to reduce toad abundance.

13.4.3 Manipulation of Toad Distribution and Population Structure

The high vagility, broad abiotic tolerances, powerful chemical defenses, and remarkable fecundity of cane toads mean that we will never eliminate them entirely from Australia. Given that we can achieve partial reductions in numbers by our new methods, should we aim for suppression of toads across their entire Australian range, or focus on specific areas where the toad's impact is highest (such as near the invasion front, and in regions that contain endangered fauna)? The latter tactic may have many benefits, by allowing persistence of predator populations in the toad-free sites, and by intensifying larval competition in the remaining areas where toads are allowed to breed unfettered by control efforts. If all of the toads in a local area breed in a single pond, either later clutches will be killed in the egg stage by existing tadpoles or (if several clutches are laid simultaneously) high levels of larval competition for food, followed by cannibalism in the metamorph stage, will result in toads largely controlling their own abundances.

Similarly, we may be able to manipulate the population structure of toads in ways that reduce recruitment. Habitat differences between adult males and females provide an opportunity for sex-biased capture. Male toads spend long periods beside water bodies calling to attract mates, whereas females move through the intervening habitat matrix acquiring food (Shine, 2010). Thus, culls conducted away from water may result in a preponderance of males among the survivors, and strong sexual conflict (Bowcock *et al.*, 2009) in a male-biased population may further reduce the survival rates of the remaining females (females that are amplexed by multiple males often drown during spawning attempts).

13.4.4 Manipulation of Toad Behavior

Evolutionary processes at work during the toads' Australian invasion have dramatically increased the rate at which the toad invasion is spreading through Australia and the invader's ability to cross barriers of unsuitable (e.g., dry) habitat, as well as the breadth of habitats into which toads will move (Shine *et al.*, 2011). Two evolutionary processes are at work. First, natural selection may favor genes for faster dispersal (reflecting trophic advantages to toads in the vanguard, where prey is plentiful and competition from conspecifics is low: Brown *et al.*, 2013a). Second, non-adaptive "spatial sorting" of dispersal-enhancing genes has also played a role (Shine *et al.*, 2011; Lindström *et al.*, 2013). Under spatial sorting, genes for more rapid dispersal accumulate at the invasion front simply because faster individuals can find their way to the invasion vanguard and, when they breed, they inevitably do so with other fast-dispersers. The end result is a sorting of genes; any gene that codes for low dispersal rates is left behind in the already-colonized range, and the invasion front accelerates because it is increasingly dominated by individuals with genes that enable rapid dispersal, even if that behavior takes the individual into a habitat where it is unlikely to survive (Shine *et al.*, 2011).

Breeding experiments have shown that this dispersal acceleration is heritable (Phillips *et al.*, 2010a). Thus, we could reset the clock on this process by simply releasing progeny of toads from long-colonized areas in Queensland in advance of the invasion front. The newly-arriving fast-dispersers would encounter an established population of sedentary toads, interbreed with them, and dilute the spatially-sorted concentration of "fast-dispersal" genes. That simple intervention could achieve three aims: (1) slow down the toad invasion (perhaps back to around 10 km per annum, instead of > 50 km per annum as currently occurs); (2) curtail the toad's ability to pass through barriers of unsuitable habitat, thus preventing further spread into isolated pockets of suitable conditions; and (3) restrict toads to the kinds of habitats they generally prefer, rather than moving out into a broader range of habitats where they will impact vulnerable native taxa. Additionally, it would enable native predators to learn toad-avoidance by encountering small (and thus, non-lethal) toads before the advent of the large toads at the invasion front (see below). Unfortunately, any proposal to release cane toads ahead of the invasion front attracts howls of outrage from community toad-busting groups, who advocate manual collection as the "solution" to the toad invasion.

13.4.5 Reduction of Predator Exposure to Lethally Large Toads

The major impact of cane toads on Australian native fauna occurs immediately after the toads arrive in a new area. Naïve predators encounter large adult toads, try to eat them, and are fatally poisoned. Smaller toads would have provided an opportunity for those predators to learn toad aversion, but that opportunity does not arise because the front is dominated by large toads. The lack of reproduction in invasion-vanguard toads may be a result of spatial sorting for dispersal-enhancing genes: a toad that foregoes reproduction, and devotes its time and energy to dispersal instead, may thereby enhance its ability to remain at the forefront of the invasion wave (Shine *et al.*, 2011; Brown *et al.*, 2013a).

The importance of the toad population structure (presence of juvenile toads, small enough to induce aversion learning without killing the predator) is highlighted by the lower and briefer impact of cane toads at the southern front, where invasion is slower because of suboptimal temperatures (Urban *et al.*, 2007). Here, toads breed soon after arriving in a new area and, as an apparent result, many predators learn to coexist with toads rather than dying (Jolly *et al.*, 2015). Studies on large varanid lizards show that vulnerable species adjust their foraging tactics to exclude toads from the menu; hence, species whose populations have been decimated as the toad wave has spread across the tropics are thriving once again in long-colonized areas of eastern Australia (Llewelyn *et al.*, 2014).

If the major impact of cane toads occurs only at the invasion front, and decreases as soon as toads begin breeding (providing aversion-inducing smaller toads), we could potentially maintain predator populations simply by protecting them from that initial onslaught. Large numbers of predators could be captured just prior to toad arrival, maintained in captivity for a year or two, then released at their sites of capture. Alternatively, the animals could be immediately re-released in an area that was invaded by toads two or three years previously, where toads have already begun to breed.

13.4.6 Increase in Predator Exposure to Non-lethal (Small) Toads

Another way to expose predators to small as well as large toads, and thus stimulate aversion learning, is to introduce small (juvenile) toads to newly-invaded areas and/or train individual predators (perhaps captive-bred, or recently captured in advance of the toad invasion front). Some proportion of predators will have the opportunity to learn taste aversion, facilitating population persistence. As in the tactic described above, the key insight is that toad impact depends upon predator behavior; predators are at risk only if they attempt to consume a toad. Thus, we can protect vulnerable predators by changing their behavior, even if we are unable to eradicate toads. We cannot possibly protect all of the predators in all areas, but even pockets of surviving (toad-averse) predators could provide nuclei for recolonization.

My proposal to release "teacher toads" for this purpose was opposed by community toad-busting groups, on the grounds that these additional toads would just add to the ecological impact of the invader rather than reduce it. However, the arguments were simplistic, without reference to the underlying rationale for the proposal. The idea of aversion-training captive-bred predators and then releasing them is less controversial, and has been adopted with great success in the case of a carnivorous marsupial, the northern quoll (*Dasyurus hallucatus*). Aversion-trained quolls survived after release, whereas their non-trained siblings did not (O'Donnell *et al.*, 2010). Capturing, aversion-training, and then releasing individuals also enhanced survival in another vulnerable predator, the northern bluetongue lizard (*Tiliqua scincoides intermedia*) (Price-Rees *et al.*, 2013). A larger-scale field test of the idea with goannas (*Varanus panoptes*) is currently underway.

13.5 The Way Forward: Integrated Pest Control

Cane toads are formidable invasion machines, and it is unlikely that any single method will ever eradicate them. Even with a combination of methods, landscape-scale extirpation is vanishingly unlikely. However, the new weapons developed out of recent ecological research on this high-profile invasive anuran provide great encouragement. We have already demonstrated that we can eradicate toad breeding in natural spawning sites (Crossland *et al.*, 2012) and that taste-aversion training can enable vulnerable predator species to persist (O'Donnell *et al.*, 2010; Price-Rees *et al.*, 2013). Current studies are assessing other (potentially more powerful) pheromone-based methods, both to disrupt embryogenesis (Crossland & Shine, 2011) and to discourage oviposition, thus concentrating toad breeding (M. R. Crossland, pers. comm.). Community groups across the toad's range in tropical Australia have adopted our tadpole-trapping methods, and have reported excellent results. The prospects for cane toad control are far brighter than was the case a few years ago, when all of the effort was targeted at killing toads rather than understanding them.

The most promising scenario includes an integrated approach, where we use a combination of methods deployed by private (community-based) conservation groups as well as employees of government at all levels (local, state, federal). Local knowledge will be invaluable to identify toad breeding sites and to evaluate the applicability of the available techniques. Part of the effort will go into collecting adult toads (as is currently the main focus), but that will increasingly be accompanied by techniques such as habitat manipulation, and prevention of toad recruitment by either suppression or collection of the toad's aquatic life-history stages. Some of my other suggestions – notably, the idea of releasing juvenile toads at the invasion front – face political obstacles, but these obstacles are fading as the futility of current toad-control methods (centered on hand-collecting adult toads) becomes increasingly apparent.

It is still far too early to claim success in the battle against the invasive cane toad in Australia. The toads have flourished despite decades of intensive effort designed to curb their numbers. We now have an arsenal of new and more effective methods to mitigate toad impact as well as reduce toad abundance, but these have yet to be deployed on a landscape scale. If that new approach succeeds, it will provide a striking example of the benefits of allocating research effort to understanding an invader before trying to eradicate it.

13.6 Acknowledgements

I thank my research group, Team Bufo, for their hard work, their innovative thinking, and their encouragement; working with them has been inspirational for me. The work was funded by the Australian Research Council.

In a nutshell

- Native to South and Central America, the large and highly toxic cane toad (*Bufo [Rhinella] marina*) has been translocated to many countries in futile attempts to control insect pests.
- Brought to Northeastern Australia in 1935, the toad has now spread across much of that continent, with devastating impacts on anuran-eating predators (lizards, snakes, marsupials, crocodiles) that are fatally poisoned when they try to consume the toxic newcomer.
- Physically removing toads has failed to reduce their numbers, or to curtail the invader's spread.
- Recent research into toad biology has provided two new approaches. The first is to release juvenile (and thus, non-lethal) toads at the current invasion front, to train native predators to avoid toads as prey. Taste aversion training can blunt the toad's impact even if we cannot eradicate the toxic invader. If sourced from long-colonized areas, these toads might also slow down toad advance by diluting invasion-front genes that code for dispersal-enhancing traits.
- The second tactic involves exploiting aspects of toad biology: for example, toad tadpoles are attracted to the toxins in newly-laid eggs, which they follow to locate and destroy future competitors. Funnel-traps baited with toad toxins can eradicate toad tadpoles from natural water bodies. Toad tadpoles also produce chemicals that disrupt embryonic development of toad eggs, providing another avenue for control.
- Cane toads in Australia provide a clear example that you need to understand an invasive species if you want to control it.

13.7 Bibliography

Alford, R. A., Brown, G. P., Schwarzkopf, L., *et al.* (2009). Comparisons through time and space suggest rapid evolution of dispersal behaviour in an invasive species. Wildlife Research, 36, 23-28.

Anonymous (1934). Toads save sugar crop. Nature, 134, 877-887.

Beckmann, C., Shine, R. (2009). Are Australia's birds at risk due to the invasive cane toad? Conservation Biology, 23, 1544-1549.

Bleach, I., Beckmann, C., Brown, G. P., *et al.* (2014). Effects of an invasive species on refuge-site selection by native fauna: the impact of cane toads on native frogs in the Australian tropics. Austral Ecology, 39, 50-59.

Bowcock, H., Brown, G. P., and Shine, R. (2009). Beastly bondage: the costs of amplexus in cane toads (*Bufo marinus*). Copeia, 2009, 29-36.

Brown, G. P., Kelehear, C., and Shine, R. (2013a). The early toad gets the worm: cane toads at an invasion front benefit from higher prey availability. Journal of Animal Ecology, 82, 854-862.

Brown, G. P., Phillips, B. L., and Shine, R. (2011). The ecological impact of invasive cane toads on tropical snakes: field data do not support predictions from laboratory studies. Ecology, 92, 422-431.

Brown, G. P., Ujvari, B., Madsen, T., *et al.* (2013b). Invader impact clarifies the roles of top-down and bottom-up effects on tropical snake populations. Functional Ecology, 27, 351-361.
Cabrera-Guzmán, E., Crossland, M. R., and Shine, R. (2011). Can we use the tadpoles of Australian frogs to reduce recruitment of invasive cane toads? Journal of Applied Ecology, 48, 462-470.
Cabrera-Guzmán, E., Crossland, M. R., and Shine, R. (2012). Predation on the eggs and larvae of invasive cane toads (*Rhinella marina*) by native aquatic invertebrates in tropical Australia. Biological Conservation, 153, 1-9.
Clarke, G., Crossland, M., Shilton, C., Shine, R. (2015a). Chemical suppression of embryonic cane toads (*Rhinella marina*) by larval conspecifics. Journal of Applied Ecology, in press. DOI: 10.1111/1365-2664.12528
Clarke, G. S., Crossland, M. R., Shine, R. (2015b). Can we control the invasive cane toad using chemicals that have evolved under intraspecific competition? Ecological Applications, in press.
Crossland, M. R., Alford, R. A., and Shine, R. (2009). Impact of the invasive cane toad (*Bufo marinus*) on an Australian frog (*Opisthodon ornatus*) depends on reproductive timing. Oecologia, 158, 625-632.
Crossland, M. R., and Shine, R. (2011). Cues for cannibalism: cane toad tadpoles use chemical signals to locate and consume conspecific eggs. Oikos, 120, 327-332.
Crossland, M. R., and Shine, R. (2012). Embryonic exposure to conspecific chemicals suppresses cane toad growth and survival. Biology Letters, 8, 226-229.
Crossland, M. R., Haramura, T., Salim, A. A., *et al.* (2012). Exploiting intraspecific competitive mechanisms to control invasive cane toads (*Rhinella marina*). Proceedings of the Royal Society B, 279, 3436-3442.
Doody, J. S., Green, B., Rhind, D., *et al.* (2009). Population-level declines in Australian predators caused by an invasive species. Animal Conservation, 12, 46-53.
Dubey, S., and Shine, R. (2008). Origin of the parasites of an invading species, the Australian cane toad (*Bufo marinus*): are the lungworms Australian or American? Molecular Ecology, 17, 4418-4424.
Florance, D., Webb, J. K., Dempster, T., *et al.* (2011). Excluding access to invasion hubs can contain the spread of an invasive vertebrate. Proceedings of the Royal Society B, 278, 2900-2908.
Freeland, W. J., Delvinquier, B. L. J., and Bonnin, B. (1986). Decline of cane toad, *Bufo marinus*, populations: status of urban toads. Australian Wildlife Research, 13, 597-601.
González-Bernal, E., Brown, G. P., Cabrera-Guzmán, E., *et al.* (2011). Foraging tactics of an ambush predator: the effects of substrate attributes on prey availability and predator feeding success. Behavioral Ecology and Sociobiology, 65, 1367-1375.
Greenlees, M. J., Brown, G. P., Webb, J. K., *et al.* (2007). Do invasive cane toads (*Chaunus marinus*) compete with Australian frogs (*Cyclorana australis*)? Austral Ecology, 32, 900-907.
Hagman, M., and Shine, R. (2006). Spawning-site selection by feral cane toads (*Bufo marinus*) at an invasion front in tropical Australia. Austral Ecology, 31, 551-558.
Hayes, R. A., Crossland, M. R., Hagman, M., *et al.* (2009). Ontogenetic variation in the chemical defenses of cane toads (*Bufo marinus*): toxin profiles and effects on predators. Journal of Chemical Ecology, 35, 391-399.
Howald, G., Donlan, C. J., Galván, J. P., *et al.* (2007). Invasive rodent eradication on islands. Conservation Biology, 21, 1258-1268.
Jolly, C. J., Shine, R., Greenlees, M. J. (2015). The impact of invasive cane toads on native wildlife in southern Australia. Ecology and Evolution, 5, 3879-3894.
Kelehear, C., Brown, G. P., and Shine, R. (2011). Influence of lung parasites on growth rates of free-ranging and captive adult cane toads. Oecologia, 165, 585-592.
Kelehear, C., Brown, G. P., and Shine, R. (2012b). Rapid evolution of parasite life-history traits on an expanding range-edge. Ecology Letters, 15, 329-337.

Kelehear, C., Cabrera-Guzmán, E., and Shine, R. (2012a). Inadvertent consequences of community-based efforts to control invasive species. Conservation Letters, 5, 360-365.

Kelehear, C., Webb, J. K., and Shine, R. (2009). *Rhabdias pseudosphaerocephala* infection in *Bufo marinus*: lung nematodes reduce viability of metamorph cane toads. Parasitology, 136, 919-927.

Lampo, M., and De Leo, G. A. (1998). The invasion ecology of the toad *Bufo marinus*: from South America to Australia. Ecological Applications, 82, 388-396.

Letnic, M., Webb, J. K., and Shine, R. (2008). Invasive cane toads (*Bufo marinus*) cause mass mortality of freshwater crocodiles (*Crocodylus johnstoni*) in tropical Australia. Biological Conservation, 141, 1773-1782.

Lever, C. (2001). The Cane Toad. The History and Ecology of a Successful Colonist. Otley, West Yorkshire: Westbury Academic and Scientific Publishing.

Lindström, T., Brown, G. P., Sisson, S. A., *et al.* (2013). Rapid shifts in dispersal behavior on an expanding range edge. Proceedings of the National Academy of Sciences of the United States of America, 110, 13452-13456.

Llewelyn, J., Schwarzkopf, L., Phillips, B. L., *et al.* (2014). After the crash: how do predators adjust following the invasion of a novel toxic prey type? Austral Ecology, 39, 190-197.

Mayer, M., Brown, G. P., Zimmerman, B., Greenleses, M. J., Shine, R. (2015). Habitat use of the introduced cane toad (*Rhinella marina*) and native frog species in tropical Australia. Journal of Tropical Ecology, 31, 531-540. DOI: 10.1017/S0266467415000474

McCallum, H. (2006). Modelling potential control strategies for cane toads. In K. L. Molloy and W. R. Henderson (Eds.), Science of Cane Toad Invasion and Control. Proceedings of the Invasive Animals CRC/CSIRO/QLD NRM&W Cane Toad Workshop, Brisbane, June 2006 (pp. 123-133). Canberra, ACT: Invasive Animals Cooperative Research Centre.

Nelson, D., Crossland, M. R., and Shine, R. (2010). Indirect ecological impacts of an invasive toad on predator-prey interactions among native species. Biological Invasions, 12, 3363-3369.

O'Donnell, S., Webb, J. K., and Shine, R. (2010). Conditioned taste aversion enhances the survival of an endangered predator imperiled by a toxic invader. Journal of Applied Ecology, 47, 558-565.

Peacock, T. (2007). Community On-Ground Cane Toad Control in the Kimberley. Review for Western Australia Department of Environment and Conservation. Canberra, ACT: Invasive Animals Cooperative Research Centre.

Phillips, B. L., and Shine, R. (2006). Allometry and selection in a novel predator-prey system: Australian snakes and the invading cane toad. Oikos, 112, 122-130.

Phillips, B. L., Brown, G. P., and Shine, R. (2010a). Evolutionarily accelerated invasions: the rate of dispersal evolves upwards during the range advance of cane toads. Journal of Evolutionary Biology, 23, 2595-2601.

Phillips, B. L., Brown, G. P., Webb, J., *et al.* (2006). Invasion and the evolution of speed in toads. Nature, 439, 803.

Phillips, B. L., Kelehear, C., Pizzatto, L., *et al.* (2010b). Parasites and pathogens lag behind their host during periods of host range-advance. Ecology, 91, 872-881.

Phillips, B., and Shine, R. (2005). The morphology, and hence impact, of an invasive species (the cane toad, *Bufo marinus*) changes with time since colonization. Animal Conservation, 8, 407-413.

Pizzatto, L., and Shine, R. (2008). The behavioral ecology of cannibalism in cane toads (*Bufo marinus*). Behavioral Ecology and Sociobiology, 63, 123-133.

Pizzatto, L., and Shine, R. (2011). The effects of experimentally infecting Australian tree frogs with lungworms from invasive cane toads. International Journal for Parasitology, 41, 943-949.

Pizzatto, L., and Shine, R. (2012). Typhoid Mary in the frogpond: can we use native frogs to disseminate a lungworm biocontrol for invasive cane toads? Animal Conservation, 15, 545-552.

Pizzatto, L., Child, T., and Shine, R. (2008). Why be diurnal? Shifts in activity time enable young cane toads to evade cannibalistic conspecifics. Behavioral Ecology, 19, 990-997.

Price-Rees, S. J., Webb, J. K., and Shine, R. (2013). Reducing the impact of a toxic invader by inducing taste-aversion in an imperilled native reptile predator. Animal Conservation, 16, 386-394.
Saunders, G., Cooke, B., McColl, K., *et al.* (2010). Modern approaches for the biological control of vertebrate pests: an update on Australian research. Biological Control, 52, 288-295.
Semeniuk, M., Lemckert, F., and Shine, R. (2007). Breeding-site selection by cane toads (*Bufo marinus*) and native frogs in northern New South Wales. Wildlife Research, 34, 59-66.
Shanmuganathan, T., Pallister, J., Doody, S., *et al.* (2010). Biological control of the cane toad in Australia: a review. Animal Conservation, 13 (Suppl. 1), 16-23.
Shine, R. (2010). The ecological impact of invasive cane toads (*Bufo marinus*) in Australia. Quarterly Review of Biology, 85, 253-291.
Shine, R., Brown, G. P., and Phillips, B. L. (2011). An evolutionary process that assembles phenotypes through space rather than time. Proceedings of the National Academy of Sciences of the United States of America, 108, 5708-5711.
Somaweera, R., and Shine, R. (2012). The (non) impact of invasive cane toads on freshwater crocodiles at Lake Argyle in tropical Australia. Animal Conservation, 15, 152-163.
Somaweera, R., Shine, R., Webb, J., *et al.* (2013). Why does vulnerability to toxic invasive cane toads vary among populations of Australian freshwater crocodiles? Animal Conservation, 16, 86-96.
Thresher, R. E., and Bax, N. (2006). Comparative analysis of genetic options for controlling invasive populations of the cane toad *Bufo marinus*. In K. L. Molloy and W. R. Henderson (Eds.), Science of Cane Toad Invasion and Control. Proceedings of the Invasive Animals CRC/CSIRO/QLD NRMandW Cane Toad Workshop, Brisbane, June 2006 (pp. 117-122). Canberra, ACT: Invasive Animals Cooperative Research Centre.
Tingley, R., Phillips, B. L., Letnic, M., *et al.* (2013). Identifying optimal barriers to halt the invasion of cane toads *Rhinella marina* in northern Australia. Journal of Applied Ecology, 50, 129-137.
Turvey, N. (2013). Cane Toads: A Tale of Sugar, Politics and Flawed Science. Sydney: University of Sydney Press.
Ujvari, B., Mun, H., Conigrave, A. D., *et al.* (2013). Isolation breeds naivety: island living robs Australian varanid lizards of toad-toxin immunity via four-base-pair mutation. Evolution, 67, 289-294.
Urban, M., Phillips, B. L., Skelly, D. K., *et al.* (2007). The cane toad's (*Chaunus marinus*) increasing ability to invade Australia is revealed by a dynamically updated range model. Proceedings of the Royal Society B, 274, 1413-1419.
van Beurden, E. K. (1981). Bioclimatic limits to the spread of *Bufo marinus* in Australia: a baseline. Proceedings of the Ecological Society of Australia, 11, 143-149.
Webb, J. K., Brown, G. P., Child, T., *et al.* (2008). A native dasyurid predator (common planigale, *Planigale maculata*) rapidly learns to avoid toxic cane toads. Austral Ecology, 33, 821-829.
White, A. W., and Shine, R. (2009). The extra-limital spread of an invasive species via "stowaway" dispersal: toad to nowhere? Animal Conservation, 12, 38-45.
Williamson, I. (1999). Competition between the larvae of the introduced cane toad *Bufo marinus* (Anura: Bufonidae) and native anurans from the Darling Downs area of southern Queensland. Australian Journal of Ecology, 24, 636-643.
Zug, G. R., and Zug, P. B. (1979). The marine toad, *Bufo marinus*: a natural history resumé of native populations. Smithsonian Contributions to Zoology, 284, 1-54.

Aníbal Pauchard, Rafael García, Sergio Zalba, Mauro Sarasola, Rafael Zenni, Silvia Ziller, Martín A. Nuñez

14 Pine Invasions in South America: Reducing Their Ecological Impacts Through Active Management

14.1 Introduction

Conifers, and specifically the Pinaceae family, have been one of the most visible and studied plant taxa in invasion biology (e.g., Richardson *et al.*, 1994; Rejmánek & Richardson, 1996; Ledgard, 2001; Essl *et al.*, 2011; Buckley *et al.*, 2005; Gundale *et al.*, 2014). Conifer invasions have several characteristics that make them appealing for ecologists, biogeographers, conservationists, and invasion biologists (Gundale *et al.*, 2014). First, conifers have been widely introduced and extensively planted as a forestry crop and ornamental species all around the world (Richardson, 2006; Simberloff *et al.*, 2010; Essl *et al.*, 2010). Second, most conifer species have attributes associated with high invasiveness such as large propagule production, small seeds, anemochoric dispersal, and broad ranges of climatic and edaphic tolerance (Rejmánek & Richardson, 1996; Essl *et al.*, 2010). Third, conifer invasions are conspicuous in the landscape and can be studied in the field with simple observational techniques (Richardson, 2006; Visser *et al.*, 2014). Fourth, conifer invasions can have severe impacts on the local biota and ecosystem processes such as changes in water and fire regimes (Simberloff *et al.*, 2010). Although conifers have become invasive across the world, the Southern Hemisphere has been especially affected by their establishment and expansion, particularly in the case of *Pinus* spp. (Simberloff *et al.*, 2010), which, with the exception of one species, did not occur naturally south of the Equator (Lusk, 2008).

Pine invasions are an important ecological and economic problem in different countries of the Southern Hemisphere (Simberloff *et al.*, 2010; Pauchard *et al.*, 2010). Many species of the Pinaceae family have been introduced throughout the world, mainly for forestry use. There are many positive outcomes of these large-scale introductions of non-native species, including timber production, pulp for paper, and job creation. These "positive" aspects of afforestation have been a matter of discussion when contrasted to changes in water regimes and soil characteristics (Jobbagy & Jackson, 2007; Little *et al.*, 2009), and also in light of their effects on human well-being and health (Guadagnin *et al.*, 2009). Additionally, many of these large-scale plantations have spread into adjacent plant communities, frequently colonizing areas of high conservation value. The characteristics that make pines a good choice for forestry, like fast growth and maturation, have also facilitated their expansion outside the planted areas. South Africa, Australia and New Zealand have historically been leading the research on tree invasions. In South America, concerns about the risk of tree invasions started much later, and systematic research has only been pursued in the last decade. The later establishment of large-scale plantations ca. 1970s, compared to the Anglo-Saxon countries, and the incipient environmental movement may have slowed the attention to tree invasions as a potential threat to biodiversity and ecosystem services (Fonseca *et al.*, 2013).

Pine invasions have been studied extensively since the 1980s, when the SCOPE report on biological invasions was published (Drake *et al.*, 1989). Myriad research projects on pine invasions were carried out in temperate and Mediterranean regions of the Southern Hemisphere, especially in South Africa and New Zealand, and largely in temperate grasslands and shrublands, as well as in the Mediterranean Fynbos (Richardson *et al.*, 1994; Higgins & Richardson, 1998; Richardson & Higgins, 1998; Simberloff *et al.*, 2010). Only recently have researchers started to pay close attention to pine invasions in lower latitudes and forest ecosystems, with most of these studies being carried out in South America (Simberloff *et al.*, 2010; Falleiros *et al.*, 2011; Zenni & Ziller, 2011; Zenni & Simberloff, 2013; Zalba & Villamil, 2002). In the Southern Hemisphere, pines were used in forestry plantations in sub-tropical and tropical regions more recently than in temperate regions (Richardson *et al.*, 2008), which probably explains the temporal and regional biases in pine invasion studies. Moreover, the paradigm that pines do not invade forests persisted until recently (Emer & Fonseca, 2010; Zenni & Simberloff, 2013).

In 2007, a group of scientists studying conifer invasions met at Bariloche, Argentina to discuss the present and future of conifer invasions in South America (Richardson *et al.*, 2008). They concluded that the relatively shorter period since massive introduction of conifers in South America was the main cause of the apparent resistance to invasion that was observed in several South American ecosystems, but that many species are showing invasive behavior on the continent (Table 14.1, Figure 14.1, Simberloff *et al.*, 2010). In 2008 and 2009 an international (Argentina, Brazil and Uruguay) project focusing on alien tree invasions in the Pampas Biome of South America highlighted the effects of large-scale plantations as focuses for pine invasion (Guadagnin *et al.*, 2009; Fonseca *et al.*, 2013). Research on conifer invasions can greatly benefit from studying processes at early stages of invasion, such as those in South America, compared to most evidence reported in the literature for areas where conifer invasions are consolidated. Furthermore, in terms of conservation and invasion control, this region still has the opportunity for prevention and adequate management actions to avoid large-scale invasions, saving resources to restore native ecosystems after invasions have occurred. In recent years, new studies have been conducted and evidence is mounting about the impacts that invasive pines are having in South American ecosystems, from microsite effects on plant diversity to large-scale changes in fire regimes. Thus, there is an urgent need for clear guidelines on how to deal with the undesired effects of invasive pines in South America.

Tab. 14.1: Invasive Pinaceae species registered as naturalized and/or invasive in different biomes of South America (modified from Simberloff *et al.*, 2010 based on current literature and records). "?" indicates there is no sufficient information to assess the invasive status.

Biome	Country	Species	Naturalized	Invasive
Tropical humid forests	Brazil	*Pinus caribaea*	no	no
		Pinus chiapensis	no	no
		Pinus elliottii	yes	yes
		Pinus maximinoi	yes	yes
		Pinus oocarpa	no	no
		Pinus patula	no	no
		Pinus taeda	yes	yes
		Pinus tecunumanii	no	no
Subtropical	Brazil	*Pinus caribaea*	no	no
	Argentina	*Pinus caribaea*	?	?
	Brazil	*Pinus chiapensis*	no	no
	Argentina	*Pinus elliottii*	yes	?
	Brazil	*Pinus elliotii*	yes	yes
		Pinus maximinoi	yes	yes
		Pinus oocarpa	yes	yes
		Pinus patula	no	no
		Pinus radiata	no	no
		Pinus serotina	?	?
	Argentina	*Pinus taeda*	yes	?
	Brazil	*Pinus taeda*	yes	yes
		Pinus tecunumanii	yes	yes
Tropical grasslands	Brazil	*Pinus caribaea*	no	no
		Pinus chiapensis	no	no
		Pinus elliottii	yes	yes
		Pinus kesiya	no	no
		Pinus maximinoi	no	no
		Pinus oocarpa	yes	yes
		Pinus patula	no	no
		Pinus pseudo-strobus	no	no
		Pinus taeda	yes	yes
		Pinus tecunumanii	yes	yes

continued **Tab. 14.1:** Invasive Pinacea species registered as naturalized and/or invasive in different biomes of South America (modified from Simberloff *et al.*, 2010 based on current literature and records). "?" indicates there is no sufficient information to assess the invasive status.

Biome	**Country**	**Species**	**Naturalized**	**Invasive**
Tropical dry forest	Argentina	*Pinus elliottii*	yes	yes
		Pinus halepensis	?	no
		Pinus taeda	yes	?
		Pinus radiata	?	no
Temperate grasslands	Argentina	*Pinus elliottii*	yes	yes
	Brazil	*Pinus elliottii*	yes	yes
	Uruguay	*Pinus elliottii*	?	?
	Argentina	*Pinus halepensis*	yes	yes
		Pinus patula	?	yes
		Pinus pinaster	yes	yes
	Uruguay	*Pinus pinaster*	yes	yes
	Argentina	*Pinus pinea*	no	no
		Pinus radiata	yes	yes
		Pinus roxburghii	?	?
		Pinus taeda	yes	?
	Brazil	*Pinus taeda*	yes	yes
Warm desert	Argentina	*Pinus ponderosa*	yes	no
Evergreen Sclerophyllous Forest	Chile	*Pinus radiata*	yes	yes
		Pseudotsuga menziesii	?	?
Temperate rain forest	Argentina	*Pinus contorta*	yes	no
	Chile	*Pinus radiata*	yes	yes
	Argentina	*Pinus radiata*	yes	?
		Pseudotsuga menziesii	yes	yes
	Chile	*Pseudotsuga menziesii*	yes	yes
Mixed mountain systems	Chile	*Pinus contorta*	yes	yes
		Pinus ponderosa	yes	no
		Pinus radiata	no	no
		Pinus sylvestris	yes	yes
		Pseudotsuga menziesii	yes	yes
Cold winter desert	Argentina	*Pinus contorta*	yes	yes
	Chile	*Pinus contorta*	yes	yes
	Argentina	*Pinus ponderosa*	yes	no
		Pseudotsuga menziesii	yes	no

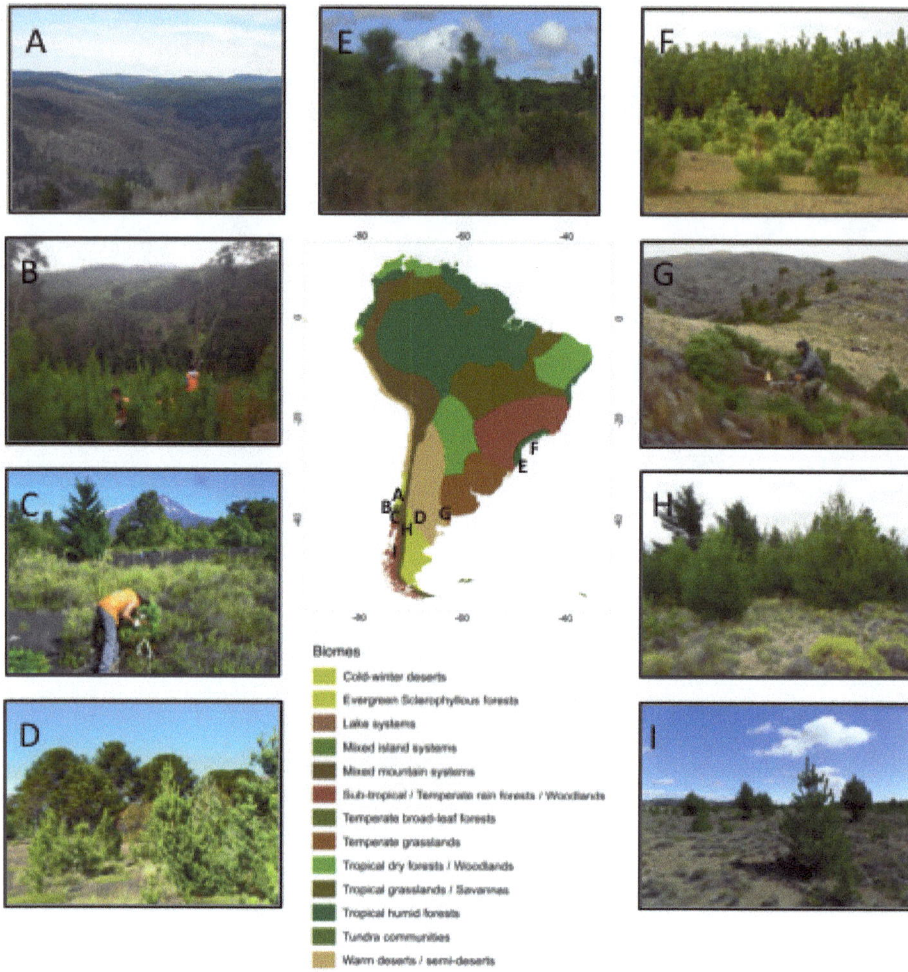

Fig. 14.1: Examples of pine invasions in South America. A) *Pinus radiata* in Maulino Forest. Cauquenes, Chile; B) University volunteers controlling *Pinus radiata* in Costal Forest. Hualpen, Chile; C) University volunteers controlling *Pseudotsuga menziesii* in Conguillio National Park, Chile; D) *Pinus contorta* in Malalcahuello National Reserve, Chile; E) and F) *Pinus taeda* in the Restinga ecosystem, Florianópolis, Brasil; G) Control of *Pinus halepensis* in Ernesto Tornquist Provincial Park. Buenos Aires, Argentina; H) *Pinus contorta* in Bariloche, Argentina; I) *Pinus contorta* in patagonian steppe. Coyhaique, Chile.

Here, we review the evidence of impacts, management, and the policy context of pine invasions in South America in order to advance in the reduction of this threat to biodiversity and ecosystem services. We will tackle these issues across biomes of South America: tropical and subtropical forests; tropical and subtropical grasslands, savannas and shrublands; Mediterranean forests and shrublands; temperate grasslands, savannas and shrublands; and temperate and sub-Antarctic forests. We advocate a more comprehensive approach to control pine invasions using prevention, early detection, containment and population management, restoration, and the inclusion of society in all steps of this process. We hope that drawing upon the experience of other countries with more advanced invasion scenarios may help to better manage the problem in South America before higher environmental, social and economical costs arise.

14.2 Pine Invasions and Their Impacts in South America

14.2.1 Tropical and Subtropical Forests

Pines were introduced to tropical and subtropical South America in the 19th century, but mostly at small scales and for horticulture. During the second half of the 20th century, governments started to import numerous pine species for silvicultural experimentation, and by the 1960s large-scale commercial pine plantations became common in many areas (Zenni & Ziller, 2011; Zenni & Simberloff, 2013). Because commercial plantations benefited from provenance trials, foresters tended to plant provenances and species in climates and soils to which they were well adapted (Zenni *et al.*, 2014). This resulted in positive genotype-by-environment interactions that promoted invasions. Currently, latitudinal patterns of pine invasion seem to match the latitudinal variation in pine native distributions. Not surprisingly, tropical pines are invasive in tropical regions, sub-tropical pines are invasive in sub-tropical regions, and temperate pines are invasive in temperate regions. Climate is a major driver limiting pine range distributions in both native and introduced ranges (e.g., Boulant *et al.*, 2009; Nuñez & Medley, 2011; McGregor *et al.*, 2012a). In the tropical Central Savannas of South America we currently see invasions by *Pinuso ocarpa* and *Pinus elliottii*, whereas in sub-tropical forests and grasslands we see invasions by *Pinus taeda* and also by *P. elliottii*. In sub-tropical forests *Pinus glabra* is also invasive, though only in one location as far we know, probably owing to the limited planting of the species (Zenni & Simberloff, 2013). We lack records of pine invasions in tropical forests (i.e., Amazon or Atlantic rainforests).

14.2.2 Tropical and Subtropical Grasslands, Savannas and Shrublands

The impacts of invasive pines in tropical and sub-tropical savannas have been studied more extensively than in other tropical and sub-tropical biomes in South America. In these savannas, *P. elliottii* densities can reach more than 12,000 plants per hectare in a period of 20 years and exclude non-woody plants from the native community. The result is a novel plant community highly dissimilar to non-invaded areas (Abreu & Durigan, 2011). However, in older pine stands, the re-establishment of native species previously excluded may occur, owing to the natural trimming process of the invasive population. Sub-tropical pines are shade-intolerant species and self-recruitment underneath dense pine invasions is limited, which results in an open canopy after pines reach full size, lowering competition for light and allowing native species to re-colonize the invaded area (Abreu *et al.*, 2011). In these cases, even though pine density in the invasive population decreases and native richness and abundance increases, the plant community remains highly dissimilar to non-invaded communities (Abreu *et al.*, 2011; Dostál *et al.*, 2013).

In tropical ecosystems, the impact caused by pine invasions on native plant communities is probably similar to the impacts observed in other biomes. The impacts are visible faster in these ecosystems, due to higher pine growth rates in warmer regions. In grassland relicts in the Atlantic rainforest, pine invasions decrease the overall richness and abundance of native grasses, forbs, and shrubs (Falleiros *et al.*, 2011). Researchers have characterized the spread of pines in South American sub-tropical forests (e.g., Emer & Fonseca, 2010, Zenni & Simberloff, 2013), but little is known about its impacts. The few studies that attempted to measure impact of pine invasions in tropical forests found that invasions tend to increase the depth of the litter layer and reduce recruitment of native species (Voltolini & Zanco, 2010).

Sub-tropical pines (e.g., *P. taeda* and *P. elliottii*) are also invasive in coastal dunes along the Atlantic coast of Brazil (Portz *et al.*, 2011; Zenni & Ziller, 2011). The invasions in dune habitats frequently form monocultures thick with pines that exclude native species and cause soil erosion. Also, most invasion reports are from coastal dunes in southern Brazil. Only preliminary assessments have been made on the status of pine invasions and their impacts on coastal dunes in lower latitudes, such as with the spontaneous spread of *P. pinaster* in southern Buenos Aires (Argentina) (Yezzi *et al.*, 2011, 2013; Cuevas & Zalba, 2011).

14.2.3 Mediterranean Forests and Shrublands

In the Mediterranean region of Chile, Spaniards and other Europeans first introduced non-native conifers in the 17th century, initially as ornamentals and later to control erosion and stabilize dunes (Peña *et al.*, 2008; Simberloff *et al.*, 2010). In the late 19th century, monospecific plantations were established where native forests had been destroyed (e.g. by mining, fires and grazing). *Pinus radiata* was introduced unintentionally in 1885 and planted with other North American conifers as an ornamental species (Lara & Veblen, 1993). Starting in the late 1960s, *P. radiata* was planted in large scale afforestation projects, heavily subsidized by the Chilean government since 1974, as the species was found to be especially suitable for the timber and pulp industries (Espinosa *et al.*, 1990; Le Maitre, 1998). At present, commercial plantations of *P. radiata* reach c.a. 1.5 million ha and are the basis for forest industry in Chile (Infor, 2013).

Pinus radiata from plantations are invading natural forests and xeric open shrublands, especially when disturbances such as harvesting and fire open the forest (Bustamante & Simonetti, 2004). There is evidence of pine invasion in open areas; however, the mechanism is not yet clear. The effect of the shading of native shrub *Lithrea caustica* on the seedling establishment of *P. radiata* is positive for recruitment and negative for seedling survival in semiarid ecosystems. This suggests that a common mechanism proposed to resist invasion in forest ecosystems such as shading is probably not sufficient to inhibit invasion in a semiarid region (Becerra & Bustamante, 2011). In coastal maulino forests, *P. radiata* is invading following deforestation and fragmentation (Bustamante & Castor, 1998; Bustamante *et al.*, 2003). In small isolated *Nothofagus* forests, surrounded by *P. radiata* plantations, the species invades protection zones, especially after disturbance that creates openings in the vegetation (Bustamante & Castor, 1998).

This invasion is relatively recent and has occurred rapidly. Bustamante and Simonetti (2005) indicated that the seeds of *P. radiata* are dispersed into the native forests, however seedling establishment occurs only at the edges and, therefore, they conclude that this exotic species is not invading less disturbed native forests. In less than a decade, it is possible to find a high proportion of young individuals and reproductive individuals growing inside fragments of native forest (Gomez *et al.*, 2011). Moreover, the pine density is negatively correlated with fragment size, while the proportion of reproductive pines is not similarly correlated (Gomez *et al.*, 2011). Due to the small size of the forest fragments present in this area along with current evidence, an increase is expected in the invasion of pine in the Mediterranean area. This situation could be reversed if active management of the invasion in native forest fragments and surrounding plantations is applied.

Pines, besides their negative impacts on biodiversity, may hold some potential for restoration of wildlife within plantations. *Pinus radiata* plantations might be a suitable habitat where the native tree *Cryptocaria alba* can regenerate (Guerrero & Bustamante 2007). Similarly, in degraded open sites where nurse plants are not available, *P. radiata* trees invading may facilitate the regeneration of native species, although studies show that facilitation produced by native trees is stronger than that produced by *P. radiata* (Becerra & Montenegro, 2013). Moreover, a well-developed understory in forestry plantations might serve as a surrogate habitat for native species and mitigate the negative effect of plantations on species richness (Simonetti *et al.*, 2013)

14.2.4 Temperate Grasslands and Savannas

Patagonia temperate grasslands or steppes in Chile and Argentina cover a much more extensive area compared with forests in Southern South America. They have also been the target of large afforestation plans, especially in Argentina. From 1970, pines have been planted mostly in these areas and nowadays, due to legal restrictions on plantations with non-native species in native forest areas, grasslands are the only ecosystem where companies and local owners can establish new pine plantations. Overall, two species were used in monospecific plantations: *Pinus ponderosa* and *Pinus contorta*. *Pinus contorta* has shown more aggressive invasive behavior, generating saplings up to 4 km from parent trees (Langdon *et al.*, 2010). When close to the source of seeds, *Pinus contorta* has invaded under different densities that depend on vegetation cover and land use, among other variables. In steppe areas where grazing has been removed, the invasion is more dense (Sarasola *et al.*, 2006). On the other hand, ponderosa pine, the species most planted, has not shown invasive potential, with saplings reaching only 50 meters from parent trees.

Extensive pine plantations have been established in areas that were formerly steppe or open woodland, where lack of fuel continuity was a major limitation to fire spread (Nuñez & Raffaele, 2007). Today, however, areas of these non-native conifers have burned and others create the potential for extensive crown fires in habitats previously characterized by surface fires and lower fuel volumes (Veblen *et al.*, 2003). Burned plantations interfere with post-fire succession to the original matorral vegetation (Nuñez & Raffaele, 2007) creating a positive feedback between fire and invasive pines. This effect of fires promoting pine expansion has also been documented for warmer pampas grasslands, where invasions by *Pinus halepensis* and *P. radiata* have been associated with fires (Zalba *et al.*, 2008). Introduced conifer plantations in Argentinean Patagonia have increased fire severity. For example, plantations of *Pinus ponderosa* and *P. contorta* have produced major conflagrations, initiated by lightning and human activity.

Replacement of native treeless vegetation with dense even-aged forests is by far the most striking impact of pine invasions in these ecosystems (Richardson *et al.*, 1994; Richardson & Higgins, 1998; Zalba & Villamil, 2002). Pine species may threaten to convert entire shrubland and grassland communities into conifer forests, with several native species in danger of at least local extirpation (Harding, 2001) even in high elevation environments (Pauchard *et al.*, 2009). In northwest Patagonia in Argentina, Lantschner *et al.*, (2008) found that the impact of pine plantations on bird communities depends on the landscape context and stand management practices. When plantations replaced steppes, the bird community was partially replaced by a new one, similar to that of ecotonal forests.

Pine invasion in Argentinean pampas reduced the diversity of native plants, displacing endemic species and promoting invasion by other non-natives (Zalba & Villamil, 2002; Cuevas & Zalba, 2009), and are also associated with changes in bird communities, including decline of obligate grassland birds and colonization by species that are less habitat-specific and colonize from forested regions (Zalba, 2000). Grazing has been found to promote plant invasion in this region (de Villalobos & Zalba, 2010, Loidy *et al.*, 2010; 2012), where cattle and feral horses reduce the biological resistance of native plant communities, creating windows of opportunity for the establishment of pine seedlings (de Villalobos *et al.*, 2011).

The light environment underneath pine plantations and invasions is critical to determine vegetation diversity. A reduction in plant diversity has been observed under young *Pinus contorta* trees (5-15 years) in the steppe of Coyhaique in the Chilean Patagonia (Pauchard *et al.*, unpublished data). Similar results were found in the steppes of Patagonia in Argentina, where dense and closed plantations with cover close to 90% see reductions of vegetation species richness from 31.7 to 10.4, and a cover reduction from 45.5% to less than 0.5% (Gyenge *et al.*, 2010). The reduction in diversity is positively correlated with pine size, and it can therefore be expected that older invasions have similar effects to those seen in plantations where almost no understory species are able to coexist with pines (Nuñez & Raffaele; 2007). Corley *et al.*, (2006; 2012) found that in dense plantation assemblages, richness of ants and beetles were modified and impoverished, with fewer ants in introduced pine plantations than in native steppe vegetation and fewer ant species in dense plantations. The impacts of pine invasions may be more intense that those associated with plantations, given the significantly higher density of stems in invaded areas (> 10 000 ind/ha) compared with the less dense plantations (ca. 1 000-1 600 ind/ha). Therefore, studies should address both the effects of plantations and invasion in these open environments.

14.2.5 Temperate and Sub-antarctic Forests

The temperate ecosystems of Chile and Argentina have been extensively planted with introduced Pinaceae species in recent decades, both for production purposes and to restore eroded and degraded land (Pauchard *et al.*, 2014). In southern Chile, the Chilean Forestry Institute (Instituto Forestal) developed a species introduction program in the early 1960s, establishing trial plots in more than 100 areas located between the semi-arid and temperate regions, including protected areas such as parks and reserves (Loewe & Murillo 2001; Pauchard *et al.*, 2014). Peña and Pauchard (2001) first warned that the introduced conifers of these trial plots were becoming invasive. *Pinus radiata* is not suitable for these harsher environmental conditions; therefore other species (*P. contorta*, *P. ponderosa*, *P. sylvestris*, and *Pseudotsuga menziesii*) are becoming widespread for plantation in more extreme, colder environments, accounting for ca. 50,000 ha. All species have been shown to be able to establish outside plantations, and *P. contorta* and *P. menziesii* are considered invasive in particular types of ecosystems (Peña *et al.*, 2007; Simberloff *et al.*, 2003). *Pinus contorta* and *P. menziesii* have become invasive in open and disturbed areas, as well as in natural vegetation, in the southernmost, colder environments (Peña *et al.*, 2008; Pauchard *et al.*, 2008). In Argentina, conifers were first introduced for forestry purposes to Isla Victoria, in the Patagonian Andes, in 1910, as an experimental project (Simberloff *et al.*, 2010). A *Pinus ponderosa* C. Lawson plantation initiated there in 1927 is one of the first conifer plantations in Latin America (Cozzo, 1987). Plantations on this island served as a government nursery for the rest of Argentina through the 1960s (references in Simberloff *et al.*, 2002), and plantation forestry in Patagonia increased rapidly in the 1970s (Schlichter & Laclau, 1998). Today, plantings of Pinaceae species are unevenly distributed across the Patagonian forest as a product of specific and localized efforts of afforestation. Those plantations include species such as *P. radiata, P. ponderosa, P. contorta, P. sylvestris, Larix decidua* and *Pseudotsuga menziesii*. Although several species are able to establish in native forests openings, *P. menziesii* is by far the most invasive conifer in forest dominated by *Austrocedrus chilensis and Nothofagus dombeyi* (Sarasola *et al.*, 2006).

Some factors seem to be slowing down the invasion of Pinaceae into forests, but these factors clearly do not provide a full barrier to the invasion (Simberloff *et al* 2010; Nuñez *et al.*, 2013). Among these factors, seed predation is playing an important role since pine seeds appear to be highly consumed and preferred by the native seed predator fauna (mainly birds and rodents) (Nuñez *et al.*, 2008; Caccia & Ballaré, 1998). Also, lack of compatible mycorrhizal fungi may be delaying the invasion in forests (Nuñez *et al.*, 2009). Although these factors may prevent invasion, other factors such as the presence of exotic mammals (deer, wild boar) can be promoting the invasion by preferring to consume their competitors – native tree species – instead of pines (Nuñez *et al.*, 2008; Relva *et al.*, 2010). In addition, exotic mammals have been shown to disperse mycorrhizal fungi needed for pine invasion, so they may be accelerating their invasion (Nuñez *et al.*, 2013). Therefore, even though some factors may be acting as a barriers to invasion in forests, there is ample evidence that pinaceae can invade in temperate and sub-Antarctic forests, producing large impacts.

Few studies have shown impacts of introduced conifers in temperate forests in South America, in either plantations or invaded areas. A decrease in plant diversity was recorded in *Araucaria araucana* forests invaded by *Pinus contorta* in the Andes of South-central Chile (Urrutia *et al.*, 2013). In forested ecosystems, invasive pines displace native species, but their effect seems to be lower than in the open temperate steppe (Pauchard *et al.*, unpublished data). Changes in fuel conditions and therefore in potential increase in fire frequency and intensity has been reported for these Araucaria forests. The main effect in young invasive stands is related to higher flammability and fuel continuity, although it is expected than in older stands the amount of fine fuel could also increase (Cóbar-Carranza *et al.*, 2014). Due to the lack of studies of the impacts of pine invasions in the biota of temperate forests, changes in animal and plant communities documented in conifer plantations can be extrapolated to invaded ecosystems. In small mixed plantations of *Pseudotsuga menziesii, Pinus radiata*, and *P. sylvestris* embedded in forests dominated by *Nothofagus dombeyi* in Northwest Patagonia, Paritsis and Aizen (2008) found decreasing species richness of understory vascular plants, epigeal beetles, and birds.

14.3 Current Policies and Management Actions on Pine Invasions

Biological invasions have only recently become part of the conservation agenda in South America. Unfortunately, many invasive species have been, and are still promoted as, important cultivars in forestry, farming and farm-fisheries (Nuñez & Pauchard, 2010). Pine plantations have been widely promoted across South America as an efficient and fast forestry cultivar, creating a strong industrial complex that uses pines as primary material for pulp, boards, and wooden furniture (Espinosa *et al.*, 1990). Government subsidies for pine plantations have been put in place in many countries, including Chile, Argentina, and Brazil. Native ecosystems and degraded areas have been widely afforested and no prevention actions have been implemented to control spread (Simberloff *et al.*, 2010).

Deficient forestry policies have increased the risk of pine invasions as much as the lack of data on invasive species has hindered control efforts. The first databases on invasive alien species in South America were online in 2003 in Argentina and in 2004 in Brazil (Box 14.1). The database structure was then disseminated to Chile, Uruguay, Paraguay, Bolivia, Peru, Ecuador, Colombia, Venezuela, and Surinam. This database was established between 2005 and 2011 by joint efforts of the Inter-American Biodiversity Information Network (IABIN) thematic network on invasive alien species (I3N), The Horus Institute for Environmental Conservation and Development in Brazil, the Universidad Nacional del Sur in Argentina, The Nature Conservancy South America Invasive Alien Species Program, and the Global Invasive Species Programme (GISP). The database led to the increased visibility of biological invasion issues at a continental scale, and to governmental concerns aligned with the guidelines and requests of the Convention on Biological Diversity.

Box 14.1. Alien Species Database and Conifer Invasion Management.

The I3N Network databases have been a substantial aid for research and a starting point for broader consultation upon building the species lists at the state or national levels. Once lists are in place, regulations are needed to define restrictions for the use of the species, and more development in legal terms, leading to changes in land use and general awareness. In Rio Grande do Sul, Brazil, where what is probably the most extensive pine (*P. taeda* and *P. elliottii*) invasion in Brazil can be seen along sand dunes and coastal grasslands, specific regulations for the use of pines have been proposed. The intention lies in reorganizing the landscape, maintaining plantations that are clearly limited and not allowed to expand over the surrounding landscape. The ornamental use of pines will be interrupted, as well as their use as shade trees or windbreaks. Production shall continue unharmed, but more responsibly while applying proper management practices to prevent and control invasions. These regulations are derived from Portaria SEMA RS 79/2013, the official list of invasive alien species published by the State Secretary of Environment.

Governments have taken considerable time to realize the importance of the threat posed by invasive species, while economic and political criteria have overridden environmental concerns. In the last few years, efforts to limit invasions have finally percolated into government and policy. Brazil, Colombia, and Uruguay developed and published official National Strategies for managing invasive alien species between 2009 and 2011, and Argentina is expected to complete one by the end of 2016. Although a rather large number of articles have been published in Latin America on biological invasions, the issue remains relatively under-studied in academic circles (Quiroz *et al.*, 2009). Pines are no exception, with most of the work already done focused on describing and understanding invasions, and less on how to control them.

Private efforts have, in some cases, moved faster than government initiatives when dealing with tree invasions. In the last decade, controlling biological invasions became a principle of the Forest Stewardship Council, first applied to pines that escape plantations, then gradually extended to other invasive species present in forest company properties. Although the principle is in place, complementary regulations are needed to make it functional. Official lists of invasive species constitute important references along with regulations that limit or even prohibit their use. The Colombian Ministry of Environment and Sustainable Development published a preliminary list of invasive species in 2011 (Ministerio de Ambiente y Desarrollo Sostenible, 2011), as well as a publication on risk analysis of introduced species in 2010 (Baptiste *et al.*, 2010). In Brazil, due to the lack of a national list, three official lists have been published at the state level in Paraná, Santa Catarina, and Rio Grande do Sul, while Rio de Janeiro and São Paulo are working on official lists at present. In Chile, FSC criteria and social awareness have pushed companies to recognize the invasive status of *Pinus radiata*. Companies are now required to control invasive pines outside plantations under these new forest certification schemes. Interestingly, forest companies are leading ambitious plans to control *P. radiata* in protection zones and forests of high conservation value.

The concern and awareness of the problem of biological invasions have grown in parallel with worldwide awareness and have been incorporated into the agenda of different agencies of national and provincial governments. This growth is the result of a maturation process over the last ten years, probably associated with pressure from environmental NGOs, as well as the increased availability of local and regional scientific information that gives certainty to the real dimension and relevance of the issue. Thus, in May 2013, a workshop funded by the national government of Argentina was held in Bariloche with the participation of international and local researchers, provincial and national decision makers, and forestry producers. Consensus was reached on the need to take action. Among the actions outlined is the inclusion of national regulations to subsidize new requirements for forestry plantations, such as measures to prevent invasions, monitoring, and control plans in future afforestation efforts, and exclusion of *Pinus contorta* among the species to be subsidized for plantations due to its high invasive potential. These regulations became effective in 2014 (MAGPNA, 2014).

In temperate and sub-antarctic forests, although substitution of native forest is no longer allowed, governments of both Chile and Argentina are still pushing for further pine plantations in eroded or deforested areas. Unfortunately, native forests take a relatively long time to recover after fire or other disturbances, especially if these are repeated over time. Thus, in an effort to reduce deforestation and increase provision of forest goods, governments have subsidized plantations in marginal conditions with limited economic results and negative environmental impacts. Forest companies, which in the past were also responsible for extensive plantations of invasive pines in these areas, are now recognizing that they have to reduce these negative impacts. Lately, these companies have not been planting some species, such as *Pinus contorta*, and have plans for the eradication of plantations of this species in Chilean Patagonia.

Efforts to control pines in South America are very limited, and pine control is only recently becoming a concern for forest companies and government agencies. In Argentinean Patagonia, the national parks administration is reducing the areas with plantations inside their domain and is controlling the spread of pines into natural areas (APN, 2000). Recently, several projects including pine invasion control and monitoring protocols in provincial forests are being conducted, funded by the national government. Their goal is to evaluate the impact of invasion and strategies to control their spread, with the ultimate goal of incorporating pine invasion control as a regular forestry practice. One incentive for these projects is the new national forest protection law that aims, among other goals, to control the presence of exotic species in national forests. South American temperate grasslands have been the setting for some of the first attempts at controlling invasive pines and restoring native communities. Mechanical control of *Pinus halepensis* and *P. radiata* has been conducted in Southern Argentinean Pampas since the early nineties, combined with studies on pine reproductive biology, seed longevity, and invasion spread, following an adaptive management strategy (Cuevas & Zalba, 2009; 2010; 2013). Considering the effects of fires on seed release from serotinous cones, and the interval between recruitment and seed set, fire has been identified both as a promoter of pine invasion but also as a potential tool for their control, providing that time between fires is less than that required for tree maturation (Cuevas & Zalba, 2010; Zalba *et al.*, 2009).

In Chile, no formal studies have been conducted to develop the best eradication techniques, but forest companies are under pressure to eradicate invasive pines in protection zones due to forest certification regulations (Pauchard *et al.*, 2014). The impact of pine invasions is slowly being considered both by private and public stakeholders, and there are currently plans to develop techniques to control them at local scales. However, it is not yet clear if large-scale management will be implemented or if specific regulations will enforce the control of pine invasions.

14.4 The Future: How to Reduce Impacts?

In technical terms, at the stand scale, pine invasions may be among the least complicated to control, especially as adult trees do not re-sprout once cut down. However, in landscapes dominated by pine plantations, propagule pressure is so high that re-invasion may occur and ecosystem processes that trigger invasion may already be changed (e.g. fire regimes). Also, because they are dispersed by wind, pines can reach places with limited accessibility in mountains and protection zones, where they are not detected until their canopies have surpassed the native vegetation.

Given the relative predictability in terms of the species and the areas than can be invaded, and how easy it is to detect and eliminate them, people managing invasive pines may have an advantage compared to other more subtle or resistant invasions (e.g. *Acacia* spp.). Nonetheless, there are several shortcomings, especially as pines are seen as an important economic resource, and therefore, cultural and economic limitations usually complicate the implementation of prevention and control strategies. Conflicts abound in governmental agencies with contrasting visions and objectives, as well as on private properties where owners wish to take advantage of invasions without taking responsibility for them, and thus do not undertake any proper forest management. The main challenges are in communication and coordination, and clear regulations are strongly needed.

New regulations and research in some key areas (e.g. restoration) are crucial in helping to prevent the impacts of these invasions throughout South America. Invasions are widespread on the continent, but appropriate management can prevent future invasions and control current ones, at least in properties associated with forestry production. A clear understanding of how biotic, abiotic, and anthropogenic factors affect pine invasion is needed for effective control (Figure 14.2). There is no universal "recipe" for pine invasion management, as the factors determining the invasion outcome vary with the species, the site, and the silvicultural and control strategies. Thus, only general guidelines of the best management practices can be addressed in this text.

Controlling pines is not only good for the environment but also for the forestry industry, given that invasions hamper future silvicultural practices, are sources of diseases, increase fire frequency, and affect the sustainability of the forestry business with negative marketing repercussions. Here, we address some of the key stages of pine invasion management:

1) Prevention and risk analysis: Despite the fact that the biological characteristics of species have not been enough to precisely predict invasion behavior by alien species, risk analysis can provide valuable inferences on invasion probability and its consequences. This information can be used in decision-making and to define management strategies and public policies (National Research Council, 2002). Risk does not only involve species biology, but also environmental damage, social and economic impacts, and public health issues, as well as the feasibility of control or eradication. Risk is the product of the likelihood of an event or process and its consequences (National Research Council, 2002). Risk analysis techniques are now an important tool to prevent plant invasions and may be used for screening new pine introductions (see Box 14.2).

Fig. 14.2: Site-specific factors determining the outcome of a pine invasion modified from Pauchard *et al.* (2014). The interaction between species traits and the invaded community determines the expected invasion risk. However, silvicultural techniques and landscape planning may reduce or increase this risk, affecting propagule pressure. For example, the use of other species wind buffers to avoid seed dispersal or the establishment of plantations in safe-sites of the landscape may reduce the risk of invasion. Control actions should target all stages of the invasion and should be coordinated with the silvicultural schemes at the stand and lanscape scales. Policies and regulations set the stage for both silvicultural and control practices and therefore are crucial to reduce pine invasions.

Box 14.2. Assessing the Risk of New Pine Introductions.

Risk assessment protocols can be used for different and complementary purposes. Originally conceived to assess the risk of introducing new species into a country, they can be applied to help define management priorities for several species in a given area according to the level of risk, select species in a commercial genus that will pose less problems in terms of invasion, provide scientific evidence of invasive capacity in the absence of field data or research, and provide impartial assessments of species to be included in official invasive species lists. Pines may also be screened using risk assessment tools (McGregor *et al.*, 2012b).

One of the most relevant variables of weed risk assessment is climate. Climate is a good predictor of conifer naturalization and invasion both at the genotype and at the species levels (Nuñez & Medley, 2011; Zenni *et al.*, unpublished). In general, climatic information from native and other invasive ranges of pines can be used to assess the risk of naturalization in another region. Climate, combined with knowledge from forestry trials (Zenni *et al.*, 2014) and intense propagule pressure, explains why commercial forestry tends to be the most frequent source of pine invasions (Essl *et al.*, 2010).

The Universidad Nacional del Sur in Argentina and the Horus Institute in Brazil designed new risk analysis protocols and adjusted existing ones to fit South American conditions through the I3N - IABIN Invasive Species Network. Results of assessments are available for 149 plants at http://www.institutohorus.org.br/index.php?modulo=inf_analise_risco. This platform can be used to evaluate new pine introductions in South America, including analyses of species pathways within the South American continent.

It is unlikely that many new conifer species will be introduced in South America in the coming years, given that a large proportion of species have already been introduced to the continent. However, introduction programs are still in place for forestry and ornamental uses. Therefore, pre-introduction screening tools (i.e., Pheloung *et al.*, 1999) should be used before allowing new introductions into South America. Predictive models that consider species mean trait values (e.g., Rejmánek & Richardson, 1996; Grotkopp *et al.*, 2002) tend to work only at larger spatial scales (Zenni & Simberloff, 2013). Instead, models of current and potential spread at the genotype level that explicitly incorporate climate, soil, belowground interactions, and commercial plantations can be highly informative. For instance, recent work has shown that genotype-environment interactions are a major driver of *Pinus taeda* invasions (Zenni *et al.*, in review). In this study, the authors found that genetic constraints limit the ability of provenances to invade in unfavorable introduced habitats. The invasive potential of provenances was largely predictable by isoclines of temperature and precipitation. The adaptive mechanism was strong enough to overcome important differences in propagule pressure.

2) Early detection: Plans to start forestry practices need to be associated with plans to control the spread of trees. Decades of pine invasion research have shed light on the process of invasion. There is strong scientific evidence available in a number of areas that can be readily used for prevention and control of invasions. There is information on the key characteristics of invaded ecosystems that make them more prone to invasion, and the characteristics of species that make them more capable of invading. For example, there is ample evidence that some areas are more invasible than others (e.g. open disturbed areas are easily invaded, while close mature forests are less so), and some species are more invasive than others (species with smaller seeds invade more than species with large seeds (Rejmanek & Richardson, 1996)). The identification of the ecological events that open opportunities for pine establishment and invasion, like fires, grazing by large ungulates, and topsoil removal, can help to prevent or lead to early detection and eradication of new foci of invasion. This information can be relatively easy to use in legislation to guide tree plantations into areas with low risk of invasion, as well as to direct the introduction of species with low probability of invasion. This may not solve the invasion problems completely, but would make it easier or cheaper to manage. Species that are notably problematic (e.g. *Pinus contorta, P. radiata, P. elliottii*) and areas more prone to invasion or sensitive zones, such as natural reserves, need to be excluded or carefully monitored.

3) Containment and population control: Regarding post-invasion management, plans have to be in place to control the spread even when all precautions selecting sites and proper species have been taken (Figure 14.2). An adaptive management strategy is usually the best way to reduce uncertainty about the best control options, and also for understanding the causes behind the invasion (Zalba, 2010; Zalba & Ziller, 2007). Pines tend to colonize new areas, even in their native range, so expecting them not to do so is unrealistic. Information on the age of cone production (i.e. the start of seed production) is key for controlling the invasion. It is important to implement control measures soon after the trees reach reproductive age and not wait long enough for invasive trees to start producing cones themselves. Knowing the dynamics of the seeds, both in the cones and in the soil bank, is key information for effective pine control. Visiting the sites every few years (depending on the area and the age of cone production) to remove all the trees colonizing plantation surroundings is critical.

When invasions reach large scales and dominate the landscape, restoration becomes essential. Otherwise, if no active restoration is conducted, new invasions by pines or by other invasive plants (with similar or higher impacts) are highly likely. In South America, we can potentially avoid reaching the levels of invasion found in other countries such as New Zealand or South Africa, but if such levels are reached, active restoration plans must be implemented. Given the cost of restoration efforts, it is highly beneficial to avoid the stage were restoration becomes a requirement.

In order to avoid invasion of areas outside plantations, along roads, and throughout the landscape, regulations need to be in place restricting the use of pines to forestry, wood, pulp and paper, and other forest products. Other common uses, such as shade, wind-breaks, or ornamental purposes, need to be banned. Plantation owners have to be made responsible for control work and forest company associations need to engage in cleaning up current invasions, taking responsibility for the wide spread of the species.

Roadsides must be prioritized and kept clean of pines. Forestry stands must be clearly visible and pines must be contained within them. Those interested in producing pines must have projects approved by environmental agencies, even if there is a simplified process for small farmers, who also have to commit to controlling spread beyond plantations. Pine seedlings must no longer be distributed by public institutions or events, except for forestry production programs with clear rules established to avoid invasions. Pine species that has shown invasive behavior, in South America or elsewhere, should be listed as invasive species so that all broader legislation referring to environmental impacts can also refer to side the effects of badly managed plantations.

4) Restoration: Pines may be relatively easy to control locally, but international experience shows that restoration is needed for achieving a true recovery of the native vegetation after prolonged invasions. New Zealand has conducted massive efforts to control pine invasions in temperate grasslands using herbicide and mechanical control, but such efforts do not necessarily imply a recovery of the natural ecosystem. Other invasive species such as shrubs and herbs may become dominant as the pine is controlled. Active restoration may be required in cases where pines have decimated the original native plant populations and where few or no propagules of the native species are present. Changes in soil chemistry, litter depth, and soil biota may hamper the recovery of native vegetation. Ad hoc restoration plans should be developed in conjunction with control measures. In some cases, passive restoration may be sufficient to restore natural vegetation, for example around isolated pine trees in protection zones. In other cases, where massive stands of pines have displaced native vegetation, active restoration needs to be planned before controlling the pine. It may even be recommended that restoration and control are gradually implemented to facilitate the establishment of the native vegetation under the protection posed by the invasive pines. Control plans should always be aimed at restoring ecosystem function, structure, and the composition of native plant communities. On the other hand, the use of pine species for restoration purposes needs to be carefully reviewed and regulated (Zalba, 2013).

5) Public awareness and regulatory frameworks: Great effort should be aimed at increasing awareness of the importance of preserving native biodiversity and the threats posed by the expansion of invasive alien species, including pines. Education actions have to be directed not just to the public in general, but also to specific actors like governmental agencies, international aid and development programmes, and professionals in the field of natural resources. For pines, the positive services that they provide when growing in plantations should be recognized, and a clear distinction should be made with those pines growing outside plantations and invading natural communities. This distinction may be difficult to comprehend for the general public, but it is crucial to avoid useless generalizations and confrontations between private and public stakeholders. Basic regulations should be approved in all countries of South America indicating that pine plantations owners should take all necessary actions to avoid the negative externality of pine invasions. Forest certification schemes have already advanced towards that goal, thus it should not be difficult to adopt those regulations into national laws.

14.5 Concluding Remarks

Pine invasions are occurring in South America across multiple biomes and landscapes. However, there is a delay to these invasions when compared to similar regions in Africa, Australia and New Zealand, because of the more recent expansion of forest afforestation. While the impacts of pine invasions vary according to the ecosystem being invaded, it is clear that they pose a risk to local and regional biodiversity by affecting ecosystem composition, structure, and function. Changes in fire regimes may be one of the most striking elements of pine invasions. Policy and legal frameworks in South America have favored pine plantations with no consideration of the negative externality caused by their invasions. Still, governments subsidize the planting of extremely invasive pines. Actions in the future should consider: 1) prevention of new introductions; 2) early detection of invasions; 3) containment and control; 4) restoration; and 5) strengthening of regulations and public awareness. South America is still on time to take the necessary actions to reduce pine invasions, and it should learn from the experience of other regions where pine invasions are now causing major ecological and economic costs.

14.6 Acknowledgements

Aníbal Pauchard and Rafael García were funded by Fondecyt 1140485, ICM P05-002 and PFB-23. Martín Nuñez was funded by NSF 0948930 and BIO Nº5. Sergio Zalba received support from the Universidad Nacional del Sur and CONICET (Consejo Nacional de Investigaciones Científicas y Técnicas).

In a nutshell

- Pines are among the best-studied groups of invasive species. In South America pines have been planted for forestry, and in many areas are becoming highly invasive.
- The impacts of pine invasions can be high and vary according to region, posing a risk to local and regional biodiversity by affecting ecosystem composition, structure, and function. Still, governments subsidize the planting of extremely invasive pines.
- It is clear that governments and stakeholders should take immediate action to prevent this problem, which is currently at a stage where its control is doable.
- The experience of other regions such Africa, Australia, and New Zealand should help South America to prevent and manage pine invasions to avoid further environmental and economic costs.

14.7 Bibliography

Abreu, R. C. R., Assis, G. B., Frison, S., *et al.*, (2011). Can native vegetation recover after slash pine cultivation in the Brazilian Savanna? Forest Ecology and Management, 262, 1452-1459.

Abreu, R. C. R., Durigan, G. (2011). Changes in the plant community of a Brazilian grassland savannah after 22 years of invasion by *Pinus elliottii* Engelm. Plant Ecology and Diversity, 4, 269-278.

APN (2000). Proyecto de manejo silvicultural para la paulatina erradicacion de especies forestales exoticas de la Isla Victoria. Administración de Parques Nacionales de la República Argentina.

Baptiste, M. P., Castaño, N., Cárdenas, D., *et al.*, (Eds.) (2010). *Análisis de riesgo y propuesta de categorización de especies introducidas para Colombia.* Bogotá, D. C., Colombia: Instituto de Investigación de Recursos Biológicos Alexander von Humboldt.

Becerra, P. I., Bustamante, R. O. (2011). Effect of a native tree on seedling establishment of two exotic invasive species in a semiarid ecosystem. Biological Invasions, 13, (12) 2763-2773.

Becerra, P. I., Montenegro, G. (2013). The widely invasive tree *Pinus radiata* facilities regeneration of native woody species in a semi-arid ecosystem. Applied Vegetation Science, 16, 173-183.

Boulant, N., Garnier, A., Curt, T., *et al.*, (2009). Disentangling the effects of land use, shrub cover and climate on the invasion speed of native and introduced pines in grasslands. Diversity and Distributions, 15, 1047-1059.

Buckley, Y.M., Brockerhoff, E., Langer, L., *et al.*, (2005). Slowing down a pine invasion despite uncertainty in demography and dispersal. Journal of Applied Ecology, 42, 1020-1030.

Bustamante, R. O., Castor, C. (1998). The decline of an endangered temperate ecosystem: the ruil (*Nothofagus alessandrii*) forest in central Chile. Biodiversity and Conservation, 7, 1607–1626.

Bustamante, R. O., Simonetti, J. A. (2005). Is *Pinus radiata* invading the native vegetation in central Chile? Demographic responses in a fragmented forest. Biological Invasions, 7, 243–249.

Bustamante, R. O., Serey, I. A., Pickett, S. T. A. (2003). Forest fragmentation, plant regeneration and invasion processes across edges in central Chile. In G. A. Bradshaw & P. A. Marquet (Eds), How Landscapes Change. Ecological Studies 162 (pp.145–60). Berlin: Springer.

Caccia, F. D., Ballaré, C. L. (1998). Effects of tree cover, understory vegetation, and litter on regeneration of Douglas-fir (*Pseudotsuga menziesii*) in southwestern Argentina. Canadian Journal of Forest Research, 28, 683-692.

Cóbar-Carranza, A., García, R. A., Pauchard, A., *et al.*, (2014). Effect of *Pinus contorta* invasion on forest fuel properties and its potential implications on the fire regime of *Araucaria araucana* and *Nothofagus antarctica* forests. Biological Invasions, 16, 2273-2291.

Corley, J. C., Sackmann, P., Rusch, V. *et al.* (2006). Effects of pine tree silviculture on the ant assemblages (Hymenoptera: Formicidae) of the Patagonian steppe. Forest Ecology and Management, 122, 162-166.

Corley, J. C., Villacide, J.M., Vesterinen, M. (2012). Can forest management lessen the impact of pine plantations on beetle and ant diversity in the Patagonian steppe? Southern Forests, 74 (3), 195–202.

Cozzo, D. (1987). Evolución de la silvicultura de plantaciones forestales en Argentina. La postulación de mayores espacios de mejoramiento silvicultura. Simposio sobre Silvicultura y Mejoramiento Genético de Especies Forestales 1, 81–95. Buenos Aires: Centro de Investigaciones y Experiencias Forestales.

Cuevas, Y. A., Zalba, S. M. (2009). Control de pinos invasores en el Parque Provincial Ernesto Tornquist (Buenos Aires): áreas prioritarias y análisis de costos. BioScriba, 2, (2) 76-89.

Cuevas, Y. A., Zalba, S. M. (2010). Recovery of native grasslands after removing invasive pines. Restoration Ecology, 18, (5) 711-719.

Cuevas, Y. A., Zalba, S. M. (2011). Forestaciones de pino y ambientes de dunas costeras: rol de la depredación de semillas. Boletín de la Sociedad Argentina de Botánica, 46, (Supl.) 68.

Cuevas, Y. A., Zalba, S. M. (2013). Efecto del tipo de corte y de tratamientos en el mantillo para la restauración de pastizales naturales invadidos por Pinus halepensis. Boletín de la Sociedad Argentina de Botánica, 48 (2), 315-329.

de Villalobos, A. E., Zalba, S. M. (2010). Continuous feral horses grazing and grazing exclusion in mountain pampean grasslands in Argentina. Acta Oecologica, 36, 514-519.

de Villalobos, A. E., Zalba, S. M., Peláez, D. (2011). Pinus halepensis invasion in mountain pampean grassland: effects of feral horses grazing on seedling establishment. Environmental Research, 111, 953-959.

Dostál, P., Müllerová, J., Pyšek, P., *et al.* (2013). The impact of an invasive plant changes over time. Ecology Letters, 16 (10), 1277-1284.

Drake, J. A., Mooney, H. A., di Castri, F., *et al.* (1989). Biological invasions: a global perspective. Chichester, UK: John Wiley & Sons.

Emer, C., Fonseca, C. R. (2010). Araucaria forest conservation: mechanisms providing resistance to invasion by exotic timber trees. Biological Invasions, 13, 189-202.

Espinosa, M., Eso, R., Drake, F. (1990). Silvicultura de las plantaciones forestales en Chile: pasado, presente y futuro. Agro-Ciencia, 6, 131–44.

Essl, F., Mang, T., Dullinger, S., *et al.* (2011). Macroecological drivers of alien conifer naturalizations worldwide. Ecography, 34, 1076-1084.

Essl, F., Moser, D., Dullinger, S., *et al.* (2010). Selection for commercial forestry determines global patterns of alien conifer invasions. Diversity and Distributions, 16, 911-921.

Falleiros, R. M., Zenni, R. D., Ziller, S. R. (2011). Invasão e manejo de Pinus taeda em campos de altitude do Parque Estadual o Pico Paraná, Paraná, Brasil. Revista Floresta, 41, 123-134.

Fonseca, C., Guadagnin, D. L., Emer, C., *et al.* (2013). Invasive alien plants in the Pampas grasslands: a tri-national cooperation challenge. Biological Invasions, 15 (8), 1751-1763.

Gómez, P., Bustamante, R., San Martin, J., *et al.* (2011). Population structure of Pinus radiata D.Don in fragments of Maulino Forest in Central Chile. Gayana Botanica, 68 (1), 97-101.

Grotkopp, E., Rejmánek, M., Rost, T.L. (2002). Toward a causal explanation of plant invasiveness: seedling growth and life-history strategies of 29 pine (Pinus) species. The American Naturalist, 159, 396-419.

Guadagnin, D. L., Zalba, S.M., Costa Gorriz, B., *et al.* (2009). Árvores e arbustos exóticos invasores no Bioma Pampa - questões ecológicas, culturais e socioeconômicas de um desafio crescente. In V. De Patta Pillar, S. C. Müller, Z. M. de Souza, A. V. Ávila Jacques (Eds.), Campos Sulinos: conservação e uso sustentable da biodiversidade (pp.300-316). Brasilia D.F.: Ministerio do Meio Ambiente.

Guerrero, P. C., Bustamante, R. O. (2007). Can native tree species regenerate in Pinus radiata plantations in Chile? Evidence from field and laboratory experiments. Forest Ecology and Management, 253, 97-102.

Gundale, M. J., Pauchard, A., Langdon, B., *et al.* (2014). Can model species be used to advance the field of invasion ecology? Biological Invasions, 16 (3), 591-607.

Gyenge, J. E., Fernández, M. E., Rusch, V., *et al.* (2010). Towards Sustainable Forestry Development in Patagonia: Truths and Myths of Environmental Impacts of Plantations with Fast-Growing Conifers. The Americas Journal of Plant Science and Biotechnology, 3, 9-22.

Harding, M. (2001). South Island Wilding Conifer Strategy. Christchurch, NZ: Department of Conservation.

Higgins, S. I., Richardson, D. M. (1998). Pine invasions in the southern hemisphere: modelling interactions between organism, environment and disturbance. Plant Ecology, 135, 79-93.

INFOR (2013). El Sector Forestal Chileno 2013. Instituto Forestal (INFOR), Ministerio de Agricultura. Chile. http://wef.infor.cl/sector_forestal/sectorforestal.php#/0 Accessed: March 2014.

Jobbagy, E. G., Jackson, R.B. (2007). Groundwater and soil chemical changes under phreatophytic tree plantations. Journal of Geophysical Research, 112, 1-15.

Langdon, B., Pauchard, A., Aguayo, M. 2010. *Pinus contorta* invasion in the Chilean Patagonia: local patterns in a global context. Biological Invasions, 12, 3961-3971.

Lantschner, M.V., Rusch, V., Peyrou, C. (2008). Bird Assemblages in Pine Plantations Replacing Native Ecosystems of N.W. Patagonia, Argentina. Biodiversity and Conservation, 17 (5), 969-989.

Lara, A., Veblen, T. T. (1993). Forest plantations in Chile: a successful model? In A. Mather (Ed.), Afforestation policies, planning and progress (pp.118-139). London, U.K: Belhaven Press.

Le Maitre D. C. (1998). Pines in cultivation: a global view. In D. M. Richardson (Ed.), Ecology and Biogeography of *Pinus* (pp.407–31). Cambridge: Cambridge University Press.

Ledgard, N. (2001). The spread of lodgepole pine (*Pinus contorta*, Dougl.) in New Zealand. Forest Ecology and Management, 141, 43–57.

Little, C., Lara, A., McPhee, J., *et al.* (2009). Revealing the impact of forest exotic plantations on water yield in large scale watersheds in South-Central Chile. Journal of Hydrology, 374, 162-170.

Loewe, V., & Murillo, P. (2001). Estudio de ensayos de introducción de especies. Santiago, Chile: Instituto Forestal (INFOR).

Loidy, A., Distel, R. A., Zalba, S. M. (2010). Large herbivore grazing and non-native plant invasions in montane grasslands of central Argentina. Natural Areas Journal, 30, (2), 148-155.

Loidy, A., Zalba, S.M., Distel, R. A. (2012). Viable seed banks under grazing and exclosure conditions in montane mesic grasslands of Argentina. Acta Oecologica, 43, 8-15.

Lusk, C. (2008). Constraints on the evolution and geographical range of Pinus. New Phytologist, 178 (1), 1-3.

MAGPNA (2014). Ministerio de Agricultura Ganadería y Pesca de la Nación Argentina. Nota 485. Dirección General de Forestación.

McGregor, K. F., Watt, M. S., Hulme, P. E., *et al.* (2012a). What determines pine naturalization: species traits, climate suitability or forestry use? Diversity and Distributions, 18, 1013-1023.

McGregor, K. F., Watt, M. S., Hulme, P. E., *et al.* (2012b). How robust is the Australian Weed Risk Assessment protocol? A test using pine invasions in the Northern and Southern hemispheres. Biological Invasions, 14, 987-998.

Ministerio de Ambiente y Desarrollo Sostenible de Colombia (2011). Plan Nacional para la Prevención, el Control y Manejo de las Especies Introducidas, Trasplantadas e Invasoras: Diagnóstico y listado preliminar de especies introducidas, Trasplantadas e invasoras en Colombia. Bogotá, D.C., Colombia: Instituto de Investigación de Recursos Biológicos Alexander von Humboldt, The Nature Conservancy –Colombia, Ministerio de Ambiente y Desarrollo Sostenible.

National Research Council (2002). Predicting invasions of non-indigenous plants and plant pests. Washington D. C., USA: National Academy Press.

Nuñez, M. A., Hayward, J., Horton, T. R., *et al.* (2013). Exotic mammals disperse exotic fungi that promote invasion by exotic trees. PLoS ONE, 8 (6), e66832.

Nuñez, M. A., Horton, T. R., Simberloff, D. (2009). Lack of belowground mutualisms hinders Pinaceae invasions. Ecology, 90, (9), 2352-2359.

Nuñez, M. A., Medley, K. A. (2011). Pine invasions: climate predicts invasion success; something else predicts failure. Diversity and Distributions, 17, 703-713.

Nuñez, M. A., Pauchard, A. (2010). Biological invasions in developing and developed countries: does one model fit all? Biological Invasions, 12 (4), 707-714.

Nuñez, M. A., Raffaele, E. (2007). Afforestation causes changes in post-fire regeneration in native shrubland communities of northwestern Patagonia, Argentina. Journal of Vegetation Science, 18 (6), 827:834.

Nuñez, M. A., Simberloff, D., Relva, M. A. (2008). Seed predation as a barrier to alien conifer invasions. Biological Invasions, 10, 1389-1398.

Paritsis, J. Aizen, M. A. (2008). Effects of exotic conifer plantations on the biodiversity of understory plants, epigeal beetles and birds in Nothofagus dombeyi forests. Forest Ecology and Management, 255, 1575-1583.

Pauchard, A., Kueffer, C., Dietz, H., *et al.* (2009). Ain't no mountain high enough: Plant invasions reaching high elevations. Frontiers in Ecology and the Environment, 7, 479-486.

Pauchard, A., Langdon, B., Jiménez, A., *et al.* (2014). Pináceas invasoras en el sur de Sudamérica: patrones, mecanismos e impactos potenciales. In F. Jaksic, S. Castro (Eds.), Ecología de Invasiones Biológicas (pp.283-308). Santiago: Editorial Universidad Católica.

Pauchard, A., Langdon, B., Peña, E. (2008). Potencial invasivo de *Pseudotsuga menziesii* (Mirb.) Franco en Bosques Nativos del Centro-Sur de Chile: patrones y recomendaciones. In: R. Mujica, H. Grosse, B. Muller-Using (Eds.) Bosques Seminaturales: una opción para la rehabilitación de bosques nativos degradados (pp.89-114). Santiago, Chile: Instituto Forestal.

Pauchard A., Nuñez, M., Raffaele, E., Bustamante, R., Legard, N., Relva, M., Simberloff, D. (2010). Introduced conifer invasions in South America: an update. Frontiers of Biogeography, 2 (2), 34-36.

Pheloung, P. C., Williams, P. A., Hallo, S. R. (1999). A weed risk assessment model for use as a biosecurity tool evaluating plant introductions. Journal of Environmental Management, 57, 239-251.

Peña, E., Hidalgo, M., Langdon, B., *et al.*, (2008). Patterns of spread of Pinus contorta Dougl. ex Loud. invasion in a Natural Reserve in southern South America. Forest Ecology and Management, 256, 1049-1054.

Peña, E., Langdon, B., Pauchard, A. (2007). Árboles exóticos naturalizados en el bosque nativo chileno. Bosque Nativo, 40, 3-7.

Peña, E., Pauchard, A. (2001). Coníferas introducidas en áreas protegidas: un riesgo para la biodiversidad. Bosque Nativo, 30, 3-7.

Portz, L., Manzolli, R. P., Saldanha, D. L., *et al.* (2011). Dispersão de espécie exótica no Parque Nacional da Lagoa do Peixe e seu entorno. Revista Brasileira de Geografia Física, 4, 33-44.

Quiroz, C., Pauchard, A., Cavieres, L. A., *et al.* (2009). Análisis cuantitativo de la investigación en invasiones biológicas en Chile: tendencias y desafíos. Revista Chilena de Historia Natural, 82, 497-505.

Rejmánek, M., Richardson, D. M. (1996). What attributes make some plant species more invasive? Ecology, 77, 1655-1661.

Relva, M. A., Nuñez, M. A., Simberloff, D. (2010). Introduced deer reduce native plant cover and facilitate invasion of non-native tree species: evidence for invasional meltdown. Biological Invasions, 12, (2), 303-311.

Richardson, D.M. (2006). Pinus: a model group for unlocking the secrets of alien plant invasions? Preslia, 78, 375–388.

Richardson, D., Higgins, S. (1998). Pines as invaders in the southern hemisphere. In D. Richardson (Ed.), Ecology and biogeography of Pinus (pp.450–473). Cambridge, UK: Cambridge University Press.

Richardson, D., Williams, P., Hobbs, R. (1994). Pine invasions in the southern hemisphere: determinants of spread and invadability. Journal of Biogeography, 21, 511-527.

Richardson, D. M., Van Wilgen, B. W., Nuñez, M. A. (2008). Alien conifer invasions in South America: short fuse burning? Biological Invasions, 10, 573-577.

Sarasola, M., Rusch, V., Schlichter, T., *et al.* (2006). Invasión de coníferas forestales en áreas de estepa y bosques de ciprés de la cordillera en la Región Andino Patagónica. Ecologia Austral, 16, 143–56.

Schlichter, T., Laclau, P. (1998). Ecotono estepa-bosque y plantaciones forestales en la Patagonia norte. Ecologia Austral, 8, 285–96.

Simberloff, D., Nuñez, M. A., Ledgard, N. J., *et al.* (2010). Spread and impact of introduced conifers in South America: lessons from other southern hemisphere regions. Austral Ecology, 35, 489-504.

Simberloff, D., Relva, A., Nuñez, M. (2002). Gringos en el bosque: introduced tree invasion in a native nothofagus/austrocedrus forest. Biological Invasions, 4, 35-53.

Simberloff, D., Relva M. A., Nuñez M. A. (2003). Introduced Species and Management of a *Nothofagus/Austrocedrus* Forest. Environmental Management, 31 (2), 263- 275.

Simonetti, J. A., Grez, A. A. Estades, C. F. (2013). Providing Habitat for Native Mammals through Understory Enhancement in Forestry Plantations. Conservation Biology, 27 (5), 1117-1121.

Urrutia, J., Pauchard, A., García, R. A. (2013). Diferencias en la composición vegetal de un bosque de *Araucaria araucana* (Molina) K.Koch y *Nothofagus antarctica* (G. Forst.) Oerst. asociadas a un gradiente de invasión de *Pinus contorta* Douglas ex Loudon. Gayana Botanica, 70 (1), 92-100.

Veblen, T. T., Kitzberger, T., Raffaele, E., *et al.* (2003). Fire history and vegetation changes in northern Patagonia, Argentina. In T. T. Veblen, W. L. Baker, G. Montenegro, *et al.* (Eds.), Fire and Climatic Change in Temperate Ecosystems of the Western Americas (pp.265–296). New York: Springer.

Visser, V., Langdon, B., Pauchard, A. *et al.* (2014). Unlocking the potential of Google Earth as a tool in invasion science. Biological Invasions, 16 (3), 513-534.

Voltolini, J. C., Zanco, L. (2010). Densidade de plântulas e jovens de espécies nativas de Floresta Atlântica em áreas com e sem o pinheiro americano (*Pinus elliottii*). Revista Biociências, 16, 102-108.

Yezzi, A., Nebbia, A. J., Zalba, S. M. (2011). Efectos de fragmentación en dunas generada por una forestación implantada. Boletín de la Sociedad Argentina de Botánica, 46, (Supl.) 79.

Yezzi, A., Nebbia, A. J., Zalba, S. M. (2013). Efectos de una plantación forestal sobre la composición de la vegetación nativa en pastizales costeros bonaerenses. Boletín de la Sociedad Argentina de Botánica, 48, (Supl.) 83.

Zalba, S. M. (2000). El pastizal pampeano, los árboles exóticos y la fauna silvestre: un problema con múltiples dimensiones. In C. Bertonatti, J. Corcuera (Eds.), Situación Ambiental Argentina 2000 (pp.332-337). Buenos Aires: Fundación Vida Silvestre Argentina.

Zalba, S. M. (2010). Controle de espécies exóticas invasoras em áreas protegidas naturais: aprender fazendo. Cadernos de Mata Ciliar, São Paulo, 3, 23-27.

Zalba, S. M. (2013). Incorporando el desafío de las invasiones biológicas a los proyectos de restauración ecológica. In D. Pérez (Ed.), Restauración ecológica en la diagonal árida de la Argentina (pp.61-72). Buenos Aires: Editorial Vazquez Mazzini.

Zalba, S. M., Cuevas, Y. A., de Villalobos, A. (2009). Lecciones aprendidas durante siete años de control de pinos invasores en pastizales naturales. In N. J. Cazzaniga, H. M. Arelovich (Eds.), Ambientes y recursos naturales del sudoeste bonaerense: Producción, contaminación y conservación. (Actas de las V Jornadas Interdisciplinarias del Sudoeste Bonaerense) (pp.325-340). Bahía Blanca: Ediuns.

Zalba, S. M., Villamil, C. B. (2002). Invasion of woody plants in relictual native grasslands. Biological Invasions, 4, 55-72.

Zalba, S. M., Ziller, S. R. (2007). Adaptive management of alien invasive species: putting the theory into practice. Natureza & Conservação, 5 (2), 86-92.

Zenni, R. D., Simberloff, D. (2013). Number of source populations as a potential driver of pine invasions in Brazil. Biological Invasions, 15, 1623-1639.

Zenni, R. D., Ziller, S. R. (2011). An overview of invasive plants in Brazil. Brazilian Journal of Botany, 34, 431-446.

Zenni, R. D., Bailey, J. K., Simberloff, D. (2014). Rapid evolution and range expansion of an invasive plant are driven by provenance-environment interactions. Ecology Letters, 17, 727-735.

Mark S. Minton, A. Whitman Miller, Gregory M. Ruiz

15 Implications of Ship Type on Delivery and Management of Ballast Water

15.1 Ships, Biological Invasions, and Ballast Water Management

Commercial shipping is a major global vector for the transport and introduction of aquatic nonindigenous species (NIS), which are associated with ballast water (BW) and hull biofouling of vessels (Carlton, 1985; Hewitt *et al.*, 2009; Ruiz *et al.*, 2000). By operating among globally and regionally interconnected ports, vessels provide the opportunity for species transfers across an extensive and intricate network of routes, breaching historical barriers to dispersal for a vast number of organisms (Drake & Lodge, 2004; Kaluza *et al.*, 2010; Keller *et al.*, 2011; NRC, 2011). The magnitude of such transfers is underscored by the scale of both shipping and the biota associated with ships. Specifically, the global fleet is estimated at more than 104,000 active commercial ships with more than 1.5 billion deadweight tons (UNCTAD, 2012), and the United States alone receives approximately 100,000 vessel arrivals per year from ports around the world (NRC, 2011). Recent research has characterized the diverse and abundant biota associated with vessels, where a single vessel can deliver millions of individual organisms and hundreds of species, ranging from microorganisms and larval invertebrates to fish and algae (Carlton, 1985; Carlton *et al.*, 1995; Davidson *et al.*, 2009; Minton *et al.*, 2005; NRC, 2011).

Prevention through vector management is generally accepted as the most cost-effective way to reduce the extent and likelihood of NIS introductions, particularly for ship-related vectors like BW and hull fouling that deliver such a diverse and abundant assemblage of organisms from around the world (Carlton *et al.*, 1995; Minton *et al.*, 2005; NRC, 2011). The goal of vector management is to interrupt the vector, reducing propagule pressure (i.e., the abundance, viability, and reproductive capacity of organisms) and thereby lowering the probability of introductions (see Chapter 1; Ruiz & Carlton, 2003). For shipping and coastal invasions, vector management has focused primarily on BW management, which has advanced at international, national, and state levels over the past several decades (see Albert *et al.*, 2013).

The most prevalent form of BW management, dating back to the early 1990s, is BW exchange (BWE), whereby a ship flushes out coastal BW (and its associated coastal biota) from its ballast tanks in open oceanic water (Box 15.1). This serves to both physically remove the coastal organisms and to create a habitat mismatch, whereby: (a) coastal organisms are unlikely to survive in the open ocean, and oceanic organisms (which can be entrained) are unlikely to survive in coastal areas upon subsequent BW discharges; and (b) organisms from low salinity coastal areas experience osmotic stress when exposed to high salinity ocean waters (Bailey *et al.*, 2011; NRC, 1996; Santagata *et al.*, 2008). The efficacy of BWE can be quite variable but on average removes approximately 90% of coastal organisms (Ruiz & Reid, 2007); however, residual organisms still persist (Minton *et al.*, 2005). While there is no doubt that BWE reduces the likelihood of introduction, this cannot be performed for all routes and operating conditions (Miller *et al.*, 2011). For this reason, and also due to some uncertainty about the likelihood of invasions from residual organisms, BWE is expected to be replaced with onboard treatment technologies to achieve lower discharge concentrations of organisms that can be applied on most vessels and routes (Hewitt *et al.*, 2009).

Box 15.1. Ballast Water Basics.

- When vessels are empty or partially in cargo, stability and trim are maintained using ballast water (BW). Thus, as vessels move around the globe, loading/unloading cargo in various ports, BW is discharged/entrained.
- As BW is entrained, the associated organisms (e.g., viruses, bacteria, phytoplankton, zooplankton, and even some fish) in the water column and sediment are drawn in as well. Similarly, during BW discharge many organisms are also discharged.
- BW exchange (BWE) the process of exchanging coastal BW (i.e., entrained within 200 nautical miles [nm] from shore) for mid-ocean water (farther than 200 nm from shore), thus flushing out most of the coastal organisms and creating osmotic stress for organisms that cannot tolerate full marine salinities. There are two types of BWE:
  - Empty-Refill: A ballast tank that contains coastal BW is emptied and refilled mid-ocean.
  - Flow-Through: A ballast tank is flushed by drawing water in through a lower valve and out through and upper opening so that 300% of the ballast tank capacity is pumped through.

In the US, BW discharge is regulated nationally by the US Coast Guard (USCG) and the US Environmental Protection Agency (US EPA); although some states have regulations, this chapter focuses on the federal regulatory framework. As mandated by the National Invasive Species Act of 1996, the USCG and the Smithsonian Environmental Research Center (SERC) established the National Ballast Information Clearinghouse (NBIC) in July 1999 to collect and analyze data on BW delivery in US waters. Nearly all commercial ships operating in US waters with the capacity to carry BW are required to submit a Ballast Water Report (BWR) to the NBIC (USCG, 2012a). BWRs contain information identifying the ship, transit details (e.g., arrival, last, and next ports), and BW history (i.e., source and discharge locations and associated dates, as well as BW management details). In July 2004, the USCG expanded management requirements mandating that all overseas BW (i.e., originating outside of the combined US and Canadian Exclusive Economic Zones [EEZ]) entrained within 200 nmi of shore be managed in an approved way.

In March 2012, the USCG published new requirements for the management of BW by ships operating in US waters (USCG, 2012b), which went into effect on 21 June 2012. These new regulations made several significant changes to BW management requirements. First, they expanded BW management requirements from only water that was entrained from coastal overseas sources to include all BW discharged by all ships approved as seagoing vessels. Second, it established US discharge standards for maximum allowed concentrations of organisms that are being phased in with specific implementation dates. Meeting the discharge standards will require the use of BW treatment systems (BWTS), since BWE cannot reliably meet the allowable concentrations. Numerous BWTS that use filtration combined with treatment (i.e., advanced oxidation, de-oxygenation, ozone, or ultra-violet) have been developed or are in development; however, none have received type approval from the USCG to date (December 2014). During the time that BWTS technology is being developed and approved for use in the US, the use of BWE will continue to be the primary BW management option. Although meeting the discharge standard will further lower the propagule pressure via the BW vector, there will remain some residual level of risk for future NIS introductions that is not yet understood (Minton *et al.*, 2005; NRC, 2011).

In this chapter, we review the current knowledge and status of BW delivery and management in the United States. We focus on overseas arrivals and discharge to characterize: (a) the magnitude of BW discharge; (b) variation by geographic source and recipient regions; and (c) differences among ship types. In addition, we consider possible future directions for BW management.

15.2 Shipping and BW Delivery to the United States

15.2.1 Overseas Arrivals

Cumulatively, the NBIC received 105,424 BWRs from overseas commercial vessel arrivals to four coastal regions of the continental US (i.e., Alaska, East, Gulf, and West coasts) during the three-year time period of 2011–2013 (Table 15.1), representing an estimated 95.5% nationwide rate of reporting for vessels overall (ranging from 94–99% among coasts). These arrivals were not distributed evenly among coastal regions ($\chi^2_3 = 47923.25$, $p < 0.001$). The East coast received 45,894 (43.5% of total ship arrivals to the US), followed by the Gulf and West coasts that received 39,440 (37.4%) and 19,486 (18.5%), respectively. Alaska received only 604 (0.6%) overseas arrivals, as most arrivals to Alaska are domestic transits from West coast ports. This distribution among coasts has been relatively stable and consistent since 2005, based on several previous analyses (Miller *et al.*, 2012; Miller *et al.*, 2011; Minton *et al.*, 2012). The disparity among coasts simply reflects historical trade patterns, driven by economics and logistical constraints (e.g., the size of the port, draft depth, type of cargo, type of equipment like cranes, supply chains, and inland distribution characteristics).

The NBIC data provide a highly detailed and nearly complete picture of vessel arrival and BW delivery characteristics, providing one of the most comprehensive and large-scale BW data sets in the world. Here, we use these data to highlight important operational aspects of vessels, BW delivery, and management patterns, and how these are distributed at various spatial scales. A critical aspect to recognize in this regard is that all ships are not the same. There are a wide variety of commercial vessel types, designed to carry different types of cargo, which affects many operational and behavioral characteristics of BW delivery (Table 15.2). Table 15.1 shows the distribution of arrivals and discharge frequency across major vessel types. Overseas arrivals to the US are dominated by bulkers, container ships, and tankers, as described previously by Verling *et al.* (2005) for the period of 1999–2002. During the more recent period (2011–2013), tankers, container ships, and bulkers accounted for 24.9, 23.8, and 17.6% of all US overseas arrivals, respectively. When compared to 1999–2002, the proportion of overseas arrivals by container ships decreased by 32%, while tanker arrivals increased by 41% and bulker arrivals increased by 26% (Verling *et al.*, 2005). Although not a large portion of all overseas arrivals, passenger ships accounted for 12.5% of arrivals in 1999–2002, but only account for 9.3% of arrivals in 2011–2013.

Tab. 15.1: Number of overseas arrivals by the nine ship types to the continental US and the percentage of those arrivals reporting discharge as reported to the NBIC for 2011–2013 (NBIC, 2014).

	Coastal Region	Bulker	Container	General Cargo	Passenger	Reefer	RoRo	Tanker	Other
Arrivals	Alaska	235	2	26	21	218	2	82	18
	East	4 305	13 223	5 876	6 954	1 346	7 672	5 592	926
	Gulf	9 383	3 402	4 606	1 903	749	893	18 056	448
	West	4 653	8 489	682	973	413	1 540	2 514	222
	Total	18 576	25 116	11 190	9 851	2 726	10 107	26 244	1 614
Percent Discharging	Alaska	93.2	0	30.8	0	61.9	0	34.1	22.2
	East	57.7	10.9	27.7	38.9	5.2	8.1	15.6	17.7
	Gulf	61.2	9.2	44.4	57.2	18.2	29.5	44.5	37.3
	West	77.8	6.9	19.2	38.7	4.8	1.1	26.8	13.5
	Total	64.9	9.3	34	42.3	13.2	8.9	36.6	22.6

Tab. 15.2: Descriptions of the various ship types.

Ship Type	Description
Bulker	Carries dry bulk cargo (e.g., ore, grain, sawdust)
Container	Containerized cargo

continued Tab. 15.2: Descriptions of the various ship types.

Ship Type	Description
General Cargo	Can carry many types of cargo (e.g., containers, bulk cargos)
Passenger	Ranging from cruise liners to ferries
Reefer	Refrigerated cargo ship
RoRo	Roll-on/Roll-off carry wheeled cargos (e.g., cars, trucks, railroad cars)
Tanker	Liquid or gas cargos (e.g., petroleum, LNG, orange juice)
Other	Includes tugs, barges, drilling rigs, offshore supply vessels, cable layers, etc.

As seen for total vessel arrivals, the traffic of various ship types is not distributed evenly among US coasts ($\chi^2_{21} = 33636.68$, $p < 0.001$). The East coast had the most homogeneous distribution of arrivals among ship types, although container ships arrived in greater numbers than other ship types (Table 15.1). In contrast, traffic to each of the other three coasts was dominated by combinations of two ship types. Nearly 70% of arrivals to Gulf coast were by tankers and bulkers, 67% of arrivals to the West coast were by containers and bulkers, and 75% of arrivals to Alaska were by bulkers and reefers. Over half of all overseas container ship arrivals were to the East coast, while the Gulf coast received at least half of all overseas arrivals by bulkers and tankers.

15.2.2 Overseas Discharge

It is important to recognize that arrival frequency alone is not a good proxy for the flux of BW, because ships do not discharge ballast in every port of call and BW discharge volumes vary significantly by ship type and location (Miller *et al.*, 2011; Minton *et al.*, 2005; NRC, 2011; Verling *et al.*, 2005). From 2011–2013, NBIC data show that different ship types discharged overseas BW at significantly different rates ($\chi^2_7 = 90.81$, $p < 0.001$), underscoring the disparity between arrivals and the frequency of BW discharged (Table 15.1). Bulkers were the only ship type that discharged BW during more than half of their overseas arrivals (64.9%), while container ships and RoRos discharged BW at less than 10% of their overseas arrivals during the 2011–2013 time period. Although some previous studies have used arrival number as the proxy for BW-mediated propagule pressure (Drake & Lodge, 2004; Kaluza *et al.*, 2010; Keller *et al.*, 2011), this may be a significant source of error, given the strong differences observed between vessel types.

The nature of commercial trade drives shipping routes and BW delivery patterns, which play out differently according to ship type. Bulkers typically transit between two ports either fully loaded in cargo or in ballast, and the destination ports can change frequently based on markets (Kaluza *et al.*, 2010). In contrast, container ships tend to make more frequent stops along regular routes, where they will load/unload only a portion of their cargo. Container ships are always in cargo, and so carry and discharge much less ballast, and at a lower frequency. In addition, 8% of container ship and 25% of RoRo overseas arrivals indicated on their BWRs that they used internal transfers between BW tanks to maintain stability during cargo operations, whereas bulkers and tankers do not report the capability to manage BW in this manner. Furthermore, it is interesting to note that the national percentages of discharging overseas arrivals per ship type have changed since BWE became mandatory in 2004. Relative to the percent of discharging overseas arrivals in 1999–2002 reported by Verling *et al.* (2005), tankers increased by 123%, general cargo ships increased by 80%, and bulkers increased by 61%; in contrast, there was a 50% decrease for RoRos, a 46% decrease for container ships, and a 23% decrease for passenger ships.

Furthermore, it is important to recognize that the behavior of each ship type (i.e., whether BW was discharged) was significantly different among coastal regions ($\chi^2_{21} = 215.5$, $p < 0.001$). These differences in behavior are likely a result of the type and the directionality of trade in which they were involved in 2011–2013. In Alaska, both bulkers and reefers were much more likely to discharge BW during overseas arrivals compared to other coasts (Table 15.1). These reefers are involved in the export of frozen fish, and bulkers are used for the export of mineral ores and timber (USCB, 2013), meaning the ships are carrying BW when arriving in Alaska and laden in cargo when departing Alaska. On the East coast, tankers are less likely than on other coasts to discharge BW during overseas arrivals. One possible reason for the lower frequency of discharge is the lack of large petroleum reserves along the East coast, so crude oil is imported (USCB, 2013). The highest frequency of overseas BW discharge by tankers is on the Gulf coast, which produces and exports large amounts of refined petroleum products. This difference in behavior of ship types among regions calls into question the ability to extrapolate from one geographic region to another, as some studies have attempted in modeling invasion dynamics (Seebens *et al.*, 2013; Xu *et al.*, 2014).

Differences in the frequency of inoculation, the size of individual inoculation events (i.e., discharge volume per arrival), the magnitude of flux (i.e., total volume of BW discharge), source regions, and management of the vector (i.e., prevalence and effectiveness of BWE) can have a large effect on the likelihood of introducing NIS and the efficacy of different management strategies. Below, we examine each of these characteristics in greater detail.

15.2.2.1 Frequency of Inoculation

The majority of overseas arrivals to the US do not discharge BW (Table 15.1), a pattern that holds true for coastwise arrivals as well (Miller *et al.*, 2012; Minton *et al.*, 2012). Since ship's ballast activities are driven by the loading and unloading of cargo, and ships may stop at multiple ports of call while transiting through US waters, determining the frequency of inoculation events (i.e., discharge events) is not simply the number of overseas arrivals multiplied by the percent of overseas arrivals that reported discharging overseas BW. Yet, the frequency of inoculation is an important factor affecting establishment (Grevstad, 1999; Hedge *et al.*, 2012; Minton & Mack, 2010; Simberloff, 1989) and should be quantified for the various ship types. Bulkers, container ships, passenger ships, and tankers reported discharge of overseas BW most frequently among ship types during 2011–2013. During this period, there were over 10,000 discharge events by tankers and 6,582 by bulkers on the Gulf coast, and 4,178 by bulkers on the West coast; these accounted for 24.7%, 16.2%, and 10.3% of all inoculations with overseas BW to the nation, respectively. On the East coast, where container ships were the most frequent arrivals (Table 15.1), there were 2931 discharge events by container ships, followed by bulkers (2,743) and passenger ships (2,262). The remaining ship type-by-coast combinations accounted for less than 30% of the overseas BW discharge events.

15.2.2.2 Volume of BW Discharge per Arrival

The magnitudes of individual inoculation events (i.e., discharge volume per discharge event) were significantly different among vessel types, and these differences were uneven among coastal regions (Figure 15.1). These differences were tested using ANOVA on the square root transformed discharge volume per discharge event ($F_{27,39373} = 1217$, $p < 0.001$) and pairwise comparisons discussed here were determined to be significant

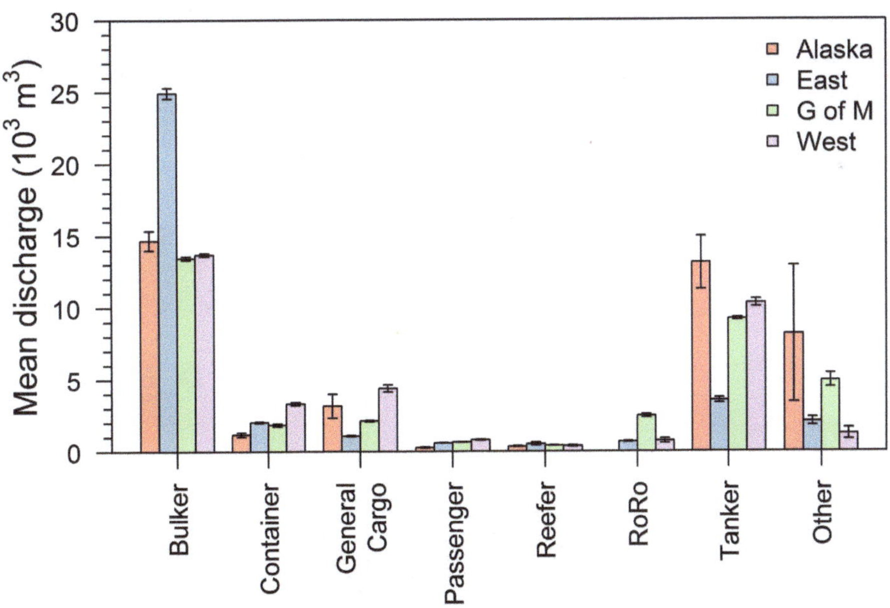

Fig. 15.1: The mean (± SE) BW discharge per discharging arrival for eight ship types to four coastal regions of the continental US. These data were reported to the NBIC for arrivals during 2011-2013 (NBIC, 2014).

based on Tukey's Honest Significant Differences at $p < 0.05$ (RoRos were excluded from the statistical tests because there were no discharges by RoRos in Alaska). Bulkers and tankers discharge significantly more BW per discharge event than the other ship types (Figure 15.1). Regional differences were observed for the three ship types that account for most arrivals (Table 15.1). Bulkers arriving to the East coast discharged, on average, 10,000 m^3 more per discharge event than did bulkers arriving to the other three coastal regions. Tankers arriving to the East coast averaged 6,000–10,000 m^3 less per discharge event than did tankers to the other coasts. Container ships on the West coast discharged significantly more per discharge event than on the East and Gulf coasts.

15.2.2.3 Overseas Discharge by Coastal Region and Ship Type

The continental US received nearly 350 million m^3 of overseas BW during 2011–2013, as reported to the NBIC (2014). The Gulf coast received more than twice as much discharge as did the East and West coasts. The Gulf coast received 54.4% of the national total (by volume), compared to 23.6%, and 20.5% for the East and West coasts, respectively, while Alaska only received 1.5%. Nationally, the overwhelming majority of overseas BW was discharged by bulkers (62.4%) and tankers (30.7%), while the remaining ship types were each responsible for less than 3% of overseas discharge. Although bulkers and tankers accounted for the majority of discharge to all four coastal regions, the distribution of discharge is not uniform among regions (Figure 15.2). Bulkers were

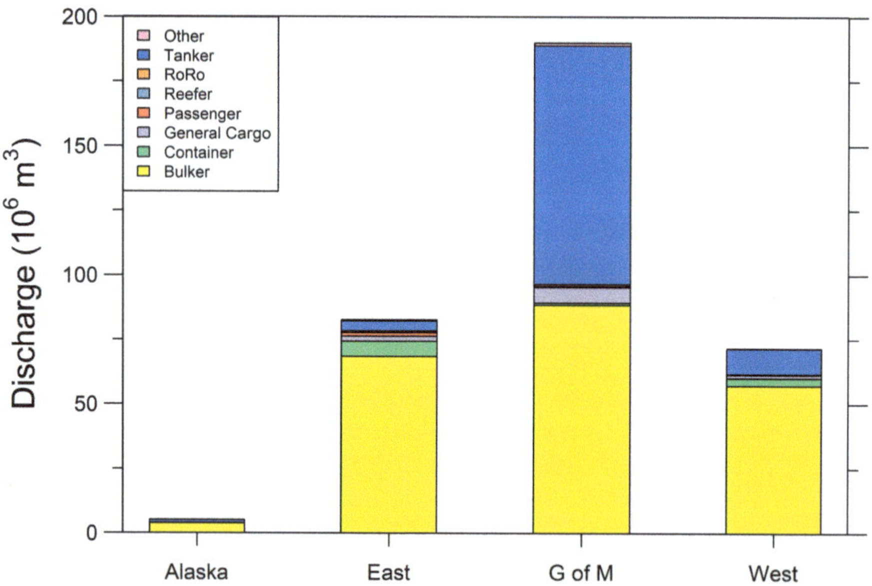

Fig. 15.2: The total volume of overseas BW discharged to the continental US by the eight vessel types. These data were reported for arrivals during 2011-2013 (NBIC, 2014).

the dominant source of overseas BW in Alaska (74%), the East (83%), and the West (80%), while tankers contributed 22%, 4.6% and 13.6% of the totals to each coast, respectively. While the Gulf coast received more overseas BW discharged by bulkers than any other region received in total, it also received 86% of all overseas BW discharged by tankers (Figure 15.2). As a result, the Gulf coast received a nearly equal amount of BW discharge by tankers and bulkers. Interestingly, the combined overseas discharge of bulkers and tankers to the Gulf coast accounted for over 51% of all overseas BW discharged to the US. The East coast was the only region where container ships discharged more BW than did tankers, contributing 7.2% of total discharge.

15.2.2.4 Overseas Discharge to US Ports

The discussion thus far has focused on BW delivery at the level of coastal regions; ships, however, aren't operating at the coastal level, but rather transit between individual ports and typically conduct much of their ballast operations in conjunction with loading and unloading cargo in these ports. BW delivery at the level of the port is even more asymmetric than it is among coastal regions. During 2011–2013, overseas BW was discharged in 211 ports on the Alaska, East, Gulf, and West coasts; however, BW discharge was concentrated in relatively few ports. The top 20 ports received 81.6% of total BW discharge, of which the top 10 ports accounted for 65.3% of total discharge. From 2011–2013 New Orleans, Louisiana received more than 54.2 million m^3 of overseas BW followed by Houston, Texas (46.7 million m^3) and Hampton Roads, Virginia (40.3 million m^3), which combined was 40.6% of all overseas discharge (Figure 15.3). The clustering of discharge by tankers in Gulf ports is striking, underscoring that the Gulf received 86% of the overseas BW discharged by tankers. Figure 15.3 demonstrates that New Orleans was the biggest recipient of overseas BW from bulkers, Houston from tankers, and the Port of New York and New Jersey from container ships.

Despite the high frequency of overseas arrivals by container ships (Table 15.1), of the top 20 discharge ports, only New York (46.7%) and Long Beach (9.7%) received greater than 5% of BW from container ships. One reason for this discrepancy between the frequency of arrivals and the volume of BW discharge is that container ships discharge much less per discharge event than do bulkers and tankers (Figure 15.1). In addition to discharging less frequently during overseas arrivals than most other ship types (Table 15.1), container ships spread their discharge over multiple ports, discharging 47% of overseas BW across multiple US ports of call after the initial overseas arrival. In contrast, the discharges by tankers and bulkers were more concentrated, discharging 90–94% of overseas BW during the overseas arrival.

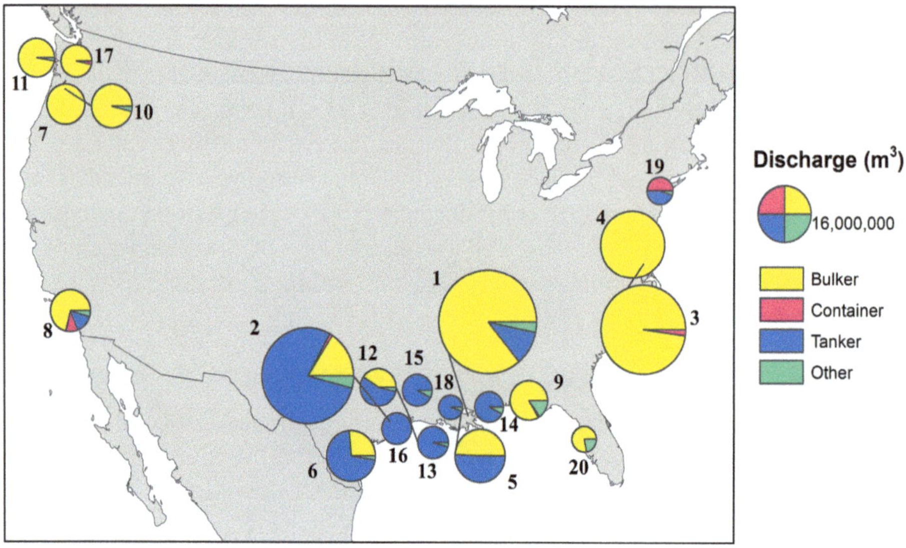

Fig. 15.3: The top 20 ports according to the total volume of overseas BW discharged from 2011–2013 (NBIC, 2014). Alaska is not included in this map since none of the top 20 ports based on discharge were located there. The symbol for each port is scaled by the volume discharged and colored according to ship type. Labels indicate the rank of each port: 1. New Orleans, LA; 2 Houston, TX; 3 Hampton Roads, VA; 4 Baltimore, MD; 5 Port of South Louisiana, LA; 6 Corpus Cristi, TX; 7 Portland, OR; 8 Long Beach, CA; 9 Mobile, AL; 10 Kalama, OR; 11 Longview, WA; 12 Port Arthur, TX; 13 Beaumont, TX; 14 Pascagoula, FL; 15 Lake Charles, LA; 16 Texas City, TX; 17 Tacoma, WA; 18 Baton Rouge, LA; 19 New York, NY; and 20 Tampa, FL.

15.2.2.5 Source of Overseas Ballast Water

The BW vector concentrates the flux from many global source locations at a few recipient locations in the US. As described above, the top 20 discharge ports received 81.6% of the overseas BW discharge in the US. In contrast, the top 40 source ports contributed only 40.5% of the overseas BW discharged in the US. In total, there were 39,024 unique source locations of overseas BW, of which 1,364 were ports or named locations and 37,660 were provided as latitude and longitude from primarily oceanic locations.

Figures 15.4–15.6 show overseas BW source volume discharged to each region in the continental US by ship type (i.e., bulker, tanker, and container). At the level of coast, the source locations reported for the time period of 2011–2013 were similar to those found by Miller *et al.* (2011) for an earlier time period of 2005–2007. It is clear from these maps that the East, Gulf, and West coasts received BW from much of the globe, but each coast and ship type had certain geographic regions that were most dominant. Overseas BW discharged on the East coast originated primarily from Northwestern Europe, the Mediterranean, the Caribbean, and Asia (China, Japan, and the Korean peninsula). The Gulf coast primarily received overseas BW from Mexico,

Central and South America, the Caribbean, northwestern Europe, and some from Asia. Sources for the West coast were largely in Asia, the western coasts of Mexico and Central America, and the Hawaiian Islands. In contrast, Alaska received the vast majority of overseas BW from Asia regardless of ship type.

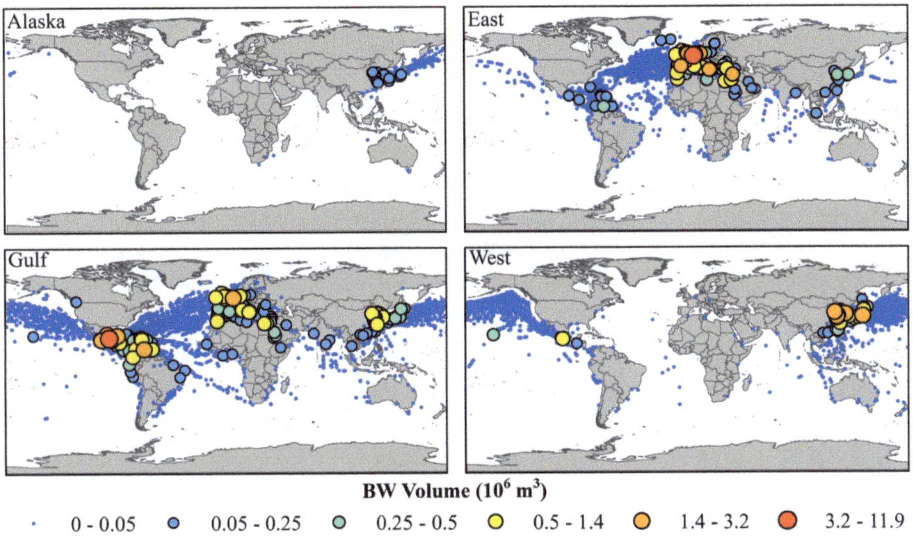

Fig. 15.4: Source locations for BW discharged by bulkers to each of the coastal regions in during 2011–2013 (NBIC, 2014).

15.2.2.5.1 Ship Type Differences - Nationally

It is also quite clear from Figures 15.4–15.6 that ship type was an important predictor of BW source location, both among and within coasts. We highlight some of these differences in source regions among bulkers, tankers, and container ships, the three dominant ship types based on BW discharge and arrival frequency. Four clear differences among these three ship types are apparent at a national scale, as follows:

1. Based on the discharge coast, bulkers had more geographically distinct BW source regions compared to tankers and container ships (Figure 15.4).
2. Tankers entrained overseas BW primarily in Central and South America and the Caribbean, regardless of the discharge coast, except for arrivals to Alaska (Figure 15.5).
3. Container ship sources were dominated by Asian sources (China, Japan, and South Korea) and Hamilton, Bermuda (Figure 15.6).
4. In contrast to bulkers and tankers, container ships entrained a substantial proportion of overseas BW beyond 200 nmi from any shore (i.e., oceanic overseas BW). Nationally, 41% of overseas BW discharge by container ships was oceanic, and on the East, Gulf, and West coasts the oceanic BW was 51.5%, 72.5%, and

59.2% of the total flux by container ships, respectively (Figure 15.6). The larger proportion of oceanic overseas BW discharged by container ships was apparent when comparing the maps of container ship source locations (i.e., larger source location symbols throughout the oceans) with those for bulkers (Figure 15.4) and tankers (Figure 15.5) each of which discharge more than 95% of overseas BW of coastal origin. By entraining oceanic BW instead of coastal BW, container ships were self-managing their ballast, thus avoiding the BWE requirement for almost 41% of the overseas BW they discharged in the US.

15.2.2.5.2 Ship Type Differences – Within Coastal Regions

When we examined the differences among bulkers, tankers, and container ships within coasts, there was significant differentiation in both the location and number of source ports. Since Alaska received relatively little overseas BW compared to the other coasts (Figure 15.2), it is not surprising that there were relatively few source locations for any of the ship types (Figure 15.4 Alaska, Figure 15.5 Alaska, Figure 15.6 Alaska). In contrast, on the East coast, container ships had the most source locations (2,871), followed closely by bulkers (2,612), and tankers (1,026). Bulkers discharging on the East coast primarily entrained BW in Northwestern Europe and the Mediterranean (e.g., Rotterdam and Amsterdam, Netherlands and Dunkirk, France) (Figure 15.4 East); whereas tankers discharging on the East coast entrained BW primarily in the Caribbean (e.g., Freeport, Bahamas; Guayanilla, Puerto Rico) with some originating in Rotterdam, Netherlands (Figure 15.5 East). The dominant source ports of container ships that discharged on the East coast (e.g., Hamilton, Bermuda; Busan, South Korea; Hong Kong; and Singapore) were also the largest source ports for container ships nationally (Figure 15.6).

On the Gulf coast, there were similar numbers of BW source locations for bulkers (3,145) and tankers (3,072), while container ships only discharged BW from 224 locations. Overseas BW discharged by bulkers was mostly from Central and South America and Northwestern Europe (e.g., Veracruz and Altamira, Mexico; Puerto Cabello, Venezuela; and Rotterdam, Netherlands) (Figure 15.4 Gulf). Tankers entrained BW primarily in Central and South America (e.g., Tuxpan and Coatzacoalcos, Mexico and Pozos Colorados, Colombia) (Figure 15.5 Gulf). Even container ships had the majority of BW entrained in Central America and the Caribbean (e.g., Guatemala; Freeport, Bahamas; and Manzanillo, Panama), and not Asia like the other coasts (Figure 15.6 Gulf).

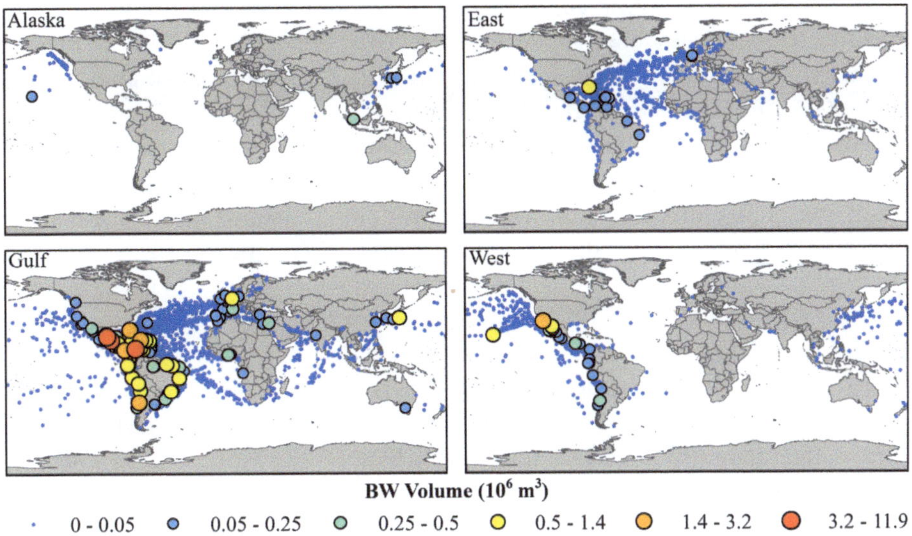

Fig. 15.5: Source locations for BW discharged by tankers to each of the coastal regions in during 2011–2013 (NBIC, 2014).

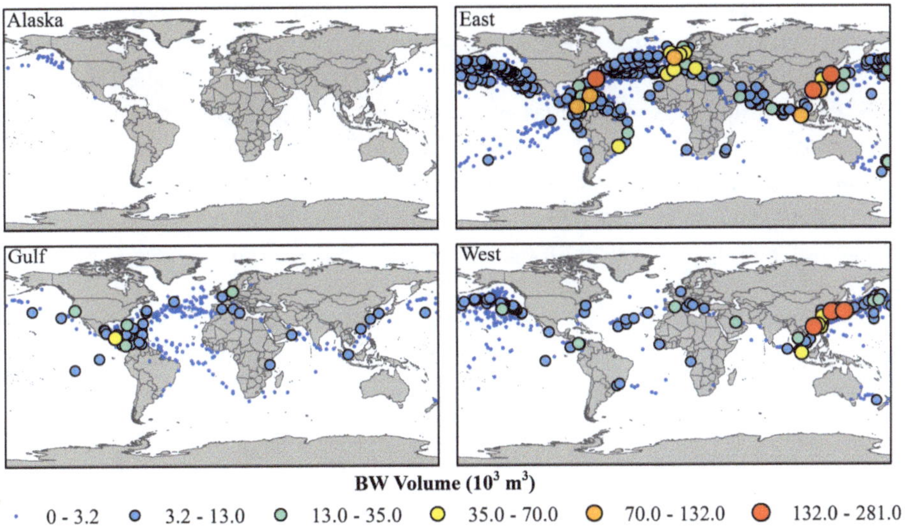

Fig. 15.6: Source locations for BW discharged by container ships to each of the coastal regions in during 2011–2013 (NBIC, 2014).

In contrast, on the West coast, the 4,357 BW source locations for bulkers was dramatically more than for container ships (1,131) or tankers (742). The vast majority of overseas bulker BW discharge was entrained in Asia (e.g., Rizhao and Xingang, China; Kashima, Japan; and Inchon, South Korea) (Figure 15.4 West). The source ports for tankers were primarily along the west coast of Mexico and in Hawaii (Figure 15.5 West), and not the eastern side of Central America or the Caribbean. Asia (e.g., Tokyo, Japan; Hong Kong; Busan, South Korea; and Yantian, China) was the main source of BW discharged by container ships on the West coast (Figure 15.6 West).

15.2.2.6 BW Exchange

The US requires the management of coastal overseas BW to reduce the likelihood of introducing new aquatic NIS into the waters of the US (USCG, 2012a), and the predominant method for managing BW at this point is BWE (Miller *et al.*, 2012; Minton *et al.*, 2012); although, effective June 2012, the management requirements were expanded to include the phased-in use of BWTS (USCG, 2012b). Since these regulatory changes took effect in the middle of the time period examined here, we will first examine the behavioral differences among ship types with respect to the regulatory framework before the newest regulations took effect.

Between 2011 and 2013 the majority of overseas BW discharged into the US was exchanged, although this varied among coasts and ship types. Nationally, 85.7% of all overseas BW by volume was exchanged, ranging from 81% on the Gulf coast to 93.8% on the West coast. Only overseas BW of coastal origin is required to undergo BWE. For all ship types, except passenger vessels, BWE was conducted on most discharged overseas BW (Figure 15.7). On the East and Gulf coasts, extensive discharge without BWE by passenger ships is largely because these vessels typically do not transit more than more than 200 nmi from shore for long enough to fully conduct a BWE, a USCG-approved reason for discharge without BWE; however, such arrivals are instructed to discharge the minimum volume necessary for safe operation.

As mentioned above, the USCG mandated that BWE be conducted more than 200 nmi from any shore for open ocean BWE. However, because of geographic constraints (e.g., the number and proximity of islands in the Caribbean), it can be difficult for ships to sail at least 200 nmi from shore for long enough to conduct a mid-ocean BWE. Miller *et al.* (2011) described two categories of overseas traffic to the United States based on the primary direction of transit, *transoceanic* arrivals that transit east-west, and *Pan-American* arrivals that transit north-south. In 2005–2007, ships conducting BWE during Pan-American transits were much more likely to exchange their BW within 200 nmi of shore (i.e., coastal BWE) than on transoceanic transits. The result was 76% of the exchanged BW from Pan-American BW sources was underwent coastal BWE, whereas only 9% of exchanged BW from transoceanic sources was exchanged in coastal waters (Miller *et al.*, 2011). The efficacy of conducting a BWE in coastal waters remains unclear and may vary depending on the location and conditions of the exchange location.

In our analysis of BWE during 2011–2013, it is clear that coastal BWE remained a significant fraction of exchanged BW discharge (28% nationally and ranging from 11% on the East and West coasts to 42% on the Gulf coast). Based on patterns of BW source locations (Figs. 15.4, 15.5, and 15.6), tankers discharging BW on the Gulf, West, and East coasts primarily fall into the Pan-American transit category, employing coastal BWE for 72%, 38%, and 34% of all discharged BW (Figure 15.7). In contrast, source locations for bulkers are mostly transoceanic for the East and West coasts, but are both Pan-American and transoceanic on the Gulf coast, which was reflected by the higher rate of coastal BWE on the Gulf coast (Figure 15.7). Even for container ships discharging on the Gulf coast, there was a larger proportion of coastal BWE than on the other coasts.

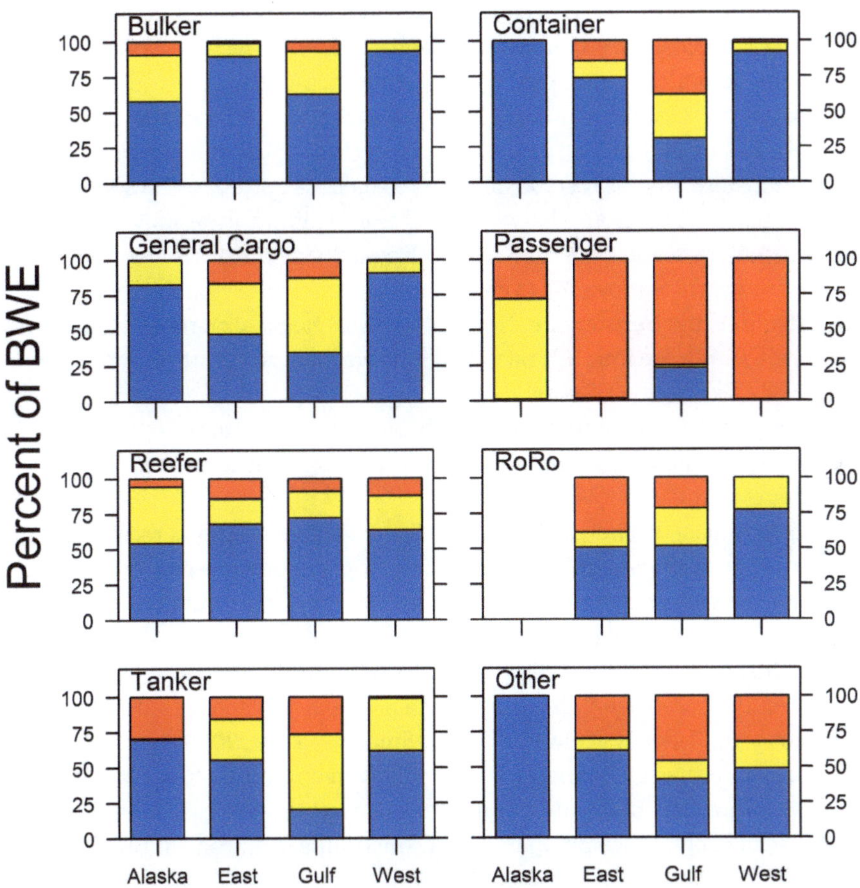

Fig. 15.7: Percent of coastal overseas BW discharge that underwent open ocean BWE (blue), coastal BWE (yellow), or was not exchanged prior to discharge (red) according to ship type and coast. Open ocean BWE occurs more than 200 nmi from any shore, while coastal BWE takes place at least partially within 200 nmi. Data were reported for the 2011–2013 time period (NBIC, 2014).

15.3 BW Management

15.3.1 The BW Vector

The movement of BW in commercial ships is a significant vector for the introduction of NIS to marine, estuarine, and freshwater systems worldwide (Carlton, 1996; Hewitt *et al.*, 2009; Hulme, 2009; NRC, 2011; Ruiz *et al.*, 2000). As discussed in Section 15.1, the most efficient efforts to reduce the rate of future introductions via a vector like ballast water, which inadvertently transports a broad suite of species and even communities, is by managing the entire vector rather than attempting management of individual ports or ship types (Drake & Lodge, 2004; NRC, 2011; Ruiz & Carlton, 2003). Successful vector management will depend on understanding how the vector operates (see Chapter 1, Carlton & Ruiz) and robust evaluations of the potential management strategies available. Thus, for vector interruption models to be most useful, they must be based on the actual operation of the vector and should capture as full a breadth of circumstances as possible. Although there is a general positive relationship between propagule pressure and successful establishment (NRC, 2011), there is also substantial evidence indicating that NIS establishment is a stochastic process, and sometimes rare events can make the difference between success and failure, especially as the number of introductions increases (Grevstad, 1999; Minton & Mack, 2010; Simberloff, 1989).

Individually, ships move BW from point to point, but the cumulative effect of this global network is the transfer of entire communities to major ports throughout much of the world. Within the four US coastal regions we analyzed, four general patterns of overseas BW flux into the US emerged:

1. The BW vector sampled from many global locations (39,024 source locations) and concentrated the inoculations into far fewer recipient locations (211 discharge locations).
2. Overseas ballast discharge was highly skewed with a few ports receiving the vast majority of the overseas BW discharge (i.e., top 20 ports received nearly 82%).
3. Despite the concentration of so much overseas ballast discharge into 20 ports, there were 191 other ports that received overseas discharge in 2011–2013. Layered upon this, but not discussed in this chapter, is the connection between all US ports via the coastwise traffic that can facilitate the secondary spread of NIS that become established (Lavoie *et al.*, 1999; Simkanin *et al.*, 2009).
4. BWE is the primary method of vector management at this time, but is not, and likely cannot, be uniformly applied to all overseas BW discharged in the US because of geographic and operational constraints (e.g., coastal BWE).

It is important to note that this analysis attempts to quantify a variety of operational characteristics of BW discharge and exchange, and uses the volume of BW discharge and the extent of BWE as coarse proxies for propagule pressure. However, given the vast amount of variability in abundance, species composition, and viability of organisms in ballast water, even the use of discharge volume is likely not a truly robust proxy (Minton *et al.*, 2005; NRC, 2011; Verling *et al.*, 2005; Wonham *et al.*, 2001). For the BW vector, BW discharge volume is undoubtedly a better proxy than raw ship arrivals and a first (albeit coarse) step toward considering propagule influx, since ships that do not discharge ballast do not pose a risk of introducing ballast-borne species. Thus, understanding the spatial and temporal patterns of BW sources, exchange, and delivery provide a starting point until experimental efforts can provide better resolution for issues such as the relationship between propagule pressure and establishment across many species and environments or the efficacy of coastal BWE (NRC, 2011).

15.3.2 Implications of Ship Type Differences

Due to the lack of comprehensive data on regional and global BW flux and management, many previous efforts to model this vector for the introduction of aquatic NIS to marine, estuarine, and freshwater systems have been limited to using ship arrivals as a proxy for propagule pressure, or have used data on BW discharge from the NBIC to make inferences beyond the US (Drake & Lodge, 2004; Kaluza *et al.*, 2010; Seebens *et al.*, 2013; Xu *et al.*, 2014). It is clear from the analyses presented here that not all ship arrivals are equal with respect to BW dynamics. Moreover, since trade determines the direction and magnitude of BW flux, inferences beyond the US based on data from the NBIC will thus have a high degree of uncertainty. Even with respect to the magnitude of BW flux in the US, one might conclude that bulkers and tankers are the BW story; however, there exists substantial variability in ship behavior among the various ship types as well as among coasts and ports, and container or other vessels types can be dominant in some individual ports. Perhaps more importantly, differences among ship types could affect the efficacy of efforts at vector management (e.g., coastal BWE) or in determining the most appropriate method for managing BW given the ship type or geographic and operational constraints. Although we highlighted differences among the three dominant ship types, this is not to imply that other vessel types are not important or that they don't require management.

15.3.2.1 Bulk Carriers

Bulkers accounted for only 17.6% of the overseas arrivals to the US in 2011–2013 but discharged 62.4% of the overseas BW, making them far and away the dominant ship type in the BW vector. Only on the Gulf coast was their dominance matched by another ship type, tankers. Bulkers not only discharged the largest amount of BW per discharge event (ranging from 13,440 ± 117 m^3 [mean ± 1 SE] on the Gulf to 24,942 ± 365 m^3 on the East) they also were responsible for 34% of the discharge events to the four coastal regions (Figure 15.1). Kaluza *et al.* (2010) noted that bulker movement was less predictable than container ships because their routes changed frequently in response to market changes. The flexibility and dynamic movement of bulkers is supported by our analysis, which found that bulkers sampled (i.e., entrained BW) more source locations (13,948) and source ports (831) globally than any of the other ship types.

Bulkers fit the profile of an effective mechanism for NIS introduction because of a) their frequent large BW discharge maximizes the inoculation dose, and b) discharge from many source locations maximizes the potential for introducing novel species when sampling from many source pools. In addition, bulker discharge was highly skewed to a few ports. Nearly 60% of bulker BW discharge was concentrated in five ports: New Orleans, LA; Hampton Roads, VA; Baltimore, MD; Portland, OR; and Kalama, WA. Consequently, bulkers were repeatedly discharging large amounts of BW (both in total and per inoculation) into the Mississippi River (New Orleans), the Chesapeake Bay (Baltimore and Hampton Roads), and the Columbia River (Portland and Kalama) from a wide array of source locations and source ports. Fortunately, bulkers also were using BWE to manage almost all overseas ballast water, but on the Gulf coast and Alaska coastal BWE accounted for a third of all exchanged discharge.

15.3.2.2 Tankers

Tankers discharged a larger amount of BW than would be expected by the number of arrivals alone, accounting for 24.9% of overseas arrivals and 30.7% of overseas BW discharge. Although tankers accounted for less than half of the total discharge of bulkers, 86% of tanker discharge was concentrated on the Gulf coast, making them the largest source of BW and most frequent discharges in this region. Unlike the other ship types, where there was geographic differentiation among discharge coasts in the BW source regions, the primary source regions for tankers (East, Gulf, and West coasts) were in the Caribbean and Central and South America. Consequently, tankers engaged primarily in Pan-American trade and a significant portion of their BW discharge underwent coastal rather than open ocean BWE.

15.3.2.3 Container Ships

Over the course of 50 years, container ships have become one of the primary ship types used to move cargo globally (Vigarie, 1999). Nationally, container ships accounted for almost as many overseas arrivals as did tankers in 2011–2013, and were the most frequent arrivals on the West (43.6%) and East coasts (28.8%). Container ships discharged much smaller volumes of BW per discharge event and frequently discharged at multiple arrival ports as they transited through US waters (i.e., initial overseas arrival and/or subsequent coastwise arrivals) in contrast to most other ship types that discharged overseas BW primarily on overseas arrivals. Although the magnitude of BW flux from container ships was less than 3% of the total flux, the inoculation frequency by container ships was relatively high on the East and West coasts, and these inoculations were globally sourced, particularly for the East coast.

Container ships have two ballast operational behaviors that separate them from most other ship types. The first is the ability to transfer BW internally among tanks to maintain stability during cargo operations, in lieu of discharging/entraining ballast water. Internal BW transfers provide a greater opportunity for containers to hold their BW without discharge, an accepted management approach that can enable them to avoid conducting BWE in many circumstances. It is not clear if this practice has become more prevalent in response to BW management regulations, but it is suggested by the decrease in the proportion of overseas container ship arrivals that discharged BW at their initial arrival since BW management was mandated in 2004 (Verling *et al.*, 2005). A second apparent behavioral change adopted by containers as an alternative to conducting BWE was to entrain BW from open ocean sources rather than coastal sources, thereby obviating the need to undergo exchange.

15.4 Where We Go from Here?

It is evident that we have come a long way in managing the BW vector via BWE. But it has been shown here and elsewhere that BWE cannot be applied uniformly by all ships en route to all ports due to geographic and operational constraints (Miller *et al.*, 2011; NRC, 2011) and BWE can still result in large introductions of NIS (Minton *et al.*, 2005). Therefore, BWE is not viewed broadly as the final solution to ballast-borne introductions and spread.

The current challenge and next frontier in managing the BW vector is achieving concentration-based discharge standards that will significantly lower propagule pressure below what is achievable with BWE (IMO, 2004; USCG, 2012b). The engineering and technical challenges of treating BW can be overcome (e.g., different ship types have different volumes of ballast both on board and discharged). These efforts are dynamic and are evolving quickly, but the timeline is uncertain. Whether we see implementation of these technologies in the next few years or more gradually over the next decade remains to be seen. Even when the technology becomes available, the time it takes to retrofit ships is significant, given the size of the global fleet. Nonetheless, this is a global experiment in invasion management that is undoubtedly reducing invasion risk and shifting the global marine invasion landscape.

In a nutshell

- Commercial ships are the major vector for global movement of NIS.
- Not all ship types are equal in their operation and behaviors. They differentially sample the world's biota and deliver BW in vastly different volumes.
- The primary effort at vector interruption has been BWE; however, BWE is imperfect because of geographic and operational constraints.
- The next frontier in efforts to manage the BW vector is achieving concentration-based discharge standards.

15.5 Bibliography

Albert, R. J., Lishman, J. M., Saxena, J. R. (2013). Ballast water regulations and the move toward concentration-based numeric discharge limits. Ecological Applications, 23 (2), 289-300, doi: 10.1890/12-0669.1.

Bailey, S. A., Deneau, M. G., Jean, L., *et al.* (2011). Evaluating efficacy of an environmental policy to prevent biological invasions. Environmental Science and Technology, 45 (7), 2554-2561, doi: 10.1021/es102655j.

Carlton, J. T. (1985). Transoceanic and interoceanic dipersal of coastal marine organisms: the biology of ballast water. Oceanography and Marine Biology, 23, 313-371.

Carlton, J. T. (1996). Pattern,process, and prediction in marine invasion ecology. Biological Conservation, 78 (1-2), 97-106.

Carlton, J. T., Reid, D. F., van Leeuwen, H. (1995). The role of shipping in the introduction of nonindigenous aquatic organisms to the coastal waters of the United States (other than the Great Lakes) and an analysis of control options. (Technical Report No. CG-D11-95). US Coast Guard: Washington, D.C.

Davidson, I. C., Brown, C. W., Sytsma, M. D., *et al.* (2009). The role of containerships as transfer mechanisms of marine biofouling species. Biofouling, 25 (7), 645-655, doi: 10.1080/08927010903046268.

Drake, J. M., Lodge, D. M. (2004). Global hot spots of biological invasions: evaluating options for ballast-water management. Proceedings of the Royal Society of London Series B-Biological Sciences, 271 (1539), 575-580.

Grevstad, F. S. (1999). Experimental invasions using biological control introductions: the influence of release size on the chance of population establishment. Biological Invasions, 1, 313-323.

Hedge, L. H., O'Connor, W. A., Johnston, E. L. (2012). Manipulating the intrinsic parameters of propagule pressure: implications for bio-invasion. Ecosphere, 3 (6), art48, doi: 10.1890/es11-000375.1.

Hewitt, C. L., Gollasch, S., Minchin, D. (2009). The vessel as a vector–biofouling, ballast water and sediments. In G. Rilov & J. A. Crooks (Eds.), Biological invasions in marine ecosystems (pp. 117-131). Heidelberg, Germany: Springer.

Hulme, P. E. (2009). Trade, transport and trouble: managing invasive species pathways in an era of globalization. Journal of Applied Ecology, 46 (1), 10-18, doi: 10.1111/j.1365-2664.2008.01600.x.

International Maritime Organization (IMO) (2004). Ballast Water Management Convention. London, UK: IMO Publication.

Kaluza, P., Koelzsch, A., Gastner, M. T., *et al.* (2010). The complex network of global cargo ship movements. Journal of the Royal Society of London Interface, 7(48), 1093-1103. doi: 10.1098/rsif.2009.0495.

Keller, R. P., Drake, J. M., Drew, M. B., *et al.* (2011). Linking environmental conditions and ship movements to estimate invasive species transport across the global shipping network. Diversity and Distributions, 17 (1), 93-102, doi: 10.1111/j.1472-4642.2010.00696.x.

Lavoie, D. M., Smith, L. D., & Ruiz, G. M. (1999). The potential for intracoastal transfer of non-indigenous species in the ballast water of ships. Estuarine Coastal and Shelf Science, 48 (5), 551-564.

Miller, A. W., Minton, M. S., Huber, T., *et al.* (2012). Status and trends of ballast water management in the United States. Fifth biennial report of the National Ballast Information Clearinghouse (January 2008 - December 2009). Edgewater: Smithsonian Environmental Research Center.

Miller, A. W., Minton, M. S., Ruiz, G. M. (2011). Geographic limitations and regional differences in ships' ballast water management to reduce marine invasions in the contiguous United States. Bioscience, 61 (11), 880-887, doi: 10.1525/bio.2011.61.11.7.

Minton, M. S., Huber, T., Ruiz, G. M., *et al.* (2012). Status and trends of ballast water management in the United States. Sixth biennial report of the National Ballast Information Clearinghouse (January 2010 - December 2011). Edgewater: Smithsonian Environmental Research Center.
Minton, M. S., Mack, R. N. (2010). Naturalization of plant populations: the role of cultivation and population size and density. Oecologia, 164 (2), 399-409, doi: 10.1007/s00442-010-1667-4.
Minton, M. S., Verling, E., Miller, A. W., *et al.* (2005). Reducing propagule supply and coastal invasions via ships: effects of emerging strategies. Frontiers in Ecology and the Environment, 3 (6), 304-308.
National Ballast Information Clearinghouse (NBIC) (2014). NBIC Online Database (doi: dx.doi.org/10.5479/data.serc.nbic). Smithsonian Environmental Research Center & United States Coast Guard. Retrieved 15 April 2014 from http://invasions.si.edu/nbic/search.html.
National Research Council (NRC) (1996). Stemming the tide: Controlling introducations of ionindgenous species by ship's ballast. Washington, D.C.: National Academy Press.
National Research Council (NRC) (2011). Assessing the relationship between propagule pressure and invasion risk in ballast water. Washington, D.C.: National Academy Press.
Ruiz, G. M., Carlton, J. T. (2003). Invasion vectors: a conceptual framework for management. In G. M. Ruiz & J. T. Carlton (Eds.), Invasive Species: Vectors and management strategies (pp.459-504). Washington, D.C.: Island Press.
Ruiz, G. M., Fofonoff, P. W., Carlton, J. T., *et al.* (2000). Invasion of coastal marine communities in North America: apparent patterns, processes, and biases. Annual Review of Ecology and Systematics, 31, 481-531.
Ruiz, G. M., Reid, D. F. (2007). Current state of understanding about the effectiveness of ballast water exchange (BWE) in reducing aquatic nonindigenous species (ANS) introductions to the Great Lakes Basin and Chesapeake Bay, USA: synthesis and analysis of existing information. (NOAA Technical Memorandum GLERL-142). Ann Arbor: U.S. Department of Commerce & National Oceanic and Atmospheric Administration.
Santagata, S., Gasiūnaite, Z. R., Verling, E., *et al.* (2008). Effect of osmotic shock as a management strategy to reduce transfers of nonindigenous species among low-salinity ports by ships. Aquatic Invasions, 3 (1), 61-76.
Seebens, H., Gastner, M. T., Blasius, B. (2013). The risk of marine bioinvasion caused by global shipping. Ecology Letters, 16 (6), 782-790, doi: 10.1111/ele.12111.
Simberloff, D. (1989). Which insect introductions succeed and which fail? In J. A. Drake, H. A. Mooney, F. di Castri, R. H. Groves, F. J. Kruger, M. Rejmanek & M. Williamson (Eds.), Biological invasions: A global perspective (pp.61-75). Chichester, England: John Wiley & Sons.
Simkanin, C., Davidson, I., Falkner, M., *et al.* (2009). Intra-coastal ballast water flux and the potential for secondary spread of non-native species on the US west coast. Marine Pollution Bulletin, 58 (3), 366-374. doi: 10.1016/j.marpolbul.2008.10.013.
United Nations Conference on Trade and Development (UNCTAD) (2012). Review of maritime transport. (UNCTAD/RMT/2012). United Nations. Retrieved 15 April 2014 from http://unctad.org/en/PublicationsLibrary/rmt2012_en.pdf.
United States Census Bureau (USCB) (2013). Total U.S. Exports. *Origin of Movement Series* Retrieved 16 May 2014, from http://www.census.gov/foreign-trade/statistics/state/data/index.html#N.
United States Coast Guard (USCG) (2012a). Ballast water management for control of non-indigenous species in waters of the United States. 33 CFR 151 Subpart D.
United States Coast Guard (USCG). (2012b). Standards for living organisms in ships' ballast water discharged in U.S. waters; Final Rule. Federal Register.
Verling, E., Ruiz, G. M., Smith, L. D., *et al.* (2005). Supply-side invasion ecology: characterizing propagule pressure in coastal ecosystems. Proceedings of the Royal Society B-Biological Sciences, 272 (1569), 1249-1256.

Vigarie, A. (1999). From break-bulk to container: the transformation of general cargo handling and trade. GeoJournal, 48, 3-7.
Wonham, M. J., Walton, W. C., Ruiz, G. M., *et al.* (2001). Going to the source: role of the invasion pathway in determining potential invaders. Marine Ecology-Progress Series, 215, 1-12.
Xu, J., Wickramarathne, T., Grey, E., *et al.* (2014). Patterns of ship-borne species spread: a clustering approach for risk assessment and management of non-indigenous species spread. arXiv preprint arXiv:1401.5407.

Part IV. **Predictions and New Tools in Biological Invasions**

João Canning-Clode, Filipa Paiva

Summary of Part IV

Invasion dynamics are being affected by global-scale changes, and new invasions have been detected in increasing numbers in several biogeographic regions, partly due to global warming but also the ability of many successful invaders to tolerate stressful conditions better than similar native species. Moreover, research in invasion science has been expanding in the last decade due to several insights. This kind of research already requires several levels of expertise in different lines of research within invasion science, particularly molecular tools to confirm the origin and history of invasions and models to anticipate future scenarios and responses to management actions. This last section of the book will contribute to different approaches in the future of biological invasions. The importance of molecular tools in examining the origin of invasive populations is discussed as well as climate change potentially promoting the spread of non-native species. Furthermore, dynamic models are presented with advanced spatiotemporal methods.

In terrestrial systems, non-native plant species pose a serious threat to the integrity of ecological systems and have detrimental economic and health impacts on society. During the last twenty years, research exploring the potential of human-induced climate change to accelerate the impact of invasive plants on native flora and ecosystems has been growing. **Chapter 16,** by Michelle Leishman and Rachael Gallagher, reviews recent investigations into the response of invasive plants to climate change drivers and to changes in temperature and rainfall patterns. The authors examine how climate change may influence the abundance and distribution of invasive plant species and draw conclusions, from manipulative experiments and from species distribution modelling, about the potential effects of climate change on several ecological aspects of plant invaders.

In addition, new spatiotemporal models have been developed over the last decade to explicitly address species responses, taking into account landscapes' heterogeneity. **Chapter 17** by Mário Santos and co-authors outlines these novel modelling techniques applied to forecasting and management, with a particular focus on biological invasions. Based on key examples of prominent biological invasions, the authors discuss the concepts, requirements, and potential outcomes of various modelling methods.

In the last two decades, DNA-based approaches have been increasingly used to study non-native species and to analyze the processes leading to introductions and invasions. Using numerous examples from different environments and taxa, **Chapter 18** by Frédérique Viard and Thierry Comtet highlights how modern molecular tools may tackle several aspects of biological invasions, such as detection of non-native species, identification of sources and routes of introductions, range expansion in invaded regions, and description of the consequences of introductions on native communities. Molecular ecology will undoubtedly have a central role in our attempts to understand the experimental evolution that is taking place accidently due to the human-assisted invasion of a multitude of species.

The impact of biological invasions in freshwater and estuarine ecosystems is well known with some of the highest rates of extinctions and endangerment of native species worldwide. In 1996 Peter Moyle and Theo Light investigated the nature of fish assemblages in highly invaded aquatic ecosystems in the context of assembly theory, and proposed twelve rules likely to regulate interactions among native and non-native species in forming new assemblages. Almost twenty years later, the same authors in **Chapter 19** of this book reassess these rules by examining them in a series of case histories of freshwater and estuarine fish assemblages with significant components of invasive species. The authors conclude with a discussion of reconciliation ecology as a framework for conservation of novel aquatic ecosystems.

Michelle R. Leishman, Rachael V. Gallagher

16 Will Alien Plant Invaders Be Advantaged Under Future Climates?

16.1 Introduction

Alien plants have successfully invaded a wide variety of habitats around the globe. There has been considerable research attention, policy, and management action directed at alien invasive plants due to their significant negative impact on terrestrial biodiversity, agricultural systems, and human health (Bridges, 1994; Pimentel *et al.*, 2005; Pyšek & Richardson, 2010). There is now increasing focus on how alien invasive plants may be affected by climate change, including changes in agricultural weed assemblages (McDonald *et al.*, 2009); responses of allergenic plants that impact human health (Ziska *et al.*, 2003; Shea *et al.*, 2008); and potential impacts of changes in the distribution, abundance, and impact of alien invasive plants on terrestrial biodiversity (Hellmann *et al.*, 2008; Beaumont *et al.*, 2009; Bradley *et al.*, 2010; Gallagher *et al.*, 2010; Bellard *et al.*, 2013).

There are a range of scales and mechanisms by which anthropogenic climate change may directly influence the distribution and abundance of alien plant species. This ranges from global-scale drivers such as elevated CO_2 resulting in changes in mean annual, minimum and maximum temperature and precipitation, to influences at regional and local scales such as increases in the frequency and severity of extreme events (e.g. floods, cyclones, and fires) (Stocker *et al.*, 2013). In addition, climate change may indirectly affect alien plant species' abundance and distribution via changes in competitive interactions, mutualisms such as pollination and seed dispersal, rates of herbivory, and seed predation (Smith *et al.*, 2000; Byers, 2002; Ward & Masters, 2007; Schweiger *et al.*, 2010). All plant species are likely to respond to these alterations in climate, but the critical question is whether alien invasive plant species are more likely to be favoured by climate change than co-occurring native plant species.

Several ideas have been proposed for why alien invasive plants are likely to be favoured by climate change (Hellmann *et al.*, 2008; Bradley *et al.*, 2010). For instance, alien invasive plant species typically possess mechanisms for long-distance dispersal and a superior ability to colonise a wide range of environments, which may allow them to move to areas of suitable climate and to colonise gaps caused by disturbance or mortality (Chown *et al.*, 2012). Alien plant species generally have large native geographic ranges (Gravuer *et al.*, 2008) and broad environmental tolerance (Gallagher *et al.*, 2011), and consequently are likely to be able to cope with a wide range of environmental variability. Finally, alien invasive plant species tend to be phenotypically plastic (Davidson *et al.*, 2011) and many are capable of rapid genetic change (Prentis *et al.*, 2008; Clements & Ditommaso, 2011), leading to high adaptive capacity that may enable persistence and responsiveness to changes in environmental conditions (e.g. Willis *et al.*, 2010).

Both experimental manipulations and functional trait approaches have shown how invaders may become more dominant under future climates. Functional traits such as specific leaf area, seed mass, or wood density are the morphological, phenological, chemical or physical attributes that relate to ecological performance of a species (McGill *et al.*, 2006). Sandel and Dangremond (2012) combined functional trait information with spatial analysis to show how warmer temperatures are likely to lead to increased dominance of exotic plant species in California's grassland flora. Experimental studies have typically found stronger biomass and reproduction responses of alien invasive compared to native plant species, suggesting a potential shift to increasing dominance by alien invasive plant species (e.g. Smith *et al.*, 2000; Tooth & Leishman, 2014).

This chapter focuses on terrestrial alien invasive plants and their potential impact on biodiversity under climate change. We first examine how climate change may influence the abundance and distribution of alien plant species, using the framework of the naturalization-invasion continuum concept (Richardson & Pysek, 2006). We go on to review evidence for the direct and indirect effects of climate change favouring alien invasive plants compared to native species. In light of this evidence, we assess the major drivers of climate change on vegetation to determine whether a shift to alien-dominated vegetation assemblages is likely under future climates. Finally, we discuss issues of management of invasive plant species under climate change and highlight knowledge gaps and potential research directions.

16.2 The Naturalization-invasion Continuum as a Framework for Understanding Potential Climate Change Influences on Alien Plants

Our understanding of the processes of alien plant arrival, persistence, and spread has been greatly influenced by the development of frameworks such as the naturalization-invasion continuum (Richardson & Pyšek, 2006; see also Catford *et al.*, 2009; Blackburn *et al.*, 2011) (Figure 16.1). The naturalization-invasion continuum conceptualises the environmental and biotic barriers that an alien species must overcome in order to become established, persist, and spread in a new environment. Several previous reviews have used this framework to consider how climate change may influence these environmental and biotic barriers (e.g. for alien species generally Hellmann *et al.*, 2008, Walther *et al.*, 2009; for invasive species in freshwater ecosystems Rahel & Olden, 2008). These reviews have resulted in a significant increase in research focus on the effect of climate change on alien invasive species. In this chapter we assess the total body of evidence, focusing particularly on differential responses of native and alien plant species, which will be crucial in determining whether shifts to alien-dominated vegetation assemblages occur under climate change.

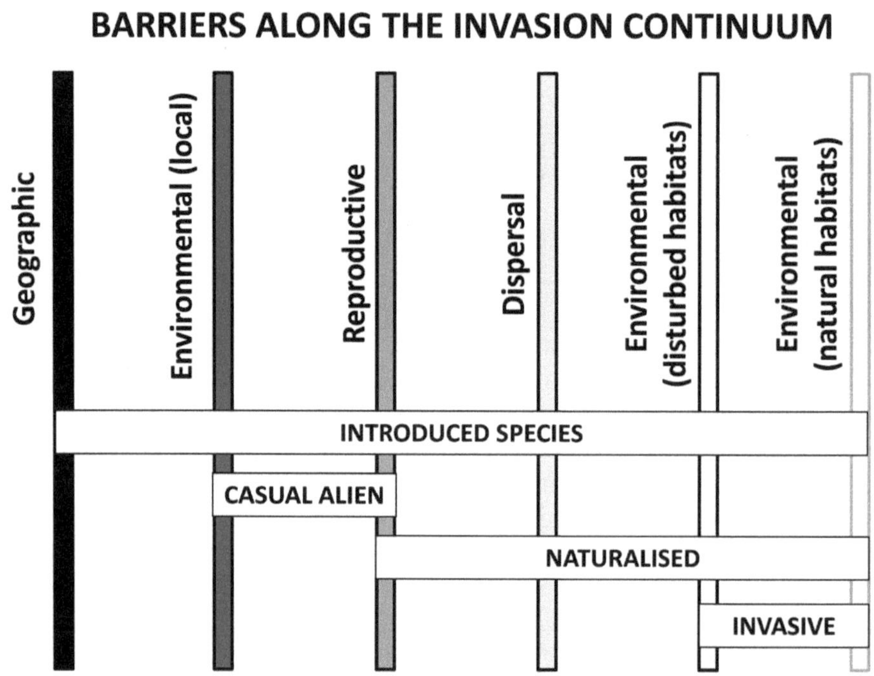

Fig. 16.1: The naturalization-invasion continuum. A schematic representation of major barriers limiting the spread of introduced plants. Climate change may affect plant invasion dynamics at each or all of these barriers. Adapted from Richardson *et al.* (2000).

16.2.1 Arrival of Propagules to New Areas

The first barriers that an alien species must overcome are those restricting the arrival of propagules into a new area. Climate change may increase the likelihood of propagule arrival (or introduction) of alien species by three main pathways. Firstly, changes in climate may result in a desire for new agricultural and horticultural species that can better cope with new climate conditions (Hellmann *et al.*, 2008; Bradley *et al.*, 2012). For example, drought-tolerant plants with low water requirements, such as desert grasses, are often promoted for garden plantings in areas experiencing reductions in rainfall. This push to plant hardier alien species and varieties may be problematic given that deliberate introductions of agricultural and horticultural species have been the largest source of alien invasive plants in many parts of the world (e.g. Australia (Groves *et al.*, 2005); United States (Reichard & White, 2001)). Secondly, increases in extreme climatic events such as floods and cyclones may result in increased propagule dispersal of alien plants (Florentine & Westbrooke, 2005; Murphy *et al.*, 2008; Diez *et al.*, 2012). Floods may carry propagules or stem fragments large distances, extending the invasion front of alien species (Sainty *et al.*, 1997; Florentine & Westbrooke, 2005) (Figure 16.2). For example, athel pine (*Tamarix aphylla*) was able to establish and spread along the Finke River system in central Australia after major floods in 1974, and is now considered one of Australia's worst weeds (Low, 2008). Thirdly, assisted colonization of native species to new areas outside their current range as a conservation action to increase a species' resilience to climate change will result in deliberate introductions of species to new areas (Mueller & Hellmann, 2008; Gallagher *et al.*, 2015). Of the three pathways, new introductions from commercial enterprises are likely to be the most important, but these have the potential to be effectively contained by Weed Risk Assessment and quarantine operations. Extreme climate events are just as likely to affect both native and alien plant species, although the superior dispersal and colonization ability of many alien invasive species may result in an advantage. Furthermore, extreme climatic events that create disturbance may be instrumental in allowing 'sleeper weeds' to transition from naturalized to invasive. Finally, assisted colonisation, whilst not likely to lead to a large number of new invaders, does have important consequences for how people perceive native, alien, and invasive plant species (see section 16.4 Issues of Management of Invasive Plants under Climate Change).

16.2.2 Establishment, Increased Abundance and Spread

Once a species has been introduced to a new area, it must overcome barriers to establishment in order to form self-sustaining populations and become naturalized. To transition from naturalized to invasive, populations must increase in abundance and spread successfully across the landscape (Richardson & Pyšek, 2006). Climate change and its drivers (e.g. elevated atmospheric CO_2 concentrations) will influence the likelihood of establishment, population growth, and spread via direct and indirect effects on demographic processes, such as seed bank persistence, seedling survival, growth rate, and reproductive output (Leishman *et al.*, 2000; Williams *et al.*, 2007). However, as native plant species will also be affected by changing climate, it is crucial to understand whether differential responses to increases in temperature and CO_2 and altered rainfall patterns will benefit invaders more than co-occurring natives. Below, we outline the evidence for the direct and indirect effect of a range of climate change factors on demographic processes of alien invasive compared to native plant species, and resultant community outcomes.

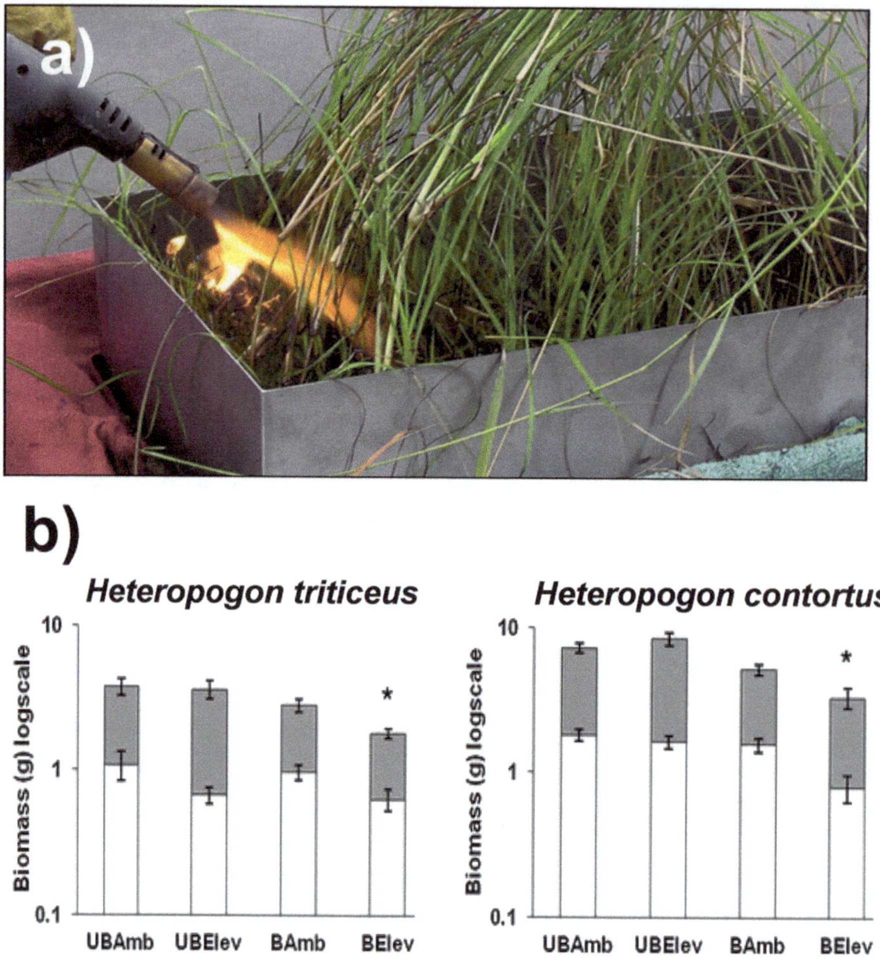

Fig. 16.2: Effects of invasion of *Nicotiana glauca* R. Graham (Solanaceae) following an extreme flooding event in early 1997 in western New South Wales, Australia. Extreme weather events of this nature are projected to increase under future climate scenarios in some regions of the globe. (a) Thick infestation of adult *N. glauca* plants (b) flower heads (c) emergence of new seedlings after flood water starts to recedes (d) mortality in co-occurring extant vegetation (e) seedling emergence. Images courtesy of S. K. Florentine (Federation University, Australia).

16.2.2.1 Elevated CO_2

Atmospheric concentrations of CO_2 have been rising steadily since the Industrial Revolution, from a concentration of ~270 ppm to current levels of ~400 ppm, and this increase is predicted to continue under a range of emission scenarios (Stocker *et al.*, 2013). It has been suggested that alien invasive species are likely to be more responsive than native species to elevated CO_2 as they tend to be capable of rapid growth when resources are not limiting due to traits such as high specific leaf area, high leaf nitrogen content, and high photosynthetic capacity (Grotkopp & Rejmanek, 2007; Leishman *et al.*, 2007; Leishman *et al.*, 2010). The ability to grow rapidly whilst exhibiting less conservative water use strategies may allow alien invasive species to benefit from both carbon fertilization and water savings under elevated CO_2 (Blumenthal *et al.*, 2013). A range of studies in growth chambers, glasshouses, open-top chambers, and FACE experiments have provided support for this (Baruch & Goldstein, 1999; Huxman *et al.*, 1999; Smith *et al.*, 2000; Belote *et al.*, 2004; Hättenschwiler & Körner, 2003; Ziska *et al.*, 2005; Dukes *et al.*, 2011; Manea & Leishman, 2011; Blumenthal *et al.*, 2013). In a meta-analysis of the responses of native and non-native species, Sorte *et al.* (2013) showed that non-native species were more responsive to elevated CO_2, although the difference was relatively weak. Therefore, whilst studies of direct fertilisation effects of elevated CO_2 on invaders show they tend to respond more strongly than native species, the difference in response between the two plant groups may not be substantial and will vary with environmental conditions.

Importantly, the influence of elevated CO_2 on alien invasive plants is likely to be the outcome of indirect effects via changes in competitive outcomes, and interactions with other drivers such as fire. For example, experiments investigating the effect of elevated CO_2 on competitive interactions between native and alien invasive plants have shown that native species became less competitive compared to invaders (Manea & Leishman, 2011). Similarly, Dukes *et al.* (2011) showed that the invasive *Centaurea solstitialis* in grassland plots responded strongly to elevated CO_2 compared to the resident native species, which responded much more weakly or not at all. In an arid system, Smith *et al.* (2001) found that aboveground biomass and seed rain of the invader *Bromus tectorum* increased much more strongly under elevated CO_2 compared to co-occurring native species. Thus it is the *relative* response of co-occurring native and alien invasive species, particularly in relation to competitive interactions, that matters.

Only a small number of experiments have examined the interaction of CO_2 with other climate change drivers, although this is an area of increasing research focus. However, from the few studies that have been conducted, the evidence is equivocal. Dukes *et al.* (2011) found no interactive effects on plant growth between CO_2 and enhanced warming, precipitation, or nitrate. Similarly, Tooth and Leishman (2013) found no consistent difference between native and alien invasive species in their re-sprouting response after fire in a temperate grassland system. However, in a similar experiment for a tropical savanna system, they found that native species' re-sprouting response was reduced under elevated CO_2, resulting in a shift to a more alien invasive-dominated community (Tooth & Leishman, 2014) (Figure 16.3). Similarly, Manea and Leishman (2014) grew mesocosms of mixed native and alien invasive grasses under ambient and elevated CO_2 combined with repeated extreme drought events and found that the alien invasive grasses were less influenced by extreme drought than native grasses under elevated CO_2. Thus, it seems that elevated CO_2 can shift the balance of plant assemblages towards greater abundance of alien invasive species, but this is strongly dependent on environmental conditions, with resource availability likely to be critical. There is clearly a need for further research in this area.

Fig. 16.3: Experimental manipulations of the interaction between fire and elevated CO_2 on native and exotic grasses. (a) Burning of competitive mixtures of native and invasive savanna grasses grown in mesocosms under glasshouse conditions of ambient and elevated CO_2 concentrations. Species in mixtures were three invasive exotic C4 grasses (Andropogon gayanus Kunth, *Cenchrus polystachios* (L.) Morrone, *Cenchrus ciliaris* L.), and three native C4 grasses (*Heteropogon triticeus* (R. Brown) Stapf ex Craib, *Heteropogon contortus* (L.) Beauv. ex Roem. & Schult., *Eriachne triseta* Steud.). (b) Graph showing total biomass (grey bars = aboveground, white bars = belowground) for two native grasses (*H. triticeus* and *H. contortus*) grown in mixed invasive and native species mesocosms under ambient and elevated CO_2. At day 154, half of the mesocosms in each glasshouse were burnt (B) and half were left unburnt (UB). These two native species had significantly less biomass under elevated compared with ambient CO_2 after burning, indicating an effect of elevated CO_2 on re-sprouting response when grown in competition with invaders. These results suggest that community composition and species interactions in this fire-prone community may alter in a high CO_2 world, shifting to a more exotic-dominated community and potentially resulting in an intensified fire frequency due to positive feedbacks. Images and graphs courtesy of I. Tooth (Royal Botanic Gardens, Sydney, Australia).

16.2.2.2 Changes in Temperature and Rainfall

Patterns of temperature and rainfall are shifting globally relative to baseline conditions. On average, combined global land and sea surface temperatures rose by 0.85°C [0.65 to 1.06] over the period from 1880 to 2012, and there have been both documented increases and declines in long-term average precipitation during this period (Stocker *et al.*, 2013). There is high confidence that the number of cold days and nights has decreased, whilst the number of warm days and nights has increased globally (Stocker *et al.*, 2013). In addition to changes in average conditions, the frequency and intensity of extreme weather events (e.g. heat-waves) has increased in some regions (Australia, Asia, Europe) and is projected to continue to increase in coming decades (Stocker *et al.*, 2013).

At the macro scale, changes in temperature and rainfall are expected to drive shifts in plant species distributions, with contractions in ranges at the equator-side boundaries and range expansions polewards and to higher elevations (Thuiller *et al.*, 2005; Kelly & Goulden, 2008). The ability to shift range will be contingent on species' ability to disperse propagules and establish populations into new regions (Corlett & Westcott, 2013). There have been a number of studies that have used species distribution modelling (SDM) approaches to examine likely responses of native plant species in large regions, such as Europe (Thuiller *et al.*, 2005), California (Loarie *et al.*, 2008), South Africa, and Western Australia (Yates *et al.*, 2010a, b). These have generally shown the potential for large species' range contractions and substantial shifts in native species assemblages. The SDM approach has also been used to project changes in suitable habitat under climate change for individual invasive plant species (e.g. Bradley *et al.*, 2009; Beaumont *et al.*, 2009; Murray *et al.*, 2012) as well as for multiple invasive species in large regions (e.g. O'Donnell *et al.*, 2012; Gallagher *et al.*, 2013; Duursma *et al.*, 2013). Generally, these studies have not found an increase in the areas of suitable habitat for alien invasive species. Instead, SDM studies show that although there is some variability among regions and taxa, areas of suitable habitat under future climate generally decline in size and shift polewards for invaders, as for native plant species. This suggests that although alien invasive plants tend to have larger native distributions with associated broader climatic ranges than native species (Milbau & Stout, 2008; Gallagher *et al.*, 2011), this does not buffer them sufficiently against the scale of climate change predicted. Instead, we can expect to see substantial shifts in species assemblages and the emergence of new colonisation opportunities created by losses of marginal populations at the edges of range boundaries of both native and alien species.

At more local scales, changes in temperature and rainfall will affect demographic processes such as seed bank mortality, seedling survival, growth rates, and reproduction. The ability of seeds to persist in the soil may be altered, via effects such as dormancy release and germination responding to changes in environmental cues, and soil pathogen activity responding to changes in soil temperature and moisture (Walck *et al.*, 2011). There have been few empirical tests on the expected effects of climate change on seed banks (but see del Cacho *et al.*, 2012; Leishman *et al.*, 2000;

Ooi *et al.*, 2009), and none have directly tested for differences in response of native and alien invasive species. Earlier work by Blaney and Kotanen (2001) found no difference in seed bank mortality due to soil fungal pathogens between native and invasive species. Whilst these authors did not test explicitly for the potential effects of climate change on seed banks, it seems reasonable that this pattern would be maintained.

Studies comparing growth responses to increased temperatures between native and alien invasive species have shown contrasting results. For example, Hou *et al.* (2014) compared the effect of extreme temperatures on seedling germination and establishment of invasive and native Asteraceae species, and showed that the invasive species coped better with extremes. However, Verlinden *et al.* (2013) grew native-invasive species pairs in ambient and elevated (+3°C) conditions and found no effect of warming on the competitive balance within pairs. In a meta-analysis incorporating results from multiple studies, Sorte *et al.* (2013) assessed whether non-native and native species responded differently to climate change factors, including warming and changes in rainfall. These authors found that there was a positive effect of warming on native terrestrial species but not on non-natives (note that this included some invertebrate species in the analysis), but no differences between natives and non-natives in response to changes in rainfall. Thus, it is likely that the effect of changes in temperature and rainfall on key plant demographic processes will be species- and system-specific, and that alien invasive plant species will not always be advantaged under changed temperature and rainfall conditions.

16.2.2.3 Changes in Disturbance Regimes Including Extreme Climatic Events

Climate change is likely to cause significant changes in disturbance regimes, such as floods, cyclones, and fire, which underpin plant demographic processes (Field *et al.*, 2012). Altered disturbance frequencies may affect germination and establishment success, regeneration and mortality, but it is unclear whether alien invasive plants will be advantaged relative to native species. More generally, invasive plants are thought to respond positively to disturbance, including physical disturbance (Burke & Grime, 1996; Daehler, 2003; Hansen & Clevenger, 2005) and nutrient enrichment (Burns, 2008; Leishman & Thomson, 2005). The most likely positive effect of changes in disturbance regimes and increased frequency of extreme climatic events on alien invasive plants will be via increased propagule input and hence colonization opportunities, increased resource availability favouring fast-growing species (Leishman *et al.*, 2010) and through reduced competition due to mortality of individuals and reduced biomass (Diez *et al.*, 2012; also see section 16.3).

16.2.2.4 Interactions Between Macro- and Local-scale Climate Change Drivers

The abundance and distribution of species is generally the outcome of a complex interplay of multiple factors. Thus, we might expect that the combined effect of multiple climate change drivers may be more important in determining ecosystem change than the effect of individual drivers such as elevated CO_2, changed temperature or rainfall, or increased frequency of extreme climatic events in isolation. Experimental work that tests hypotheses on the effects of multiple drivers on mixed assemblages of native and alien invaders is now emerging (e.g. Dukes *et al.*, 2011; Tooth & Leishman, 2013; Tooth & Leishman, 2014; Manea & Leishman, 2014). Sheppard *et al.* (2012) found that species introduced into experimental European meadow communities did not show responses to any of the combinations of warming and extreme drought or deluge consistent with their native/alien provenance. Similar results were found by Godfree *et al.* (2013), who compared responses of a native and alien grass species in Australia under ambient and simulated 2050 conditions of drought, temperature, and elevated CO_2, and by Tooth and Leishman (2013), who compared re-sprouting responses after fire under ambient and elevated CO_2 for mixed temperate grassland assemblages. However, other studies have found evidence for a shift towards more alien-dominated communities under combinations of increased CO_2 and fire or drought (Dukes *et al.*, 2011; Tooth & Leishman, 2014; Manea & Leishman, 2014) (Figure 16.3). It seems likely that the effect of multiple climate change drivers on vegetation assemblages will not consistently favour alien invasive plants, but that the outcomes may be dependent on the response of key species (Thuiller *et al.*, 2007) to specific combinations, such as elevated CO_2 and drought, or warming and increased fire frequency.

16.3 Will There Be a Shift to Alien-dominated Vegetation Assemblages under Climate Change?

So far we have examined the evidence for differences in the responses of native and alien invaders to the direct effects of climate change factors, either individually or in combination, in order to assess how likely it is that there will be a shift to aliendominated vegetation as climate change progresses. However, changes in plant species assemblages under climate change are likely to come about due to a combination of the following drivers: (1) species-level responses to changed climate conditions, such as temperature and rainfall; (2) changes in competitive interactions between species under changed conditions of CO_2, temperature, rainfall, extreme events, and disturbance regimes; (3) reduced biomass or mortality of individuals resulting in the creation of colonization opportunities; and (4) changes in interactions between plant species and enemies (including pathogens, herbivores, and seed predators) or mutualists (including mycorrhizae, pollinators, and seed dispersers). We assess the evidence for each driver below and explore the interactions between them.

For the first driver (species-level responses to changed climate conditions such as temperature and rainfall) it is apparent from SDM approaches that both native and alien invasive species will have shifts in areas of climatic suitability towards the poles and to higher elevations, associated for many species with substantial contractions in size or connectivity between populations. Critical determinants for how these shifts in climate suitability play out in terms of changes in species' distributions are the dispersal capacity of a species and interactions among newly arriving species and the recipient community (driver 2). Several glasshouse-based experimental studies have shown that competitive interactions between native and alien invasive plant species shift to increase the relative advantage of the alien invaders under conditions such as increased CO_2 (Manea & Leishman, 2011), increased temperature (Verlinden & Nijs, 2010, but see Verlinden *et al.*, 2013), and extreme temperature (Song *et al.*, 2010). Similarly, field-based experiments have shown that warming (Chuine *et al.*, 2012) and elevated CO_2 (Dukes *et al.*, 2011) result in increased abundance of the invader in a native-dominated recipient community. These results suggest that competitive interactions will be an important determinant of vegetation composition under climate change. Thus, alien plant invaders may become more dominant in the novel vegetation assemblages created by changing climatic conditions due to their greater competitive ability, in combination with their greater capacity to disperse long distances in order to keep up with their changing climate space.

The third driver of changes in plant assemblages (reduced biomass or mortality of individuals resulting in the creation of colonization opportunities) arises due to reduced population viability on species' range margins with climate change, as well as from increased frequency and intensity of extreme climatic events (Diez *et al.* 2012). These factors are likely to result in reduced resilience of vegetation assemblages. Alien invasive plants may be best able to take advantage of these colonisation and growth opportunities in low resilience communities due to their capacity for dispersal into new areas (Pyšek & Richardson, 2007) and for rapid growth when resources are not limiting (Grotkopp & Rejmanek, 2007; Leishman *et al.*, 2010). Thus the reduced competitive ability of natives under changed climate conditions and disturbance-induced mortality creating opportunities for colonisation and growth, in combination with the typical invader traits of long-distance dispersal and fast growth, may result in clear advantages for alien invaders and resultant shifts to more alien-dominated species assemblages under future climates.

Finally, the fourth driver (changes in interactions between plant species and enemies or mutualists) is also likely to affect species' distributions and community composition. However, our knowledge of the impact of climate change on these interactions, and particularly on differential impacts on native and alien invasive species, is quite limited. It seems likely that climate change will impact interactions between plant species and enemies or mutualists in many complex and unpredictable ways.

In summary, the combination of climate change drivers 1, 2, and 3 is likely to result in a shift to alien-dominated vegetation assemblages, with a reduction in vegetation resilience (driver 3 - reduced biomass or mortality of individuals, resulting in the creation of colonization opportunities) in combination with typical invader traits of good dispersal ability and capacity for fast growth, being the most important. This has clear implications for management options, including increasing vegetation resilience and monitoring, and early eradication efforts in response to disturbance events such as fire, storms, and floods (as outlined below in section 16.4).

16.4 Issues of Management of Invasive Plants under Climate Change

Climate change poses a distinct set of challenges for the management of invasive plants. In many instances, the application of established techniques for controlling populations may become increasingly ineffective, requiring the development of new approaches. A failure to incorporate the potential effects of climate change into prevention measures that target key stages of the invasion continuum may lead to poor management outcomes.

16.4.1 Weed Risk Assessment

Pre-border Weed Risk Assessment (WRA) protocols aim to identify potentially problematic species prior to their introduction from foreign locations. Screening typically involves the use of a standardised set of questions on the biology and invasive behaviour of the species for which permission to import is being sought, as well as an assessment of the match between the climate in the native range of the species and of the recipient region (Pheloung, 2001; Groves *et al.*, 2001). Whilst current WRA protocols have been shown to be highly effective in preventing new weed establishment (Gordon *et al.*, 2008), they do not address the potential for changes in climate over coming decades to alter patterns of weed establishment (Beaumont *et al.*, 2014). Given increasing evidence that climate change may affect the growth (Smith *et al.*, 2000), spread (Crossman *et al.*, 2011), and extent of suitable habitat of weeds (Murray *et al.*, 2012; Gallagher *et al.*, 2013; Duursma *et al.*, 2013), the omission of clear questions about climate change in the WRA process requires attention. To do this, models which integrate spatial information about weed spread in the landscape and biological traits have been developed (e.g. Crossman *et al.*, 2011), but are yet to be widely adopted into formal WRA decision tools.

16.4.2 Controlling Weeds

Methods for controlling established weeds may also suffer from a lack of an integration of knowledge about the drivers of climate change, such as elevated atmospheric CO_2. For instance, the tolerance of some C_4 grasses to glyphosate-based herbicides has been shown to increase under elevated CO_2 levels (Manea *et al.*, 2011). This increased tolerance was attributed to a dilution effect from increased biomass production, also reported for a C_3 species – Canada Thistle (*Cirsium arvense* (L.) Scop.) (Ziska *et al.*, 2004). Whilst chemical control will still be feasible for weed species under elevated CO_2, dosage and frequency of application may need to increase in order to maintain a regulatory effect on population growth. Chemical-free methods, such as biological control, are also predicted to become less effective in some regions under climate change due to changes in host-plant and control-agent interactions (Hellmann *et al.*, 2008). Biocontrol agents, in particular insects, may use environmental cues like air temperature to precisely time life-cycle events, such as overwintering or emergence. Where climate change alters these signals, peak periods of host-plant growth or seed set and insect abundance may become decoupled, reducing the potential for control (Lu *et al.*, 2013). However, this effect will be site-specific and may lead to increased efficacy in regions where changing temperatures bring the phenology of host-plant and agent into better synchrony (Hellmann *et al.*, 2008; Gerard *et al.*, 2013).

16.4.3 Management Following Extreme Climatic Events

As discussed above (section 16.2.1), climate change is projected to lead to an increase in the frequency and magnitude of extreme climatic events (Field *et al.*, 2012). In addition to facilitating new arrivals, extreme climatic events may facilitate already-established alien plants by the creation of colonisation opportunities associated with disturbances such as cyclones, floods, and wildfires. For example, non-native vines may form thick infestations in canopy gaps and forest edges following cyclones (Elmqvist *et al.*, 1994; Murphy *et al.*, 2008). Strategies for targeted weed eradication and management following disturbance would be a valuable way of limiting the impact of invaders on ecosystems as climate changes. Pre-emptive monitoring in areas projected to provide suitable habitat for multiple species under future climate scenarios may also be a useful tool for detecting new weed incursions and fast-tracking local eradication (O'Donnell *et al.*, 2012).

16.4.4 Weed Management in the Wider Context of Biodiversity Planning

Whilst planning to limit the effects of climate change on ecosystems is an essential part of biodiversity conservation, some adaptation measures may inadvertently promote the dispersal and establishment of invasive plants. For instance, actions which increase the connectivity of the landscape through the introduction or reinstatement of dispersal corridors may facilitate the spread of invasive plants (Simberloff & Cox, 1987; Minor & Gardner, 2011; however see Noss, 1991 for counter-arguments). Similarly, deliberate translocation of native species deemed at-risk from climate change beyond their current range limits (assisted colonisation) may lead to an increased risk of native plants showing invasive behaviour (Gallagher *et al.*, 2015). That is, native plants that become decoupled from co-evolved pests or pathogens following translocation may become highly abundant, potentially displacing other native species in recipient communities. Whilst translocation for climate change is still in its infancy, there is substantial evidence of native plants becoming invasive when introduced outside historical range boundaries (e.g. acacias in Australia; Richardson *et al.*, 2011) that should be taken into account when designing translocation programs. In addition, defining which populations are native and which are invasive will become an increasingly difficult task for those species that shift their range without human assistance (Walther *et al.*, 2009). Balancing the need to allow species to colonise new areas that feature optimal conditions for their growth and survival against the potential for invasive behaviour will be a challenging management issue in coming decades.

16.5 What Research Is Needed to Inform Successful Management of Invasive Plants under Climate Change?

Whilst substantial progress has been made towards understanding potential synergies between plant invasions and climate change over the last two decades, research is still lacking in key areas. These areas range across a need for pure ecological research into factors such as accurately predicting dispersal dynamics and demography, to more practical questions of how to better engage the agricultural and horticultural sectors to reduce new species introductions, and novel ways to control weeds once they are established in a high CO_2 world. Below, we outline major topics where research is needed in order to effectively plan for the prevention and control of invasive plants in coming decades. Whilst we do not advocate for one research area over another, we have attempted to prioritise these topics in terms of their influence on reducing the impact of invasions under climate change.

Preventing invasions: The development of proactive weed management strategies that capitalise on the short window of opportunity following extreme events such as cyclones and floods, and large-scale disturbances such as fires, are needed to limit the damage associated with weed encroachment in affected areas, before weeds colonise and spread. A proactive approach to weed management will help to achieve better eradication outcomes at local scales and help to direct the typically finite funds available for weed management towards actions with the highest likelihood of long-term success.

Multiple drivers of invasion: The combined effects and interactions between multiple drivers and consequences of climate change on invasive plants (e.g. elevated CO_2, rising temperatures, increased drought conditions, increased extreme climatic events) are relatively poorly understood. Despite a rise in the amount of research on this topic, we still lack a clear theoretical framework based on a mechanistic understanding for predicting how interactions between climate change factors may affect alien plant populations and vegetation more generally. Similarly, more research is needed into the interactive effects of propagule pressure, resource availability, and competition between natives and invaders in order to increase our predictive capacity.

Mutualists and enemies: Little is known about how the interactions between invasive plants and mutualists, or equally, enemies such as herbivores and pathogens, may be altered by changing climate regimes. Filling this knowledge gap, particularly in relation to epidemiology of plant diseases and pests, will have important consequences for the continued efficacy of biocontrol as a means to suppress weed populations. However, beneficial relationships between plants and mutualists (e.g. N-fixing bacteria, soil-borne microorganisms) may also be affected by climate change in ways that may favour alien invasive species over co-occurring native species or, conversely, increase native plant community resistance to invasion.

Reproductive biology: The impact of climate change on the reproductive biology of invasive plants and the native species they co-occur with requires research attention. Establishment is a key demographic phase governed by multiple factors such as seed availability (linked to seed output and adult population dynamics, and dispersal – including biotic, abiotic, and human-mediated) and seed quality (linked to maternal provisioning and rates of seed-set). Without an understanding of how changes in temperature, rainfall, and CO2 concentration may affect seed production and storage within seed banks, the ability to predict how invaders may be advantaged by changing climate will be limited. This type of information on key demographic stages could also be coupled to spatial models to better predict how key invasive plants may respond to climate change.

16.6 Acknowledgments

We thank Anthony Manea for his input into sections on the impact of elevated CO_2 on invasive species. Ifeanna Tooth and Singarayer Florentine kindly provided access to figures and photographs of their study species, for which we are grateful.

In a nutshell

- The last two decades have seen an upsurge in research into the potential synergies between invasive species and climate change, with evidence emerging of increased invader success under climate change. All stages along the naturalization-invasion continuum are likely to be affected, from the introduction and establishment of alien species to their spread and transition to serious invaders. A key question is whether alien plants will have a relative advantage under climate change conditions.
- So far, evidence for differential responses of alien invasive and native species to climate change drivers (elevated CO_2) and outcomes (increasing temperature, changing rainfall patterns, changes in disturbance regimes) is mixed. Although alien invasive plants appear to be more responsive to elevated CO_2 than many native species, plant response to elevated CO_2 and other climate change components is dependent on environmental conditions and resource availability. Similarly, correlative modelling of species-climate relationships has not revealed clear evidence that invasive plants are likely to be able to increase the extent of suitable habitat under future climates any more than their native counterparts.
- We suggest that the most important driver of a shift to alien-dominated vegetation under climate change will be the superior capacity of alien invasive plants to take advantage of colonisation opportunities arising from climate change, such as extreme climatic events, changes in disturbance regimes, and widespread reduction in vegetation resilience as range margin populations decline.
- There are substantial challenges ahead for managing invasive plants under future climates. Weed risk assessment and management approaches must incorporate consideration of future climatic conditions. Most importantly, we will need a shift in management approaches away from a focus on the control of undesirable alien plant species to building resilience of resident vegetation assemblages, in association with targeted monitoring and early eradication of alien plant species.

16.7 Bibliography

Baruch, Z., Goldstein, G. (1999). Leaf construction cost, nutrient concentration, and net CO_2 assimilation of native and invasive species in Hawaii. Oecologia, 121, 183-192.

Beaumont, L. J., Gallagher, R. V., Downey, P. O., *et al.* (2009). Modelling the impact of *Hieracium* spp. on protected areas in Australia under future climates. Ecography, 32, 757-764.

Beaumont, L. J., Gallagher, R. V., Leishman, M. R., *et al.* (2014). How can knowledge of the climate niche inform the weed risk assessment process? A case study of *Chrysanthemoides monilifera* in Australia. Diversity and Distributions, 20, 613–625.

Bellard, C., Thuiller, W., Leroy, B., *et al.* (2013). Will climate change promote future invasions? Global Change Biology, 19, 3740–3748.

Belote, R. T., Weltzin, J. F., Norby, R. J. (2004). Response of an understory plant community to elevated [CO_2] depends on differential responses of dominant invasive species and is mediated by soil water availability. New Phytologist, 161, 827-835.

Blackburn, T. M., Pyšek, P., Bacher, S., *et al.* (2011). A proposed unified framework for biological invasions. Trends in Ecology and Evolution, 26, 333-339.
Blaney, C. S., Kotanen, P. M. (2001). Effects of fungal pathogens on seeds of native and exotic plants: a test using congeneric pairs. Journal of Applied Ecology, 38, 1104-1113.
Blumenthal, D. M., Resco, V., Morgan, J. A., *et al.* (2013). Invasive forb benefits from water savings by native plants and carbon fertilization under elevated CO_2 and warming. New Phytologist, 200, 1156-1165.
Bradley, B. A., Blumenthal, D. M., Wilcove, D. S., *et al.* (2010). Predicting plant invasions in an era of global change. Trends in Ecology and Evolution, 25, 310-318.
Bradley, B. A., Oppenheimer, M., Wilcove, D. S. (2009). Climate change and plant invasions: restoration opportunities ahead? Global Change Biology, 15, 1511-1521.
Bridges, D. C. (1994). Impact of weeds on human endeavors. Weed Technology, 8, 392-395.
Burke, M. J., Grime, J. P. (1996). An experimental study of plant community invasibility. Ecology, 77, 776-790.
Burns, J. H. (2008). Demographic performance predicts invasiveness of species in the Commelinaceae under high nutrient conditions. Ecological Applications, 18, 335–346.
Byers, J. E. (2002). Impact of non-indigenous species on natives enhanced by anthropogenic alteration of selection regimes. Oikos, 97, 449-458.
Catford, J. A., Jansson, R., Nilsson, C. (2009). Reducing redundancy in invasion ecology by integrating hypotheses into a single theoretical framework. Diversity and Distributions, 15, 22-40.
Chown, S. L., le Roux, P. C., Ramaswiela, T., *et al.* (2012). Climate change and elevational diversity capacity: do weedy species take up the slack? Biology Letters, 20120806, doi: 10.1098/rsbl.2012.0806.
Chuine, I., Morin, X., Sonié, L., *et al.* (2012). Climate change might increase the invasion potential of the alien C4 grass *Setaria parviflora* (Poaceae) in the Mediterranean Basin. Diversity and Distributions, 18, 661–672.
Clements, D. R., Ditommaso, A. (2011). Climate change and weed adaptation: can evolution of invasive plants lead to greater range expansion than forecasted? Weed Research, 51, 227-240.
Corlett, R. T., Westcott, D. A. (2013). Will plant movements keep up with climate change? Trends in Ecology and Evolution, 28, 482-488.
Crossman, N. D., Bryan, B. A., Cooke, D. A. (2011). An invasive plant and climate change threat index for weed risk management: integrating habitat distribution pattern and dispersal process. Ecological Indicators, 11, 183-198.
Daehler, C. C. (2003). Performance comparisons of co-occurring native and alien invasive plants: implications for conservation and restoration. Annual Review of Ecology, Evolution and Systematics, 34, 183-211.
Davidson, A. M., Jennions, M., Nicotra, A. B. (2011). Do invasive species show higher phenotypic plasticity than native species and, if so, is it adaptive? A meta-analysis. Ecology Letters, 14, 419-431.
del Cacho, M., Saura-Mas, S., Estiarte, M., *et al.* (2012). Effect of experimentally induced climate change on the seed bank of a Mediterranean shrubland. Journal of Vegetation Science, 23, 280-291.
Diez, J. M., D'Antonio, C. M., Dukes, J. S., *et al.* (2012). Will extreme climatic events facilitate biological invasions? Frontiers in Ecology and the Environment, 10, 249-257.
Dukes, J. S., Chiariello, N. R., Loarie, S. R, *et al.* (2011). Strong response of an invasive plant species (*Centaurea solstitialis* L.) to global environmental changes. Ecological Applications, 21, 1887-1894.
Duursma, D. E., Gallagher, R. V., Roger, E., *et al.* (2013). Next-generation invaders? hotspots for naturalised sleeper weeds in Australia under future climates. PLoS ONE 8, e84222.

Elmqvist, T., Rainey, W. E., Pierson, E. D., *et al.* (1994). Effects of tropical cyclones Ofa and Val on the structure of a Samoan lowland rain forest. Biotropica, 26, 384-391.

Field, C.B., V. Barros, T.F. Stocker, *et al.* (Eds.) (2012). Managing the Risks of Extreme Events and Disasters to Advance Climate Change Adaptation. A Special Report of Working Groups I and II of the Intergovernmental Panel on Climate Change. Cambridge, UK, New York, USA: Cambridge University Press.

Florentine, S. K., Westbrooke, M. E. (2005). Invasion of the noxious weed *Nicotiana glauca* R. Graham after an episodic flooding event in the arid zone of Australia. Journal of Arid Environments, 60, 531-545.

Gallagher, R. V., Duursma, D. E., O'Donnell, J., *et al.* (2013). The grass may not always be greener: projected reductions in climatic suitability for exotic grasses under future climates in Australia. Biological Invasions, 15, 961-975.

Gallagher, R. V., Hughes, L., Leishman, M. R., *et al.* (2010). Predicted impact of exotic vines on an endangered ecological community under future climate change. Biological Invasions, 12, 4049-4063.

Gallagher, R. V., Leishman, M. R., Miller, J. T., *et al.* (2011), Invasiveness in introduced Australian acacias: the role of species traits and genome size. Diversity and Distributions, 17, 884–897.

Gallagher, R.V., Hancock, N., Makinson, R.O., *et al.* (2015). Assisted colonisation as a climate change adaptation tool. Austral Ecology, 40 (1), 12-20.

Gerard, P. J., Barringer, J. R. F., Charles, J. G., *et al.* (2013). Potential effects of climate change on biological control systems: case studies from New Zealand. BioControl, 58, 149-162.

Godfree, R. C., Robertson, B. C., Gapare, W. J., *et al.* (2013). Nonindigenous plant advantage in native and exotic Australian grasses under experimental drought, warming, and atmospheric CO_2 enrichment. Biology, 2, 481-513.

Gordon, D. R., Onderdonk, D. A., Fox, A. M., *et al.* (2008). Consistent accuracy of the Australian weed risk assessment system across varied geographies. Diversity and Distributions, 14, 234-242.

Gravuer, K., Sullivan, J. J., Williams, P. A., *et al.* (2008). Strong human association with plant invasion success for Trifolium introductions to New Zealand. Proceedings of the National Academy of Sciences, 105, 6344-6349.

Grotkopp, E., and Rejmánek, M. (2007). High seedling relative growth rate and specific leaf area are traits of invasive species: phylogenetically independent contrasts of woody angiosperms. American Journal of Botany, 94, 526-532.

Groves, R. H., Lonsdale, M., Boden, R. (2005). Jumping the garden fence: invasive garden plants in Australia and their environmental and agricultural impacts. Canberra: WWF-Australia.

Groves, R. H., Panetta, F. D., Virtue, J. G. (Eds.) (2001). Weed Risk Assessment. Clayton: CSIRO Publishing.

Hansen, M. J., Clevenger, A. P. (2005). The influence of disturbance and habitat on the presence of non-native plant species along transport corridors. Biological Conservation, 125, 249-259.

Hättenschwiler, S., Körner, C. (2003). Does elevated CO2 facilitate naturalization of the non-indigenous *Prunus laurocerasus* in Swiss temperate forests? Functional Ecology, 17, 778-785.

Hellmann, J. J., Byers, J. E., Bierwagen, B. G., *et al.* (2008). Five potential consequences of climate change for invasive species. Conservation Biology, 22, 534-543.

Hou, Q. Q., Chen, B. M., Peng, S. L., *et al.* (2014). Effects of extreme temperature on seedling establishment of nonnative invasive plants. Biological Invasions, 16, 2049-2061.

Huxman, T. E., Hamerlynck, E. P., Smith, S. D. (1999). Reproductive allocation and seed production in *Bromus madritensis* ssp. *rubens* at elevated atmospheric CO_2. Functional Ecology, 13, 769-777.

Kelly, A. E., Goulden, M. L. (2008). Rapid shifts in plant distribution with recent climate change. Proceedings of the National Academy of Sciences, 105, 11823-11826.

Leishman, M. R., Haslehurst, T., Ares, A., *et al.* (2007). Leaf trait relationships of native and invasive plants: community-and global-scale comparisons. New Phytologist, 176, 635-643.

Leishman, M. R., Masters, G. J., Clarke, I. P., *et al.* (2000). Seed bank dynamics: the role of fungal pathogens and climate change. Functional Ecology, 14, 293-299.
Leishman, M. R., Thomson, V. P. (2005). Experimental evidence for the effects of additional water, nutrients and physical disturbance on invasive plants in low fertility Hawkesbury Sandstone soils, Sydney, Australia. Journal of Ecology, 93, 38-49.
Leishman, M. R., Thomson, V. P., Cooke, J. (2010). Native and exotic invasive plants have fundamentally similar carbon capture strategies. Journal of Ecology, 98, 28-42.
Loarie, S. R., Carter, B. E., Hayhoe, K., *et al.* (2008). Climate change and the future of California's endemic flora. PLoS ONE 3, e2502.
Low, T. (2008). Climate Change and Invasive Species: A Review of Interactions. Canberra: Biological Diversity Advisory Committee.
Lu, X., Siemann, E., Shao, X., *et al.* (2013). Climate warming affects biological invasions by shifting interactions of plants and herbivores. Global Change Biology, 19, 2339-2347.
Manea, A., Leishman, M. R. (2011). Competitive interactions between native and invasive exotic plant species are altered under elevated carbon dioxide. Oecologia, 165, 735-744.
Manea, A., Leishman, M. R. (2014). Leaf area index drives soil water availability and extreme drought-related mortality under elevated CO_2 in a temperate grassland model system. PLoS ONE 9, e91046.
Manea, A., Leishman, M. R., Downey, P. O. (2011). Exotic C4 grasses have increased tolerance to glyphosate under elevated carbon dioxide. Weed Science, 59, 28-36.
McDonald, A., Riha, S., DiTommaso, A., *et al.* (2009). Climate change and the geography of weed damage: analysis of US maize systems suggests the potential for significant range transformations. Agriculture, Ecosystems and Environment, 130, 131-140.
McGill, B. J., Enquist, B. J., Weiher, E., *et al.* (2006). Rebuilding community ecology from functional traits. Trends in Ecology and Evolution, 21, 178-185.
Milbau, A., Stout, J. C. (2008). Factors associated with alien plants transitioning from casual, to naturalized, to invasive. Conservation Biology, 22, 308-317.
Minor, E. S., Gardner, R. H. (2011). Landscape connectivity and seed dispersal characteristics inform the best management strategy for exotic plants. Ecological Applications, 21, 739-749.
Mueller, J. M., Hellmann, J. J. (2008). An assessment of invasion risk from assisted migration. Conservation Biology, 22, 562-567.
Murphy, H. T., Metcalfe, D. J., Bradford, M. G., *et al.* (2008). Recruitment dynamics of invasive species in rainforest habitats following Cyclone Larry. Austral Ecology, 33, 495-502.
Murray, J. V., Stokes, K. E., Klinken, R. D. (2012). Predicting the potential distribution of a riparian invasive plant: the effects of changing climate, flood regimes and land-use patterns. Global Change Biology, 18, 1738-1753.
Noss, R. F. (1991). Landscape connectivity: different functions at different scales. In W. E. Hudson (Ed.), Landscape Linkages and Biodiversity (pp.27-39). Washington, DC, USA: Island Press.
O'Donnell, J., Gallagher, R. V., Wilson, P. D., *et al.* (2012). Invasion hotspots for non-native plants in Australia under current and future climates. Global Change Biology, 18, 617-629.
Ooi, M. K., Auld, T. D., Denham, A. J. (2009). Climate change and bet-hedging: interactions between increased soil temperatures and seed bank persistence. Global Change Biology, 15, 2375-2386.
Pheloung, P. C. (2001). Weed risk assessment for plant introductions to Australia. In R. H. Groves, F. D. Panetta, J.G. Virtue (Eds.), Weed Risk Assessment (pp. 83-92). Collingwood: CSIRO Publishing.
Pimentel, D., Zuniga, R., and Morrison, D. (2005). Update on the environmental and economic costs associated with alien-invasive species in the United States. Ecological Economics, 52, 273-288.
Prentis, P. J., Wilson, J. R., Dormontt, E. E., Richardson, D. M., and Lowe, A. J. (2008). Adaptive evolution in invasive species. Trends in Plant Science, 13, 288-294.

Pyšek, P., Richardson, D. M. (2007). Traits associated with invasiveness in alien plants: where do we stand? In W. Nentwig (Ed.), Biological Invasions (pp.97-125). Berlin Heidelberg: Springer.
Pyšek, P., Richardson, D. M. (2010). Invasive species, environmental change and management, and health. Annual Review of Environment and Resources, 35, 25-55.
Rahel, F. J., Olden, J. D. (2008). Assessing the effects of climate change on aquatic invasive species. Conservation Biology, 22, 521-533.
Reichard, S. H., White, P. (2001). Horticulture as a pathway of invasive plant introductions in the United States most invasive plants have been introduced for horticultural use by nurseries, botanical gardens, and individuals. BioScience, 51, 103-113.
Richardson, D. M., Carruthers, J., Hui, C., *et al.* (2011). Human-mediated introductions of Australian acacias–a global experiment in biogeography. Diversity and Distributions, 17, 771-787.
Richardson, D. M., Pyšek, P. (2006). Plant invasions: merging the concepts of species invasiveness and community invasibility. Progress in Physical Geography, 30, 409-431.
Sainty, G., McCorkelle, G., Julien, M. (1997). Control and spread of Alligator Weed *Alternanthera philoxeroides* (Mart.) Griseb., in Australia: lessons for other regions. Wetlands Ecology and Management, 5, 195-201.
Sandel, B., Dangremond, E. M. (2012). Climate change and the invasion of California by grasses. Global Change Biology, 18, 277-289.
Schweiger, O., Biesmeijer, J. C., Bommarco, R., *et al.* (2010). Multiple stressors on biotic interactions: how climate change and alien species interact to affect pollination. Biological Reviews, 85, 777-795.
Shea, K. M., Truckner, R. T., Weber, R. W., *et al.* (2008). Climate change and allergic disease. Journal of Allergy and Clinical Immunology, 122, 443-453.
Sheppard, C. S., Alexander, J. M., Billeter, R. (2012). The invasion of plant communities following extreme weather events under ambient and elevated temperature. Plant Ecology, 213, 1289-1301.
Simberloff, D., Cox, J. (1987). Consequences and costs of conservation corridors. Conservation Biology, 1, 63-71.
Smith, S. D., Huxman, T. E., Zitzer, S. F., *et al.* (2000). Elevated CO_2 increases productivity and invasive species success in an arid ecosystem. Nature, 408(6808), 79-82.
Song, L., Chow, W. S., Sun, L., *et al.* (2010). Acclimation of photosystem II to high temperature in two *Wedelia* species from different geographical origins: implications for biological invasions upon global warming. Journal of Experimental Botany, 61, 4087-4096.
Sorte, C. J., Ibanez, I., Blumenthal, D. M., *et al.* (2013). Poised to prosper? A cross-system comparison of climate change effects on native and non-native species performance. Ecology Letters, 16, 261-270.
Stocker, T. F., Dahe, Q., Plattner, G. K. (Eds.) (2013). Climate Change 2013: The Physical Science Basis. Working Group I Contribution to the Fifth Assessment Report of the Intergovernmental Panel on Climate Change. Summary for Policymakers. Cambridge, UK, New York, USA: Cambridge University Press.
Thuiller, W., Lavorel, S., Araújo, M. B., *et al.* (2005). Climate change threats to plant diversity in Europe. Proceedings of the National Academy of Sciences of the United States of America, 102, 8245-8250.
Thuiller, W., Richardson, D. M., Midgley, G. F. (2007). Will climate change promote alien plant invasions? In W. Nentwig (Ed.), Biological invasions (pp. 197-211). Berlin Heidelberg: Springer.
Tooth, I. M., Leishman, M. R. (2013). Post-fire resprouting responses of native and exotic grasses from Cumberland Plain Woodland (Sydney, Australia) under elevated carbon dioxide. Austral Ecology, 38, 1-10.

Tooth, I. M., Leishman, M. R. (2014). Elevated carbon dioxide and fire reduce biomass of native grass species when grown in competition with invasive exotic grasses in a savanna experimental system. Biological Invasions, 16, 257-268.

Verlinden, M., Nijs, I. (2010). Alien plant species favoured over congeneric natives under experimental climate warming in temperate Belgian climate. Biological Invasions, 12, 2777-2787.

Verlinden, M., Van Kerkhove, A., Nijs, I. (2013). Effects of experimental climate warming and associated soil drought on the competition between three highly invasive West European alien plant species and native counterparts. Plant Ecology, 214, 243-254.

Walck, J. L., Hidayati, S. N., Dixon, K. W., *et al.* (2011). Climate change and plant regeneration from seed. Global Change Biology, 17, 2145-2161.

Walther, G. R., Roques, A., Hulme, P. E., *et al.* (2009). Alien species in a warmer world: risks and opportunities. Trends in Ecology and Evolution, 24, 686-693.

Ward, N. L., Masters, G. J. (2007). Linking climate change and species invasion: an illustration using insect herbivores. Global Change Biology, 13, 1605-1615.

Williams, A. L., Wills, K. E., Janes, J. K., *et al.* (2007). Warming and free-air CO_2 enrichment alter demographics in four co-occurring grassland species. New Phytologist, 176, 365-374.

Willis, C. G., Ruhfel, B. R., Primack, R. B., *et al.* (2010). Favorable climate change response explains non-native species' success in Thoreau's woods. PloS One, *5*, e8878.

Yates, C. J., Elith, J., Latimer, A. M., *et al.* (2010a). Projecting climate change impacts on species distributions in megadiverse South African Cape and Southwest Australian Floristic Regions: Opportunities and challenges. Austral Ecology, 35, 374-391.

Yates, C. J., McNeill, A., Elith, J., *et al.* (2010b). Assessing the impacts of climate change and land transformation on Banksia in the South West Australian Floristic Region. Diversity and Distributions, 16, 187-201.

Ziska, L. H., Faulkner, S., Lydon, J. (2004). Changes in biomass and root: shoot ratio of field-grown Canada thistle (*Cirsium arvense*), a noxious, invasive weed, with elevated CO_2: implications for control with glyphosate. Weed Science, 52, 584-588.

Ziska, L. H., Gebhard, D. E., Frenz, D. A., *et al.* (2003). Cities as harbingers of climate change: common ragweed, urbanization, and public health. Journal of Allergy and Clinical Immunology, 111, 290-295.

Ziska, L. H., Reeves, J. B., Blank, B. (2005). The impact of recent increases in atmospheric CO2 on biomass production and vegetative retention of Cheatgrass (*Bromus tectorum*): implications for fire disturbance. Global Change Biology, 11, 1325-1332.

Mário Santos, Rita Bastos, Joana Vicente, Uta Berger, Britaldo Silveira Soares Filho, Hermann Rodrigues, Joaquim Alonso, Carlos Guerra, João Martins, João Honrado, João Alexandre Cabral

17 Anticipating Invasions and Managing Impacts: A Review of Recent Spatiotemporal Modelling Approaches

17.1 Introduction

The world's landscapes are influenced by a set of anthropogenic pressures that put at risk their sustainability and weaken their ecological functions and services (Kubiszewski *et al.*, 2013). This problem has led to an emphasis on developing accurate assessments of ecosystems status (e.g. Petchey & Belgrano, 2010). In this context, the need for rapid, standardized, and cost-saving assessment methodologies is crucial, namely to predict how anthropogenic environmental changes will affect the integrity of ecosystems (Pejchar & Mooney, 2009). Species invasions are one of the major causes of ecosystem changes, with serious consequences for economic and social systems (Pejchar & Mooney, 2009). Spatiotemporal patterns of this phenomenon are difficult to predict, considering that the structure and functioning of ecosystems are also substantially changing alongside the invasion (e.g. Santos *et al.*, 2011). At present, we have no universal framework to predict how the ecological effects of a particular invader will change over time (Catford *et al.*, 2012). By understanding the long-term feedbacks between invasive species and the native communities, we will be able to identify targets for early assessments by using alternative management tools that support decision-making for eradication or minimizing of severe ecological impacts (Catford *et al.*, 2012).

Scientific models contain the fundamental features that are of interest for solving a problem, and biologists have improved ecological studies by creating quantitative models that simultaneously attempt to capture the structure and composition of ecosystems (Jørgensen, 2008). Following the rapid development of computing technology, and high performance computation in particular, detailed simulations have become cheaper and more available. The wide application of advanced survey technologies makes it possible to collect large-scale and high resolution spatial data. Facilitated by this computing power and spatial data, spatially-explicit approaches have been rapidly developed and applied in ecosystem modelling (Chen *et al.*, 2011). Quantitative models might be divided into mathematically- (black-box models such as Species Distribution Models) and ecologically-driven models (system dynamics, individual-based, discrete-event and others) (Chen *et al.*, 2011). Black-box models, although used for simplifying modelling protocols, are usually considered unable to describe, in a comprehensible way, the structural changes when ecosystem conditions are substantially altered (Kumar & Zhao, 2011). On the other hand, the strength of ecologically-driven models lies in the ability to take into account the individual/ systemic and evolving nature of inter-related activities, showing the interactions between principal drivers in an invasion process (Shackelford *et al.*, 2013). Even though ecological models are usually data-intensive and frequently over-parameterized, they have been used in a wide variety of applications related to ecosystem functioning (e.g. Evans *et al.*, 2013). Modellers should also address the issue of the applicability of the models in data-poor conditions, especially when multitudes of field parameters, which are necessary for empiric model calibration, are not available (e.g. Filgueira *et al*, 2013). Since species invasions are far too complex to be comprehensively quantified, most models focus on a small subset of processes occurring in an ecological system (e.g. Santos *et al.*, 2013). Consequently, the determination of the appropriate scale and parameters is usually the critical and most challenging part of determining the most suitable modelling approach for each case (Santos *et al.*, 2013). This chapter highlights the applicability of different spatially-explicit models focused on predicting invasive species dynamics at distinct scales and objectives, in which main constrains and advantages of the different modelling paradigms are discussed.

17.1.1 Individual-Based Models (IBMs)

Whereas the occurrence of invasive species and their subsequent negative effects on the native community have been often documented, information about the success or failure of the invasion processes themselves is still fragmented. The most successful mechanisms of invasive plants, for example, are allelopathy, hybridisation, changes of soil chemistry, and water or radiation budget (Callaway & Aschehoug, 2000; Hierro & Callaway, 2003). Their dominance and interplay, however, depends on the given species constellation and environmental conditions. A detailed understanding of the mechanisms is thus essential for a confident prediction of distribution patterns (Ren & Zhang, 2009). Mechanistic models (or process-based models, "bottom-up models") such as IBM actively address the underlying biological processes at the organism level that lead to invasion (Morin & Thuiller, 2009). Due to the complexity of the processes to be described, model development and data acquisition for individual-based models is relatively extensive (Morin & Thuiller, 2009), but important features like demographic variability, spatial heterogeneity (Jongejans *et al.*, 2008), seed dispersal (Nathan *et al.*, 2011), migration, as well as intra- and interspecific competition (e.g., Berger *et al.*, 2008; Lin *et al.*, 2012) can be explicitly addressed. In that way, critical issues defining the success or failure of an invasion process like dispersal limitations and timing of climatic events, among others, are inherent parts of the virtual system mimicking random effects and uncertainties occurring in nature.

There are numerous examples demonstrating the suitability of IBMs for studying the spread of invasive organisms (e.g. Goslee *et al.*, 2006; Nehrbass & Winkler, 2007; Peters, 2002; Rebaudo *et al.*, 2011; Shackelford *et al.*, 2013; Tonini *et al.*, 2013). The majority of established forest simulators, including individual-based models, describe a static community composition and are considered unsuitable for modelling invasion (Martin *et al.*, 2009). In this review, we found two forest IBMs, which share a grid-based design. One cell is equal to the size of an adult canopy (Marco *et al.*, 2002). Local interactions (competition, inhibition etc.) are not considered, neither between trees occurring in neighbourhood cells nor within the same cell, since it is limited to only one tree.

This also accounts for allelopathy, which occurs between two or more species (one species emitting and another affected by allelopathic substances). Allelopathy has been identified as a key mechanism in plant invasions by empirical studies (Hierro & Callaway, 2003; Ren & Zhang 2009), but only a few IBMs address this issue explicitly (see, e.g. Peters, 2002; Travis *et al.*, 2011). This is surprising, since IBMs are technically suitable for it (Berger *et al.* 2008). So far, most models addressing allelopathy follow a simple mathematical approach outlining toxic effects in a generic single-species system (e.g., An *et al.*, 2002), or in interacting two-species systems (Martines *et al.*, 2009) in homogeneous environments (Fassoni & Martins, 2014). The individual-based models for grasslands (Goslee *et al.*, 2001) are spatially explicit and able to consider environmental heterogeneity. Based on simulations experiments using the latter two, general insight could be achieved about the coexistence of species or their extinction in invaded communities. In contrast, landscape models like FORECAST (Kimmins *et al.*, 1999) are more applied. Originally developed to simulate stresses between different layers in a forest ecosystem, FORECAST was then applied to study allelopathic effects of ericaceous understory in Canadian conifer forests (Blanco, 2007), and their impacts on stand level. Since the model itself is spatially implicit, it is still unsuitable to simulate the driving forces in heterogeneously-structured forest ecosystems, which could be modified by allelopathy within the same layer.

Following Blanco's (2007) suggestion for an ideal allelopathic forest ecosystem model, we suggest that a simulation model suitable to describe the mechanisms of species invasions should consider different types of disturbances and their effects on organisms' growth and mortality, because they create windows of opportunity (Myster, 1993) for the establishment of native and invasive species. Individuals' variability (e.g., age- and size class-dependent demographic performances), species composition, as well as species autoecology, including allelochemical production and sensibility, dispersal, and viability, since all of them affect the interactions between native and alien species, and subsequently influence invasion success.

In principle, spatially explicit, individual-based models can fulfil these criteria. They are designed to inherit complex life cycles, different types of neighbourhood interactions above- and belowground, disturbances, and allelopathy. Empirical data and logistic difficulties, however, seem to form a bottleneck for a more extended application of IBMs for studying species' invasions (see also Berger *et al.*, 2008). As long as this problem cannot be solved, correlative and statistical modelling approaches coupled with mechanistic approaches are still needed, which we will introduce and discuss in the following.

17.1.2 Stochastic Dynamic Methodology (StDM)

The Stochastic Dynamic Methodology (StDM) is a mechanistic and dynamic framework for understanding ecological processes based on parameter estimation methods founded on the premise that general statistical patterns of ecological phenomena are emergent indicia of complex ecological processes (e.g. Santos & Cabral, 2004; Santos *et al.*, 2013). In this context, the main objective of applying the StDM refers to the final use of management models based on medium-long-term databases collected in the scope of programs to measure and monitor ecological attributes, minimizing the constraints of conventional dynamic modelling procedures, such as construction/conceptualization, parameterization, structural complexity, and criteria of variable selection (Santos *et al.*, 2011). The StDM has been successfully applied, tested, and validated in several types of systems and contexts, namely in the scope of invasive processes, interactions, and effects (e.g. Bastos *et al.*, 2012; Santos *et al.*, 2011). To improve the applicability of this framework, the StDM can be coupled with Geographic Information Systems (GIS) to produce simulations that allow the interactive creation of spatially dynamic ecological patterns (Bastos *et al.*, 2012; Santos *et al.*, 2013).

To illustrate the potential implementation of the spatially explicit StDM framework for forecasting invasions and managing invasive species, following the research developed by Bastos *et al.* (2012), we present a case study applied to an endemic bird of São Miguel island (Azores Archipelago, Portugal), the Azores bullfinch (*Pyrrhula murina*). This species is currently threatened by two of the major causes of biodiversity loss worldwide: invasion of native habitats by exotic plants, and habitat destruction by land use changes (Ramos, 1996). The main goal of this research was to relate bird population trends to changes in the surrounding habitats, and predict the species responses to scenarios of native forest management. For this the basic principle of the StDM methodology was applied to the interaction between the Azores bullfinch (i.e. state variable, measured in number of individuals) and habitat features, mainly resulting from invasion processes acting through the Azorean Laurel Forest. The invasion process represents key local effects with influence on the species abundance. This is based on the relationships detected in the Multi-Model Inference statistical analysis applied to datasets that include whole regional gradients of the studied habitat changes (Bastos *et al.*, 2012). Furthermore, a structural-dynamic model concerning the invasion of laurel forest by exotic plants, along with additional structural features of the surrounding habitats, was designed in order to predict the local influence of habitat dynamics in the species response (Bastos *et al.*, 2012). Local dynamic trends of Azores bullfinch abundance were extrapolated to the regional level by submitting the resultant StDM independent output (i.e. for each study unit, represented by an area of 50 ha) to a geostatistical interpolation, in order to foresee changes in the Azores bullfinch spatio-temporal patterns (Bastos *et al.*, 2012). To illustrate species response to invasion processes acting at the remaining fragments of natural forest, we focus on the lily-of-the-valley tree (*Clethra arborea*), which is one of the main plant invaders driving native biodiversity losses in the Azores laurel forest (Heleno *et al.*, 2009). In this way, two habitat classes concerning pure fragments of laurel forest and invaded laurel forest by *C. arborea* were taken into account (i.e. by aggregation of habitat classes comprehending invasion by *C. arborea* in target laurel forest habitats, considered in Bastos *et al.*, 2012). The spatial projections were carried out for the central area of the Azores bullfinch distribution, coincident with the core of the Special Protection Area of "Pico da Vara/Ribeira do Guilherme", allowing the dynamic model to simulate the trends of invasive plant species in the absence of management actions. On the other hand, hypothetical management actions were considered in an area of 300 ha (i.e. 6 study units), by restoring native laurel vegetation through *C. arborea* removal. Management actions were simulated between the 21st and 23rd years (i.e. 100 ha intervened per year) from a simulation period of 50 years, considering an average intensity of 70% for exotics removal, a realistic performance taking into account logistical constraints (Bastos *et al.*, 2012). The predicted changes in the habitats and in the Azores bullfinch distribution, according to the scenarios considered, are shown in Figure 17.1. Furthermore, the respective dynamic trends expressed in percentage cover of each habitat class and in bullfinch abundances throughout the simulation period are displayed in Figure 17.2.

Overall, the sustainability of the Azores bullfinch population seems to be guaranteed by the central range distribution, retaining the qualities of a favourable breeding area over the next 50 years, even with the invasion of *C. arborea* (Bastos *et al.*, 2012). In fact, since mixed laurel habitats with *C. arborea* represent an important food resource for the Azores bullfinch in winter (Ramos, 1995; 1996), the range of this exotic plant expansion inside the SPA does not seem to represent the same disruptive effect on the species distribution as the one caused by more aggressive invaders (i.e. *Pittosporum undulatum*) that occur at lower altitudes, thus not affecting the bullfinch core area (Ceia *et al.*, 2011b; Ramos, 1996). These projections are in agreement with other authors that consider the Azores bullfinch a species resilient to habitat invasion by *C. arborea*, notably when present in a matrix of native forest (Ceia *et al.*, 2011a). As a consequence of management actions, the spatial projection reveals higher local abundances/densities towards the central range of the species occurrence, which is confirmed by a significant increasing of birds foraging in managed areas (Ceia *et al.*, 2011b).

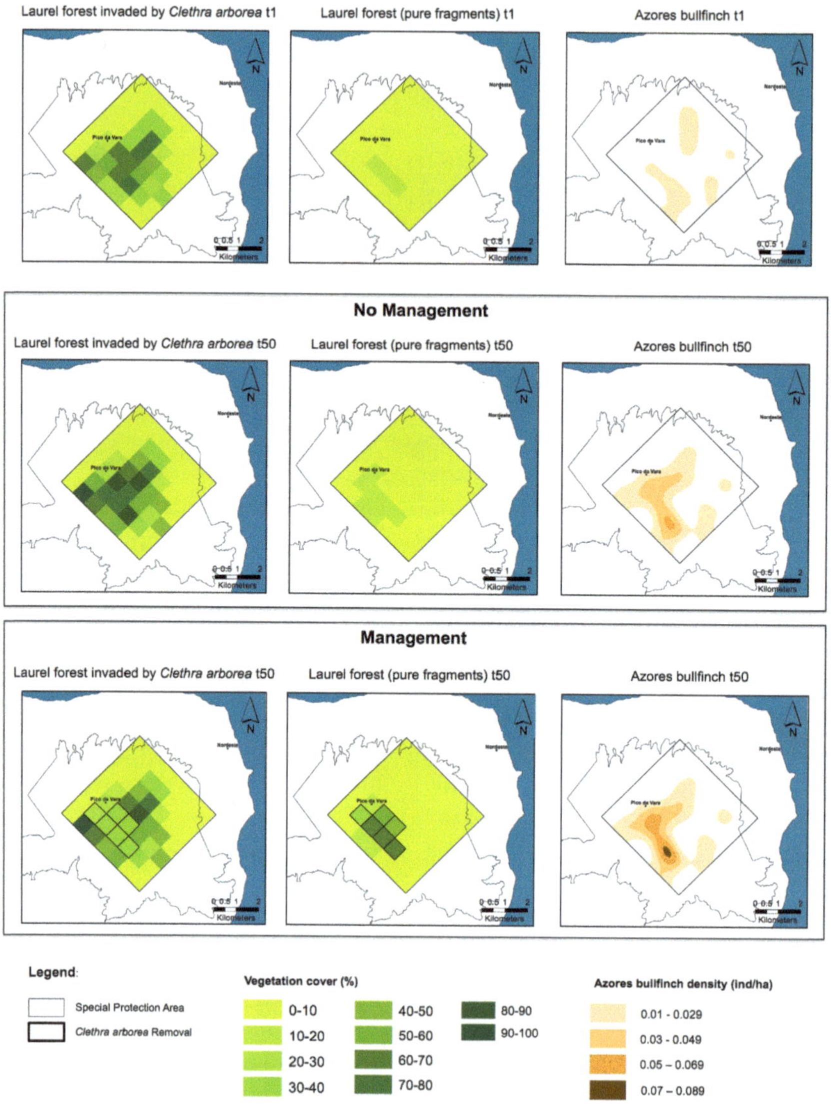

Fig. 17.1: Spatial projection of the Azores bullfinch main habitats (i.e. pure fragments of laurel forest and invaded laurel forest by *C. arborea*) and its distribution in the center of the Special Protection Area of "Pico da Vara/Ribeira do Guilherme", considering scenarios without and with management of *C. arborea* removal in target areas (i.e. 300 ha). The spatial patterns were projected using a continuous distribution function based on a simple kriging and its temporal variation from t = 1 to t = 50. The habitats were represented by 10% cover classes of each vegetation category. The selected Azores bullfinch density classes are expressed in bird numbers per study unit area (in hectares) based on the values simulated by the StDM framework.

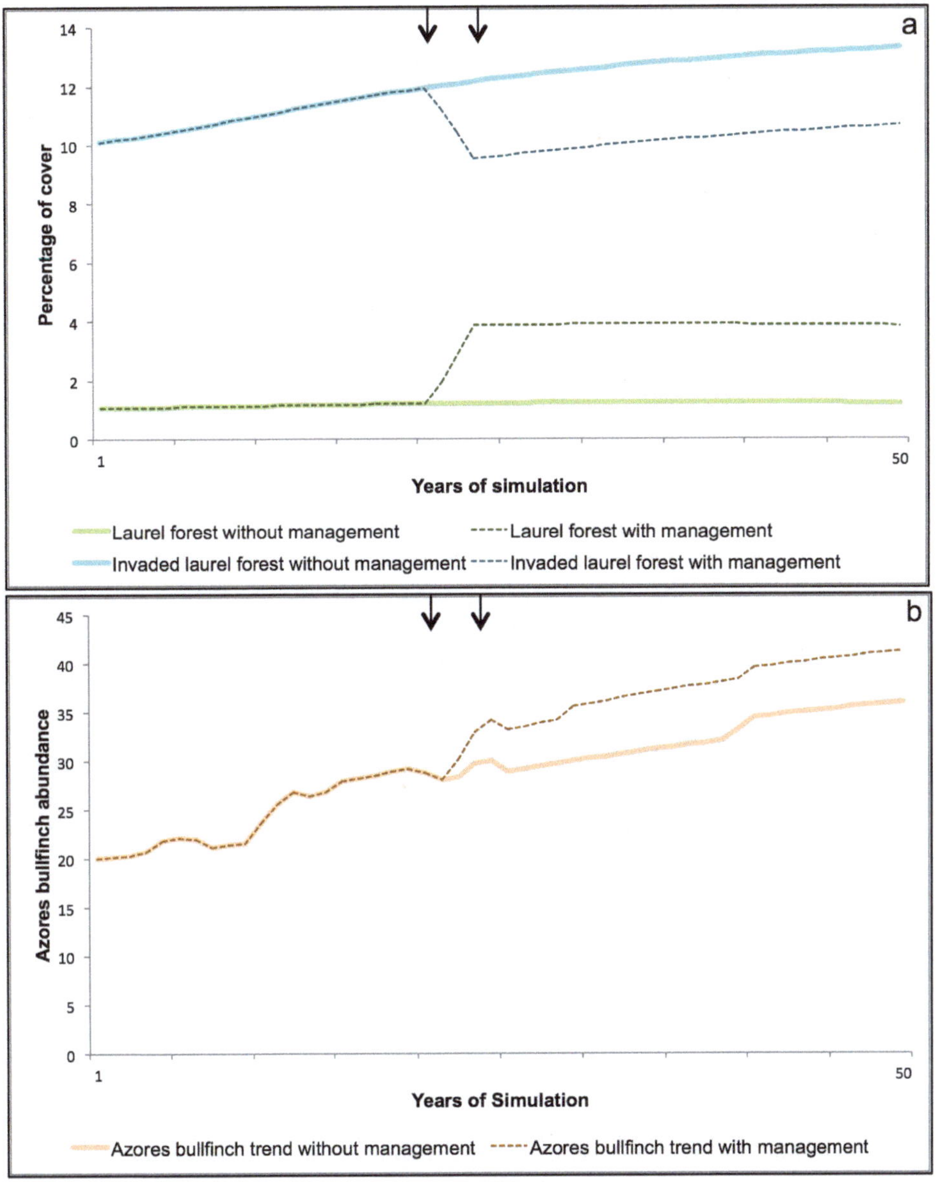

Fig. 17.2: StDM simulation results concerning (a) the habitat dynamics (i.e. between pure fragments of laurel forest and invaded laurel forest by *C. arborea*) and (b) the response trends of Azores bullfinch abundances in the centre of the Special Protection Area of "Pico da Vara/Ribeira do Guilherme". The scenarios without and with management of *C. arborea* removal in target areas (i.e. 300 ha) are discriminated. Arrows indicate the time interval where the management actions took place through the simulation period.

The StDM is still very recent, with the first results published in 2003/2004 (e.g. Santos & Cabral, 2004). Meanwhile, ecologists applied, tested, and validated StDM applications in several types of scenarios. StDM models, which are produced in the form of rules, are transparent, easily understood by experts, and straightforwardly interpretable in order to allow decision-makers to incorporate pertinent qualitative data before simulations. The StDM exhibits these structural qualities, but also provides simple, suitable, and intuitive outputs, easily interpreted by non-experts (ranging from resource users to senior policy makers) (Bastos *et al.*, 2012; Santos *et al.*, 2013). Although conceptually simple, the StDM models capture the stochastic complexity of some holistic ecological trends, including relevant temporal and spatial gradients of stochastic environmental characteristics, which allows the simulation of structural changes when habitat and environmental conditions are substantially changing. The obtained results seem to demonstrate the StDM reliability in capturing the dynamics of the studied ecosystems by predicting the behavioural pattern for the key components selected under very complex and variable environmental scenarios, namely when conditions relatively unaffected by human activities were changed by man-induced disturbances. Another goal when developing methodologies for assessing changes in the ecological integrity of ecosystems is the feasibility of application and extent to which the results are applicable in other contexts (Santos *et al.*, 2013). Considering that StDM is easily applicable to data from natural, semi-natural, and artificial ecosystems affected by gradients of changes, this approach will facilitate the development of more global techniques in the scope of invasion ecology.

17.1.3 Species Distribution Models (SDMs)

Species Distribution Models (SDMs) are statistical frameworks that relate the geographical distribution of species to spatial variations in environmental conditions (Guisan & Zimmermann, 2000; Guisan & Thuiller, 2005; Elith & Leathwick, 2009). SDMs thus explore the statistical relations between environmental conditions potentially shaping the distribution of a given species (i.e. predictors) and the distribution data for that same species, identifying the range of conditions that allow its presence – i.e. modelling the ecological niche *sensu* Hutchinson (1957). The *fundamental niche* is defined as the *n*-dimensional hypervolume created by all the points corresponding to a state of the environment that allows for a given species to exist indefinitely. The *realized niche* corresponds to a subset of the *fundamental niche* constrained by biotic interactions and competitive exclusion. SDMs, being based on empirical datasets, likely predict the realized niche (Guisan & Zimmermann, 2000). Another assumption on which SDMs are based is that the whole niche of the species is captured in the distribution data, i.e. the distribution dataset is representative of the whole range of conditions that the species may occupy.

The statistical development of SDMs should be underpinned by a sound conceptual model, based on ecological processes and taking into account the objectives of the model (Guisan & Zimmermann, 2000). The subsequent workflow involved in SDM development is represented in Figure 17.3. Distribution data (usually obtained by surveying the target area following a statistical sampling design) and environmental variables (which should be chosen based on the theoretical model and preliminary statistical tests) (Guisan & Zimmermann, 2000) are used to fit a statistical model relating the presence or absence of the species with the environmental conditions (i.e. the predictors). It is of utmost importance that the performance of the model is assessed by evaluating its predictive power on a dataset (either independent or obtained by resampling the training data) (Elith & Leathwick, 2009). When addressing only the relation between environmental factors and the species distribution, we are modelling the species distribution in the environmental space captured by the training data (Elith & Leathwick, 2009). However, if the final goal is to obtain spatial predictions of distribution, the whole environmental space must be projected into geographic space (i.e. latitude, longitude, elevation), based on spatial data of the environmental variables for the whole study area. This projection may either be made for unsampled sites in the study area where the original distribution was sampled (model-based interpolation) or for other geographical locations (extrapolation or spatial projection), and even different time frames (past — backcasting, or future — forecasting) e.g. as part of model-based scenario analysis routines (Elith & Leathwick, 2009) .

In the field of biological invasions, SDMs play a particularly important role since invasions represent a growing threat to biodiversity (Pimentel *et al.*, 2000) and once alien invasive species are established, they often become difficult to eradicate (Genovesi, 2005). Thus, anticipating and preventing future invasions is the most cost-effective form of managing biological invasions. SDMs have been used (Figure 17.4.) to analyse and explain current patterns of invasion of various organisms (Peterson & Vieglais, 2001; Peterson, 2003; Peterson & Robins, 2003; Guisan & Thuiller, 2005; Broennimann & Guisan, 2008; Petitpierre *et al.*, 2012).

Modelling invasive species is nonetheless a challenging task. For example, sometimes it violates the assumption of the whole niche being captured, as the invading species may not yet occupy all areas of the invaded region where environmental conditions are suitable for its growth and reproduction. Another assumption of SDMs is that the environmental niche of a species is conserved across space and time (niche conservatism) (Petitpierre *et al.*, 2012). This assumption may have important effects not only on the prediction of the invasion extent under current conditions (spatial projections) (Petitpierre *et al.*, 2012), but also on predictions of future invasions under environmental change scenarios (temporal projections) (Broennimann & Guisan, 2008).

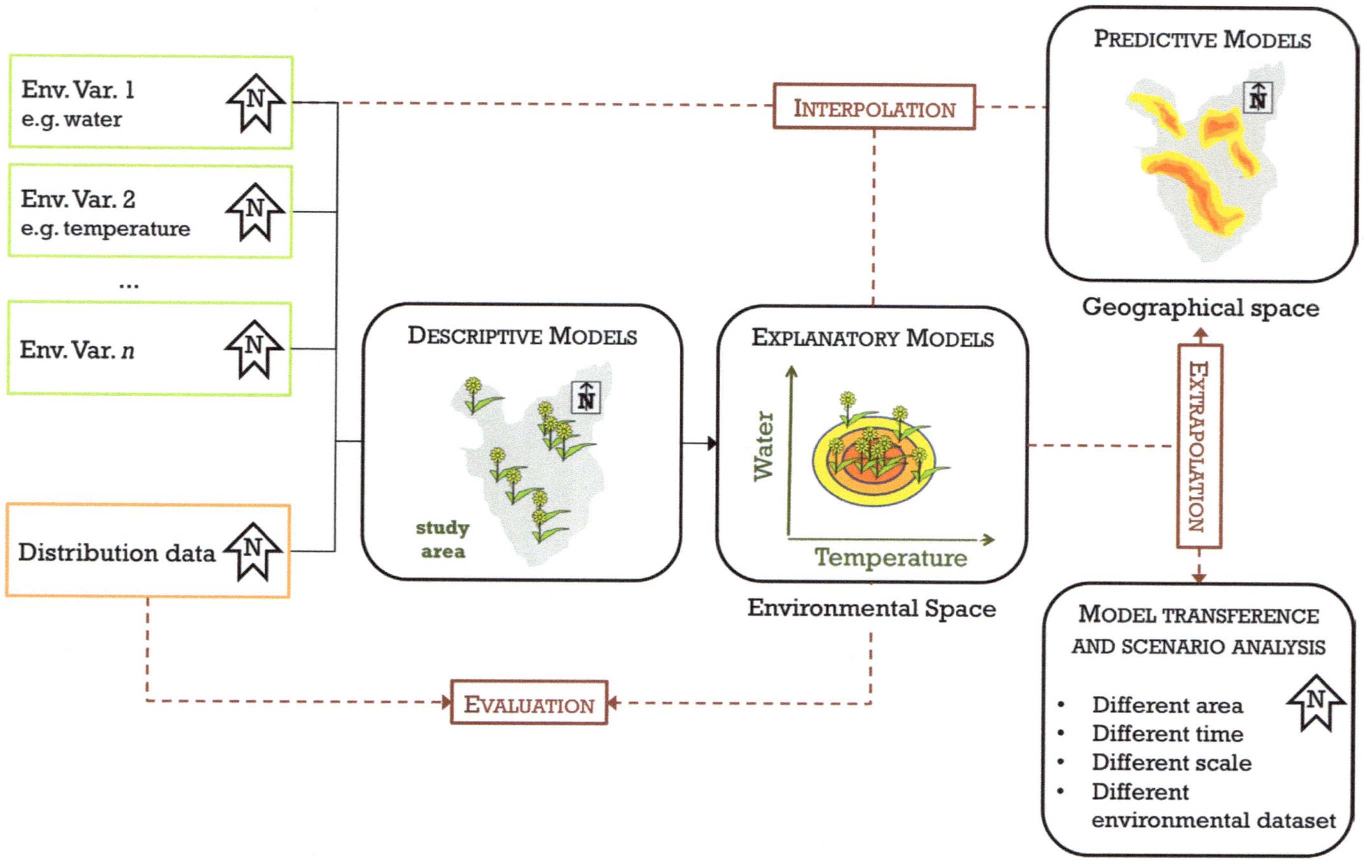

Fig. 17.3: Simplified representation of a general workflow for the statistical development of a Species Distribution Model.

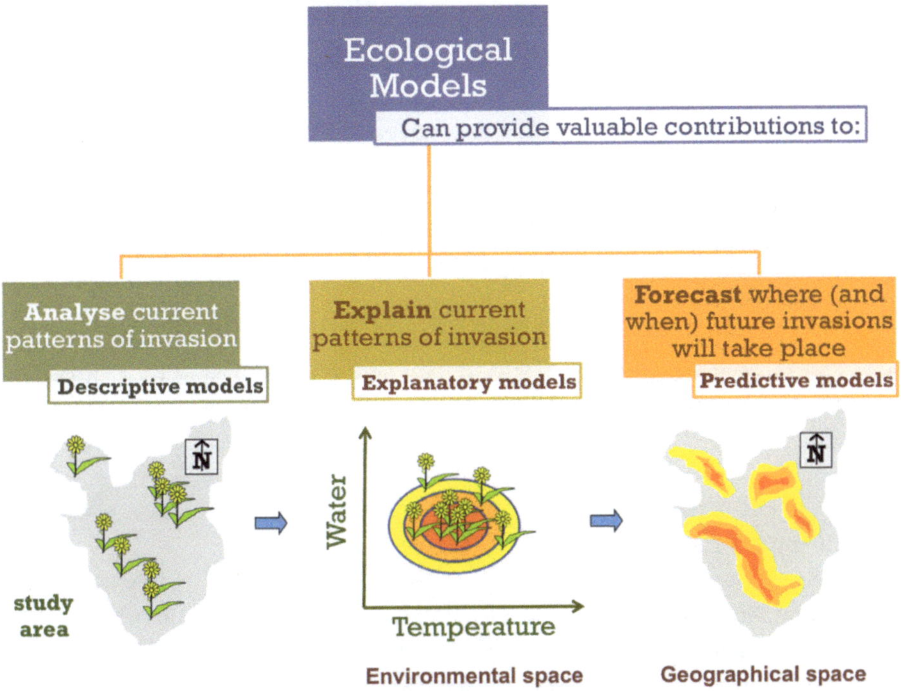

Fig. 17.4: Species Distribution Models have been used to tackle many important topics in invasion ecology, from research to management. Graphs courtesy of A. Guisan.

An example of the usefulness of the ability to make predictions for different geographical locations is presented in Thuiller *et al.* (2005), who used SDMs for 96 species native to South Africa but invasive in other parts of the world, to evaluate the risk of invasions at a global scale. SDMs were calibrated with data from the native range (i.e. South Africa) and spatially projected over the world, in order to identify areas potentially vulnerable to invasion by each individual species. The individual species predictions were then combined to obtain a cumulative probability surface, obtaining a broad estimate of invasion risk by South African species.

The ability to predict potential future distribution makes SDMs very useful for conservation planning considering the impacts of invasive species, due to the prediction of future distributions that allow the prioritization of preventive actions. For example, Gassó *et al.* (2012) applied current distribution data for the 78 major plant invaders species in Spain using climatic, geographic, and land use variables as predictors, to predict their current potential distribution. The potential distribution maps were used as an early warning tool to guide control and eradication plans. The work of Vicente *et al.* (2011) provides another example of the use of model-based forecasting to advise conservation and management. In their work, the authors applied a new modelling approach to obtain current and future potential distributions for an invasive (*Acacia dealbata*) and a rare (*Ruscus aculeatus*) plant species in the northwest of Portugal. The new modelling framework allowed the identification of areas where both species may potentially be present (considered as potential conflict areas), thereby offering insights for the management of potential impacts of the invader on the conservation of the rare species. The new modelling framework was found to be more informative than traditional models, as it combines predictors according to the geographic scale (regional or local) at which they are expected to influence species distributions. The idea of using SDMs to identify conflict areas and prioritize management actions was also applied in Vicente *et al.* (2013), who developed a framework allowing for the identification of potential conflict areas between alien invasive plants and selected ecosystem services, by modelling invasive species richness and overlaying their spatial projection with maps of the relative importance of ecosystem service provision.

17.1.4 Dinamica EGO

Land-use and land-cover changes (LUCC) have been the focus of predictive modelling in recent years (Geist & Lambim, 2002; Soares-Filho *et al.*, 2006) because they represent the most evident manifestations of human activities on natural systems, and thus a leading matter of discussion for regional planning (Nepstad *et al.*, 2009;Stickler *et al.*, 2009), sustainable development (Merry *et al.*, 2009;), and mitigation (Soares-Filho *et al.*, 2010), among other areas. LUCC models are by nature spatially explicit representations of interactions between humans and the environment, and as such are used to simulate the patterns of change on the landscape in response to coupled human-ecological dynamics. In this way, they represent a heuristic device useful to perform ex ante evaluation of the outcomes of a variety of scenarios, translated as different socioeconomic, political, and environmental frameworks.

In this context, the Dinamica software project goal is to develop tools and methods for LUCC and environmental modelling in support of sound policy development, and to freely disseminate this technology to students and researchers interested in its use. From a spatially explicit simulation model of landscape dynamics (Soares-Filho *et al.*, 2002), Dinamica software has evolved to Dinamica EGO (Soares-Filho *et al.*, 2013). EGO stands for Environment for Geoprocessing Objects. Dinamica EGO is an environmental modelling platform for the design of analytical and space-time models. Its main features include nested interactions, multi-transitions, dynamic feedbacks, multi-region and multi-scale approaches, and decision processes for bifurcating and joining execution pipelines. Dinamica EGO 64-bit architecture includes multiprocessor computing, smart handling of large raster datasets, dynamic compilation of logical and mathematical equations, and a library of algorithms for the analysis and simulation of space-time phenomena, such as cellular automata transition functions, as well as calibration and validation methods (Soares-Filho *et al.*, 2002; Soares-Filho *et al.*, 2010; Soares-Filho *et al.*, 2013). The software environment, written in C++ and Java, contains a series of algorithms called "functors" (See list at csr.ufmg.br/dinamica/wiki). Each functor performs an operation. A special class of functor is the "container" that may envelop a series of functors and other containers. Functors, including containers, are sequenced in graph form to establish a visual data flow (Fig. 17.5.). With help of a graphical interface, a user can create models by dragging and connecting functors via their ports; each port represents a connector to a data element, such as a map, table, matrix, mathematical expression, or constant. Thus, a model can be designed as a diagram whose execution follows a data flow chain. Models can also be converted into submodels and stored in the functor library as new functors. In addition, Dinamica EGO contains the most common spatial analysis algorithms available in commercial GIS, and a wizard tutorial for preparing a model interface for end-users. The graphical interface allows for creative design of spatial models that are saved in Extensible Markup Language or EGO script language; the latter format enables script writing using a text editor, which can be converted to EGO graphical diagrams and vice-versa.

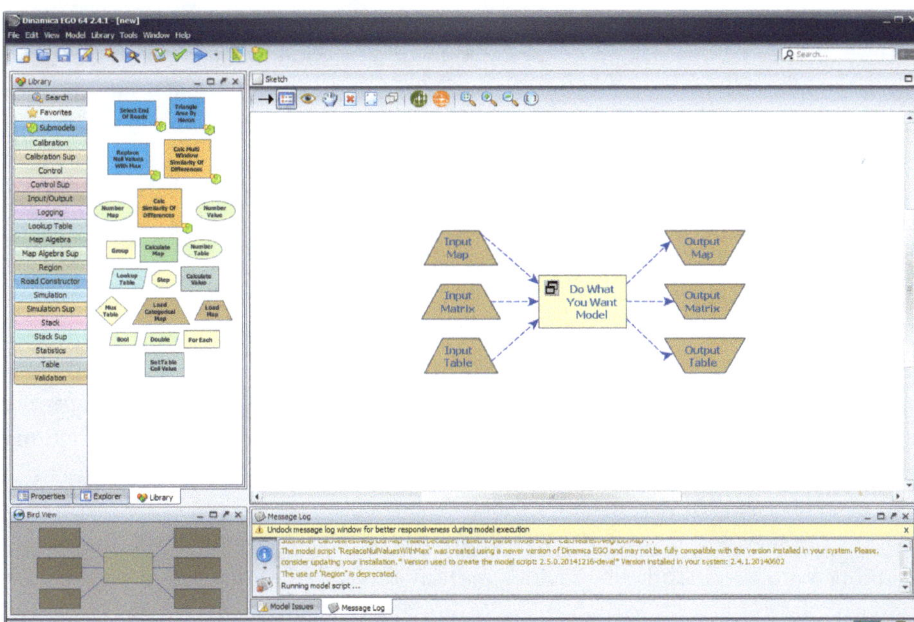

Fig. 17.5: Graphical Interface of Dinamica EGO.

Dinamica EGO innovative modelling techniques provide a complete solution for calibrating, running, and validating space-time models. For modelling transition probabilities for LUCC models or calibrating environmental models, Dinamica EGO uses an adapted version of the Bayesian method of conditional probability (Bonham-Carter, 1994), known as the Weights of Evidence—WOFE (Soares-Filho *et al.*, 2004; Soares-Filho *et al.*, 2009; Soares-Filho *et al*, 2010,). In addition, Dinamica EGO provides a hybrid heuristic-analytical calibration approach, which takes advantage of both the WOFE method and Genetic Algorithm (GA). The GA tool is flexible enough to embrace a variety of models as well as their specific fitness functions, thus offering a practical way to optimize the performance of a wide variety of models. For model validation, Dinamica EGO employs a map comparison method named Reciprocal Similarity (Almeida *et al.*, 2008; Soares-Filho *et al.*, 2009; Soares-Filho *et al*, 2013). Soares-Filho *et al.* (2013) evaluated several map comparison methods and found that Reciprocal Similarity dealt the best with comparison of changes. Thanks to its set of cellular automata transition functions, Dinamica EGO handles issues related to both spatial prediction and replication of patterns of change. Worthy of mention, these functions replicate the expanding and contracting landscape elements, thereby simulating the evolving structure of dynamic landscapes. To convey its powerful modelling toolbox to students and other professionals interested in environmental modelling, Dinamica EGO comes with a comprehensive guidebook containing a series of exercises with increasing level of complexity (http://csr.ufmg.br/dinamica/dokuwiki/doku.php?id=tutorial:start). In summary, the Dinamica EGO platform allows for the design of very simple static spatial models to very complex dynamic ones, favouring usability, performance, and flexibility. Such features are particularly important as models become increasingly complex and are fed with greater volumes of abundance of data.

17.2 Discussion

The modelling strategies shown in this work might be used to improve our prediction of invasions, but our ability to reliably forecast invasion processes is still very limited. These limitations result from the fact that most types of modelling languages and concepts are either exclusively focused on holistic “top-down” predictions (e.g. SDMs) or on incorporating “bottom-up” processes (e.g. IBMs) (Gallien *et al.*, 2010, Santos *et al.*, 2013).

Individual-based models (IBMs) allow the study of system-level properties emerging from the adaptive behaviour of individuals, but also the influence of the system on the individuals (Grimm *et al.*, 2006). IBMs consider features disregarded in most analytical models: variability among individuals, local interactions, complete life cycles, and even adaptation to a changing internal and external environment (Railsback & Grimm, 2011). The possibility of implementing competing hypotheses about the functioning of the particular ecological system provides a unique chance to detect the most plausible mechanisms and contributes to a comprehensive understanding. The potential of IBMs is also associated with complexity in structure and dependence on powerful computers. A subsequent drawback of these models is that they are more difficult to analyse, understand, and communicate than traditional analytical models (Railsback & Grimm, 2011). New standards for model descriptions like the ODD protocol (Grimm *et al.*, 2006, 2010) have been developed to overcome this limitation (Table 17.1).

Tab. 17.1: Comparison of the different spatio-temporal methodologies to model species invasions.

Method	Strengths	Limitations	Where to start?
Individual based models	Variability among individuals; Local interactions; Adaptation to changing environments	Difficult to analyse; Difficult to parameterize; Difficult to communicate; Tends to be limited by software and hardware requirements	Railsback & Grimm, 2011
Stochastic Dynamic Methodology	Based on information theoretic approaches and systems dynamics; Easy parameterization; Easy interpretation; Less complex then other spatially dynamic approaches	Specific gradients in databases are required; Spatial interpolation not fully developed; Software requirements; Work-intensive when implementing over a growing number of simulations	Santos & Cabral, 2004 Bastos *et al.*, 2012 Santos *et al.*, 2013
Species Distribution Models	Based on information theoretic approaches; Highly adaptable; Easy to interpret and communicate; Large scale processes.	Highly dependent on theoretic assumptions; Temporal transitions not fully developed and limited; Hardware requirements	Elith & Leathwick, 2009
Celular autómata coupled with Geographic information systems (Dinamica EGO)	Several interactions at different scales might be imposed; Easy to interpret and communicate; Highly adaptable	Transition rules; Complexity; Hardware requirements; Difficult to parameterize	Soares-Filho *et al.*, 2002 Soares-Filho *et al.*, 2009

The Stochastic Dynamic Methodology (StDM) combines statistical and dynamic modelling with geostatistical techniques to address complex spatially-explicit emergent problems, from the individual habitat patch to the whole landscape context (Santos *et al.*, 2013). The StDM is a sequential modelling process initiated by the analysis of landscape and habitat composition, which defines the convenient parameters that contextualize the physical and biotic descriptors at the study unit level. This procedure involves the use of robust information and theoretic approaches based on generalized linear models in order to establish the interaction criteria between the construction of the dynamic model and the resulting stochastic dynamic simulations for each study unit (Bastos *et al.*, 2012; Santos *et al.*, 2013). These simulations, when projected into a geographic space and submitted to an appropriate geostatistical interpolation, create an integrative picture, in space and time, of the responses to the gradients of habitat changes, namely considering management options and invasive species responses (Table 17.1).

Species distribution models (SDMs) are numerical tools that combine observations of species occurrence or abundance with environmental estimates, and are used to predict distributions across landscapes (e.g. Lomba *et al.*, 2010). SDMs are currently applied across all ecosystems, and model robustness seems to be influenced by the selection of relevant predictors and modelling method, scale, and the extent of extrapolation. In many situations the link between SDM practice and ecological theory is weak, hindering progresses (Benito *et al.*, 2013). Some drawbacks have been identified and several authors proposed methods to improve the method's capabilities, namely for: modelling of presence-only data; model selection and evaluation; accounting for biotic interactions; and assessing model uncertainty (Benito *et al.*, 2013) (Table 17.1).

Dinamica EGO (EGO stands for Environment for Geoprocessing Objects) is an environmental modelling platform based on cellular automata (CA) models coupled with powerful Geographic Information System data. Dinamica Ego models consist of a simulation environment represented by a gridded space (raster), in which a set of transition rules determine the attribute of each given cell, taking into account the attributes of cells in its vicinities (Soares-Filho *et al.*, 2013). These models have been very successful in view of their operability, simplicity, and ability to embody logics—as well as mathematics-based transition rules, and have also been enhanced by the incorporation of decision-support tools strongly enabled by the linkages between CA and GIS (Soares-Filho *et al.*, 2009). Dinamica EGO presents a wide-open possibility for design, from very simple to very complex space-time models. The Dinamica EGO graphical interface allows the design of a model by simply dragging and connecting operators that perform calculations (Table 17.1).

IBMs, StDM, SDMs, and Dinamica EGO have all proved successful in a number of works associated with ecosystem dynamics, but none represent a "best" approach or provide a universal procedure. The idea of modelling species distributions on the basis of large-scale holistic relationships ("top-down") while at the same time considering the most important processes ("bottom-up") has recently led to the development of so-called hybrid models. In holistic "top-down" approaches, such as SDMs, many reductionist details are not considered, such as individual variability, adaptation, local idiosyncrasies, and highly dynamic ecological phenomena. Although this simplification enables the understanding of whole-system processes such as resilience, resistance, persistence, regulation, and density dependence, it is difficult to follow at other scale approaches, namely when local properties are crucial to global system behaviour. This is one of the major advantages of "bottom-up" models, such as individual-based models (IBMs) and, partially, the stochastic dynamic methodology (StDM) (Grimm, 1999; Santos *et al.*, 2013): to enable information crossing at a local scale, potentiating a better understanding of ecosystem dynamics, and leading to emergence and similar system characteristics. StDM and DinamicaEGO are able to partially combine the different approaches, and may pave the way to more research in this scope. We believe that the combination of both modelling approaches will result in promising future outcomes, allowing a better understating of invasion ecology, which will make methods more instructive and credible to decision-makers and environmental managers.

17.3 Acknowledgements

We are indebted to Martin Schubert for the valuable discussion associated with the individual-based models, and to Jaime Ramos for the data associated with the Azores bullfinch. The present study was supported by FEDER funds through the Operational Programme for Competitiveness Factors – COMPETE, and by National Funds through FCT – Foundation for Science and Technology under the project PTDC/AAG-MAA/4539/2012 / FCOMP-01-0124-FEDER-027863 (IND_CHANGE) FCT (Portuguese National Board of Scientific Research) and PEst-OE/290AGR/UI4033/2014.

In a nutshell

- The mobility of our societies accelerates the global redistribution of species. As a consequence, biological invasions increasingly contribute to environmental changes affecting biodiversity, ecosystem functioning, and subsequent services. Assessing and forecasting the consequences, as well as the development of effective adaptation or mitigation strategies, requires a comprehensive understanding of the involved processes, including introduction pathways, invasiveness of exotics, and the vulnerability of ecosystems and landscapes.
- The complex nature of these processes and their interactions often result in non-linear and emergent properties of the ecosystems, which cannot decoded by classical approaches. Conventional dynamic models, for example, are useful to understand temporal changes but the underlying assumption of spatial uniformity in environmental conditions and processes does not correspond to real ecosystems, and thus limits their suitability for studying biological invasions.
- New spatiotemporal approaches have been developed over the last decade which explicitly address species responses, taking into account landscape heterogeneity. They include individual-based models (IBMs), stochastic dynamic models (StDM), statistical species distribution models (SDMs), and complex cellular automata models (CA) coupled with geographic information systems (GIS).
- The advantages and/or limitations of these modelling approaches strongly depend on the ecological contexts, scales, target organisms, guilds and/or communities under study, or the availability and quality of available of datasets, among others.
- This chapter reviews these recent spatiotemporal modelling approaches applied to forecasting and management, with a focus on biological invasions. Based on case samples of prominent biological invasions, we discuss the concepts, requirements, and potential outcomes of such approaches and describe the ecological relevance of applications, along with specific advantages and drawbacks. This review will guide ecologists and conservation biologists in selecting adequate tools when facing the multiplicity of ecological conditions and/or scenarios that would be difficult or impossible to otherwise understand.

17.4 Bibliography

Almeida, C. M., Gleriani, J. M., Castejon, E. F., *et al.* (2008). Using neural networks and cellular automata for modelling intra-urban land-use dynamics. International Journal of Geographical Information Science, 22 (9), 943-963.

An, M., Johnson, I. R. & Lovett, J. V. (2002). Mathematical modelling of residue
allelopathy: the effects of intrinsic and extrinsic factors. Plant and Soil, 246, 11- 22.

Bastos, R., Santos, M., Ramos, J.A., *et al.* (2012). Testing a novel spatially-explicit dynamic modelling approach in the scope of the laurel forest management for the endangered Azores bullfinch (*Pyrrhula murina*) conservation, Biological Conservation 147: 243–254.

Benito, B.M., Cayuela, L. & Albuquerque, F.S. (2013). The impact of modelling choices in the predictive performance of richness maps derived from species-distribution models: guidelines to build better diversity models. Methods in Ecology and Evolution, 4: 327-335.
Berger, U., Piou, C., Schiffers, K., *et al.* (2008). Competition among plants: Concepts, individual-based modelling approaches, and a proposal for a future research strategy. Perspectives in Plant Ecology, Evolution and Systematics. 9, 121-135.
Blanco, J. A. (2007). The representation of allelopathy in ecosystem-level forest models.
Ecological Modelling, 209 (2-4), 65-77.
Bonham-Carter, G. F. (1994). Geographic information systems for geoscientists: modelling with GIS. New York: Elsevier.
Broennimann, O., & Guisan, A. (2008). Predicting current and future biological invasions: both native and invaded ranges matter. Biology Letters, 4 (5), 585-589.
Callaway, R. M. & Aschehoug, E. T. (2000). Invasive Plants Versus Their New and Old Neighbors: A Mechanism for Exotic Invasion. Science, 290, 520-523.
Catford, J.A., Daehler, C.C., Murphy, H.T., *et al.* (2012). The intermediate disturbance hypothesis and plant invasions: Implications for species richness and management. Perspectives in Plant Ecology, Evolution and Systematics, 14, 231-241.
Ceia, R.S., Ramos, J.A., Heleno, R.H., *et al.* (2011a). Status assessment of the critically endangered Azores Bullfinch. Bird Conservation International, 21, 477–489.
Ceia, R.S., Sampaio, H.L., Perejo, *et al.* (2011b). Throwing the baby out with the bathwater: does laurel forest restoration remove a critical winter food supply for the critically endangered Azores bullfinch? Biological Invasions, 13, 93–104.
Chen, Q., Han, R., Ye, F., *et al.* (2011). Spatio-temporal ecological models. Ecological Informatics, 6 (1), 37-43.
Elith, J. & Leathwick, J.R. (2009). Species Distribution Models: Ecological Explanation and Prediction Across Space and Time. Annual Review of Ecology, Evolution, and Systematics, 40, 677–697.
Evans, M. R., Grimm, V., Johst, K., *et al.* (2013). Do simple models lead to generality in ecology? Trends in Ecology and Evolution, 28 (10), 578-583.
Fassoni, A. C., Martins. M.L. (2014). Mathematical analysis of a model for plant invasion mediated by allelopathy. Ecological Complexity, 18, 49-58, doi.org/10.1016/j.ecocom.2013.03.004
Filgueira, R., Grant, J., Stuart, R., *et al.* (2013). Ecosystem modelling for ecosystem-based management of bivalve aquaculture sites in data-poor environments. Aquaculture Environment Interactions, 4 (2), 117-133.
Gallien, L., Münkemüller, T., Albert, C. H., *et al.* (2010). Predicting potential distributions of invasive species: where to go from here? Diversity and Distributions, 16 (3), 331-342.
Gassó, N., Thuiller, W., Pino, J., Vilà, M. (2012). Potential distribution range of invasive plant species in Spain. NeoBiota, 12, 25-40.
Geist, H. J., Lambin, E. F. (2002). Proximate Causes and Underlying Driving Forces of Tropical Deforestation: Tropical forests are disappearing as the result of many pressures, both local and regional, acting in various combinations in different geographical locations. BioScience, 52 (2), 143-150.
Genovesi, P. (2005). Eradications of invasive alien species in Europe: a review. Biological Invasions, 7, 127–133.
Goslee, S.C., Peters, D.P.C. & Beck, K.G. (2006). Spatial Prediction of Invasion Success Across Heterogeneous Landscapes using an Individual-Based Model. Biological Invasions, 8, 1387-3547.
Grimm, V., Berger, U., Bastiansen, F., *et al.* (2006). A standard protocol for describing individual-based and agent-based models, Ecological Modelling, 198, 115-126.
Grimm, V., Berger, U., DeAngelis, D.L., *et al.* (2010). The ODD protocol: A review and first update. Ecological Modelling, 221, 2760-2768

Guisan A., & Thuiller W. (2005). Predicting species distribution: offering more than simple habitat models. Ecology Letters, 8, 993–1009.
Guisan A., & Zimmermann N.E. (2000). Predictive habitat distribution models in ecology. Ecological Modelling, 135, 147–186.
Heleno, R. H., Ceia, R. S., Ramos, J. A., *et al.* (2009). Effects of Alien Plants on Insect Abundance and Biomass: a Food-Web Approach. Conservation Biology, 23 (2), 410-419.
Hierro, J. L. & Callaway, R. M. (2003). Allelopathy and exotic plant invasion. Plant and Soil, 256, 29-39.
Hutchinson, G. E. (1957). Concluding remarks. Cold Spring Harbor Symposia on Quantitative Biology, 22, 415-427.
Jongejans, E., O. Skarpaas, & Shea, K. (2008). Dispersal, demography and spatial population models for conservation and control management. Perspectives in Plant Ecology, Evolution and Systematics, 9 (3-4), 153-170.
Jørgensen, S. E. (2008). Overview of the model types available for development of ecological models. Ecological Modelling, 215 (1), 3-9.
Kimmins, J. P., Mailly, D. & Seely, B. (1999). Modelling forest ecosystem net primary production: the hybrid simulation approach used in FORECAST. Ecological Modelling, 122 (3), 195-224.
Kubiszewski, I., Costanza, R., Franco, C., *et al.* (2013). Beyond GDP: Measuring and achieving global genuine progress. Ecological Economics, 93, 57-68.
Kumar, J. L. G., & Zhao, Y. Q. (2011). A review on numerous modeling approaches for effective, economical and ecological treatment wetlands. Journal of Environmental Management, 92 (3), 400-406.
Lin, Y., Berger, U., Grimm, V., *et al.* (2012). Differences between symmetric and asymmetric facilitation matter: exploring the interplay between modes of positive and negative plant interactions. Journal of Ecology, 100 (6), 1482-1491, doi: 10.1111/j.1365-2745.2012.02019.x.
Lomba, A., Pellissier, L., Randin, C., *et al.* (2010). Overcoming the rare species modelling paradox: a novel hierarchical framework applied to an Iberian endemic plant. Biological Conservation, 143 (11), 2647-2657.
Marco, D. E., Sergio, A., & Cannas, S. A. (2002). Species Invasiveness in Biological Invasions: A Modelling Approach. Biological Invasions, 4, 193-205.
Martin, P. H., C. D. Canham, & Marks, P. L. (2009). Why forests appear resistant to exotic plant invasions: intentional introductions, stand dynamics, and the role of shade tolerance. Frontiers in Ecology and the Environment, **7**, 142-149.
Martines, I. P., Kojouharov, H. V. & Grover, J. P. (2009). A chemostat model of resource competition and allelopathy. Applied Mathematics and Computation, 215, 573 -582.
Merry, F., Soares-Filho, B., Nepstad, D., *et al.* (2009). Balancing conservation and economic sustainability: the future of the Amazon timber industry. Environmental Management, 44 (3), 395-407.
Morin, X. & Thuiller, W. (2009). Comparing niche- and process-based models to reduce prediction uncertainty in species range shifts under climate change. Ecology, 90 (5), 1301-1313
Myster, R. W. (1993). Tree invasion and establishment in old fields at Hutcheson Memorial Forest. The Botanical Review, 59 (4), 251-272.
Nathan, R., Katul, G., Bohrer, A., *et al.* (2011). Mechanistic models of seed dispersal by wind. Theoretical Ecology, 4 (2), 113-132.
Nehrbass, N. & Winkler, E. (2007). Is the Giant Hogweed still a threat? An individual-based modelling approach for local invasion dynamics of *Heracleum mantegazzianum*. Ecological Modelling, 201, 377-384.
Nepstad, D., Soares-Filho, B., Merry, F., *et al.* (2009). The end of deforestation in the Brazilian Amazon. Science, 326 (5958), 1350-1351.
Pejchar, L., Mooney, H. A. (2009). Invasive species, ecosystem services and human well-being. Trends in Ecology and Evolution, 24 (9), 497-504.

Petchey, O. L. & Belgrano, A. (2010). Body-size distributions and size-spectra: universal indicators of ecological status? Biology letters, 6 (4), 434-437.

Peters, D.P.C. (2002). Plant species dominance at a grassland – shrubland ecotone: an individual-based gap dynamics model of herbaceous and woody species. Ecological Modelling, 152, 5-32.

Peterson, A.T. & Vieglais, D.A. (2001). Predicting Species Invasions Using Ecological Niche Modeling: New Approaches from Bioinformatics Attack a Pressing Problem. BioScience, 51, 363.

Peterson A.T. (2003). Predicting the geography of species' invasions via ecological niche modeling. The Quarterly Review of Biology, 78, 419–433.

Petitpierre B., Kueffer C., Broennimann O., *et al.* (2012). Climatic niche shifts are rare among terrestrial plant invaders. Science, 335, 1344–1348.

Pimentel D., Lach L., Zuniga R., *et al.* (2000). Environmental and Economic Costs of Nonindigenous Species in the United States. BioScience, 50, 53.

Railsback, S. F., & Grimm, V. (2011). Agent-based and individual-based modeling: A practical introduction. New Jersey: Princeton University Press.

Ramos, J. A. (1995). The diet of the azores bullfinch *Pyrrhula murina* and floristic variation within its range. Biological Conservation, 71 (3), 237-249.

Ramos, J. A. (1996). Introduction of exotic tree species as a threat to the Azores bullfinch population. Journal of Applied Ecology, 33 (4), 710-722.

Ren, M.-X. & Zhang, Q.-G. (2009). The relative generality of plant invasion mechanisms and predicting future invasive plants. Weed Research, 49, 449-460.

Rebaudo, F., Crespo-Pérez, V., Silvain, J.-F., *et al.* (2011). Agent-Based Modeling of Human-Induced Spread of Invasive Species in Agricultural Landscapes: Insights from the Potato Moth in Ecuador. Journal of Artificial Societies and Social Simulation, 14, (3) 7.

Santos, M. & Cabral, J. A. (2004). Development of a stochastic dynamic model for ecological indicators' prediction in changed Mediterranean agroecosystems of north-eastern Portugal. Ecological Indicators, 3 (4), 285-303.

Santos, M., Bastos, R. & Cabral, J. A. (2013). Converting conventional ecological datasets in dynamic and dynamic spatially explicit simulations: Current advances and future applications of the Stochastic Dynamic Methodology (StDM). Ecological Modelling, 258, 91-100.

Santos, M., Freitas, R., Crespí, A. L., *et al.* (2011). Predicting trends of invasive plants richness using local socio-economic data: An application in North Portugal. Environmental Research, 111 (7), 960-966.

Shackelford, N., Renton, M., Perring, M. P. *et al.* (2013). Modeling disturbance-based native invasive species control and its implications for management. Ecological Applications, 23 (6), 1331-1344.

Soares-Filho, B. S., Coutinho Cerqueira, G. & Lopes Pennachin, C. (2002). DINAMICA—a stochastic cellular automata model designed to simulate the landscape dynamics in an Amazonian colonization frontier. Ecological modelling, 154 (3), 217-235.

Soares-Filho, B. S., Rodrigues, H. & Costa, W. (2009). Modeling environmental dynamics with Dinamica EGO. Centro de Sensoriamento Remoto, IGC/UFMG, www.csr.ufmg.br/dinamica/tutorial/Dinamica_EGO_guidebook.pdf (accessed on 7.10.2013).

Soares-Filho, B., Alencar, A., Nepstad, D., *et al.* (2004). Simulating the response of land-cover changes to road paving and governance along a major Amazon highway: the Santarém–Cuiabá corridor. Global Change Biology, 10 (5), 745-764.

Soares-Filho, B., Moutinho, P., Nepstad, D., *et al.* (2010). Role of Brazilian Amazon protected areas in climate change mitigation. Proceedings of the National Academy of Sciences, 107 (24), 10821-10826.

Soares-Filho, B., Rodrigues, H. & Follador, M. (2013). A hybrid analytical-heuristic method for calibrating land-use change models. Environmental Modelling & Software, 43, 80-87.

Stickler, C., Nepstad, D., Coe, M., *et al.* (2009). The potential ecological costs and co-benefits of REDD: a critical review and case study from the Amazon region. Global Change Biology, 15, 2803–2824.

Thuiller W., Richardson D.M., Pyšek P., *et al.* (2005). Niche-based modelling as a tool for predicting the risk of alien plant invasions at a global scale. Global Change Biology, 11, 2234–2250.

Tonini, F., Hochmair, H.H., Scheffrahn, R.H., *et al.* (2013). Simulating the spread of an invasive termite in an urban environment using a stochastic individual-based model. Environmental Entomology, 42, 412-423

Vicente J.R., Fernandes R.F., Randin C.F., *et al.* (2013). Will climate change drive alien invasive plants into areas of high protection value? An improved model-based regional assessment to prioritise the management of invasions. Journal of Environmental Management, 131, 185-195.

Travis, J.M.J., Harris, C.M., Park, K.J., *et al.* (2011). Improving prediction and management of range expansions by combining analytical and individual-based modelling approaches, Methods in Ecology and Evolution, 2 (5), 477- 488

Vicente J.R., Pinto A.T., Araújo M.B., *et al.* (2013). Using Life Strategies to Explore the Vulnerability of Ecosystem Services to Invasion by Alien Plants. Ecosystems, 16, 678–693.

Vicente, J., Randin, C. F., Gonçalves, J., *et al.* (2011). Where will conflicts between alien and rare species occur after climate and land-use change? A test with a novel combined modelling approach. Biological Invasions, 13 (5), 1209-1227.

Frédérique Viard, Thierry Comtet

18 Applications of DNA-based Methods for the Study of Biological Invasions

18.1 Introduction

Numerous conceptual frameworks and methodologies have been developed for the study of the invasion of an area by a non-indigenous (syn. non-native) species (NIS) (Williamson, 1996; Shigesada & Kawasaki, 1997; Richardson, 2011). The issues addressed may be classified into three main research themes: 1) history of the invasion processes and subsequent dispersal (e.g. date(s) and location(s) of the introduction(s); source(s) and vector(s); pathway(s) of colonization); 2) biological impact of the invaders and factors favoring new species in colonizing new territories (e.g. ecological consequences for the native community; invasion dynamics; life history traits favoring foreign species); 3) predictions and control of biological invasions. Modeling, population dynamics, and field- or laboratory-based community ecology are scientific fields that have been extensively used to investigate these issues since the seminal book by Elton (1958).

Until the end of the 1990s, only a few studies based on genetic data specifically addressed non-native species. These early studies demonstrated the insights such approaches may offer, across a wide range of taxa and environments:

- to identify the invader (e.g. Geller *et al.*, 1997 for crabs)
- to determine the geographical origin of the NIS (e.g. Goff *et al.*, 1992 in algae; O'Foighil *et al.*, 1998 in the Portuguese oyster)
- to test for hypotheses of founder events or recurrent introductions (e.g. Davies *et al.*, 1999a for the medfly; Suarez *et al.*, 1999 for the Argentine ant) and examine dispersal strategies (e.g. Wilson *et al.*, 1999 for freshwater mussels)
- to investigate the genetic consequences of introductions (Stone & Sunnucks, 1993 in a gall wasp) or to look for hybridization with native species (e.g. Goodman *et al.*, 1999 in deer)
- to demonstrate changes in behavior (Tsutsui *et al.*, 2000 in ants) and reproductive systems in plants (Eckert *et al.*, 1996; Daehler, 1998) between the native and colonized areas.

With the emergence of DNA tools in the last two decades, population genetics and related scientific fields, like phylogeography or DNA barcoding (see definitions below), have been increasingly used to tackle these issues across many taxa, regions, and habitats. The number of papers dramatically increased, and several reviews summarize the main findings, highlight the usefulness of inferences from genetic data, provide warnings and recommend specific analytical designs, and link genetic data with the evolutionary potential of non-native species (e.g. Holland, 2000; Darling & Blum, 2007; Roman & Darling, 2007; Estoup & Guillemaud, 2010; Geller *et al.*, 2010; Dormontt *et al.*, 2011; Fitzpatrick *et al.*, 2012; Lawson Handley *et al.*, 2011; Rius & Darling, 2014 ; Rius *et al.*, 2015). This book chapter does not aim at covering all the issues addressed in these reviews, and we encourage readers to benefit from these papers for their own study designs. Our aim is to point out some of the major applications of DNA-based approaches, each of them focused on a specific step of the invasion dynamics (Figure 18.1): 1) the primary introduction, where molecular tools are powerful for early detection of NIS; 2) the history of the introduction process (routes, pathways and spread); and 3) the study of the evolutionary and ecological consequences after establishment.

18.2 A Need for Powerful Non-Native Species Identification Tools

An accurate and rapid identification of non-native species is a pre-requisite for successful survey, monitoring, and management schemes (Pyšek *et al.*, 2013). This is particularly true at the early stages of an introduction (i.e. transport and primary introduction) for attempts to eradicate the new species and prevent its future successful proliferation (Simberloff *et al.*, 2013). To address this issue, molecular (DNA-based) tools have proven to be particularly relevant, having numerous advantages over traditional approaches (Darling & Blum, 2007; Le Roux & Wieczorek, 2009). These methods (1) are fast, (2) do not require expertise in different taxonomic fields, (3) can be applied to fragments of specimens or to particular life history stages for which morphological diagnostic characters are lacking, and (4) can be applied to complex matrices (e.g. environmental samples). Choosing among the various methods available depends on the sample to be analyzed and the questions to be answered (Darling & Blum, 2007; Bott *et al.*, 2010) (Table 18.1).

Many methods have been developed to identify a single target species and are used to confirm the identity of pre-identified specimens or to detect the presence of the target species: this is true for PCR-RFLP (e.g. Darling & Tepolt, 2008), PCR with species-specific primers (e.g. Harvey *et al.*, 2009), *in situ* hybridization (e.g. Le Goff-Vitry *et al.*, 2007; Mountfort *et al.*, 2007), sandwich hybridization assay (e.g. Harvey *et al.*, 2012), real-time quantitative PCR, or the new hybridization coupled with light transmission spectroscopy method (Egan *et al.*, 2013). Alternative approaches using DNA sequences, in which various DNA sequences from unknown specimens are compared to those in databases (e.g. Genbank), not only allowed confirmation of the identity of specimens, but also offered the possibility of detecting the presence of non-native species without any *a priori*, for example during regular surveys. This will be detailed below.

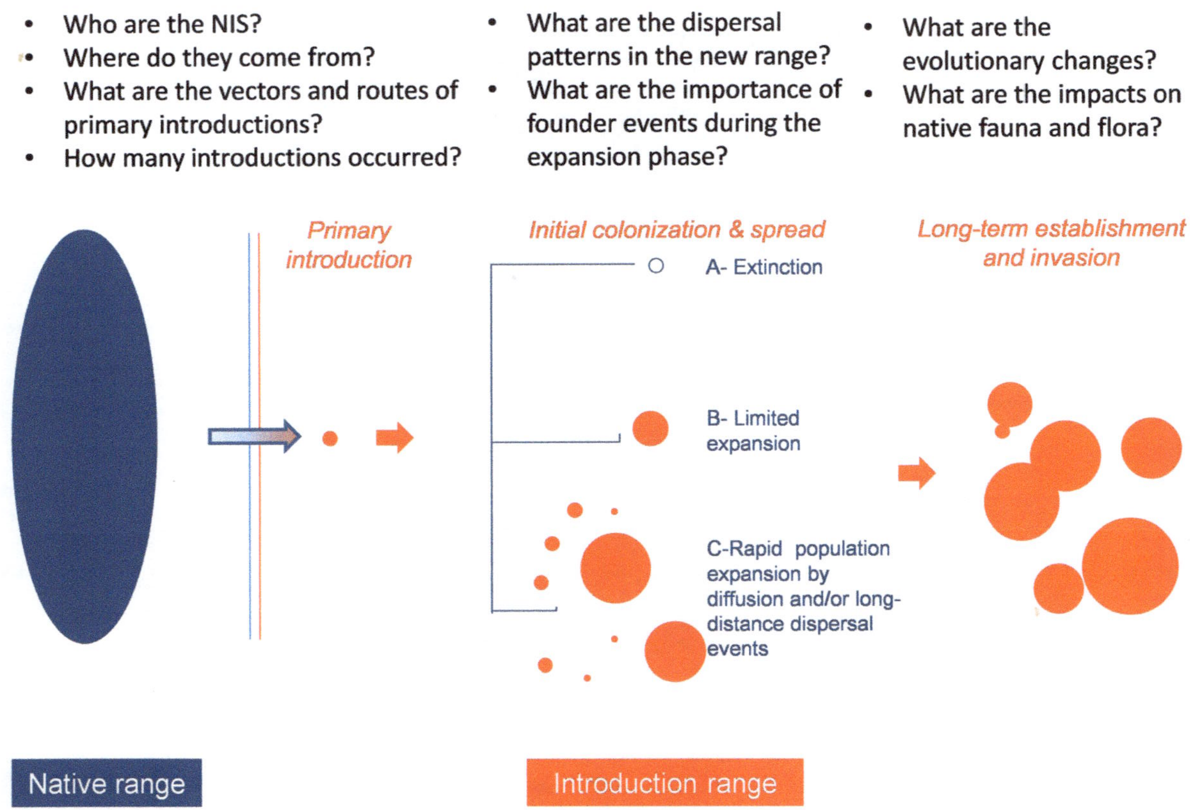

Fig. 18.1: Main issues of biological invasions commonly addressed by using genetic data. These issues are typically related to different stages of the introduction/invasion process. The blue ellipse depicts populations in the native range, and red circles represent populations in the introduction range. Ellipse and circle size are proportional to the population size.

Tab. 18.1: Overview of popular molecular approaches and tools with their most relevant applications (*, **, *** stand for low to high relevancy) in the study of biological invasions.

	Approach	**Species-specific marker**	**DNA barcoding**	**DNA metabarcoding**	**Phylogeography**	**Population genetics**
	Example of tools	**PCR-RFLP**	**Sanger sequencing with e.g. *coxI* for metazoans; *rbcL* for plants**	**Illumina or 454 sequencing**	**Organellar or nuclear DNA sequencing**	**Microsatellites, SNPs, RAD-Seq**
Theme	Question					
NIS identification	Targeting a specific NIS	***	***			*
	NIS inventories		***	***		
	Looking for cryptic NIS				***	
Pathways	Source identification				***	***
	Testing for single vs. repeated introductions				***	***
	Patterns of spread in the new range(s)				*	***
Evolutionary changes	Testing for selection on standing genetic variation					*** (many markers needed)
	Looking for admixture				** (if nuclear)	***
	Looking for founder events				*	***
	Studying hybridization				** (if nuclear)	***
Community level	Food web	***	***	***		
	Host-parasite interactions		***	***	*** (co-evolution studies)	** (co-evolution studies)
	Biodiversity assessment		* (time consuming)	***		

18.2.1 Molecular Barcoding: A Popular Approach for Aliens' Identification

The identification of NIS culminated in the use of DNA barcoding (Floyd *et al.*, 2002; Hebert *et al.*, 2003), which has proven to be particularly relevant in the context of biological invasions (e.g. Armstrong & Ball, 2005; Cross *et al.*, 2011; Comtet *et al.*, 2015). The main novelty that DNA barcoding has brought is the standardization of the identification process. DNA barcoding relies on short standard DNA sequence(s), called DNA barcodes, which can be applied to a wide range of taxa (such as *coxI* for most metazoans; *ITS* for fungi; *rbcL*, *matK*, and *ITS* for plants), whose design requires upstream research based on taxonomy and phylogeny. The power of DNA barcoding is based on the principle that the intraspecific barcode polymorphism is lower than the interspecific divergence, the difference between the two being known as the barcoding gap (Meyer & Paulay, 2005; Wiemers & Fiedler, 2007; Figure 18.2). The wider the barcoding gap is, the better the species discrimination is. Thanks to standardization, a lot of efforts have been and are made to feed international reference databases composed of barcodes recovered from voucher specimens identified morphologically. For example, the BoLD database (Ratnasingham & Hebert, 2007), with its wide taxonomic coverage, is the most comprehensive international database dedicated to DNA barcoding.

Such initiatives lead DNA barcoding to be particularly well-suited to the identification of species out of their native range, for which no *a priori* on the geographic origin is available. It is particularly useful in some taxonomic groups that are very difficult to identify due to a paucity of diagnostic morphological criteria, for example in algae (Geoffroy *et al.*, 2012) or ascidians (Callahan *et al.*, 2010; Bishop *et al.*, 2013). In these taxa, many non-native species may be overlooked because of poor taxonomy at the species level and poor knowledge of their biogeographic status (Bishop *et al.*, 2013). A further advantage of standardization is the possibility of using DNA barcoding in routine analyses in any laboratory where basic molecular equipment is available, but also in governmental agencies in charge of biosecurity and management strategies (Darling & Mahon, 2011). It is thus routinely used by government agencies in New Zealand and Australia for survey purposes, for example for the detection of the invasive ascidians *Didemnum vexillum* and *D. perlucidum* (Smale & Childs, 2012) or to control the presence at the border of high-risk insect species (Armstrong, 2010).

DNA barcoding applications are numerous. It first allows the discovery of new non-native species during regular surveys. For example, a solitary styelid tunicate was repeatedly observed along the coasts of Brittany during surveys in marinas. DNA barcoding showed it belonged to *Asterocarpa humilis*, a species that has never been reported and probably remained unnoticed for years (Bishop *et al.*, 2013). Similarly, the alien macroalga *Gracilaria vermiculophylla* was detected in British Columbia by using DNA barcoding (Saunders, 2009). Second, when applied to historical samples (e.g. museum specimens), DNA barcoding may reveal former misidentifications. For example, non-native calyptraeid gastropods from California, putatively identified as *Crepidula fornicata*, were later (20 years) identified through DNA barcoding as *C. convexa*, a species with different dispersal abilities, with potential consequences in management strategies (McGlashan *et al.*, 2008). Finally, because DNA barcoding can be used on early life history stages, it can be used to detect invaders during the introduction step, for example during quar-

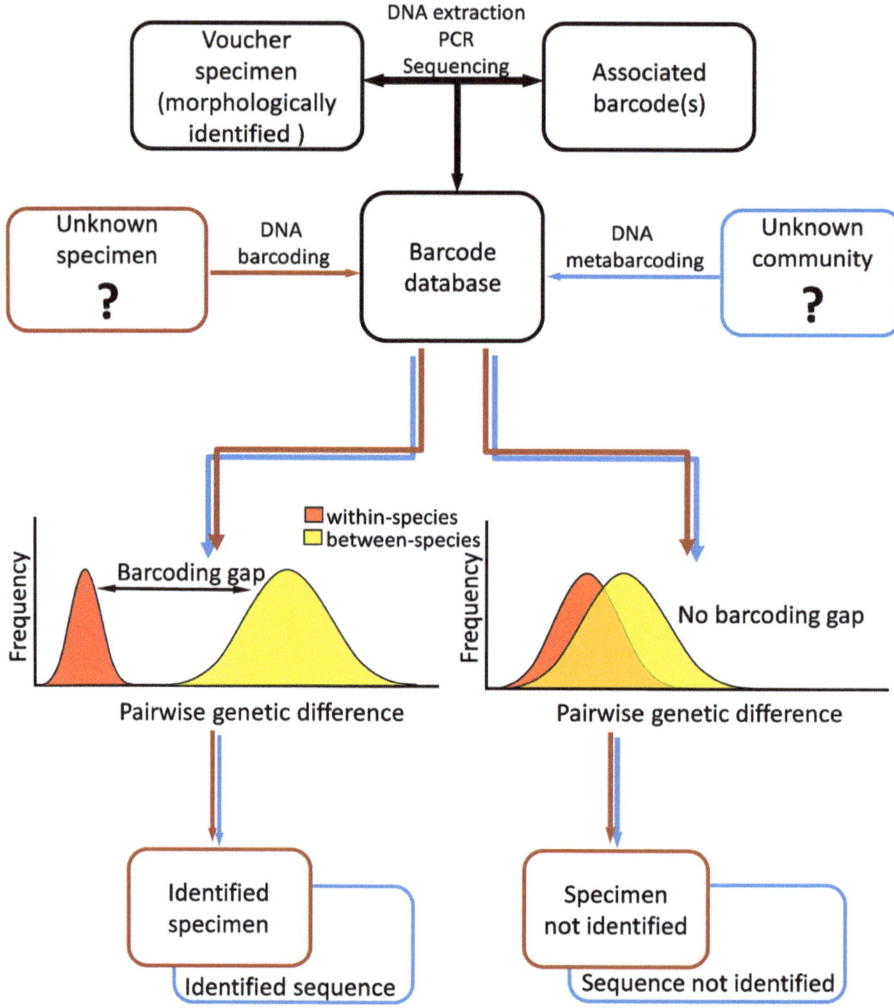

Fig. 18.2: Main steps of the DNA barcoding and metabarcoding approaches for the identification and inventory of non-native species. Barcoding (in red) and metabarcoding (in blue) rely on the availability of reference barcodes recorded in international databases. Such databases are established from upstream research (in black). The power of these approaches is conditioned by the existence of a barcoding gap (i.e. the lack of overlap between within-species and between-species polymorphism), which allows unambiguous identification of species. In case of such an overlap (i.e. no barcoding gap), identification to the species level will fail. However, identification at higher taxonomic levels is still valuable. Whereas barcoding consists in the identification of specimens based on their barcode, metabarcoding identifies sequences obtained from a mixture of many species (i.e. specimens are not observed).

antine procedures at the border (e.g. Armstrong & Ball, 2005), or through examination of introduction vectors (ships, planes...). In this context, DNA barcoding was used successfully to inventory invertebrates living in ships' ballast sediments, focusing on the diapausing eggs of rotiferans and crustaceans (Briski *et al.*, 2011).

18.2.2 Early Detection: A Challenging But Critical Task

One of the challenges of early detection of non-native species is the detection of a low number of specimens hidden within the local species pools. All the above molecular methods of identification are particularly relevant in that context, being typically more sensitive than traditional methods involving sorting and counting. For example, PCR-RFLP or PCR with species-specific primers allow the detection of single alien invertebrate larvae in plankton (Darling & Tepolt, 2008; Harvey *et al.*, 2009). Ultimately, alien species may be detected in the form of molecular imprints, i.e. free or particle-bound DNA molecules released by organisms (environmental DNA, eDNA), which may still be detected several weeks after the species has been removed (Ficetola *et al.*, 2008; Jerde *et al.*, 2011; Dejean *et al.*, 2012).

It is expected in the future that early detection of aliens will be further enhanced thanks to the development of next-generation sequencing techniques (NGS), which typically provide billions of sequence reads in a single run, quickly and at low cost. In particular, NGS allows assessing biodiversity from complex environmental samples through DNA metabarcoding, an extension of traditional barcoding (Shokralla *et al.*, 2012; Taberlet *et al.*, 2012; Cristescu, 2014) (Figure 18.2). DNA metabarcoding relies on the same principle as traditional barcoding, differing by the sequencing depth, which theoretically allows recovering the whole diversity of the sample, and offers the possibility to simultaneously analyze several samples. A recent study focusing on the detection of marine aliens showed that DNA metabarcoding allowed the detection of a single *Asterias amurensis* (an invasive seastar in New Zealand) larva in water and sediment samples containing a large array of environmental eukaryotes (Pochon *et al.*, 2013).

18.2.3 Revealing New Cryptic NIS

Identifying non-native species with molecular tools like DNA barcoding requires that their taxonomy is well understood, so that the developed markers indeed identify single species. However, in some taxa, taxonomy at the species level remains unclear, and many species are in fact species complexes composed of several species which looks identical, called cryptic species. The taxa for which molecular approaches are needed to help identification (because of the paucity of diagnostic characters) are often also those for which taxonomic status is unclear. As a result, many non-native species may be overlooked because of cryptism. Within-species sequencing approaches, like phylogeography approaches (see 18.3.1), are particularly relevant to reveal the existence of non-native cryptic species by showing unexpected large molecular divergence between some lineages. Many examples exist in the marine realm where taxa that were considered as worldwide invasive 'species' were recently shown to be comprised of several cryptic lineages, possibly corresponding to several sister-species or sub-species, like in the bryozoan *Watersipora subtorquata* (Mackie *et al.*, 2012) and the ascidians *Botryllus schlosseri* (Bock *et al.*, 2012) and *Ciona intestinalis* (Zhan *et al.*, 2010). Revealing cryptic non-native species is crucial because they may differ in their life history traits or their invasion histories and pathways. For example, the occurrence of two cryptic lineages of the non-native amphipod *Grandidierella japonica* suggested the existence of two independent introduction events on the Pacific coasts of North America (Pilgrim *et al.*, 2013).

18.3 Tracing Back Introduction and Expansion Processes

Tracing back the origin of an introduction to identify the dispersal routes and pathways (e.g. vectors, number of introduction events) and studying the spread of NIS in their new range are key issues to better understand the ecology and biology of an invader and to propose management strategies. DNA-based studies are widely used to tackle these issues.

18.3.1 Routes and Pathways of Colonization

One central feature of introduction processes is the disruption of natural dispersal pathways with new populations often established far away from their native range (Wilson *et al.*, 2009). For example, the Japanese kelp *Undaria pinnatifida*, native to the Northwest Pacific, has colonized new areas in the Northern Atlantic and Western Pacific within less than 30 years (Voisin *et al.*, 2005, and references therein). Such a rapid global spread would not have been possible through natural routes of dispersal, especially in a species characterized by low dispersal abilities. Determining routes, vectors or numbers of introduction events is difficult based only on direct observations or using logbooks of ships and planes or shellfish importation registers: the information is often difficult to get or incomplete, and such approaches are usually inadequate to document repeated and cryptic introductions. Investigating the routes and pathways of introductions has been central in many genetic studies of NIS (Holland, 2000; Estoup & Guillemaud, 2010; Geller *et al.*, 2010). Both phylogeography-based methods (Avise *et al.*, 1987; Avise, 2000; Hickerson *et al.*, 2010) and population genetics (Hartl & Clark, 1989; Weir, 1990; Hedrick, 2000) are commonly used to tackle this issue, using different molecular tools (Table 18.1): sequencing of mitochondrial or nuclear genes, microsatellites (SSRs), and single nucleotide polymorphisms (SNPs). Molecular markers are most often neutral, for examining migration and drift processes, and polymorphic enough in the native range at the population level to capture traces left during and after the introduction.

Despite some success in analyzing general features of the introduction processes (e.g. testing the propagule pressure hypothesis, testing competing introduction scenarios (Lawson Handley *et al.*, 2011; Geller *et al.*, 2010), many studies failed to identify the sources of the target invaders, because of rather restrictive conditions to be fulfilled (Figure 18.3). One critical aspect for determining the source is the sampling effort, with all putative sources being included (Muirhead *et al.*, 2008; Geller *et al.*, 2010) (see case II as compared to case I in Figure 18.3). Determining the source is particularly challenging for species lacking genetic structure in their native range (case II in Figure 18.3), like the numerous marine invertebrates characterized by a long-lived dispersal stage and large effective population size.

Even with a significant genetic structure in the native range (a distinct genetic footprint of each putative source population), determining the source of NIS can be difficult (Geller *et al.*, 2010). This may be due to the difficulty of getting a comprehensive sampling, and the post-introduction changes in the genetic composition of the introduced or native populations through genetic drift, if the time elapsed since the introduction is long. Attempts to determine the geographic origins of an introduction thus require an important sampling effort, ideally in the early stages of the process. If the precise origin cannot be determined, important information can nevertheless be obtained about introduction pathways and processes. For instance, Voisin *et al.* (2005) showed that populations of *U. pinnatifida* introduced in Europe were far less polymorphic and genetically similar to populations cultivated in Asia (native range) than populations introduced in Australasia. This supported earlier hypotheses explaining the different introduction pathways of this species in Europe (through aquaculture) and Australasia (through shipping).

Gene sequencing and parsimony networks (Posada & Crandall, 2001) are commonly used in invasion studies in light of phylogeography approaches (which examine the geographical distribution of gene lineages (Avise *et al.*, 1987)). For species in their natural distribution range, the geographical distribution of gene lineages is analyzed to understand the historical and natural demography and migratory changes (Avise, 2000; Maggs *et al.*, 2008). Common interpretation keys (e.g. star-like networks featuring demographic expansion) are, however, misleading in non-native species, as the observed patterns cannot be associated to natural long-term vicariant, demographic or migratory processes, given the short time elapsed since the introduction, and the disruption of natural migration routes due to human-mediated activities. Parsimony networks are nevertheless useful as descriptive tools to highlight genetic mixing between evolutionary divergent lineages in introduced populations, an indication of repeated introductions from several sources (e.g., Kolbe *et al.*, 2004; Simon-Bouhet *et al.*, 2006). Besides, unexpected patterns observed in phylogeography studies of a given species across its supposed native range can reveal unreported introductions. For example, in two annelid species, Jolly *et al.* (2006) explained the presence of haplotypes typical of Northern European clades in Southern Europe by human-mediated colonization.

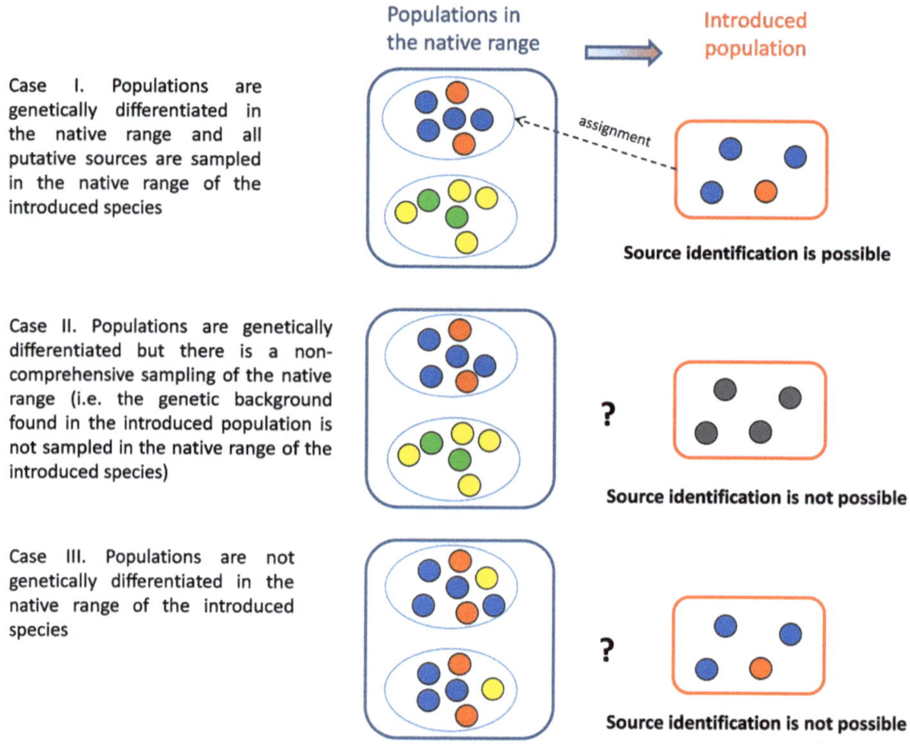

Fig. 18.3: Tracing back the sources: success and failure in assigning sources of introduction. Figures illustrate conditions under which a correct identification of the source may or may not be possible. Blue rectangles and ellipses (left) feature the native range and populations (here 2 populations), respectively, of the introduced species. Red rectangles (right) feature an introduced population of this species. Filled circles represent individuals; each color refers to a different genetic background.

Assignment analyses, based on multilocus genetic data and maximum likelihood calculations, have been largely used to trace back introductions. They provide a means of assigning individuals to particular putative sources and assessing to some extent population structure (Cornuet *et al.*, 1999; Davies *et al.*, 1999b; Manel *et al.*, 2005; Broquet & Petit, 2009). Using such methods, Davies *et al.* (1999a) demonstrated the difficulty of controlling the infestation by the medfly *Ceratitis capitata*, by showing that a single individual captured in California was most likely an immigrant from an unnoticed re-introduction rather than a remnant individual produced locally. New analytical methods, like Approximate Bayesian Computations (ABC) have also brought much to the study of colonization history (Cornuet *et al.*, 2008; Lawson Handley *et al.*, 2011). Their main advantage is to allow a probabilistic approach of different competing introduction scenarios. Using a simulation-based approach and a case study with the western corn rootworm, *Diabrotica virgifera virgifera*, Guillemaud *et al.* (2010) showed the higher efficiency of ABC compared to traditional genetic distance-based methods (e.g. F_{st}), or assignment-likelihood statistics, in testing the hypothesis that two invasive populations have the same origin. The usefulness of these methods was also pointed out by Rius *et al.* (2012), who showed that the colonization pathways of the ascidian *Microcosmus squamiger* over the whole introduction range followed the historical taxonomic records.

Altogether, despite the above limitations inherent to the use of DNA-based methods (Geller *et al.*, 2010; Fitzpatrick *et al.*, 2012), many studies shed light on the routes and pathways of invasions and their associated vectors or processes. These studies showed that for a given non-native species, the introduction pathways can differ between regions of the introduction range, as exemplified above in *U. pinnatifida* (Voisin *et al.*, 2005). They also documented that both single and multiple introductions can be observed in successful introduced species (Dlugosch & Parker, 2008). In terrestrial plants (Dormontt *et al.*, 2011) and marine species (Roman & Darling, 2007; Rius *et al.*, 2015), a genetic diversity in the introduced populations equal or higher than in the native ones was the most frequently observed pattern. It is explained by multiple and repeated introductions that may be unnoticed based only on field observations, i.e. cryptic introductions (Geller *et al.*, 2010). New statistical methods, like ABC, and molecular tools, like SNPs and DNA-seq, which deliver an enlarged set of markers, are now available for deeper investigations of colonization histories under complex scenarios (Estoup & Guillemaud, 2010).

18.3.2 Spread Dynamics and Temporal Changes

Population genetics studies are analyzing the way genetic diversity is distributed in space or time, to determine the relative importance of various evolutionary forces (migration, genetic drift, selection, mutation) in shaping the evolutionary trajectories of populations. Invasive species may, however, be challenging when using population genetics approaches (Fitzpatrick *et al.*, 2012), for instance because introduced populations did not yet reach equilibrium. Some of the basic assumptions of classical population genetics are thus expected to be violated as, for instance, the relationship between *Fst*, an estimator of the genetic differentiation between populations, and *Nm*, the effective number of migrants. Note, however, that such deviations from the model assumptions also hold for many species in their native range, as pointed out by Whitlock & McCauley (1999). An increasing number of population genetics studies of non-native species thus go beyond the usual measurements of genetic distance, like *Fst*-estimates, using for instance ABC analyses or assignment tests, which are less or non-sensitive to equilibrium assumptions.

One issue commonly addressed with DNA-based studies is the way NIS spread into their new range. Patterns of dispersal may follow several models (Figure 18.1), from 'simple' diffusion from the primary site(s) of the introduction, up to long-distance, jump or saltatory dispersal events with or without diffusion around the new site (Shigesada & Kawasaki, 1997). DNA-based studies have been used to discriminate these scenarios, to assess the importance of long-distance dispersal events, or to estimate dispersal rates. Based on the compilation of several studies of non-native insects, Lawson Handley *et al.* (2011) pointed out that the spread of non-native species is often a combination of short- and long-distance events, a pattern named 'stratified dispersal' (Shigesada *et al.*, 1995). The importance of long-distance and jump dispersal can often be related to human activities, particularly for species with low dispersal abilities, like many algae with short-lived spores or tunicate species with short-lived gametes and larvae. For example, using microsatellites, Lacoursière-Roussel *et al.* (2012) showed the importance of regional recreational boating in spreading the invasive colonial tunicate *Botryllus schlosseri* from its primary introduction sites located in commercial ports. Genetic studies also demonstrated that mechanisms involved in secondary spread (i.e. within the introduction range) of non-native species may differ from those involved in primary introduction (i.e. from the native area). For instance, using assignment tests with microsatellites, Grulois *et al.* (2011) investigated the genetic structure of populations of the kelp *Undaria pinnatifida*, first introduced deliberately in Brittany (France, English Channel) for aquaculture, and which subsequently established sustainable populations in the wild. The lack of isolation-by-distance patterns and the significant genetic divergence between wild and cultivated populations suggested that although farming initiated the escape into the wild, it was likely not the main source of the long-term establishment and population renewal of this species. Long-distance dispersal, through drifting thalli or fouling on buoys, ships, etc., more likely explained its rapid regional spread. The comparison between modeling of the natural dispersal (e.g. oceanic currents for larval dispersal in marine species) and genetic data can help disentangle natural and human-mediated patterns of spread of aliens (e.g. Viard *et al.*, 2006; Dupont *et al.*, 2007).

Invasion processes are obviously dynamic and fast-evolving processes. As such, monitoring the changes in the genetic composition of an invader along its spreading range can shed light on the vectors of spread (see above) and the invasion dynamics (e.g. additional propagule pressure), and help assess the efficiency of management actions. In the western corn rootworm, *Diabrotica virgifera virgifera*, Ciosi *et al.* (2011) analyzed established populations from the center towards the edge of its expansion route: they observed an interesting genetic pattern characterized by an increase of the genetic variation that was unexpected under the hypothesis of serial founder events during the expansion. They suggested that this may result from control measures in the center of the invasion zone, which could have significantly reduced the population size with consequences on both the demography and genetic composition of the primary introduced populations (i.e. demographic and genetic bottlenecks). Surprising patterns have also been evidenced, in particular 'gene surfing' patterns, where singular clines in allele frequencies have been observed in relation to genetic drift events along the colonization wave (Excoffier & Ray, 2008; Excoffier *et al.*, 2009). Temporal analyses are another way to follow the invasion dynamics; for example, by using herbarium or museum specimens, it is possible to compare modern and historical specimens. In this context, Saltonstall (2002) showed the high invasive potential of a specific non-native lineage of the common reed *Phragmites australis*, which invaded and even displaced previous lineages over 50 years in the USA. Alternative approaches consist of repeated sampling over time at the same location. Such an approach allowed Pérez-Portela (2012) to show a sequential loss of genetic diversity in one population of the introduced ascidian *Perophora japonica*, which suggested either serial bottlenecks or selection effects.

The study of invasion dynamics in both space and time has greatly benefited from DNA-based methods. The studies mentioned above, however, highlighted the importance of combining several approaches like modeling, field observations, and demography studies with genetic data to better describe the expansion wave, to evaluate the effectiveness of management strategies, and to investigate the likely evolution of non-native species.

18.4 Long-Term Establishment and Consequences on the Ecosystems

18.4.1 Insights about the Evolutionary Potential of the Invaders

Genetic studies are increasingly used to help understand the evolution of the invaders and their invasiveness (Figure 18.4). Studying invasive species adaptation and evolution is, however, still a very challenging issue (Sax *et al.*, 2007; Keller & Taylor, 2008; Dormontt *et al.*, 2011; Rius *et al.*, 2015).

18.4.1.1 Evidence in Favor of the Emergence of Rapid Adaptation

Evidence of rapid adaptive evolution associated with phenotypic shifts has been documented in several invasive species. In the cane toad *Bufo marinus*, for example, the rate of progress of the invasion front was related to changes in toad morphology (e.g. length of the legs; Phillips *et al.*, 2006) although other phenotypic changes were not adaptive but likely determined by stochastic processes (Shine, 2012). Although rapid adaptation may occur, other mechanisms may explain the sudden invasion by non-native species, in particular the introduction of new lineages from new sources. This is exemplified by the sudden expansion of the European crab *Carcinus maenas* in the Canadian Maritimes, where water temperatures are sub-optimal for this crab, although it was established there for a long time. This sudden expansion was first explained by a putative adaptation of the crab to cold-water conditions, combined with an increase in local water temperatures. A cryptic introduction by a new genetic lineage originating from Northern Europe was later revealed based on genetic analyses (Roman, 2006), suggesting that individuals better adapted to cold conditions could have been introduced. Among the approaches available to examine the emergence of adaptive evolution is the joint analyses of the genetic structure at neutral markers (*Fst* measures) and at quantitative traits (*Qst* measures), provided that a comprehensive examination of the traits exist over the native and invasive ranges (Keller & Taylor, 2008). Following such an approach, Xu *et al.* (2010) showed that phenotypic shift in the perennial herb *Phyla canescens* was due to selective rather than stochastic processes.

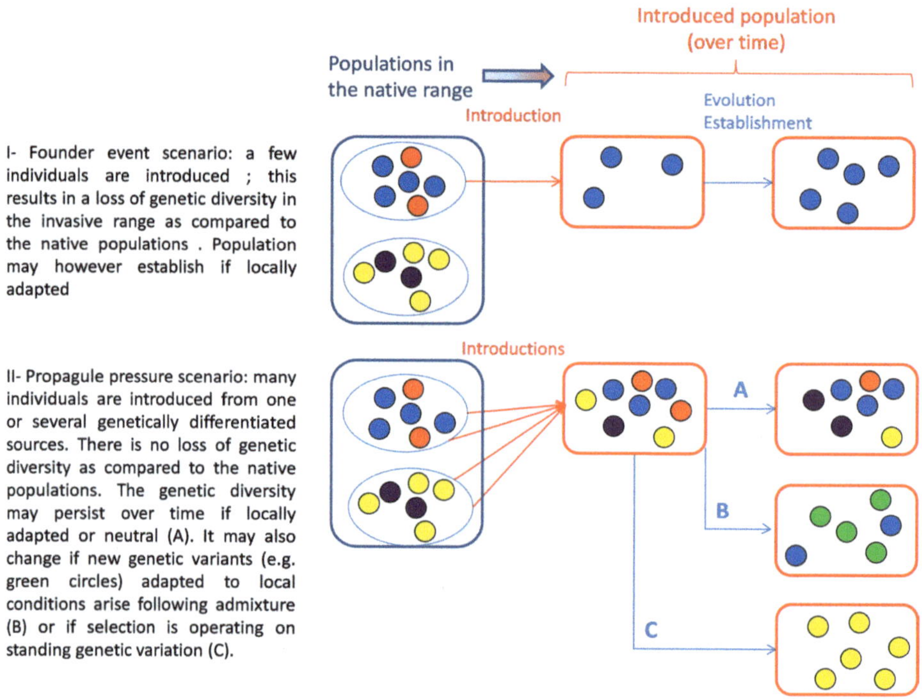

Fig. 18.4: Schematic diagram illustrating the relationships between introduction patterns, genetic diversity, and possible evolutionary trajectories of invader non-native species. Blue rectangles and ellipses depict the native range and populations, respectively. Red rectangles represent an introduced population (i.e. introduction range). Filled circles depict individuals; each color represents a different genetic background.

18.4.1.2 Studying Mechanisms That May Promote Rapid Adaptation

Two major mechanisms have been proposed as drivers of rapid evolutionary changes in invaders at the within-species level: selection of standing genetic variation (Barrett & Schluter, 2008; Figure 18.4-II-C) and genetic admixture (Rius & Darling, 2014; Figure 18.3.II-B). Many studies showed that introduced populations have similar or even higher genetic diversity than native populations (Dormontt *et al.*, 2011; Roman & Darling, 2007; Rius *et al.* 2015). This genetic diversity offers a large basis on which selection may operate over short time scales, as compared to selection on new mutations (Barrett & Schluter, 2008). Genome scan analyses, in which outlier loci (i.e. loci showing atypical patterns because they are influenced by selection processes) are looked for, are classical ways to study adaptation (Beaumont & Balding, 2004), but should be used and discussed with care (Bierne *et al.*, 2011). In European introduced populations of the Pacific oyster *Crassostrea gigas*, based on outlier identification, Rohfritsch *et al.* (2013) found two groups of populations established in fjord-like environments, a result which may reveal parallel adaptations in similar environments. Conversely, Riquet *et al.* (2013) did not find any outliers between native American and introduced European populations of the slipper limpet *Crepidula fornicata*, while their study revealed outliers between populations in the native range. They suggested that genome scans may not be always efficient for identifying selection between native and introduced ranges because of the short time elapsed since the introduction. Rapid evolution through genetic admixture is supported by observations of repeated introductions of many individuals from several genetically-diversified sources. Through reproduction between individuals with different genetic background, genetic admixture may lead to evolutionary novelties that confer selective advantage in the new environments. Examples supporting this mechanism are still scarce (Rius & Darling, 2014); a particularly convincing example was shown through the invasion by the freshwater snail *Melanoides tuberculata* (Facon *et al.*, 2008). Genetic admixture may, however, be rare in the wild because of the underlying costs, like outbreeding depression due to reproduction between individuals with evolutionary-divergent genetic backgrounds.

It is possible that the genetic diversity of invaders is an important component of their invasiveness and adaptive potential. And yet, invaders suffering genetic bottlenecks (with associated loss of genetic diversity) have been successful (Figure 18.4-I). In addition, adaptive shifts require a substantial amount of adaptive genetic diversity, i.e. genetic variation on which selection may act and that determines fitness traits, and in most of the examples cited in the preceding sections, neutral genetic diversity is measured. There are thus debates about 1) the importance of founder events for preventing, or conversely enhancing, the rapid adaptation of non-native species to their new environments; and 2) the relevance of neutral genetic diversity as a proxy for adaptive genetic diversity (Sax *et al.*, 2007; Dlugosch & Parker, 2008; Dormontt *et al.*, 2011). The emergence of -omics technologies (Hohenlohe *et al.*, 2010; Rius & Darling, 2014) may be helpful in the future for a better characterization of the genomic architecture of the traits under selection as well as detailed investigations of how recombination is operating between genetic lineages in contact after their introductions.

18.4.2 Linking Hybridization and Invasiveness

Hybridization can be viewed as an extreme case of genetic admixture extended to different species when they are not fully reproductively isolated. Such events are common in plants, and hybridization has long been investigated as the main mechanism explaining their invasiveness (Abbott, 1992; Ellstrand, 2000). It has been studied more recently in invasive animals (but see Hedrick, 2013, and references therein).

Hybridization may occur even between highly divergent cryptic species, as exemplified by the successful matings observed between *Ciona robusta* and *Ciona intestinalis*; these two tunicate species were shown to hybridize in the English Channel, a sympatric area where *Ciona robusta* was putatively introduced recently (Nydam & Harrison, 2010). Introgression of *Carcinus maenas* into *C. aestuarii* in Japan, recently documented by Darling (2011), also shed light on the importance of these processes in the history of the introduction. In this specific case, the most parsimonious explanation of the observed genetic patterns, especially a similar genetic diversity in all the studied populations and no genetic structure between them, was a single introduction event into Japan from a source where hybridization between the two species occurred earlier. Adaptive introgression was recently emphasized as a major mechanism by which hybridization may favor the establishment success of an introduced species with potential risks for native species (Fitzpatrick *et al.*, 2010). Adaptive introgression is defined by the transfer of native species alleles into the genome of the introduced species, thus providing the latter with genes that are locally-adapted. Such processes may lead to a mosaic genome made of native and introduced species genes. Like for genetic admixture and selection on standing genetic variation, next-generation sequencing approaches can be useful to examine the way genes are transferred from one species to another, and how the whole genome architecture is modified in the donor and recipient species. To what extent hybridization is widespread in nature and may facilitate invasions is one important direction for future research on invasions to take (Dormontt *et al.*, 2011).

18.4.3 Examining Species Interactions

Once a non-native species has been established, it becomes part of the local community and thus interacts with native species. In particular, it enters the food web both as a potential consumer and as potential prey. Molecular methods are one of the existing approaches available to decipher the trophic links within ecosystems (e.g. King *et al.*, 2008; Traugott *et al.*, 2013). As such, various DNA-based approaches were used to assess the trophic role of aliens, from PCR with species-specific primers to identify target prey (Sheppard *et al.*, 2004; Gorokhova, 2006) to barcoding-like approaches (Kasper *et al.*, 2004). In this latter example, the inventory of prey of one native and one non-native social wasp by molecular identification (16S mtDNA sequences) allowed the determination of prey overlap between the two species. As in all other potential applications, metabarcoding approaches (see 18.2.1) may enhance the assessment of the diet of non-native species, and then their role into the food web (e.g. Pompanon *et al.*, 2012).

DNA barcoding and metabarcoding may also help elucidate the role of parasites in the introduction and invasion processes (Roy & Lawson Handley, 2012), allowing the identification of parasites and pathogens of aliens. For example, 18S rDNA sequences allowed the identification of larvae of the sea anemone *Edwardsiella*, a parasite of the highly invasive marine ctenophore *Mnemiopsis leidyi* in the Northeast Atlantic (Selander *et al.*, 2010). Parasitic anemone larvae are common in the native habitat, suggesting that this may result from a co-introduction. However, the discriminatory power of 18S is not sufficient to identify sea anemones at the species level, so the parasite may also be a local species (Selander *et al.*, 2010). This first occurrence of an endoparasite in *M. leidyi* in its introduced area might help control its establishment and spread, and has been proposed as a biological control agent.

18.4.4 Monitoring Community Diversity and Structure

Invasions by non-native species greatly impact biodiversity, sometimes with drastic consequences on the structure and functioning of the recipient communities, like community displacement or extinction of native species (Nichols *et al.*, 1990; Simberloff *et al.*, 2013). Documenting the changes in the composition of a community after one alien species has been introduced requires temporal surveys. In this context, molecular tools may help identify the NIS as reported in section 18.2.1. In some cases, both alien and native species may be difficult to identify, and both may benefit from molecular identification tools (e.g. soil and marine sediment meiofauna, pelagic protists, insect larvae). A few multiple-taxa surveys using DNA barcoding were applied to soil fauna (Porco *et al.*, 2013) and lepidopteran communities (deWaard *et al.*, 2009), but such approaches are time-consuming and expensive. For example, in the latter study, 190 species, including 31 non-natives, were identified from more than 900 barcoded individual insects. Next-Generation Sequencing technologies would allow us to overcome these drawbacks, being time- and cost-efficient. They offer the possibility of assessing the composition of the whole community through DNA metabarcoding (Cristescu, 2014), including rare species (Pochon *et al.* 2013; Zhan *et al.* 2013). In addition to species richness, DNA metabarcoding may also allow estimation of the relative abundances of different taxa (Porazinska *et al.*, 2010; Hajibabaei *et al.*, 2011; Comtet *et al.*, 2015), even if it may sometimes fail (Porazinska *et al.*, 2009), and the development of new technologies would certainly help improve these estimates. DNA metabarcoding will thus help with monitoring of the changes that occur after the introduction of new species, or after the implementation of management procedures.

18.5 Acknowledgements

We are grateful to João Canning-Clode for inviting us to write this book chapter following stimulating discussions through various research projects. Some ideas developed in this chapter benefited from insights and outcomes from results obtained from the Interreg IVa Marinexus and ANR HySea (ANR-12-BSV7-0011) programmes.

In a nutshell: about DNA-based studies of non-native species

- In the last two decades, molecular tools have become very popular for addressing questions relating to biological invasions.
- Knowing earlier is better: DNA-based identifications are critical for early and accurate detection of non-native species.
- An ever-increasing number of non-native cryptic species are recognized; revealing these cryptic invasions is a prerequisite to trace back introduction pathways.
- When studying colonization and spread by non-native species, particular care has to be given to sampling design and the statistical framework. The study design is guided by the question addressed (e.g. assignment tests to determine the geographic origin require a comprehensive coverage of all putative sources). New statistical frameworks (e.g. Approximate Bayesian Computations, Maximum-likelihood methods) have to be considered.
- The success in tracing back introduction processes, particularly for assigning sources, depends on the time elapsed since the introduction (the shorter the better) because of post-introduction changes in the genetic composition of native and invasive populations. In this context, the knowledge of the life history traits of the NIS is also crucial.
- Despite their inherent limitations, DNA-based studies shed light on introduction pathways and processes; for instance, by showing that the same NIS can invade different regions through different routes and vectors.
- DNA-based studies have shown that introduced populations can be genetically as or even more diversified than native ones. This underlined the importance of propagule pressure and recurrent events in biological invasions.
- Genetic admixture between different genetic pools has been documented in non-native species. Such a process may lead to evolutionary novelties in sexually reproducing species.
- Hybridization between native and non-native species could facilitate adaptive introgression and thus long-term establishment of the invader.
- Evolutionary outcomes of invasions in both non-native and native species have yet to be examined in detail, particularly the effects of genetic admixture at the species level and hybridization processes.
- Next-Generation Sequencing technologies offer new opportunities to study biological invasions, from the development of huge numbers of polymorphic markers for population genetics, to biodiversity assessments through DNA metabarcoding. New statistical and theoretical frameworks are, however, required.

18.6 Bibliography

Abbott, R.J. (1992). Plant invasions, interspecific hybridization and the evolution of new plant taxa. Trends in Ecology and Evolution, 7 (12), 401-405.

Armstrong, K. (2010). DNA barcoding: a new module in New Zealand's plant biosecurity diagnostic toolbox. EPPO Bulletin, 40, 91-100.

Armstrong, K.F. & Ball, S.L. (2005). DNA barcodes for biosecurity: invasive species identification. Philosophical Transactions of the Royal Society of London B. Biological Sciences, 360, 1813-1823.

Avise, J.C. (2000). Phylogeography: the history and formation of species. Cambridge, MA, Harvard: University Press.

Avise, J.C., Arnold, J., Ball, R.M., Bermingham, E., Lamb, T., Neigel, J.E., Reeb, C.A. & Saunders, N.C. (1987). Intraspecific phylogeography: the mitochondrial DNA bridge between population genetics and systematics. Annual Review of Ecology Evolution and Systematics, 18, 489–522.

Barrett, R.D. & Schluter, D. (2008). Adaptation from standing genetic variation. Trends in Ecology and Evolution, 23, 38-44.

Beaumont, M.A. & Balding, D.J. (2004). Identifying adaptive genetic divergence among populations from genome scans. Molecular Ecology, 13 (4), 969-980.

Bierne, N., Welch, J., Loire, E., Bonhomme, F. & David, P. (2011). The coupling hypothesis: why genome scans may fail to map local adaptation genes. Molecular Ecology, 20 (10), 2044-2072.

Bishop, J.D.D., Roby, C., Yunnie, A.L.E., Wood, C.W., Lévêque, L., Turon, X. & Viard, F. (2013). The Southern Hemisphere ascidian *Asterocarpa humilis* is unrecognised but widely established in NW France and Great Britain. Biological Invasions, 15 (2), 253-260.

Bock, D.G., MacIsaac, H.J. & Cristescu, M.E. (2012). Multilocus genetic analyses differentiate between widespread and spatially restricted cryptic species in a model ascidian. Proceedings of the Royal Society of London, B, 279, 2377-2385.

Bott, N.J., Ophel-Keller, K.M., Sierp, M.T., Herdina, Rowling, K.P., McKay, A.C., Loo, M.G.K., Tanner, J.E. & Deveney, M.R. (2010). Toward routine, DNA-based detection methods for marine pests. Biotechnology Advances, 28, 706-714.

Briski, E., Cristescu, M.E., Bailey, S.A. & MacIsaac, H.J. (2011). Use of DNA barcoding to detect invertebrate invasive species from diapausing eggs. Biological Invasions, 13, 1325-1340.

Broquet, T. & Petit, E.J. (2009). Molecular estimation of dispersal for ecology and population genetics. Annual Review of Ecology Evolution and Systematics, 40, 193-216.

Callahan, A.G., Deibel, D., McKenzie, C.H., Hall, J.R. & Rise, M.L. (2010). Survey of harbours in Newfoundland for indigenous and non-indigenous ascidians and an analysis of their cytochrome c oxidase I gene sequences. Aquatic Invasions, 5 (1), 31-39.

Ciosi, M., Miller, N.J., Toepfer, S., Estoup, A. & Guillemaud, T. (2011). Stratified dispersal and increasing genetic variation during the invasion of Central Europe by the western corn rootworm, *Diabrotica virgifera virgifera*. Evolutionary Applications, 4 (1), 54-70.

Comtet, T., Sandionigi, A., Viard, F. & Casiraghi, M. (2015). DNA (meta)barcoding of biological invasions: a powerful tool to elucidate invasion processes and help managing aliens. Biological Invasions, 17 (3), 905-922

Cornuet, J.-M., Piry, S., Luikart, G., Estoup, A. & Solignac, M. (1999). New methods employing multilocus genotypes to select or exclude populations as origins of individuals. Genetics, 153 (4), 1989-2000.

Cornuet, J.M., Santos, F., Beaumont, M.A., Robert, C.P., Marin, J.M., Blading, D.J., Guillemaud, T. & Estoup, A. (2008). Inferring population history with DIY ABC: a user-friendly approach to approximate Bayesian computation. Bioinformatics, 24, 2713-2719.

Cristescu, M.E. (2014). From barcoding single individuals to metabarcoding biological communities: towards an integrative approach to the study of global biodiversity. Trends in Ecology & Evolution, 29 (10), 566-571.

Cross, H.B., Lowe, A.J. & Gurgel, C.F.D. (2011). DNA barcoding of invasive species. In D.M. Richardson (Ed.) Fifty years of invasion ecology: the legacy of Charles Elton (pp.289-299). Chichester: Blackwell Publishing, Ltd.

Daehler, C.C. (1998). Variation in self-fertility and the reproductive advantage of self-fertility for an invading plant (*Spartina alternifora*). Evolutionary Ecology, 12, 553-568.

Darling, J.A. (2011). Interspecific hybridization and mitochondrial introgression in invasive *Carcinus* shore crabs. PloS ONE, 6 (3), e17828.
Darling, J.A. & Blum, M.J. (2007). DNA-based methods for monitoring invasive species: a review and prospectus. Biological Invasions, 9 (7), 751-765.
Darling, J.A. & Tepolt, C.K. (2008). Highly sensitive detection of invasive shore crab (*Carcinus maenas* and *Carcinus aestuarii*) larvae in mixed plankton samples using polymerase chain reaction and restriction fragment length polymorphisms (PCR-RFLP). Aquatic Invasions, 3 (2), 141-152.
Darling, J.A. & Mahon, A.R. (2011). From molecules to management: adopting DNA-based methods for monitoring biological invasions in aquatic environments. Environmental Research, 111, 978-988.
Davies, N., Villablance, F.X. & Roderick, G.K. (1999a). Bioinvasions of the medfly *Ceratitis capitata*: source estimation using DNA sequences at multiple intron loci. Genetics, 153, 351-360.
Davies, N., Villablanca, F.X. & Roderick, G.K. (1999b). Determining the source of individuals: multilocus genotyping in nonequilibrium population genetics. Trends in Ecology and Evolution, 14 (1), 17-21.
Dejean, T., Valentini, A., Miquel, C., Taberlet, P., Bellemain, E. & Miaud, C. (2012). Improved detection of an alien invasive species through environmental DNA barcoding: the example of the American bullfrog *Lithobates catesbeianus*. Journal of Applied Ecology, 49, 953-959.
deWaard, J.R., Landry, J.-F., Schmidt, C., Derhousoff, J., McLean, J.A. & Humble, L.M. (2009). In the dark in a large urban park: DNA barcodes illuminate cryptic and introduced moth species. Biodiversity and Conservation, 18 (14), 3825-3839.
Dlugosch, K.M. & Parker, I.M. (2008). Founding events in species invasions: genetic variation, adaptive evolution, and the role of multiple introductions. Molecular Ecology, 17 (1), 431-449.
Dormontt, E.E., Lowe, A.J. & Prentis, P.J. (2011). Is rapid evolution important in successful invasions? In D.M. Richardson (Ed.), Fifty years of invasion ecology - The legacy of Charles Elton (pp. 175-193). Chichester: Blackwell Publishing Ltd.
Dupont, L., Ellien, C. & Viard, F. (2007). Limits to gene flow in the slipper limpet *Crepidula fornicata* as revealed by microsatellite data and a larval dispersal model. Marine Ecology-Progress Series, 349, 125-138.
Eckert, C.G., Manicacci, D. & Barrett, S.C.H. (1996). Genetic drift and founder effect in native *versus* introduced populations of an invading plant, *Lythrum salicaria* (Lythraceae). Evolution, 50 (4), 1512-1519.
Egan, S.P., Barnes, M.A., Hwang, C.-T., Mahon, A.R., Feder, J.L., Ruggiero, S.T., Tanner, C.E. & Lodge, D.M. (2013). Rapid invasive species detection by combining environmental DNA with Light Transmission Spectroscopy. Conservation Letters, 6 (6), 402-409.
Ellstrand, N. (2000). Hybridization as a stimulus for the evolution of invasiveness in plants. Proceedings of the National Academy of Sciences, USA, 97 (13), 7043-7050.
Elton, C.S. (1958). The ecology of invasion by animals and plants. London: Methuen.
Estoup, A. & Guillemaud, T. (2010). Reconstructing routes of invasion using genetic data: why, how and so what? Molecular Ecology, 19, 4113-4130.
Excoffier, L. & Ray, N. (2008). Surfing during population expansions promotes genetic revolutions and structuration. Trends in Ecology & Evolution, 23 (7), 347-351.
Excoffier, L., Foll, M. & Petit, R.J. (2009). Genetic consequences of range expansions. Annual Review of Ecology, Evolution, and Systematics, 40 (1), 481-501.
Facon, B., Pointier, J.-P., Jarne, P., Sarda, V. & David, P. (2008). High genetic variance in life-history strategies within invasive populations by way of multiple introductions. Current Biology, 18 (5), 363-367.
Ficetola, G.F., Miaud, C., Pompanon, F. & Taberlet, P. (2008). Species detection using environmental DNA from water samples. Biology Letters, 4, 423-425.

Fitzpatrick, B., Fordyce, J., Niemiller, M. & Reynolds, R. (2012). What can DNA tell us about biological invasions? Biological Invasions, 14 (2), 245-253.

Fitzpatrick, B.M., Johnson, J.R., Kump, D.K., Smith, J.J., Voss, S.R. & Shaffer, H.B. (2010). Rapid spread of invasive genes into a threatened native species. Proceedings of the National Academy of Sciences of the United States of America, 107 (8), 3606-3610.

Floyd, R., Abebe, E., Papert, A. & Blaxter, M. (2002). Molecular barcodes for soil nematode identification. Molecular Ecology, 11 (4), 839-850.

Geller, J.B., Darling, J.A. & Carlton, J.T. (2010). Genetic perspectives on marine biological invasions. Annual Review in Marine Sciences, 2, 401-427.

Geller, J.B., Walton, E.D., Grosholz, E.D. & Ruiz, G.M. (1997). Cryptic invasions of the crab *Carcinus* detected by molecular phylogeography. Molecular Ecology, 6, 901-906.

Geoffroy, A., Le Gall, L. & Destombe, C. (2012). Cryptic introduction of the red alga *Polysiphonia morrowii* Harvey (Rhodomelaceae, Rhodophyta) in the North Atlantic Ocean highlighted by a DNA barcoding approach. Aquatic Botany, 100, 67-71.

Goff, L.J., Liddle, L., Silva, P.C., Voytek, M. & Coleman, A.W. (1992). Tracing species invasion in *Codium*, a siphonous green alga, using molecular tools. American Journal of Botany, 79 (11), 1279-1285.

Goodman, S.J., Barton, N., Swanson, G., Abernethy, K. & Pemberton, J. (1999). Introgression through rare hybridization: a genetic study of hybrid zone between red and sika deer (genus *Cervus*) in Argyll, Scotland. Genetics, 152, 355-371.

Gorokhova, E. (2006). Molecular identification of the invasive cladoceran *Cercopagis pengoi* (Cladocera: Onychopoda) in stomachs of predators. Limnology and Oceanography: Methods, 4, 1-6.

Grulois, D., Lévêque, L. & Viard, F. (2011). Mosaic genetic structure and sustainable establishment of the invasive kelp *Undaria pinnatifida* within a bay (Bay of St-Malo, Brittany). Cahiers De Biologie Marine, 52 (4), 485-498.

Guillemaud, T., Beaumont, M.A., Ciosi, M., Cornuet, J.M. & Estoup, A. (2010). Inferring introduction routes of invasive species using approximate Bayesian computation on microsatellite data. Heredity, 104 (1), 88-99.

Hajibabaei, M., Shokralla, S., Zhou, H., Singer, G.A.C. &, Baird, D.J. (2011). Environmental barcoding: a next-generation sequencing approach for biomonitoring applications using river benthos. PloS ONE, 6 (4), e17497.

Hartl, D.L. & Clark, A.G. (1989). Principles of population genetics. Sunderland: Sinauer Associates Inc.

Harvey, J.B.J., Hoy, M.S. & Rodriguez, R.J. (2009). Molecular detection of native and invasive marine invertebrate larvae present in ballast and open water environmental samples collected in Puget Sound. Journal of Experimental Marine Biology and Ecology, 369, 93-99.

Harvey, J.B.J., Ryan, J.P., Marin, R., III, Preston, C.M., Alvarado, N., Scholin, C.A. & Vrijenhoek, R.C. (2012). Robotic sampling, *in situ* monitoring and molecular detection of marine zooplankton. Journal of Experimental Marine Biology and Ecology, 413, 60-70.

Hebert, P.D.N., Cywinska, A., Ball, S.L. & deWaard, J.R. (2003). Biological identifications through DNA barcodes. Proceedings of the Royal Society of London B, 270, 313-321.

Hedrick, P.W. (2000). Genetics of Populations. Boston: Jones and Bartlett Publishers, Inc.

Hedrick, P.W. (2013). Adaptive introgression in animals: examples and comparison to new mutation and standing variation as sources of adaptive variation. Molecular Ecology, 22 (18), 4606-4618.

Hickerson, M.J., Carstens, B.C., Cavender-Bares, J., Crandall, K.A., Graham, C.H., Johnson, J.B., Rissler, L., Victoriano, P.F. & Yoder, A.D. (2010). Phylogeography's past, present, and future: 10 years after. Molecular Phylogenetics and Evolution, 54 (1), 291-301.

Hohenlohe, P.A., Phillips, P.C. & Cresko, W.A. (2010). Using population genomics to detect selection in natural populations: key concepts and methodological considerations. International Journal of Plant Sciences, 171 (9), 1059-1071.

Holland, B.S. (2000). Genetics of marine bioinvasions. Hydrobiologia, 420, 63-71.
Jerde, C.L., Mahon, A.R., Chadderton, W.L. & Lodge, D.M. (2011). "Sight-unseen" detection of rare aquatic species using environmental DNA. Conservation Letters, 4 (2), 150-157.
Jolly, M.T., Viard, F., Gentil, F., Thiébaut, E. & Jollivet, D. (2006). Comparative phylogeography of two coastal polychaete tubeworms in North East Atlantic supports shared history and vicariant events. Molecular Ecology, 15, 1841-1855.
Kasper, M.L., Reeson, A.F., Cooper, S.J.B., Perry, K.D. & Austin, A.D. (2004). Assessment of prey overlap between a native (*Polistes humilis*) and an introduced (*Vespula germanica*) social wasp using morphology and phylogenetic analyses of 16S rDNA. Molecular Ecology, 13, 2037-2048.
Keller, S.R. & Taylor, D.R. (2008). History, chance and adaptation during biological invasion: separating stochastic phenotypic evolution from response to selection. Ecology Letters, 11 (8), 852-866.
King, R.A., Read, D.S., Traugott, M. & Symondson, W.O.C. (2008). Molecular analysis of predation: a review of best practice for DNA-based approaches. Molecular Ecology, 17, 947-963.
Kolbe, J.J., Glor, R.E., Rodriguez Schettino, L., Chamizo Lara, A. & Losos, J.B. (2004). Genetic variation increases during biological invasion by a Cuban lizard. Nature, 431, 177-181.
Lacoursière-Roussel, A., Bock, D.G., Cristescu, M.E., Guichard, F., Girard, P., Legendre, P. & McKindsey, C.W. (2012). Disentangling invasion processes in a dynamic shipping–boating network. Molecular Ecology, 21 (17), 4227-4241.
Lawson Handley, L.J., Estoup, A., Evans, D., Thomas, C., Lombaert, E., Facon, B., Aebi, A. & Roy, H. (2011). Ecological genetics of invasive alien species. BioControl, 56 (4), 409-428.
Le Goff-Vitry, M.C., Chipman, A.D. & Comtet, T. (2007). *In situ* hybridization on whole larvae: a novel method for monitoring bivalve larvae. Marine Ecology Progress Series, 343, 161-172.
Le Roux, J. & Wieczorek, A.M. (2009). Molecular systematics and population genetics of biological invasions: towards a better understanding of invasive species management. Annals of Applied Biology, 154, 1-17.
Mackie, J.A., Darling, J.A. & Geller, J.B. (2012). Ecology of cryptic invasions: latitudinal segregation among *Watersipora* (Bryozoa) species. Scientific Reports, 2, 871.
Maggs, C.A., Castilho, R., Foltz, D., Hensler, C., Jolly, M.T., Kelly, J., Olsen, J., Perez, K.E., Stam, W., Vainola, R., Viard, F. & Wares, J. (2008). Evaluating signatures of glacial refugia for North Atlantic marine organisms? Ecology, 89 (Supplement 11), S108–S122.
Manel, S., Gaggiotti, O.E. & Waples, R.S. (2005). Assignment methods: matching biological questions with appropriate techniques. Trends in Ecology & Evolution, 20 (3), 136-142.
McGlashan, D., Ponniah, M., Cassey, P. & Viard, F. (2008). Clarifying marine invasions with molecular markers: an illustration based on mtDNA from mistaken calyptraeid gastropod identifications. Biological Invasions, 10, 51-57.
Meyer, C.P. & Paulay, G. (2005). DNA barcoding: error rates based on comprehensive sampling. PLoS Biology, 3 (12), e422.
Mountfort, D., Rhodes, L., Broom, J., Gladstone, M. & Tyrrell, J. (2007). Fluorescent *in situ* hybridisation assay as a species-specific identifier of the northern Pacific seastar, *Asterias amurensis*. New Zealand Journal of Marine and Freshwater Research, 41, 283-290.
Muirhead, J.R., Gray, D.K., Kelly, D.W., Ellis, S.M., Heath, D.D. & Macisaac, H.J. (2008). Identifying the source of species invasions: sampling intensity *vs.* genetic diversity. Molecular Ecology, 17 (4), 1020-1035.
Nichols, F.H., Thompson, J.K. & Schemel, L.E. (1990). Remarkable invasion of San Francisco Bay (California, USA) by the Asian clam *Potamocorbula amurensis*. II. Displacement of a former community. Marine Ecology Progress Series, 66, 95-101.
Nydam, M.L. & Harrison, R.G. (2010). Introgression despite substantial divergence in a broadcast spawning marine invertebrate. Evolution, 65 (2), 429-442.

O'Foighil, D., Gaffner, P., Wilbur, A. & Hilbish, T. (1998). Mitochondrial cytochrome oxydase I gene sequences support an Asian origin for the Portuguese oyster *Crassostrea angulata*. Marine Biology, 131, 497-503.

Pérez-Portela, R., Turon, X. & Bishop, J. (2012). Bottlenecks and loss of genetic diversity: spatio-temporal patterns of genetic structure in an ascidian recently introduced in Europe. Marine Ecology Progress Series, 451, 93-105.

Phillips, B.L., Brown, G.P., Webb, J.K. & Shine, R. (2006). Invasion and the evolution of speed in toads. Nature, 439 (7078), 803-803.

Pilgrim, E.M., Blum, M.J., Reusser, D.A., Lee II, H. & Darling, J.A. (2013). Geographic range and structure of cryptic genetic diversity among Pacific North American populations of the non-native amphipod *Grandidierella japonica*. Biological Invasions, 15, 2415-2428.

Pochon, X., Bott, N.J., Smith, K.F. & Wood, S.A. (2013). Evaluating detection limits of next-generation sequencing for the surveillance and monitoring of international marine pests. PloS ONE, 8 (9), e73935.

Pompanon, F., Deagle, B.E., Symondson, W.O.C., Brown, D.S., Jarman, S.N. & Taberlet, P. (2012). Who is eating what: diet assessment using next generation sequencing. Molecular Ecology, 21, 1931-1950.

Porazinska, D.L., Giblin-Davis, R.M., Faller, L., Farmerie, W., Kanzaki, N., Morris, K., Powers, T.O., Tucker, A.E., Sung, W. & Thomas, W.K. (2009). Evaluating high-throughput sequencing as a method for metagenomic analysis of nematode diversity. Molecular Ecology Resources, *9*, 1439-1450.

Porazinska, D.L., Sung, W., Giblin-Davis, R.M. & Thomas, W.K. (2010). Reproducibility of read numbers in high-throughput sequencing analysis of nematode community composition and structure. Molecular Ecology Resources, 10, 666-676.

Porco, D., Decaëns, T., Deharveng, L., James, S.W., Skarżyński, D., Erséus, C., Butt, K.R., Richard, B. & Hebert, P.D.N. (2013). Biological invasions in soil: DNA barcoding as a monitoring tool in a multiple taxa survey targeting European earthworms and springtails in North America. Biological Invasions, 15, 899-910.

Posada, D. & Crandall, K. (2001). Intraspecific gene genealogies: trees grafting into networks. Trends in Ecology and Evolution, 16 (1), 37-45.

Pyšek, P., Hulme, P.E., Meyerson, L.A., Smith, G.F., Boatwright, J.S., Crouch, N.R., Figueiredo, E., Foxcroft, L.C., Jarošik, V., Richardson, D.M., Suda, J. & Wilson, J.R.U. (2013). Hitting the right target: taxonomic challenges for, and of, plant invasions. AoB PLANTS, 5, plt042; doi:10.1093/aobpla/plt042.

Ratnasingham, S. & Hebert, P.D.N. (2007). BOLD: The Barcode of Life Data System (www.barcodinglife.org). Molecular Ecology Notes, 7, 355-364.

Richardson, D.M. (Ed.) (2011). Fifty years of Invasion Ecology - The legacy of Charles Elton. Chichester: Blackwell Publishing Ltd.

Riquet, F., Daguin-Thiébaut, C., Ballenghien, M., Bierne, N. & Viard, F. (2013). Contrasting patterns of genome-wide polymorphism in the native and invasive range of the marine mollusc *Crepidula fornicata*. Molecular Ecology, 22 (4), 1003-1018.

Rius, M. & Darling, J.A. (2014). How important is intraspecific genetic admixture to the success of colonising populations? Trends in Ecology & Evolution, 29 (4), 233-242.

Rius, M., Turon, X., Ordonez, V. & Pascual, M. (2012). Tracking invasion histories in the sea: facing complex scenarios using multilocus data. PloS ONE, 7 (4), e35815

Rius, M., Turon, X., Bernard, G., Volckaert, F. & Viard, F. (2015). Marine invasion genetics: from spatio-temporal patterns to evolutionary outcomes. Biological Invasions, 17 (3): 869-885

Rohfritsch, A., Bierne, N., Boudry, P., Heurtebise, S., Cornette, F. & Lapegue, S. (2013). Population genomics shed light on the demographic and adaptive histories of European invasion in the Pacific oyster, *Crassostrea gigas*. Evolutionary Applications, 6 (7), 1064-1078.

Roman, J. (2006). Diluting the founder effect: cryptic invasions expand a marine invader's range. Proceedings of the Royal Society B, *273* (1600), 2453-2459.
Roman, J. & Darling, J.A. (2007). Paradox lost: genetic diversity and the success of aquatic invasions. Trends in Ecology & Evolution, 22 (9), 454-464.
Roy, H.E. & Lawson Handley, L.-J. (2012). Networking: a community approach to invaders and their parasites. Functional Ecology, 26, 1238-1248.
Saltonstall, K. (2002). Cryptic invasion by a non-native genotype of the common reed, *Phragmites australis*, into North America. Proceedings of the National Academy of Sciences, USA, 99 (4), 2445-2449.
Saunders, G.W. (2009). Routine DNA barcoding of Canadian Gracilariales (Rhodophyta) reveals the invasive species *Gracilaria vermiculophylla* in British Columbia. Molecular Ecology Resources, 9 (Suppl. 1), 140-150.
Sax, D.F., Stachowicz, J.J., Brown, J.H., Bruno, J.F., Dawson, M.N., Gaines, S.D., Grosberg, R.K., Hastings, A., Holt, R.D., Mayfield, M.M., O'Connor, M.I. & Rice, W.R. (2007). Ecological and evolutionary insights from species invasions. Trends in Ecology & Evolution, 22 (9), 465-471.
Selander, E., Møller, L.F., Sundberg, P. & Tiselius, P. (2010). Parasitic anemone infects the invasive ctenophore *Mnemiopsis leidyi* in the North East Atlantic. Biological Invasions, 12, 1003-1009.
Sheppard, S.K., Henneman, M.L., Memmott, J. & Symondson, W.O.C. (2004). Infiltration by alien predators into invertebrate food webs in Hawaii: a molecular approach. Molecular Ecology, 13, 2077-2088.
Shigesada, N. & Kawasaki, K. (1997). Biological Invasions: Theory and Practice. Oxford: Oxford University Press.
Shigesada, N., Kawasaki, K. & Takeda, Y. (1995). Modeling stratified diffusion in biological invasions. American Naturalist, 146, 229–251.
Shine, R. (2012) Invasive species as drivers of evolutionary change: cane toads in tropical Australia. Evolutionary Applications, 5 (2), 107-116.
Shokralla, S., Spall, J.L., Gibson, J.F. & Hajibabaei, M. (2012). Next-generation sequencing technologies for environmental DNA research. Molecular Ecology, 21, 1794-1805.
Simberloff, D., Martin, J.-L., Genovesi, P., Maris, V., Wardle, D.A., Aronson, J., Courchamp, F., Galil, B., García-Berthou, E., Pascal, M., Pyšek, P., Sousa, R., Tabacchi, E. & Vilà, M. (2013). Impacts of biological invasions: what's what and the way forward. Trends in Ecology & Evolution, 28 (1), 58-66.
Simon-Bouhet, B., Garcia-Meunier, P. & Viard, F. (2006). Multiple introductions promote range expansion of the mollusc *Cyclope neritea* (Nassariidae) in France: evidence from mitochondrial sequence data. Molecular Ecology, 15, 1699-1711.
Smale, D.A. & Childs, S. (2012). The occurrence of a widespread marine invader, *Didemnum perlucidum* (Tunicata, Ascidiacea) in Western Australia. Biological Invasions, 14, 1325-1330.
Stone, G.N. & Sunnucks, P. (1993). Genetic consequences of an invasion through a patchy environment - the cynipid gallwasp *Andricus quercuscalicis* (Hymenoptera : Cynipidae). Molecular Ecology, 2, 251-268.
Suarez, A.V., Tsutsui, N.D., Holway, D.A. & Case, T.J. (1999). Behavorial and genetic differentiation between native and introduced populations of the Argentine ant. Biological Invasions, 1, 43-53.
Taberlet, P., Coissac, E., Pompanon, F., Brochmann, C. & Willerslev, E. (2012). Towards next-generation biodiversity assessment using DNA metabarcoding. Molecular Ecology, 21, 2045-2050.
Traugott, M., Kamenova, S., Ruess, L., Seeber, J. & Plantegenest, M. (2013). Empirically characterising trophic networks: what emerging DNA-based methods, stable isotope and fatty acid analyses can offer. Advances in Ecological Research, 49, 177-224.
Tsutsui, N.D., Suarez, A.V., Holway, D.A. & Case, T.J. (2000). Reduced genetic variation and the success of an invasive species. Proceedings of the National Academy of Sciences of the United States of America, 97 (11), 5948-5953.

Viard, F., Ellien, C. & Dupont, L. (2006). Dispersal ability and invasion success of *Crepidula fornicata* in a single gulf: insights from genetic markers and larval-dispersal models. Helgoland Marine Resarch, 60 (2), 144-152.

Voisin, M., Engel, C. & Viard, F. (2005). Differential shuffling of native genetic diversity across introduced regions in a brown alga: aquaculture vs. maritime traffic effects. Proceedings of the National Academy of Sciences of the United States of America, 102 (15), 5432-5437.

Weir, B.S. (1990). Genetic data analysis. Sunderland: Sinauer Associates Inc.

Whitlock, M.C. & McCauley, D.E. (1999). Indirect measures of gene flow and migration: $Fst \neq 1/(4Nm+1)$. Heredity, 82, 117-125.

Wiemers, M. & Fiedler, K. (2007). Does the DNA barcoding gap exist? – a case study in blue butterflies (Lepidoptera: Lycaenidae). Frontiers in Zoology, 4, 8.

Williamson, M. (1996). Biological invasions. London, Chapman & Hall.

Wilson, A.B., Naish, K.-A. & Boulding, E.G. (1999). Multiple dispersal strategies of the invasive quagga mussel (*Dreissena bugensis*) as revealed by microsatellite analysis. Canadian Journal of Fisheries and Aquatic Sciences, 56, 2248-2261.

Wilson, J.R.U., Dormontt, E.E., Prentis, P.J., Lowe, A.J. & Richardson, D.M. (2009). Something in the way you move: dispersal pathways affect invasion success. Trends in Ecology and Evolution, 24 (3), 136-144.

Xu, C.-Y., Julien, M.H., Fatemi, M., Girod, C., Van Klinken, R.D., Gross, C.L. & Novak, S.J. (2010). Phenotypic divergence during the invasion of *Phyla canescens* in Australia and France: evidence for selection-driven evolution. Ecology Letters, 13 (1), 32-44.

Zhan, A., MacIsaac, H.J. & Cristescu, M.E. (2010). Invasion genetics of the *Ciona intestinalis* species complex: from regional endemism to global homogeneity. Molecular Ecology, 19 (21), 4678-4694.

Zhan, A., Hulák, M., Sylvester, F., Huang, X., Adebayo, A.A., Abbott, C.L., Adamowicz, S.J., Heath, D.D., Cristescu, M.E. & MacIsaac, H.J. (2013). High sensitivity of 454 pyrosequencing for detection of rare species in aquatic communities. Methods in Ecology and Evolution, 4, 558-565.

Theo Light, Peter Moyle

19 Assembly Rules and Novel Assemblages in Aquatic Ecosystems

19.1 Introduction

Freshwater and estuarine ecosystems are the most altered and most invaded ecosystems worldwide (Leidy & Moyle, 1997; Dudgeon *et al.*, 2006; Rahel, 2002). Not surprisingly, many freshwater organisms are in danger of extinction. In the best documented group, fishes, an estimated 40% are seriously threatened worldwide (Moyle & Leidy, 1992; Helfman, 2007). In California alone, of 129 native species, 5% are extinct; 26% are in danger of extinction in the near future; and 26% are on a trajectory towards extinction if present trends continue (Moyle *et al.*, 2011). With the addition of environmental change as the result of climate change, the percentage of native fishes threatened with extinction rises to 82% (Moyle *et al.*, 2013). The causes of the declines are multiple and interactive, but boil down to three major factors: watershed (catchment) alteration, introductions of non-native species, and competition between humans and aquatic organisms for water, with humans winning. Increasingly, aquatic ecosystems contain non-native species that are integrated into them and may be favored by the altered environments (e.g., Marchetti *et al.*, 2001; Moyle & Marchetti, 2006). The severity of the problem indicates that saving aquatic biodiversity requires new approaches to conservation, incorporating native species into permanently altered waterways, heavily used by humans, that contain large numbers of non-native species. This new reality is labeled "novel ecosystems", where mixtures of native and non-native species form new assemblages living in irreversibly altered environments (Hobbs *et al.*, 2009, 2013; Moyle, 2013). We think that the best hope for many native species is integration into novel ecosystems that are intensely managed to favor native species. This requires understanding of how stable (persistent) communities within ecosystems are assembled and structured.

Moyle and Light (1996a) reviewed the nature of fish assemblages in highly invaded aquatic ecosystems, in the light of assembly theory. They proposed 12 generalizations ("rules") likely to regulate interactions among native and non-native species in forming new assemblages, six for the establishment stage of invasions of new species and six for the integration stage, in which more predictable (stable) assemblages become established (Table 19.1). In this chapter, we revisit these generalities, especially for the integration stage, to see how well they hold up under our expanded knowledge of the nature of contemporary fish assemblages that contain non-native species, which are increasingly recognized as being part of novel ecosystems (Moyle, 2013). The basic questions we address are:

1. Have advances in assembly theory been useful for understanding the nature of novel assemblages?
2. Are the original rules in Moyle and Light (1996a) useful in predicting the nature of novel assemblages, using a biogeographic perspective?
3. Can the driver/passenger conceptual model of interactions among habitat changes, invasions, and native species (Didham *et al.*, 2005) shed light on how new freshwater fish assemblages are created and maintained?
4. Can significant conservation of native fishes and other species be accomplished when the native species are part of novel assemblages/ecosystems?

We examine these questions in a series of case histories of freshwater and estuarine fish assemblages with significant components of non-native species. We conclude with a discussion of reconciliation ecology as a framework for conservation of novel aquatic ecosystems (Rosenzweig, 2003; Moyle, 2013).

Tab. 19.1: Some proposed rules for biotic invasions into freshwater and estuarine systems during two major stages of invasion, initial establishment and long-term integration into the existing fish community (from Moyle & Light, 1996a).

Establishment	Integration
Most invasions fail	Most successful invasions are accommodated without major community effects
All aquatic systems are invasible	Major community effects are most often observed where species richness is low
Piscivores and detritivore/omnivores are most likely to be successful in systems with low levels of human disturbance	Piscivores are most likely to alter invaded communities; omnivores/detritivores least likely do so
Any species with the right physiological and morphological traits can invade, given the opportunity	Long-term success depends on a close physiological match between the invader and the system being invaded
Successful invasions are most likely when native assemblages are depleted or disrupted	Long-term success is most likely in aquatic systems highly altered by human activity
Invasibility of aquatic systems is related to interactions among environmental variability, predictability, and severity	Invaders are much more likely to extirpate native species in aquatic systems with either extremely high or extremely low variability or severity

19.2 Assembly Theory and Novel Assemblages

Moyle and Light (1996a,b), following Lodge (1993), looked to community assembly theory (e.g., Keddy, 1992; Weiher *et al.*, 2011) as a possible source of theoretical and predictive insight for invasion biology. On the one hand, there has been broad adoption within the invasion biology literature of certain terms and concepts that are common (though not exclusive) to assembly theory. For example, the ideas of species pools, environmental and biotic filters, and empty niches/niche opportunities are commonly used to structure discussions of the invasion process (Shea & Chesson, 2002; Rahel & Olden, 2008; Leprieur *et al.*, 2008; Lapointe & Light, 2012), and the importance of appropriate null hypotheses and neutral assembly (Weiher *et al.*, 2011) has been emphasized with reference to propagule pressure (Colautti *et al.*, 2006; Daleo *et al.*, 2009). Nevertheless, community assembly theory and invasion biology have developed in the past 15-20 years seemingly in isolation, with major reviews in assembly theory (e.g., Weiher *et al.*, 2011) rarely citing the invasion biology literature and vice versa for reviews of invasion ecology (e.g., Lockwood *et al.*, 2013; but see Catford *et al.*, 2008). A notable exception is provided by Olden and colleagues (2010); they draw on both the assembly and invasion literature in order to highlight major research challenges in the conservation of fishes, including biotic homogenization and the emergence of novel assemblages. This work identifies two major challenges facing the study of novel fish assemblages: (1) the need to understand functional homogenization—the replacement of specialist native species by generalist non-natives; and (2) how to account for ongoing and future habitat and climate change in predicting and managing current and future assemblages. Still, it is not clear how assembly theory per se will contribute to addressing these challenges; Olden *et al.* (2010) conclude with a call for more basic research to improve taxonomic and biogeographic data as well as better data management and systems for global data sharing. One problem with assembly theory is that, despite reference to environmental filters, it largely fails to incorporate abiotic factors into community models. We argue that, in addition to basic research and data sharing, improved statistical modeling and adaptive ecosystem experiments that focus on the interactions among habitat alteration, invasions, and conservation outcomes are more likely to lead to advances.

19.3 Assembly Rules and Invasions: A Biogeographic Perspective

The 'rules' in Moyle and Light (1996a) were derived as generalizations from a number of detailed, more or less long-term studies of individual aquatic systems, mainly in California. We marshal similar evidence to reflect on them here. Given the difficulty of obtaining experimental data at the scales needed for testing assembly rules, especially those related to the integration stage, the other major source of data for testing invasion hypotheses is biogeographic. An increasing number of regional- (Gido *et al.*, 2004; Marchetti *et al.*, 2004; Maceda-Veiga *et al.*, 2010; Hermoso *et al.*, 2011; Lapointe *et al.*, 2012), continental- (Gido & Brown, 1999; Blanchet *et al.*, 2009), and global-scale (Leprieur *et al.*, 2008) studies of fish invasion patterns have been published in the last 20 years. These studies have examined patterns of non-native fish species richness, usually in relation to various measures of anthropogenic disturbance and natural environmental variability, and in some cases to measures of impact on native fish communities (Table 19.2). Here, we examine insights from some of these biogeographical studies as they relate to the rules of Table 19.1 and related hypotheses of invasion establishment and integration.

Tab. 19.2: Association of non-native species richness of watersheds with various measures of human activity and disturbance, as well as features of the natural environment, including native richness. + indicates a positive relationship, – a negative relationship, ns a non-significant relationship, and blank cells indicate the variable was not tested in that study. NPP = Net Primary Productivity, precip = precipitation.

	Urbanization	Agriculture	Hydrologic alteration	Population density	Affluence (GDP)	Colonization pressure	Native richness	Spatial heterogeneity	Other
Region									
Pennsylvania, USA[1]	+	+					ns	+ (basin area)	
Mississippi Basin, USA[2]			ns				+	+ (basin area)	
Great Plains, USA[3]			+				+	+ (basin area)	– (precip)
California, USA[4]	+	+	+				+	+ (basin area)	
Iberian Peninsula[5]	ns	ns	+						
Mid-Atlantic, USA[6]	+	marginal		+		+	+	+ (altitudinal range)	
Continental or biogeographic realm									
North America[2]			+				–, +[9]	+ (basin area)	– (precip)
Nearctic[7]	+			+	+		+	+ (altitudinal range)	ns (NPP)
Palearctic[7]	+			+	+		+	+ (basin area)	+ (NPP)
Australian[7]	+			+	+		ns	ns	ns (NPP)
Oriental[7]	ns			ns	ns		+	+ (basin area)	ns (NPP)
Neotropical[7]	ns			ns	ns		ns	+ (altitudinal range)	– (NPP)
Afrotropical[7]	ns			ns	ns		ns	+ (altitudinal range)	ns (NPP)
Global[8]	+			+	+		ns	+ (basin area, altitudinal range)	ns (NPP)

[1]Argent & Carline, 2004; [2]Gido & Brown, 1999; [3]Gido *et al.*, 2004; [4]Marchetti *et al.*, 2004, Light & Marchetti, 2007; [5]Hermoso *et al.*, 2011; [6]Lapointe *et al.*, 2012, Lapointe & Light, 2012; [7]Blanchet *et al.*, 2009; [8]Leprieur *et al.*, 2008; [9]negative for all introduced species, but positive for species exotic to North America.

19.3.1 Biotic Resistance

Biotic resistance, the idea that diverse communities should be more resistant to invasion (e.g., Moyle & Light, 1996a, Shea & Chesson, 2002), is the assembly-related invasion hypothesis that has received the most attention and broadest testing; however, this hypothesis has its roots in Elton (1958) and indeed Darwin (Ludsin & Wolfe, 2001). Moyle and Light (1996a) hypothesized that diversity would not hinder establishment ('all aquatic systems are invasible') but that it might limit impacts of invaders at the integration stage (Table 19.1). At the establishment stage, most studies of fish invasion patterns actually suggest a positive association between native and non-native diversity, particularly at regional scales (Table 19.2). Although limited smaller-scale studies, experimental studies, and modeling suggest a negative relationship between native and non-native diversity, supporting biotic resistance (Shea & Chesson, 2002; Levine *et al.*, 2004; Fridley *et al.*, 2007), most of this work has been done on plant invasions, and we are not aware of such small-scale or experimental studies for freshwater fish. The positive relationship between native and non-native diversity observed in many taxa at medium to large scales has been termed the 'biotic acceptance hypothesis' (Stohlgren *et al.*, 2006) and has been variously attributed to facilitation (Bulleri *et al.*, 2008), spatial heterogeneity (Davies *et al.*, 2005), or correlation of both native and non-native diversity with other extrinsic factors such as human population density or research effort (McKinney, 2001, Lapointe *et al.*, 2012). Overall, the biogeographic evidence suggests that high native diversity does not hinder establishment for fish invasions.

Relatively few studies have tested whether native diversity mitigates or limits the impacts of invasions at the integration stage (Ricciardi *et al.*, 2013). For California fishes (Light & Marchetti, 2007), multivariate models weakly suggest that successful invasions do not alter the strong positive association between native fish richness and number of declining fishes in a watershed. This would be expected as a null hypothesis: the more species, the more can potentially be declining. However, in one of the only experimental studies of biotic resistance for freshwater fishes, Carey and Wahl (2010) showed that greater native diversity in mesocosms limited the effects of introduced common carp *(Cyprinus carpio)* on native fish growth and zooplankton abundance, giving limited support to biotic resistance at the local scale.

19.3.2 Human Activity and Propagule/Colonization Pressure

The idea that human disturbance enhances invasibility of ecosystems and long-term persistence of invaders (Table 19.1) also has its roots in Elton (1958), and has been examined extensively with reference to establishment and long-term persistence of freshwater fishes. The effects of human activity in and around aquatic systems are many and various, and changes in fish communities are typically due to an interacting and sometimes contradictory suite of impacts, including flow regime changes that both increase and decrease natural variability, water abstraction, sedimentation and sediment reduction, water chemistry changes, drastic channel modification, loss of upstream-downstream connectivity, impoundments, removal or alteration of riparian vegetation, and many more. Most measures of human disturbance in medium- to larger-scale biogeographic studies are somewhat indirect (e.g., road densities), so physical and chemical changes to aquatic environments are tied to multiple interacting human activities. Such activities are expected to increase with human population density and affluence, and these are simultaneously linked to increasing propagule and colonization pressure (Leprieur *et al.*, 2008; Lockwood *et al.*, 2009), making it difficult to distinguish between habitat disturbance and increased introductions as explanations for invasion success. Most studies find that non-native species richness increases with one or more measures of human activity in a catchment, though the details vary among studies (Table 19.2). However, Blanchet *et al.*, (2009) found that measures of human disturbance, population density, and affluence were mainly important in the more developed regions of the world, while measures of the natural environment such as basin area and altitudinal range were the main predictors of non-native species richness in the less developed biogeographic realms (Table 19.2). Colonization pressure was positively linked to the number of non-natives established per watershed in the only study examining this separately from other human activity variables (Lapointe *et al.*, 2012; Table 19.2). We expect it also contributes to the links between human activity and non-native richness in other regions, but good data on introduction effort and failed invasions are lacking in most cases.

19.3.3 Are Non-Native Species the Drivers or Passengers of Native Species Declines?

Underlying several of the rules in Table 19.1 is the question of whether and how habitat degradation interacts with non-native species invasions to produce community change, particularly the loss of native species diversity. Since invasions, habitat degradation, and native species declines are correlated in most systems, it is not always clear whether non-native species are directly causing community change or are simply 'passengers' of more fundamental ecosystem changes driven by habitat changes and other human activities that disfavor native species and favor non-natives (Didham *et al.*, 2005; MacDougall & Turkington, 2005). Distinguishing between the so-called driver and passenger models (Figure 19.1) is difficult in freshwater systems because of the long history of habitat modification of most systems and ubiquitous presence of certain invaders, many of which are closely associated with modified habitats. For example, dam and reservoir construction is nearly always accompanied by deliberate introductions of reservoir-tolerant sport and forage fishes; when native fishes decline in abundance or disappear, it is generally not clear whether the altered habitat or invaders are responsible.

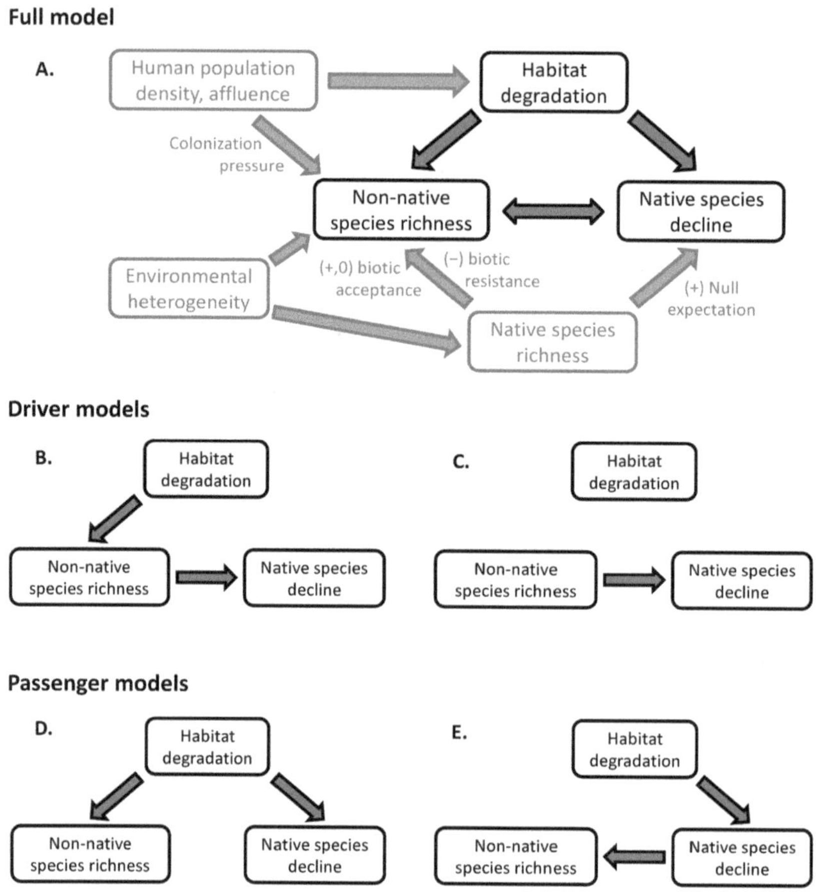

Fig. 19.1: Three general models of the relationships among habitat degradation, invasions, and native species losses (modified from Hermoso *et al.*, 2011; Light & Marchetti, 2007). Model A also includes external factors likely to affect the invasion process. The full or driver model (A or B) best fit the California and Iberian Peninsula data (Light & Marchetti, 2007; Hermoso *et al.*, 2011). Driver model C describes a situation where non-native species lead to native species loss even in the absence of significant habitat degradation (as in the Cosumnes River, California), while the passenger model (D) best fits the Pennsylvania data (Light, unpublished; analysis of data from Argent, 2000). An alternative conception of the driver model (E) posits that non-native species are indirect beneficiaries of habitat disturbance through its suppressing of native species (e.g., Didham *et al.*, 2005).

Light and Marchetti (2007) explicitly contrasted the driver and passenger models of fish invasions and native fish declines in California's watersheds using path analysis. Their analysis indicated that non-native fish richness in California's watersheds was a better predictor of native fish declines than measures of land use alteration or hydrologic modification; in the most parsimonious path model, the habitat modification variables were associated with native fish declines only indirectly, by way of non-native richness. The California fish data as a whole, then, best fit the driver hypothesis, in which non-native fish richness is directly linked to native fish declines, while the most important effect of habitat and hydrologic alteration is that it facilitates invasion by non-natives (Figure 19.1). This scenario is also supported by numerous case studies showing that, for example, altered habitats such as California's mid-elevation reservoirs initially supported abundant native fish populations (Sacramento sucker, *Catostomus occidentalais;* Sacramento pikeminnow, *Ptychocheilus grandis*; hardhead, *Mylopharodon conocephalus*), which only began to decline following the introduction of non-native predators such as largemouth bass, *Micropterus salmoides* (Moyle, 2002). The inverse scenario is playing out in the Cosumnes River (below), where the invasion of redeye bass (*M. coosae)* has led to extirpation or near-extirpation of most native species even in the absence of significant habitat or hydrologic alteration (Moyle *et al.*, 2003).

19.4 Case Histories

Guadiana River. The Guadiana River flows through one of the largest catchments on the Iberian Peninsula, and has been extensively dammed and diverted. The region has a Mediterranean climate, so the river has a flow regime with high seasonal and interannual flow variability (Hermoso *et al.*, 2011). As in California's drainages, the native fish fauna has low diversity (14 species), high endemicity (79%), high levels of conservation concern (64%), and is highly invaded (13 species). Hermoso *et al.*, (2011) made 170 fish collections by electrofishing in 2002–06, and measured 33 environmental variables at the site and basin scale. The detailed nature of this study allowed abundance of non-native individuals to be examined at each site, and response variables included both native species richness and a measure of native biotic integrity. Non-native fish abundance in each reach was the strongest predictor of both biotic integrity and native richness, while measures of habitat degradation were only weakly associated with native richness in structural equation models (SEM, a more general form of path analysis). Non-native abundance, however, was positively associated with the presence of and proximity to reservoirs. These results partly reflect that non-native fishes can invade upstream and downstream areas that have fairly natural characteristics (except perhaps flow regime), even if they are unable to establish self-sustaining populations, with negative effects on the native fishes. Hermoso *et al.*, (2011) therefore concluded that their study supports the driver conceptual model (Figure 19.1B).

Martis Creek. Martis Creek, California, is a small montane stream whose fish populations were studied for 30 years (Moyle & Vondracek, 1985; Kiernan & Moyle, 2012). The study monitored four stations that differed in habitat in a 3 km reach of the creek between its mouth and a flood control dam, constructed in 1972. The dam was operated so that it only diminished extreme high flow events. The creek also flowed along the edge of a large gravel mining operation that began about 5 years into the study. Over the years, eight native species and six non-natives were found in the creek, although only 5 natives and 2 non-natives were consistently members of the overall fish assemblage. The two non-native species, rainbow trout (*Oncorhynchus mykiss)* and brown trout (*Salmo trutta*), were typically the most abundant fishes, having replaced the native cutthroat trout (*O. clarki)* throughout its range decades previously.

Over the 30-year period, fish assemblages were persistent, but individual species exhibited marked year to year variability in biomass and numbers. The native fishes generally declined and only non-native rainbow trout increased in both density and biomass at all sites over time. Annual variation in fish assemblage structure was largely explained by extent of high flows in winter, amount of pool habitat, and extent of gravel beds. The contribution of non-native species to the total fish assemblage was inversely related to mean annual streamflow, maximum discharge, and the frequency of springtime high flow events.

Overall, the novel fish assemblage in Martis Creek behaved like an integrated assemblage with some intrinsic stability in structure, species with similar responses to flows, and an ability to resist invasions of new species, including those abundant in the upstream reservoir. The high variability in flows, both within and among years, was partly responsible for assemblage persistence; only species capable of living under such conditions could persist, although the reduction in peak flows caused by the dam may have permitted the non-native trout species to occur in larger numbers than they might otherwise. The assemblage appears to be structured following the predictions of Moyle and Light (1996a) and native fishes continue to coexist with the non-native trout species, despite the piscivorous nature of the brown trout. Only the once-common native species, Lahontan redside (*Richardsonius egregious),* became rare during the study, but the reason seems to be reduction in its favored pool habitat through lack of high flood flows; such flows maintain pools through scouring.

The model that best fits the structuring of the assemblage is the driver conceptual model (Figure 19.1B), although the initial invasion success of brown and rainbow trout did not rely on habitat degradation; both species can drive cutthroat trout to extinction through aggressive interactions, even in relatively pristine streams (Moyle, 2002). However, the slow change in habitat taking place from reduced extreme high flow events has apparently caused the decline of at least one native species, fitting the passenger conceptual model (Figure 19.1D).

Putah Creek. Lower Putah Creek, Yolo-Solano County, California, bears only superficial resemblance to the creek that once existed there, a creek with multiple channels that meandered through vast prairies and marshes (Kiernan *et al.*, 2012). Today it flows 30 km through a productive agricultural region; its single channel is deeply incised between levees constructed in the 19th century. Flows to the creek were largely cut off in 1957, following construction of two large dams that divert the water for agricultural and urban use. Fortuitously, flows remained in 1-2 km of stream below the lowest dam, to satisfy some riparian water rights; this reach served as a refuge for native fishes. Non-native fishes were present in the reservoirs upstream, in some large pools created by gravel mining downstream, in sewage effluent, and similar situations. A 1996 lawsuit resulted in flows being restored to the lower creek in 2000. The new water was released from the dam according to a flow regime designed to favor native fishes. The fish assemblages were monitored in multiple stations annually both before and after the new flow regime was instituted. The studies found that much of the creek became dominated by native fishes (Kiernan *et al.*, 2012). The expansion of native fishes was facilitated by creation of favorable spawning and rearing conditions (e.g., elevated springtime flows), cooler water temperatures, maintenance of minimum flows over the length of the creek, and displacement of non-native species by high-discharge events. Native fishes were especially favored by manipulating stream flows at biologically important times of the year. This only required a small increase in the total volume of water delivered downstream during most years.

During the study, 13 native and 22 non-native fish species were found. Non-native species continued to be found throughout the creek but were abundant only in the lowermost reaches where the water was warmer and flowing slowly through many large pond-like pools. The assemblage of fishes at each sampling station was fairly consistent through time, with highest variability in the assemblages dominated by non-native species. Overall, the flow regime and its related temperature regime were the 'master variables' that determined the structure, persistence, and resilience of fish assemblages. It appears that in most of the creek, abundant native fishes, which are adapted to the 'natural' flow regime, could then further suppress non-native species, which mostly spawn later and in warmer water. The fish assemblages are still novel because non-native species are always present and because the stream channel is highly altered, although it is increasingly managed to provide natural habitat for native fishes (e.g., shade, structure). The persistence of the assemblages dominated by native fish, however, *does* depend on the continued release of water from the dams in a designed flow regime. During a period of extended drought, when all but minimum flows can be eliminated, it is likely that non-native fishes will invade and dominate large sections of creek now devoted to native fishes.

It is worth noting that studies of riparian plants, terrestrial vertebrates, and butterflies along the creek find patterns similar to those observed in the fishes. Assemblages are made up of mixed native and non-native species, but natives tend to dominate more in upstream areas (M. Truan, unpublished data).

Overall, the Putah Creek data shows a fish assemblage that was originally generated by extreme habitat degradation, following the passenger model. Once a more natural flow regime was re-established, the passenger model continued to hold, as native species abundance seemed driven more by environmental conditions than by interactions with non-native species. However, in the lowermost reaches of the creek, where habitat recovery was less pronounced (lower flows, higher temperatures, etc.), non-native fishes remain abundant and native fishes have had a harder time becoming re-established, suggesting the driver model as being more explanatory.

Suisun Marsh. Suisun Marsh is part of the San Francisco Estuary (SFE), California. The SFE was dealt with extensively in Moyle and Light (1996a) because it is a highly altered and highly invaded system. Suisun Marsh, at 405 km^2, is the largest tidal marsh on the west coast of North America. While located in the middle of the SFE, so subject to regular incursions of salt water, considerable effort is made to keep the water as fresh as possible, to favor duck hunting clubs and wildlife areas (Moyle *et al.*, 2014). There is nevertheless considerable variability, within and between years, in salinity and temperature. Regular monthly sampling at fixed locations throughout the Marsh began in 1980 to develop an understanding of the fish assemblages in the Marsh, which contain a mixture of native and non-native species, as do the macroinvertebrate assemblages (Moyle *et al.*, 1986).

During the 34-year sampling period, 50 species of fish were collected, although only 28 were abundant enough to be considered part of the assemblage (Moyle *et al.*, 2012). Of these species, half were non-natives. During the study, two new non-native fish species invaded the Marsh, as did several macroinvertebrates. The most abundant fishes of the marsh were euryhaline and most of the rare species were stenohaline, sometimes becoming locally abundant when conditions (salinity in particular) stabilized in a small area. Over the period, the assemblage composition was fairly stable, although abundance ranks of species changed somewhat (Table 19.3). There was also considerable regional variability in assemblage composition in different parts of the Marsh, reflecting local environmental conditions (Matern *et al.*, 2002, Moyle *et al.*, 2012, Moyle unpublished data). Populations of native and non-native species tended to vary concordantly, although non-native species showed wider fluctuations in populations from year to year (Moyle, unpublished data). This concordance has persisted despite the invasion of the overbite clam (*Potamocorbula amurensis*), which has caused a major shift in the fish-oriented food web of the SFE from planktonic- to benthic-based (Moyle *et al.*, 2012).

Suisun Marsh is a good example of a system that fits the full conceptual model, with different non-native species being drivers and passengers. On the one hand, the major players in the fish assemblage are an equal mixture of native and non-native species; these species are all euryhaline and respond in similar ways to environmental variability in an environment that is highly altered, including altered tidal and riverine flow patterns and regimes. Invaders such as shimofuri goby (*Tridentiger bifasciatus*) can become very abundant, but have no apparent effect on other fishes, even ecologically and morphologically similar species such as prickly sculpin (*Cottus asper*). In contrast, the invasion of overbite clam has caused the decline of at least two native plankton-feeding fishes.

Cosumnes River. Moyle and Randall (1998) rated the Cosumnes River as one of the most important watersheds to protect in the Sierra Nevada in California because it is the region's largest river without a major dam. Although relatively unstudied at the time, it was assumed the natural flow regime of the river would favor assemblages of native fishes. A study of the fish assemblages, however, revealed that most of the river contained mainly non-native, predatory redeye bass (Moyle *et al.*,

2003). Native fishes continued to exist mainly in stream reaches above natural barriers to bass invasion. The headwaters were also found to contain mainly non-native brown trout and brook trout (*Salvelinus fontinalis*). Redeye bass were introduced into California streams because they are a game species that lives in similar streams in the Southwestern USA. They appear to have eliminated 4–5 species of native fish through predation and the ability of different life history stages to occupy different habitats (Moyle *et al.*, 2003). This example runs contrary to the other examples, because the bass had a major impact on a fish assemblage in a river with relatively low human disturbance. This example fits the driver model, only without habitat degradation driving species invasions (Figure 19.1D). Presumably one of the reasons for this was a close ecological match of the Cosumnes River to the native habitats of the bass.

Tab. 19.3: Top ten fish species caught in otter trawls in monthly samples in Suisun Marsh in the first ten years of study (1980-89) and ten-year period, 2004–2013. In 2004–2013, longfin smelt dropped from the top 10, reflecting an estuary-wide drop in abundance, while white catfish and black crappie became more abundant because a series of wetter years reduced overall salinities in the marsh. Species with asterisk (*) are non-native species. For scientific names see Matern *et al.* (2002). Moyle, unpublished data.

Rank	First ten years 1980-89	Last ten years 2003-14
1	Striped bass*	Striped bass*
2	Threespine stickleback	Splittail
3	Tule perch	Tule perch
4	Splittail	White catfish*
5	Longfin smelt	Yellowfin goby*
6	Prickly sculpin	Shimofuri goby*
7	Yellowfin goby*	Prickly sculpin
8	Common carp*	Threespine stickleback
9	Sacramento sucker*	Common carp*
10	Shimofuri goby*	Black crappie*

Pennsylvania streams. In contrast to California and the Iberian Peninsula, Eastern North America has relatively high native fish diversity and fewer non-natives. For example, the Susquehanna River drainage historically supported 117 fish species and has 33 non-natives, only a few of which are widespread (Snyder 2005). Argent and Carline (2004) examined a dataset of paired fish collections in 200 small streams (first through fourth order) in Pennsylvania, USA, with sites in the state's three major drainages: Delaware (34 sites), Susquehanna (93), and Ohio (72). The streams were sampled once in the 1950s through early 1970s, and a second time in the late 1970s–90s; land cover, but not measures of hydrologic modification, were available for the catchments containing each site. The original work examined an outcome of overall community change (as measured by the Jaccard coefficient of similarity between sampling periods) and found that fish community change was associated with urban/developed uses in the Delaware drainage and with agricultural land uses in the Susquehanna and Ohio drainages. A re-analysis of this dataset (Argent, 2000 & pers. comm.) allowed us to test several invasion hypotheses and compare the outcomes with our California results (Light, unpublished).

Pennsylvania stream sites averaged 2.0 (± 1.6, 1SD) non-native and 11.8 (± 5.5, 1SD) native fish species per site, aggregating across sampling periods. Between the two sampling periods, mean native richness declined by 1.3 species per site (95%CI: 0.8, 1.8), and the average non-stocked native fish lost 17% (95% CI: 10, 24) of its range. Meanwhile, non-native richness increased by 0.6 species per site (95% CI: 0.4, 0.8) and the most widespread non-natives increased their ranges by an average of 122% (95% CI: 63, 180). Most of the common non-natives were stocked sport fishes, mainly piscivores (Table 19.4), and nearly all sites contained at least one of these species. However, there was no significant association between increased non-native richness and any measure of native fish decline, even when the analysis was limited to the two more heavily invaded drainages (Susquehanna and Delaware). Total non-native richness was weakly associated with number of native species lost in univariate tests ($r = 0.19$, $P = 0.009$), but both variables were strongly associated with measures of watershed size, native diversity, and altered land uses (agriculture and urbanization), and in multivariate analyses the relationship disappeared. Path analysis indicated that while both non-native increases and native declines were associated with altered land use, there was no significant direct effect of non-native fish richness on native losses in the Pennsylvania dataset (Light, unpublished). Thus, in contrast to the situation in California and the Iberian Peninsula, the Pennsylvania study supports the passenger model, with both native species losses and non-native increases more directly tied to measures of land use change, population density, and habitat alteration than to one another (Figure 19.1D).

Fit with Moyle and Light (1996a). The 'rules' of Moyle and Light (1996a) proved to have only limited usefulness for the integration stage of invasions in these seven case histories (Table 19.5). The examples do show, however, the not-surprising result that successful invasions are most likely in assemblages with low numbers of species in highly altered systems. The effect of piscivores on native fish diversity in many of these systems is still poorly understood, although their impact has been strongly negative in the Cosumnes River and other California rivers (Moyle, 2002). On the other hand, no clear evidence exists that the broader distribution of stocked piscivores in recent fish collections in Pennsylvania has led to native fish declines beyond what can be explained by land use changes. This may reflect a fauna already adapted to the presence of many piscivorous fishes.

19.5 Discussion

Moyle and Light (1996a) noted that making predictions about aquatic invasions and their impacts requires good understanding of both the invader and the system being invaded. While such information is steadily improving, solid predictive models have not emerged, especially from assembly theory. This is partly because of the idiosyncratic nature of invasions and invaded systems, although stream fish assemblages in Mediterranean climate areas are becoming increasingly homogenized worldwide (Marr *et al.*, 2013), and many of the same widespread fish species are common invaders in Pennsylvania as well (Table 19.4). The generalities of Moyle and Light (1996a) still mostly hold, but they have not proven to be very useful as predictors, given all the idiosyncrasies. However, empirical and conceptual models are becoming increasingly useful for at least understanding patterns and process in aquatic invasions (e.g., Moyle & Marchetti, 2006; Light & Marchetti, 2007; Hermoso *et al.*, 2011).

The majority of invasion studies continue to be case studies of single systems. While these are still essential to build our knowledge base, we think that fundamental insights are most likely to be derived from a combination of long-term studies that follow the trajectories of native and non-native assemblages over time, along with larger scale biogeographic studies. The most useful scale for these larger studies, we believe, is the regional scale (e.g., Marchetti *et al.*, 2004; Lapointe *et al.*, 2012), across which native assemblages and human pressures are fairly comparable. Smaller scale studies may lack variation, while larger scale studies may have so much region-to-region variation as to obscure crucial processes and relationships (e.g., Blanchet *et al.*, 2009). Comparisons among regions with distinctly different natural environments, invasion histories, and levels or types of human disturbance will also be fertile ground for elucidating larger patterns (Blanchet *et al.*, 2009, this chapter).

If there is any universal truth to come out of the many studies of aquatic invasions, it is that the more a system is altered by humans, the more non-native species are likely to be abundant, and the more native species are likely to be in decline. Even here, the degree to which this is true depends on the system and the scale at which the phenomenon is being investigated (e.g., Table 19.2). There is growing evidence (e.g., Putah Creek example, above) that to some extent, non-native species can be controlled through environmental manipulations that favor native species, at least in streams. This of course requires both understanding of factors that regulate fish populations and the ability to manage streams intensively to create the desired conditions. But such management is encouraged by the growing realization that species invasions do not have to result in extinctions of native species, at least in the short term, and that novel assemblages, with mixtures of native and non-native species, can develop that have characteristics of 'natural' co-evolved assemblages.

Tab. 19.4: Number of sites (of 200 sampled streams) occupied by the most widespread non-native fishes and widespread native fishes showing the greatest declines in distribution in three drainage basins in Pennsylvania, USA (Argent, 2000; Argent & Carline, 2004). Many fish that are native in the Ohio River drainage, which is part of the Mississippi basin, are non-native in the Atlantic-coastal Susquehanna and Delaware River drainages. No native, non-stocked fishes increased their distributions by more than 26%. Basins: Del = Delaware, Sus = Susquehanna. Trophic and tolerance designations are for the Northeastern U.S., following Barbour *et al.* (1999): P = piscivore, G = generalist, H = herbivore, I = insectivore; tolerance (to sedimentation and other water quality reductions): I = intolerant, M = moderately tolerant, T = tolerant. For scientific names see Barbour *et al.* (1999). Time 1 = 1950-74, Time 2 = 1975-95.

Common name	Basin	Time 1	Time 2	Total	Increase/ decrease	Trophic designation	Tolerance designation
Widespread non-natives (10+ sites)							
brown trout	all	88	148	155	68%	P	I (M)1
rainbow trout	all	21	55	68	162%	P	I (M)
smallmouth bass	Del, Sus	21	36	40	71%	P	M
rock bass	Del, Sus	21	26	37	24%	P	M
bluegill	Del, Sus	8	25	29	213%	G	M
largemouth bass	Del, Sus	8	14	21	75%	P	M
green sunfish	Del, Sus	5	13	15	160%	G	T
common carp	all	4	12	14	200%	G	T
Widespread natives (10+ sites) showing the greatest decreases in range							
central stoneroller	all	82	56	86	-32%	H	T (M)
bluntnose minnow	all	63	26	67	-59%	G	T
fantail darter	Ohio, Sus	43	28	48	-35%	I	M
johnny darter	Ohio	39	27	47	-31%	I	M
redside dace	Ohio, Sus	25	14	29	-44%	I	I
spotfin shiner	Ohio, Sus	19	4	22	-79%	I	T (M)
pearl dace	all	18	5	21	-72%	G	M
satinfin shiner	Ohio, Sus	18	5	20	-72%	I	I
silverjaw minnow	Del, Sus	19	4	20	-79%	I	T (M)
blackside darter	Ohio	14	8	17	-43%	I	M
striped shiner	Ohio	15	0	15	-100%	I	T (M)
banded killifish	all	10	6	11	-40%	I	T
swallowtail shiner	Ohio, Sus	10	4	11	-60%	I	I
mimic shiner	Ohio	10	1	11	-90%	G	M (I)

[1] Tolerance designation in parentheses indicates most common U.S. designation outside of the Northeast region.

Tab. 19.5: Fit to 'rules' (Moyle & Light, 1996a) for integration for fish assemblages in the case studies discussed above. "Yes" indicates a good fit. "No" indicates a poor fit.

Integration	Guadiana River	Martis Creek	Putah Creek	Suisun Marsh	Cosumnes River	California drainages	Pennsylvania streams
Most successful invasions are accommodated without major community effects.[1]	No	Yes	Yes	Yes	No	No	Yes
Major community effects are most often observed where species richness is low.	Yes (14 natives)	Yes (6 natives)	Yes (12 natives)	Yes (15 natives)	Yes (5 natives)	Yes (2-12 natives per stream)	Yes (206 natives, 2–28 per site)
Piscivores are most likely to alter invaded communities; omnivores/ detritivores least likely do so.	Not tested	Yes	Yes	Yes	Yes	Not tested	No (most non-natives are piscivores)
Long-term success depends on a close physiological match between the invader and the system being invaded.	Not tested	Yes	Yes	Yes	Yes	Not tested	Yes[2]
Long-term success is most likely in aquatic systems highly altered by human activity.	Yes	Yes	Yes	Yes	No	Yes	Yes
Invaders are much more likely to extirpate native species in aquatic systems with either extremely high or extremely low variability or severity.[3]	Yes	Yes?	Yes?	Yes?	Yes	Yes	Yes?

[1] Predatory fishes such as brown trout and striped bass were often introduced before studies, so initial impacts are unknown.

[2] The most successful invaders are from nearby drainages (centrarchids) or from similar environments in Eurasia (brown trout, common carp).

[3] Example systems here would be classified as having moderate variability, with the possible exception of the Cosumnes River.

19.5.1 Conclusions

(1) *Have advances in assembly theory been useful for understanding the nature of novel assemblages?* Unfortunately, assembly theory has contributed little to our understanding of invasion biology of aquatic ecosystems. The rapid increase in knowledge of aquatic invasions, however, has led to better conceptual models and improved understanding of the idiosyncratic ways that non-native species affect ecosystems, including fish assemblages.
(2) *Are the original rules in Moyle and Light (1996a) useful in predicting the nature of novel assemblages, using a biogeographic perspective?* Most still hold up under improved knowledge of fish assemblages, but their utility for making predictions is limited.
(3) *Can the driver/passenger conceptual model of interactions among habitat changes, invasions, and native species (Didham et al., 2005) shed light on how new freshwater fish assemblages are created and maintained?* Limited application by us and others (Hermoso *et al.*, 2011) indicates that this model is useful for providing a basic understanding of how invasions and habitat degradation work together to create observed conditions, leading to novel ecosystems. It was useful, for example, in pointing out similarities and differences among our case histories. The key insight of this model is the importance of disentangling the correlated effects of invasions and habitat changes, and we highlight the utility of methods such as path analysis and SEM for distinguishing among competing models (e.g., Figure 19.1) when only observational data are available. However, future studies of systems actively managed to either directly remove non-native species or to return habitats (e.g., flow regimes) to a more natural state should continue to shed light on the relative roles of invaders as drivers or passengers of ecosystem change, particularly if these manipulations are designed and followed as the large-scale experiments they have the potential to be (Olden *et al.*, 2014).
(4) *Can significant conservation of native fishes and other species be accomplished when the native species are part of novel assemblages/ecosystems?* We think that this approach has considerable promise in making approaches to managing invasions and invaded systems more realistic. At the same time, it has to be recognized that some invaders are very harmful and cannot easily be part of a novel ecosystem that also supports a significant contingent of native species. Such invasions either need to be prevented or be subject to eradication or control programs during the early invasion stage. Since we often still cannot predict which invaders will be harmful in which systems, it is vital that we also continue to advocate and vigorously support prevention and early eradication of new invasions when these are feasible (Simberloff, 2013).

19.5.2 Management Implications

Freshwater ecosystems that are not heavily influenced by humans are increasingly rare, if they exist at all. Introduction of non-native species into aquatic systems by multiple pathways is a worldwide phenomenon that is not likely to slow down soon.

In addition, global climate change is further stressing aquatic ecosystems, often making them more invasible (Moyle *et al.*, 2013). Thus novel ecosystems—altered systems containing mixtures of native and non-native species—are increasingly going to be the focus of management. This reality makes it imperative to understand how biotic communities in fresh water are assembled and structured by both biotic and abiotic factors. This understanding must include knowing the idiosyncratic requirements of local environments and fishes. Our examples suggest that understanding the ecological requirements of native species in particular can allow environmental manipulations that favor them and discourage non-native species, or at least allow stable assemblages to develop in which native species are an integral part. Managers and ecologists have to admit that we are now in charge of assembling communities in many of our aquatic ecosystems. We wish that there was solid theory on which to base such management but so far it has not developed much beyond Moyle and Light (1996a, b), although several recent reviews (Catford *et al.*, 2009; Olden *et al.*, 2010; Ricciardi *et al.*, 2013) have the potential to help focus questions and research efforts. The growing attention being paid to novel ecosystems, however, suggests that help is on the way (Hobbs *et al.*, 2013). But there will be no substitute for local knowledge and dedication to conservation for creating or protecting conditions that favor native species. We favor the reconciliation ecology approach to management that works to integrate biodiversity conservation in human-dominated landscapes (Rosenzweig, 2003; Moyle, 2013), but urge managers not to give up on prevention and eradication when feasible.

In a nutshell: about DNA-based studies of non-native species

- Freshwater and estuarine ecosystems are highly altered, highly invaded, and have some of the highest rates of extinction and endangerment of native species worldwide. Along with global climate change, this means that conservation will increasingly take place in the context of novel assemblages in highly modified ecosystems.
- Although assembly theory seems to show promise for advancing understanding of these novel assemblages, we find that more progress is likely from basic research and improved modeling of patterns derived from large datasets.
- In particular, we believe that regional scale surveys, larger-scale comparative studies, and long-term, detailed studies of individual systems are needed. These show promise in advancing both our understanding of invasion dynamics and our ability to manage invaded systems to favor native species.
- Most invaders cannot be eliminated, few aquatic systems can be returned to a fully pristine state, and humans will continue to compete with aquatic species for water. Therefore, we favor a reconciliation ecology approach to management of freshwater systems.
- However, we also strongly advocate policies and management practices that reduce the rate of invasions, and, where possible, eradicate new and potentially harmful invaders.

19.6 Bibliography

Argent, D. G., & Carline, R. F. (2004). Fish assemblage changes in relation to watershed landuse disturbance. Aquatic Ecosystem Health & Management, 7, 101-114.

Argent, D. G., (2000). Fish distribution patterns in Pennsylvania: anthropogenic influences and the status of and rare fishes. Ph.D. Thesis. The Pennsylvania State University, University Park, PA.

Barbour, M. T., Gerritsen, J., Snyder, B. D., & Stribling, J. B. (1999). Rapid bioassessment protocols for use in streams and wadeable rivers. USEPA, Washington.

Blanchet, S., Leprieur, F., Beauchard, O., Staes, J., Oberdorff, T., & Brosse, S. (2009). Broad-scale determinants of non-native fish species richness are context-dependent. Proceedings of the Royal Society B: Biological Sciences, 276 (1666), 2385-2394.

Bulleri, F., Bruno, J. F., & Benedetti-Cecchi, L. (2008). Beyond competition: incorporating positive interactions between species to predict ecosystem invasibility. PLoS Biology, 6 (6), e162.

Carey, M. P., & Wahl, D. H. (2010). Native fish diversity alters the effects of an invasive species on food webs. Ecology, 91, 2965-2974.

Catford, J. A., Jansson, R., & Nilsson, C. (2009). Reducing redundancy in invasion ecology by integrating hypotheses into a single theoretical framework. Diversity and Distributions, 15, 22-40.

Colautti, R. I., Grigorovich, I. A., & MacIsaac, H. J. (2006). Propagule pressure: a null model for biological invasions. Biological Invasions, 8, 1023-1037.

Daleo, P., Alberti, J., & Iribarne, O. (2009). Biological invasions and the neutral theory. Diversity and Distributions, 15, 547-553.

Davies, K. F., Chesson, P., Harrison, S., Inouye, B. D., Melbourne, B. A., & Rice, K. J. (2005). Spatial heterogeneity explains the scale dependence of the native-exotic diversity relationship. Ecology, 86, 1602-1610.

Didham, R. K., Tylianakis, J. M., Hutchison, M. A., Ewers, R. M., & Gemmell, N. J. (2005). Are invasive species the drivers of ecological change? Trends in Ecology and Evolution, 20, 470-474.

Dudgeon, D., Arthington, A.H, Gessner, M.O., Kawabata Z-I., & Knowler, D.J., *et al.*, (2006). Freshwater biodiversity: importance, threats, status and conservation challenges. Biological Reviews, 81, 163–182.

Elton, C. S. (1958). The ecology of invasions by plants and animals. London: Methuen.

Fridley, J. D., Stachowicz, J. J., Naeem, S., Sax, D. F., Seabloom, E. W., Smith, M. D., *et al.*, (2007). The invasion paradox: reconciling pattern and process in species invasions. Ecology, 88, 3-17.

Gido, K. B., & Brown, J. H. (1999). Invasion of North American drainages by alien fish species. Freshwater Biology, 42, 387-399.

Gido, K. B., Schaefer, J. F., & Pigg, J. (2004). Patterns of fish invasions in the Great Plains of North America. Biological Conservation, 118, 121-131.

Helfman,, G. S. (2007). Fish Conservation. Covelo, CA: Island Press.

Hermoso, V., Clavero, M., Blanco-Garrido, F., & Prenda, J. (2011). Invasive species and habitat degradation in Iberian streams: an analysis of their role in freshwater fish diversity loss. Ecological Applications, 21, 175-188.

Hobbs, R.J., Higgs, E., & Harris, J.A. (2009). Novel ecosystems: implications for conservation and restoration. Trends in Ecology and Evolution, 24, 599-605.

Hobbs, R.J., Higgs, E.S., & Hall, C.M. (Eds.) (2013). Novel ecosystems: Intervening in the new ecological world order. Oxford: Wiley-Blackwell.

Keddy, P. A. (1992) Assembly and response rules: two goals for community ecology. Journal of Vegetation Science, 3, 157-64.

Kiernan, J. D., & Moyle, P. B. (2012). Flows, droughts, and aliens: factors affecting the fish assemblage in a Sierra Nevada, California, stream. Ecological Applications, 22, 1146-1161.

Kiernan, J.D., Moyle, P.B. & Crain, P.K. (2012). Restoring native fish assemblages to a regulated California stream using the natural flow regime concept. Ecological Applications, 22,1472-1482.
Lapointe, N. W. & Light, T. (2012). Landscape-scale determinants of non-native fish communities. Diversity and Distributions, 18, 282-293.
Lapointe, N. W., Thorson, J. T., & Angermeier, P. L. (2012). Relative roles of natural and anthropogenic drivers of watershed invasibility in riverine ecosystems. Biological Invasions, 14, 1931-1945.
Leidy, R. A. & Moyle, P.B. (1997). Conservation status of the world's fish fauna: an overview. In P. A. Fiedler and P. M. Karieva (Eds.), Conservation biology for the coming decade (pp.187-227). New York: Chapman and Hall.
Leprieur, F., Beauchard, O., Blanchet, S., Oberdorff, T., & Brosse, S. (2008). Fish invasions in the world's river systems: when natural processes are blurred by human activities. PLoS Biology, 6 (2), e28.
Levine, J. M., Adler, P. B., & Yelenik, S. G. (2004). A meta-analysis of biotic resistance to exotic plant invasions. Ecology Letters, 7 (10), 975-989.
Light, T., & Marchetti, M. P. (2007). Distinguishing between invasions and habitat changes as drivers of diversity loss among California's freshwater fishes. Conservation Biology, 21, 434-446.
Lockwood, J. L., Cassey, P., & Blackburn, T. M. (2009). The more you introduce the more you get: the role of colonization pressure and propagule pressure in invasion ecology. Diversity and Distributions, 15, 904-910.
Lockwood, J. L., Hoopes, M. F., & Marchetti, M. P. (2013). Invasion ecology. New York: John Wiley & Sons.
Lodge, D.M. (1993). Species invasions and deletions: community effects and responses to climate and habitat change. In P.M. Kareiva *et al.* (Eds.), Biotic interactions and global change (pp. 367-387). Sunderland, MA: Sinauer.
Ludsin, S. A., & Wolfe, A. D. (2001). Biological invasion theory: Darwin's contributions from The Origin of Species. Bioscience, 51, 780-789.
MacDougall, A. S., & Turkington, R. (2005). Are invasive species the drivers or passengers of change in degraded ecosystems? Ecology, 86, 42-55.
Maceda-Veiga, A., Monleon-Getino, A., Caiola, N., Casals, F. & De Sostoa, A. (2010). Changes in fish assemblages in catchments in north-eastern Spain: biodiversity, conservation status and introduced species. Freshwater Biology, 55, 1734–1746.
Marchetti M. P., Light, T., Feliciano. J., Armstrong, T., Hogan, Z, Moyle, P.B. (2001). Homogenization of California's fish fauna through abiotic change. In J. L Lockwood & M.L. McKinney (Eds.), Biotic Homogenization (pp.269-288), New York: Kluwer/Academic Press.
Marchetti, M.P., T. Light, P.B. Moyle and J. Viers. 2004. Fish invasions in California watersheds: testing hypotheses using landscape patterns. Ecological Applications, 14 (5), 1507–1525.
Marr, S. M., Olden, J. D., Leprieur, F., Arismendi, I., Ćaleta, M., Morgan, D. L., & García-Berthou, E. (2013). A global assessment of freshwater fish introductions in mediterranean-climate regions. Hydrobiologia, 719, 317-329.
Matern, S. A., Moyle, P.B. & Pierce, L.C. (2002). Native and alien fishes in a California estuarine marsh: twenty-one years of changing assemblages. Transactions of the American Fisheries Society, 131, 797-816.
McKinney, M. L. (2001). Effects of human population, area, and time on non-native plant and fish diversity in the United States. Biological Conservation, 100, 243-252.
Moyle, P. B. (2002). Inland fishes of California. Berkeley: University of California Press.
Moyle, P. B., & Randall, P.J. (1998). Evaluating the biotic integrity of watersheds in the Sierra Nevada, California. Conservation Biology, 12, 1318-1326.
Moyle, P. B., &. Vondracek, B. (1985). Persistence and structure of the fish assemblage in a small California stream. Ecology, 66, 1-13.

Moyle, P. B., and Light, T. (1996a). Biological invasions of fresh water: empirical rules and assembly theory. Biological Conservation, 78, 149-162.
Moyle, P. B., and Light. T. (1996b). Fish invasions in California: do abiotic factors determine success? Ecology, 77, 1666-1670.
Moyle, P. B., Crain, P. K., Whitener, K., & Mount, J.F. (2003). Alien fishes in natural streams: fish distribution, assemblage structure, and conservation in the Cosumnes River, California, USA. Environmental Biology of Fishes, 68, 143-162.
Moyle, P. B., Daniels, R.A., Herbold,B., & Baltz, D.M. (1986). Patterns in distribution and abundance of a non-coevolved assemblage of estuarine fishes in California. Fishery Bulletin, 84, 105-117.19.
Moyle, P. B., Hobbs, J. and O'Rear, T. (2012). Fishes. In A. Palaima (Ed.), Ecology, conservation and restoration of tidal marshes: The San Francisco Estuary (pp.161-173). Berkeley: University of California Press.
Moyle, P.B. & Marchetti, M.P. (2006). Predicting invasion success: freshwater fishes in California as a model. Bioscience, 56, 515-524.
Moyle, P.B. (2013). Novel aquatic ecosystems: the new reality for streams in California and other Mediterranean climate regions. River Research and Applications, 30 (20), 1335-1344, DOI: 10.1002/rra.2709.
Moyle, P.B., Katz, J.V. & Quiñones, R.M. (2011). Rapid decline of California's native inland fishes: a status assessment. Biological Conservation, 144, 2414-2423.
Moyle, P.B., Kiernan, J.D., Crain, P.K. & Quiñones, R.M. (2013) Climate change vulnerability of native and alien freshwater fishes of California: a systematic assessment approach. PloS One, 8 (5), e63883., DOI:10.1371/journal.pone.0063883.
Moyle, P.B., Manfree, A.D. & Fiedler, P.L. (2014). Suisun Marsh: Ecological History and Possible Futures. Berkeley: University of California Press.
Moyle, P.M. & Leidy, R.A. (1992). Loss of biodiversity in aquatic ecosystems: evidence from fish faunas. In P.L. Fiedler and S.K. Jain (Eds.). Conservation biology – the theory and practice of nature conservation, preservation and management (pp.127-169). London: Chapman and Hall.
Olden, J. D., Kennard, M. J., Leprieur, F., Tedesco, P. A., Winemiller, K. O., & García-Berthou, E. (2010). Conservation biogeography of freshwater fishes: recent progress and future challenges. Diversity and Distributions, 16, 496-513.
Olden, J. D., Konrad, C. P., Melis, T. S., Kennard, M. J., Freeman, M. C., Mims, M. C. *et al.*, (2014). Are large-scale flow experiments informing the science and management of freshwater ecosystems? Frontiers in Ecology and the Environment, 12, 176-185.
Rahel,, F. J. (2002). Homogenization of freshwater faunas. Annual Review of Ecology and Systematics, 33, 854-856.
Rahel,, F. J., & Olden, J. D. (2008). Assessing the effects of climate change on aquatic invasive species. Conservation Biology, 22, 521-533.
Ricciardi, A., Hoopes, M. F., Marchetti, M. P., & Lockwood, J. L. (2013). Progress toward understanding the ecological impacts of nonnative species. Ecological Monographs, 83, 263-282.
Rosenzweig, M.L. (2003). Reconciliation ecology and the future of species diversity. Oryx, 37, 194-205.
Shea, K., & Chesson, P. (2002). Community ecology theory as a framework for biological invasions. Trends in Ecology and Evolution, 17, 170-176.
Simberloff, D. (2009). The role of propagule pressure in biological invasions. Annual Review of Ecology, Evolution, and Systematics, 40, 81-102.
Simberloff, D. (2013). Biological invasions: What's worth fighting and what can be won? Ecological Engineering, 65, 112-121.

Snyder, B. D. (2005). The Susquehanna River fish assemblage: surveys, composition, and changes. American Fisheries Society Symposium, 45, 451-470.
Stohlgren, T. J., Jarnevich, C., Chong, G. W., Evangelista, P. H., Pyšek, P., Kaplan, Z., & Richardson, D. M. (2006). Scale and plant invasions: a theory of biotic acceptance. Preslia, 78, 405-426.
Weiher, E., Freund, D., Bunton, T., Stefanski, A., Lee, T., & Bentivenga, S. (2011). Advances, challenges and a developing synthesis of ecological community assembly theory. Philosophical Transactions of the Royal Society B: Biological Sciences, 366 (1576), 2403-2413.

List of Figures

List of Tables

Index

O

P

Q

R

S

X

Y

Z

Kolibri 360 GmbH
Ferdinand-Jühlke-Straße 7
99095 Erfurt, Deutschland
produktsicherheit@kolibri360.de